Audio and Video Systems

Audio and Video Systems

M.L. Anand
Consultant Engineer

CRC Press
Taylor & Francis Group
Boca Raton London New York

CRC Press is an imprint of the
Taylor & Francis Group, an **informa** business

Manakin
PRESS

First published 2025
by CRC Press
4 Park Square, Milton Park, Abingdon, Oxon, OX14 4RN

and by CRC Press
2385 NW Executive Center Drive, Suite 320, Boca Raton FL 33431

CRC Press is an imprint of Informa UK Limited

British Library Cataloguing-in-Publication Data
A catalogue record for this book is available from the British Library

ISBN: 9781032867588 (hbk)
ISBN: 9781032867618 (pbk)
ISBN: 9781003529064 (ebk)

DOI: 10.4324/9781003529064

Typeset in Times New Roman
by Manakin Press, Delhi

Manakin
PRESS

Preface

The book is an outcome of my long teaching experience. It is written for B.Tech as well as Diploma students and covers the syllabi of major universities and institutions.

The book is student-friendly, thought-provoking, and stimulating. It helps in clearing the cobwebs of the mind. Although a book cannot replace teachers, the reader will feel the physical presence of the teacher.

Chapters 1, 2, and 3 deal with basic concepts, Audio Engineering, Audio and video amplifiers, and Sinusoidal oscillators.

Chapters 4, 5, and 6 deal with Microphones, Loudspeakers, P.A. systems, Audio recording, Data storage devices, hi-fi stereo, VCR, VCD, and DVD.

Chapter 7 and 8 deal with Communication systems, radio receivers, and Antenna systems.

Chapters 9, 10, and 11 deal with Monochrome and color TVs, as well as Fax.

Chapter 12 deals with Radar systems.

Chapter 13 deals with Telegraphy and telephony.

Chapter 14 deals with Satellite communication.

Chapter 15 deals with Transmission lines and waveguides.

Chapter 16 deals with Optical fiber.

Chapter 17 deals with Consumer electronics.

E.g., digital calculator, microwave oven, A.V.R, inverter, U.P.S., mobile, e-mail, Internet, computers, memory systems, medical instruments, etc...

Experiments, Appendices etc, etc.

Errors might have crepts in, in spite of utmost care to avoid them. I will be grateful if the same are pointed out along with suggestion for the improvement of the book.

I am thankful to the management and the editorial team of the Company for cooperation and support in the publication of the editor.

Author

Contents

1

Basic Concepts: Audio Engineering

The Audio Video System is the study of electronic devices used in our day to life. Basically this is *consumer electronics*.

1.1 AUDIO SYSTEMS

An audio system is a system which involves an audio signal. In other words, Audio Systems "process" sound. The examples of audio systems/equipment/ components are:

1. Radio
2. Tape Recorder/Player/Stereo
3. Microphone
4. Loud speaker
5. Telephone.

1.2 AUDIO VIDEO SYSTEM

An audio/video system is that process sound as well as picture signals simultaneously. The examples of AV systems are:

(*a*) Human AV systems. Our two ears and two eyes act as AV systems.

(*b*) Man made AV systems

1. Television
2. VCP
3. VCR

4. Cinema Projector
5. Radar
6. Picture telephones etc.

1.2.1 Importance of A.V. Systems

Every body knows the important of Radio, Tape recorder, Telephone, TV, VCR etc. These have been the necessity of every home.

The AV Systems have very much entered also in the field of education. Video cassettes have been prepared and these are displayed in schools and colleges. Lectures of eminent professors/ educationists are prepared on all subjects and can be viewed on TV at home. There is no need to join computer classes. All computer languages can be learnt just sitting in the bed room watching a TV.

Now it has become a fashion to prepare video films on all occasions. It may be a birth day party, or a any procession, we can keep our memories afresh everytime. In old age, we can see our childhood.

AV systems have entered also in the world of crime. Video cameras (C.C.T.V.) can be installed in any place or house. All activities of smuggling/killings can be recorded on a video. The same càn be presented in the courts as a witness. How a criminal will deny!

Now in medical sciences, say a patient is being operated upon in the operation theatre, A video cassette (V.C.D.) can be prepared and patient can see at any time later on. Further by using close circuit TV,. The relatives of the patient sitting outside can see what is happening inside the operation theatre.

1.3 AUDIO ENGINEERING

Audio Engineering is the branch of engineering dealing with the study of audio (sound) energy.

1.3.1 Definition of Audio (Sound)

Audio (sound) may be defined as generation, transmission and reception of energy in the form of vibrations, e.g., when a spring is stretched, the air is compressed, vibrations are produced by their molecules. This results in the oscillatory vibrations and thereby acoutistic waves are generated and transmitted.

1.3.2 Terminology Related to Audio Engineering

1. *Acoustics.* The word "acoustics", is a term associated with sound waves, or with individual media, phenomena, apparatus, etc. It is the science of sound waves.

   Acoustical. The word "acoustical", is a term related with acoustics.

2. *Pressure*

   (*i*) *Static Pressure* (P_0). This is the pressure in the medium that would exist at a point with no sound waves present. Microbar (μB) is the unit of pressure used in acoustic. One μB is equal to 0.1 Newtons per m^2.

   (*ii*) *Instantaneous pressure* (p). This is the incremental change from the static pressure at a given instant caused by the presence of sound waves. The unit is μB.

   (*iii*) *Sound pressure* (P) is the RMS value of the instantaneous sound pressure over a time interval at that point.

3. *Velocity*

   (*i*) *Instantaneous particle velocity.* This is the velocity at a point of a given infinitesimal part of the medium at a given instant due to sound waves only. The unit is m/s.

   (*ii*) *Effective particle velocity* at a point is the RMS value of instantaneous particle velocity. The unit is *meter per second* (m/s).

   (*iii*) *Instantaneous volume velocity.* This is the velocity due to sound waves only and is the rate of flow of the medium perpendicularly through a specified area. This is equal to the product of instantaneous particle velocity and the surface area in consideration. The unit is *cubic meters per second.*

4. *Impedence*

   (*i*) *Acoustic Impedence* (Z_A). The acoustic impedence at a given surface is the complex ratio (this has the same meaning as the complex ratio of voltages and currents in ac circuits involving j factor) of average sound pressure (effective) over the surface to the effective volume velocity through it. The surface may be either a hypothetical, surface of an acoustic medium or the moving surface of a mechanical device. The units are *acoustic ohms*.

(*ii*) *Specific acoustic impedence* (Z_S). This is the complex ratio of the effective sound pressure at a point to the effective particle velocity at that point. The unit is Newton Seconds/m^2 or *Rayl.*

(*iii*) *Characteristic impedence* (ρ_c). This is the ratio of the effective sound pressure at a given point to the effective particle velocity at that point in a progressive sound wave. This is equal to the product of the density of the medium and the speed of sound in the medium. It is analogous to the characteristic impedence of an infinite long dissipation loss transmission line. The unit is Newton-sec/m^3 or MKS rayl. For air, this value is 407 MKS rayls or 40.7 rayls at 22°C temperature and 0.76 m mercury pressure.

5. *Sound Energy Density (D)*. This is the ratio of sound energy in a given part of the gas to the volume of that part of the gas. This has units of watt seconds per cubic meter.
6. *Sound Intensity (I)*. The sound intensity measured in a spacific direction at a point is the average rate at which the sound energy is transmitted through a unit area perpendicular to the specific direction at the point considered. The unit is Watt per square meter.
7. *Acoustic Intensity Level (AIL)*. This is the ratio of two sound intensities in a logarithmic form. The unit is decibel.

$$\text{AIL} = 10 \log_{10} \frac{I_1}{I_2}.$$

8. *Sound Pressure Level (SPL)*. Sound pressure level of a sound in decibels is 20 times the logarithmic to the base of 10 of the ratio of measured effective sound pressure (P) of the sound to a reference effective sound pressure (P_{ref}).

$$\text{SPL} = 20 \log_{10} \frac{P}{P_{ref}}$$

The reference pressure (P_{ref}) is taken as 0.1 newton/m^2 (or 1 μB).

9. *Acoustic Power Level (APL)*. The acoustic power level of a sound source in decibals is 10 times the logarithm to the base 10 of the ratio of the acoustic power radiated by the source (W) to a reference acoustic power (W_{ref}).

$$\text{APL} = 10 \log_{10} (W/W_{ref})\ db.$$

The reference power is taken as 10^{-13} watts. That is, a source radiating one acoustic watt has a power level of 130 db.

Table 1.1: Pressure and intensity of sound

S. No.	Sounds	Pressure		Intensity	
		Pa (*N/m*²)	μ Bars	Watt/m²	db
1.	Threshold of hearing	20×10^{-6}	200×10^{-6}	10^{-11}	4
2.	Threshold of pain	65	620	10	128
3.	Thunder	2	20	10^{-2}	98
4.	Heavy traffic	0.2	1.9	10^{-4}	78
5.	Light traffic	18×10^{-3}	200×10^{-3}	10^{-6}	58
6.	Ordinary conversation	65×10^{-4}	650×10^{-4}	10^{-7}	52
7.	Whispering	20×10^{-5}	20×10^{-5}	10^{-9}	18

1.4 TYPES OF VIBRATIONS

The vibrations are of following types:

(*i*) Having a frequency range between 20 Hz to 20 kHz, which have audible range for a common man, *i.e.,* we can hear these vibrations.

(*ii*) Having a frequency range above 20 kHz called is ultrasonics, which are not audible but have many engineering applications like: drilling, flaw detection in metals, cold welding, ultrasonic mixing and sonar (sound navigation & ranging) .

Dogs can hear these high frequency sounds. The birds can hear upto 50,000 Hz. Similarly, bats produce ultrasonics in the darkness to find presence of obstacles in their way.

(*iii*) Having a frequency range below 20 kHz called *infrasonics,* which are also not audible by man.

We should note that the vibrations (oscillations) may be periodic (sinusoidal) as produced by a tuning fork [Fig. 1.1 (*a*)] or may be non periodic (non-sinusoidal) as produced by a noise [Fig. 1.1 (*b*)].

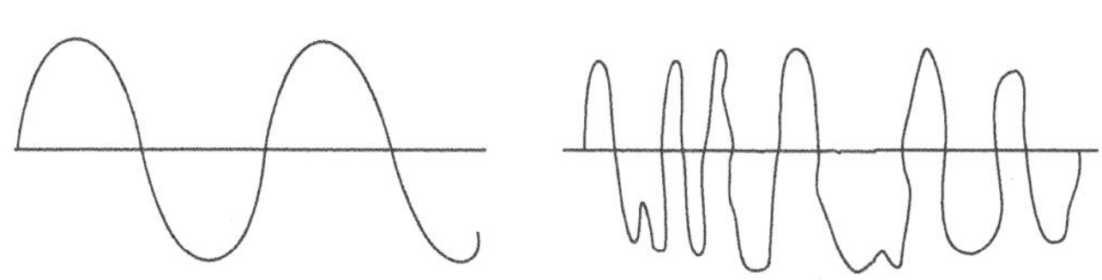

(*a*) (*b*)

Fig. 1.1

1.5 SIMPLE HARMONIC MOTION (SHM)

The particles of every medium are bound by some elastic forces. When a particle of the medium is disturbed by giving some energy, it is displaced from its mean position. The disturbed particle begins to vibrate with simple harmonic motion. The energy of this particle is transferred to the next particle due to elasticity of the medium. The second particle also starts vibrating about its mean position and hands over the disturbance to the next particle and so on. In this way, the disturbance is transmitted from particle to particle in the medium, whereas the particles do not change their position and continue to vibrate about their mean position.

All material objects when struck properly produce SHM due to their vibrations. The common examples are a simple pendulum [See Fig. 1.2 (*a*)] and a tuning fork [Fig. 1.2 (*b*)].

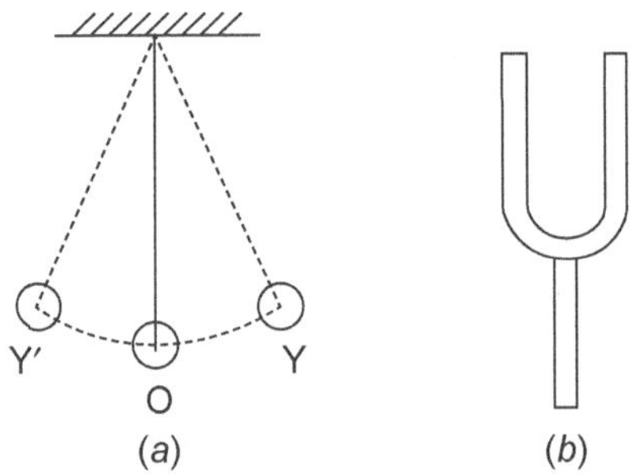

Fig. 1.2

The graphic representation of a SHM is given in [Fig. 1.3 (*a*) and 1.3 (*b*)], which clearly shows the position (phase) of the vibrating particle at different positions. The displacement of the particle at any time is given by the expression.

$$AM = y = a \sin \theta = a \sin \omega t.$$

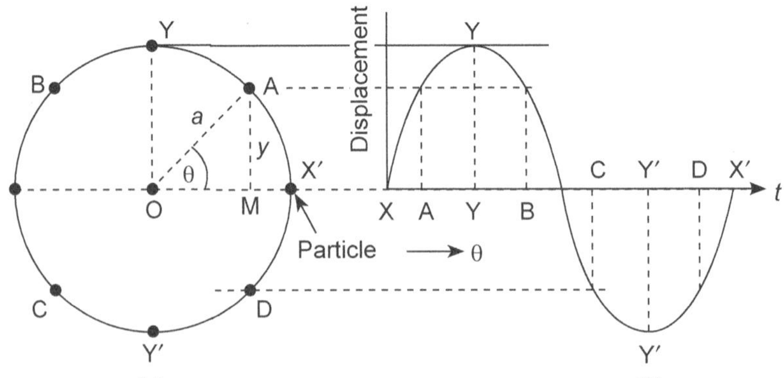

Fig. 1.3

1.5.1 Definition of SHM

A particular type of periodic motion (which repeats itself over and over again after regular intervals of time) is known as Simple Harmonic Motion (SHM). Sometimes their vibration is a combination of two or more than two SHMs.

The SHM may therefore be defined as a motion, which satisfies two conditions:

1. It accelerates always towards a certain fixed point and
2. Its acceleration is proportional to its distance from the fixed point.

Alternative Definition

The SHM is a motion executed by a particle subjected to a force that is proportional to the displacement of the particle but opposite in sign.

1.5.2 Bodies executing SHM

The various shapes of bodies executing SHM may be classified as : strings, bars, pipes, plates and diaphragms etc. Whenever any of them emits sound waves, it is found to be in a state of rapid vibrations.

(*i*) *Vibrations of a string* [Fig. 1.4 (*a*)]. If a string stretched between two fixed points is plucked in the middle, a sound is produced. On touching the string, vibrations disappear and sound ceases.

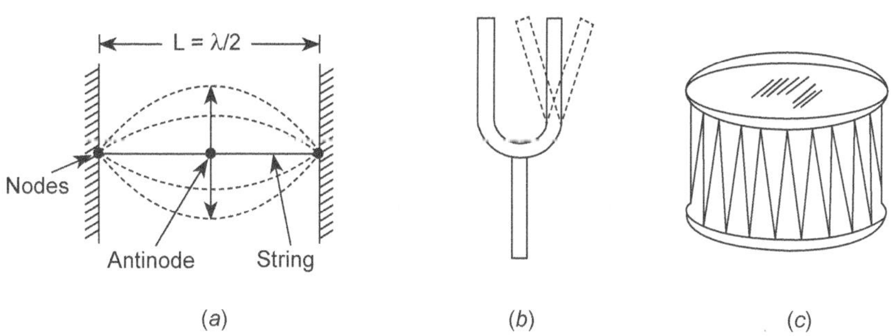

Fig. 1.4

(*ii*) *Vibrations of a bar* [Fig. 1.4 (*b*)]. The most common example of bar is a tuning fork. When it is struck at one of its arms, it vibrates and produces sound.

(*iii*) *Vibrations of a diaphragm* [Fig. 1.4 (*c*)]. Sounding of a drum is an example of vibrations of a diaphragm.

Note. It should be noted that oscillatory motion is always periodic but every periodic motion is not oscillatory. Examples of periodic motions are :

1. Change of seasons
2. Motion of earth around the sun
3. Motion of planets and comets
4. Change of phases by moon
5. Human heart beat.

1.6 VIBRATIONS OF STRETCHED MEMBRANES, (DIAPHRAGMS) AND THIN PLATES

So far, we have considered vibrations of a stretched string which was an example of one dimensional system. A *membrane* is a two dimensional system. The analysis of the vibrations of a stretched membrane is helpful in the study of electro acoustical transducers (microphone, loudspeakers etc.). There are two such cases of vibrations:

1. *Vibrations of stretched circular membrane* in which case, restoring force due to stiffness is negligible in comparison to the tension *e.g.,* drum heads, diaphragm of a *condenser microphone*.
2. *Vibrations of stretched thin circular plates*, where stiffness cannot be neglected *e.g.,* diaphragms of telephone receivers.

   The Fig. 1.5 show examples of stretched membranes (diaphragms).

   The transverse vibrations of a string (or bar) are sinusoidal, but the transverse vibrations of membranes are not sinusoidal, therefore analysis of membranes is bit complicated. A membrane if regarded as an assembly of parallel strings, it becomes easier to analyse.

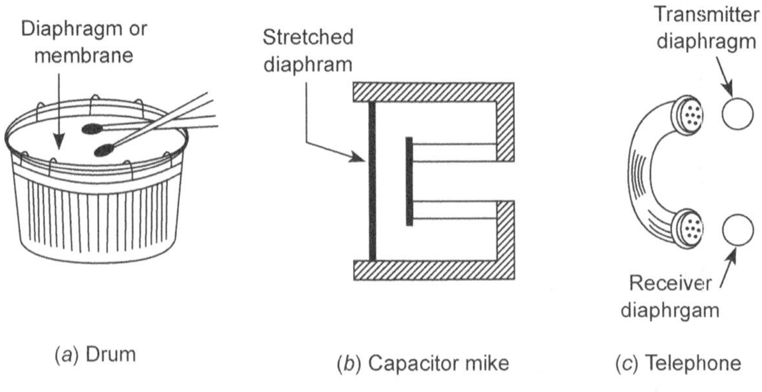

(*a*) Drum (*b*) Capacitor mike (*c*) Telephone

Fig. 1.5

Let us take an example of a drum (or a capacitor microphone), which have a membrane stretched tightly over the open ends of a vessel. As the drum head is made to vibrate, the air inside the vessel passes through compression and rarefaction alternately. The velocities of sound waves are lesser in the membrane than in the air but the sound pressure is uniform over entire membrane.

The Fig. 1.6 (*a*) shows vibration modes of a rectangular and Fig. 1.6 (*b*) shows of a circular membrane.

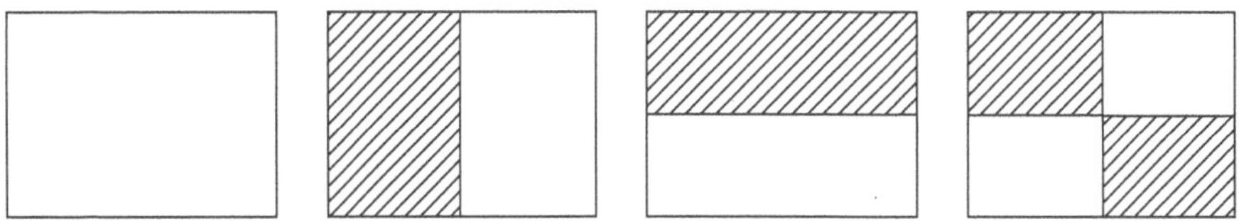

(*a*)

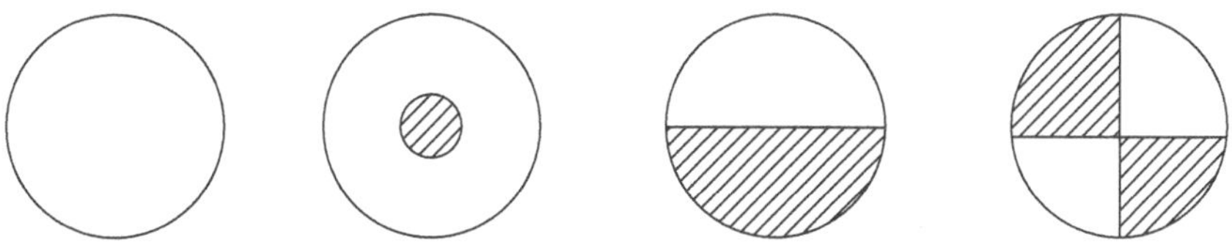

(*b*)

Fig. 1.6

There is a little difference in vibrations in thin plates and in membranes. That is in former, the restoring force is entirely due to the stiffness of the diaphragm, no tension is applied, where as in later the restoring force is entirely due to the tension applied to the membrane.

1.7 TYPES OF WAVES

(*a*) Depending upon the direction of propagation of waves and of the medium particles, the waves are of two types :

1. *Transverse Waves.* In these waves, the particles of the medium vibrate perpendicular to the direction of propagation of waves. See Fig. 1.7.

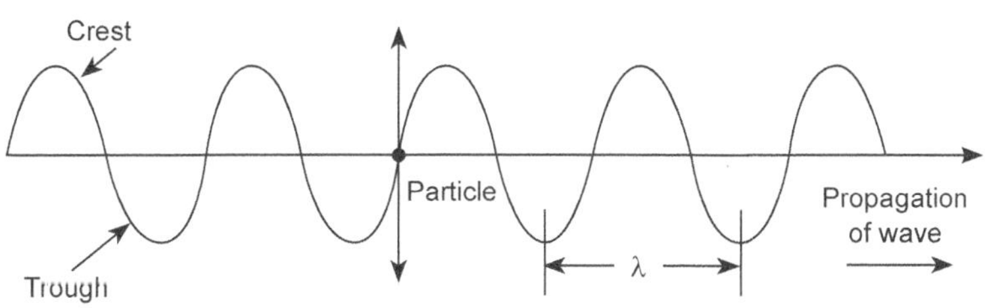

Fig. 1.7

The waves produced in a stretched string are transverse waves. These waves travel in the form of *crests* and troughs, which are parts of the medium above and below the mean position respectively.

2. *Longitudinal Waves.* In these waves, the particles of the medium vibrate parallel to the direction of propagation of waves. See Fig. 1.8 (*a*).

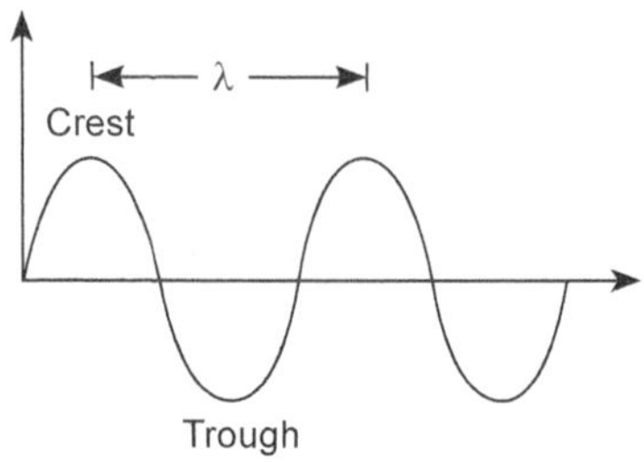

Fig. 1.8 (*a*)

The sound waves in air are longitudinal waves. These waves travel in the form of *compressions* and *rarefactions*, which are alternate regions of high and low density of the medium respectively.

The distance between two adjacent crests or troughs (or compressions and rarefaction) is called wavelength (λ) [Fig. 1.8 (*b*)].

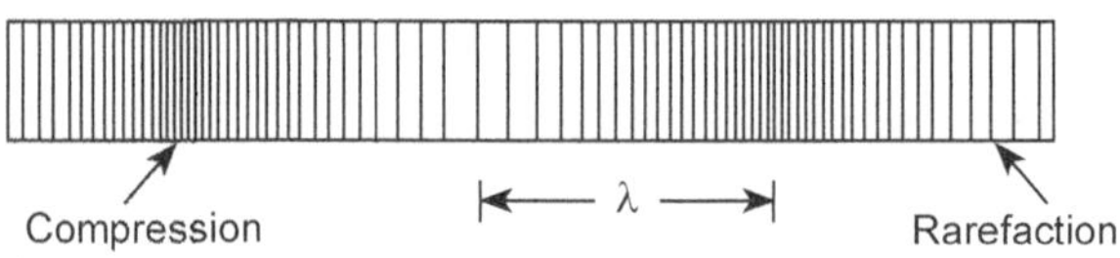

Fig. 1.8 (*b*)

Table 1.2 Difference between Transverse and Longitudinal Waves

Transverse Waves	Longitudinal Waves
1. These are the waves in which vibrations occur perpendicular to the direction of propagation of waves.	In these waves, vibrations occur along the direction of propagation of the waves.
2. These waves travel in the form of crests and troughs.	These waves travel in the form of compressions and rarefactions.
3. The transverse waves can travel through solids only,	Longitudinal waves can travel through solids, liquids and even gases.
4. These waves can be polarised.	These waves cannot be polarised.

(*b*) On the basis of propagation of energy, the waves can be of two types :

1. *Progressive or Travelling Waves.* If there is a continuous transfer of energy in one direction across any cross section of the medium, the waves are called *progressive or travelling waves*.

2. *Stationary or Standing Waves.* If the net transfer of energy across any cross section of the medium is zero, the waves are called stationary or standing waves.

   These waves are produced due to the superposition of a wave and its reflected wave.

Table 1.3 Difference between Progressive and Stationary Waves

Progressive waves	Stationary waves
1. The amplitude of each particle is same every where.	The amplitude of the particle varies from zero at *nodes* to maximum at *antinodes*
2. The phase of two consecutive particles is not same.	The phase of all particles between two nodes is same and is opposite to that of the particles in the segments on either side.
3. No particle is permanently at rest.	The particles at the *nodes* are permanently at rest.
4. Particles pass through mean position at different instants.	All particles cross the mean position simultaneously.
5. All particles have same maximum velocity.	The velocity is maximum (largest) for the particles at *antinodes* and it is zero for the particles at the *nodes*.
6. Energy is continuously transferred across every cross-section of the string.	No net transportation of energy takes place across any cross-section of the string.

When two progressive waves of the same type (transverse or longitudinal), of same wavelength and of same amplitude travel with the same velocity in opposite directions through a medium, they superpose on each other to form a *stationary wave*.

In stationary waves, crests and troughs (or compressions or rarefactions) donot move forward through the medium. They simply appear and disappear alternately at the same positions. No energy transfer takes place in the medium.

Stationary waves are further of two types :

(*i*) *Transverse stationary waves.* These are the waves, which are formed due to the super-position of two progressive transverse waves. For example, waves produced in a *sonometer* are transverse stationary waves.

(*ii*) *Longitudinal stationary waves.* These are the waves, which are formed due to super-position of two progressive longitudinal waves. For example, waves produced in *organ pipes* or in a resonance apparatus are longitudinal stationary waves.

1.8 VELOCITY OF SOUND

Sound waves are produced due to the vibratory motion, which may (or sometimes) may not be visible. Sound is a longitudinal wave motion in a material medium. For propagation of sound, a medium is a must. Sound cannot propagate in vacuum. The material medium required may be solid, liquid or a gas.

The velocity of sound depends upon the

1. the elasticity of the medium, and
2. the inertia of the medium (*i.e.,* density/mass of the medium).

The velocity of sound in a medium is given by :

$$V = \sqrt{\frac{\text{Elasticity}}{\text{Density}}} = \sqrt{\frac{E}{\rho}}$$

This is the Newton's formula for velocity of sound. The same formula holds good for all mediums.

1.8.1 Velocity of Sound in Solids

For finding velocity of sound in solids, the elasticity (E) is replaced by young modulus of elasticity (Y).

Hence, $$V = \sqrt{\frac{Y}{\rho}}$$

1.8.2 Velocity of Sound in Liquids

In case of liquids, E is replaced by bulk modulus of elasticity (K)

$$V = \sqrt{\frac{K}{\rho}}$$

1.8.3 Velocity of Sound in Gases (or air)

Newton found that coefficient of elasticity of a gas is equal to its pressure. While providing this, he assumed that sound travelled through air under *isothermal conditions*. He gave the formula for velocity of sound in air as under :

$$V = \sqrt{\frac{P}{\rho}}$$

where P is the pressure of air at normal temperature and pressure (NTP) condition, and ρ is the density of air.

Pressure of air at NTP, $P = 0.76 \times 13.6 \times 10^3 \times 9.8 \text{ N/m}^2$

and density $\rho = 1.293 \text{ kg/m}^3$

Hence velocity of air at NTP

$$V = \sqrt{\frac{0.76 \times 13.6 \times 10^3 \times 9.8}{1.293}} = 280 \text{ m/s}$$

But the velocity of sound in air has been found as 332 m/s, therefore the above formula needs corrections.

(a) Laplace's Corrections

The French Mathematician Laplace found that sound travels through air under *adiabatic, and not isothermal conditions*, as newton presumed. According to Laplace :

1. Temperature rises slightly in the region of *compression* and it decreases slightly in *rarefaction.*
2. Air is a bad conductor of heat and compressions or rarefactions follow each other so rapidly that heat generated in compression or absorbed during rarefaction is not transferred, thus the changes are *adiabatic*.

   He gave the following formula for velocity of sound in air :

$$V = \sqrt{\frac{\gamma P}{\rho}}$$

Where γ = Ratio of two specific heats of air (or any gas) i.e. at constant pressure and volume $\gamma = C_P/C_V$

P = Pressure of the air or gas

ρ = Density of air or gas.

Using this formula for air, for which, $\gamma = 1.41$

$$V = \sqrt{\frac{1.41 \times \left(0.76 \times 13.6 \times 10^3 \times 9.8\right)}{1.293}}$$

$$= 333.5 \text{ m/s}$$

This value agrees with the experimental value of velocity of sound in air at NTP. Hence, Laplace correction is valid.

(b) Factors affecting velocity of sound in gases

1. **Effect of density:** The velocity of sound in two gases at the same pressure, is inversely proportional to the square root of their densities :

$$\frac{v_1}{v_2} = \sqrt{\frac{\rho_2}{\rho_1}}$$

Density of oxygen is 16 times to that of hydrogen *i.e.*,

$$\frac{v_h}{v_0} = \sqrt{\frac{16}{1}} = 4$$

or, velocity of sound in hydrogen (V_h) is 4 times to that in oxygen (V_o).

2. *Effect of pressure.* There is no effect of pressure on the velocity of sound.
3. *Effect of moisture.* The velocity of sound is more in the moist air than in the dry air.
4. *Effect of temperature.* The velocity of sound is proportional to the square root of absolute temperature

   i.e., $$V \propto \sqrt{T} \text{ or } \frac{V_1}{V_2} = \sqrt{\frac{T_1}{T_2}}$$

   It can be easily shown that increase in the velocity per deg. centigrate rise in temperature is 0.62 m/s.
5. *Effect of wind.* When the wind blows in the direction of sound, velocity of sound is increased and in the opposite case, it is decreased. If velocity of sound is V_S and of the wind is V_W, the resultant velocity of sound is,

   $$V = V_S \pm V_W$$

1.8.4 Velocity of Sound in Water

It is more important to find velocity of sound in sea water than in fresh water. The reason being that this has practical importance in all methods of finding distances and depths, locating ships, and submarines under sea water.

For this, sound is recorded by listening it through a hydrophone at a distance of 150 to 200 km. A radio signal is also transmitted at the same instant and arrival of both is recorded and compared.

The velocity in the sea water is found to increase with temperature and the salinity (quantity of salt) of water. The following expression is used for sound in m/s in sea.

$$V = 1410 + 4.21\,\theta - 0.237\,\theta^2 + 1.4S$$

Where θ is the temperature of sea water in °C and S its salinity in parts per thousand.

Table 1.4. Velocity of sound in different medium

S. No.	Medium	Velocity m/s	S. No.	Medium	Velocity m/s
1.	Steel	5100	6.	Mercury	1470
2.	Cast iron	4700	7.	Water (4°C)	1432
3.	Brick	3650	8.	Hydrogen (0°C)	1296
4.	Copper	3586	9.	Air (0°C)	332
5.	Brass	3500	10.	Oxygen (0°C)	317

Example 1.1: *Calculate velocity of sound in hydrogen at 363° K. The density of hydrogen at NTP is 0.089 kg/m³. Assume $\gamma = 1.41$.*

Solution: We know that NTP = 273°K and 0.76 m of mercury pressure (or $P = 0.76 \times 13.6 \times 10^3 \times 9.81 = 1.013 \times 10^5$ N/m²) and density $\rho = 0.089$

Now velocity at 273°K, $V_{273} = \sqrt{\gamma . P / \rho}$

$$= \sqrt{1.41 \times 1.013 \times 10^5 / 0.089} = 1267 \text{ m/s}$$

Now
$$\frac{V_{363}}{V_{273}} = \sqrt{\frac{363}{273}} \times 1267$$

or
$$V_{363} = 1.15 \times 1267 = 1457 \text{ m/s. } \textbf{Ans.}$$

Example 1.2: *At what temperature, the velocity of sound in air is double the velocity of sound in air at 0°C.*

Solution: Let at temperature t, the velocity of sound is double of that of at 0°C

$$\frac{V_t}{V_0} = \sqrt{\frac{T}{T_o}} = \text{or} \sqrt{\frac{273+t}{273+0}} = 2$$

or
$$273 + t = 4 \times 273 = 1092 \text{ or } t = 819°\text{C. } \textbf{Ans.}$$

Example 1.3: *An observer sets his watch by the sound of a gun fired from a tower. Find the distance between tower and the observer, if his watch was slow by 5 seconds. Temperature of air is 15°C.*

Solution: We know that velocity changes at the rate of 0.62 m/s per °C change in temperature and velocity of sound at 0°C is 340 m/s.

Now, the velocity of sound at 15°C.

$$V_{15} = 340 + (0.62 \times 15)$$
$$= 340 + 9.30 = 349.3 \text{ m/s.}$$

The distance of the tower from the observer is the distance travelled by sound in 5 seconds, *i.e.*

$$= 349.3 \times 5 = 1747.5 \text{ meters. } \textbf{Ans.}$$

Example 1.4: *A body vibrating with a certain frequency sends 2.5 m long waves through a medium A and 2.0 m long waves through a medium B. The velocity of waves in A is 120 m/s. Find the velocity in medium B.*

Solution: Suppose frequency of vibrating body be n. We know that frequency remains constant irrespective of the medium.

Medium A. Wave length $\lambda_1 = 1.5$ m

Velocity $V_1 = 120$ m/s

$$n = \frac{V_1}{\lambda_1} = \frac{120}{1.5} = 80 \text{ Hz.}$$

Medium B. Frequency, $n = 80$ Hz

Wave length $\lambda_2 = 2.0$ m

Velocity $V_2 = n \,.\, \lambda_2 = 80 \times 2 = 160$ m/s. **Ans.**

1.9 TYPES OF OSCILLATIONS

There are following types of oscillations :

1.9.1 Free Oscillations

When a body is set into vibrations and allowed to vibrate freely under the influence of its own elastic or gravitational forces then it vibrates with a certain time period called the *natural period*. Such vibrations are called *free vibrations* or oscillations and the frequency with which the body vibrates without any external influence *i.e.* when left to itself is called *natural frequency*. For example when a pendulum is displaced from its mean position, and it vibrates with a natural time period given by

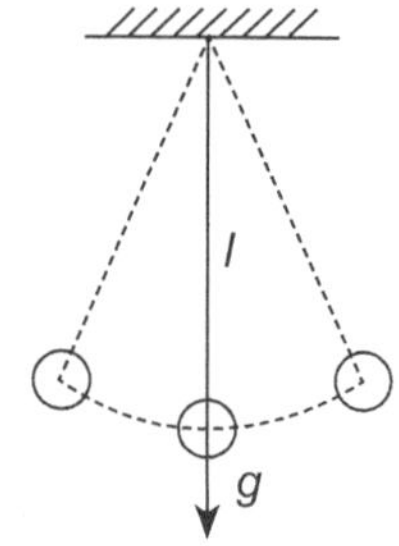

Fig. 1.9

$$T = 2\pi\sqrt{\frac{l}{g}}$$

Where l is length of the pendulum and g is the gravity (see Fig. 1.9).

Other examples of free vibrations are: vibrations of tuning fork, strings of musical instruments, air columns in organ pipes etc.

1.9.2 Damped Oscillations

The oscillations in which amplitude decreases with time are called as damped oscillations or vibrations. The wave form of these oscillations are shown in

Fig. 1.10. The decrease in the amplitude is due to the losses taking place, which may be due to friction of the surface or air friction, and no means are provided to compensate these losses. It is to be noted that the amplitude of oscillations decreases but the frequency remains the same, since in the later depends upon the parameters of the system.

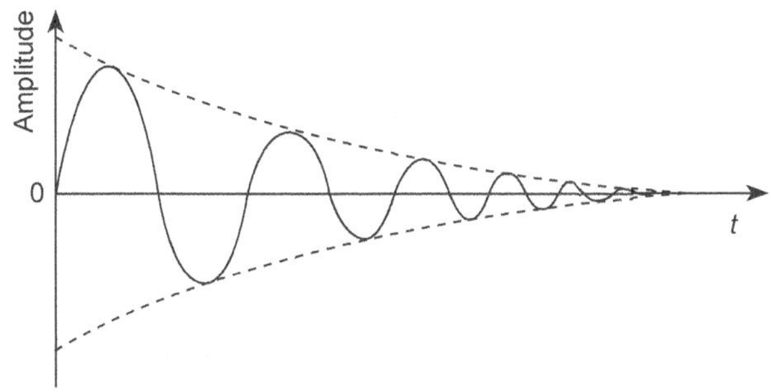

Fig. 1.10

The oscillations of a freely oscillating pendulum are damped as they go on decreasing with time.

It is to be mentioned here that friction forces are of many types but they always result in the damping (decreasing) amplitude of the oscillations.

The most important frictional force is the *resistance to motion* of the oscillator (e.g. pendulum) and this resistance may arise from *radiation of sound waves* and from the viscous fluid present in the surrounding. This frictional force is proportional to the velocity of the system and can be expressed as

$$F_r = -R_m \frac{dx}{dt} \qquad \left[\text{velocity} = \frac{dx}{dt}\right]$$

Where F_r is the frictional force, R_m is mechanical resistance of the system. If R_m is also included, the equation of motion of a simple oscillator becomes :

$$m\frac{d^2x}{dt^2} + Rm\frac{dx}{dt} + kx = 0$$

Where m is the mass of the oscillatory system, R_m is the mechanical resistance of the system and k is the stiffness constant.

1.9.3 Undamped (Forced) Oscillations

The oscillations in which amplitude does not decrease with time (and remains constant) are called undamped or forced oscillations or vibrations. The wave form is shown in Fig. 1.11. These oscillations are produced by a system in which some means are provided to compensate for the losses, *i.e.,* the loss of energy after each oscillation is provided by some external means. We are more interested in these oscillations, as these are used in audio video systems.

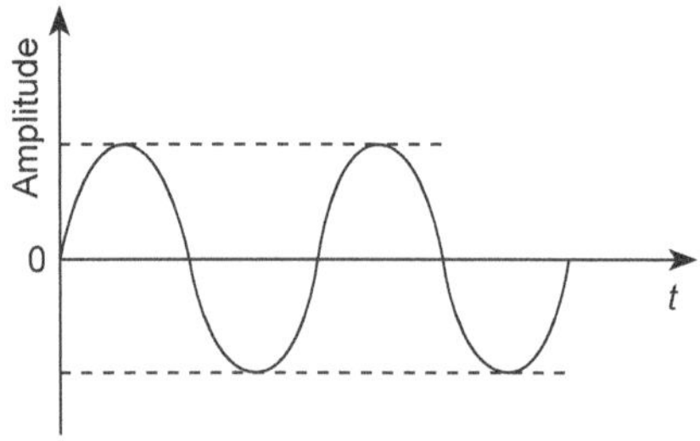

Fig. 1.11

In this case, the oscillatory system is set into vibrations with the help of a strong external periodic force having a frequency different from its natural frequency. If a pendulum instead of allowing free oscillations is moved by hands it can be made to oscillate with the frequency, we like. In other words, it will then not vibrate with its own natural frequency but with the frequency of the motion of our hands. It is obvious that now the vibrations of the pendulum will be *forced* and not free. Similarly melody of stringed instruments is the result of forced vibrations of their sounding boxes.

1.10 ELECTRICAL OSCILLATORS

1.10.1 Damped Electrical Oscillator

The Fig. 1.12 shows a damped oscillator. The dissipation of energy occurs in resistance (R) of the circuit. Due to loss of energy, the amplitude of oscillations goes on decreasing and ultimately they disappear.

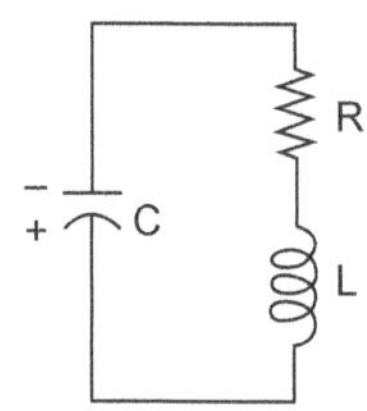

Fig. 1.12

Electrical damping is of three types (Fig. 1.13)

1. *Heavy damping.* In this case, the charge on the capacitor (c) decays to zero in minimum possible time.

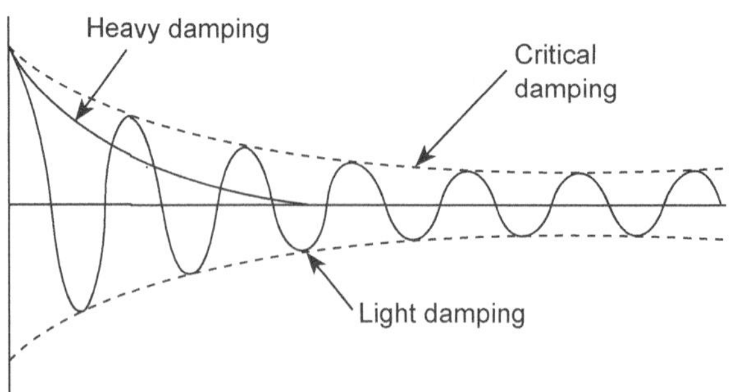

Fig. 1.13

2. *Critical damping.* In this case, the discharge of the capacitor is non oscillatory. The rate of discharge depends upon the damping constant or the resistance R. The behaviour of the oscillator is said to be *dead beat.*

3. *Light damping.* The discharge is oscillatory and takes place slowly. The amplitude of oscillations decreases with time, i.e. reduces exponentially to zero. The oscillations cease almost during the same time, in which the critically damped oscillator returns to the initial state.

The damped oscillations are not simple harmonic in the sense that the maximum displacement on either side of the mean position goes on decreasing with time. The motion cannot be termed as periodic, as it does not repeat itself exactly as each swing being smaller than that of the preceeding one. However when the damping is small, the motion tends to be an SHM as well as periodic.

1.10.2 Undamped (Or Forced) Electrical Oscillator

An oscillator, to which a continuous excitation is provided by some external agency is called an undamped or forced oscillator. It consists of

1. *The driven system i.e.,* damped oscillator.
2. *The external driving system.* In case of a *mechanical oscillator*, it is a source of mechanical force (such as a spring in old mechanical wall clocks) and in case of *electrical system*, it may be an external emf. Such an electrical oscillator consists of a RLC series circuit with an external source of EMF (V). See Fig. 1.14.

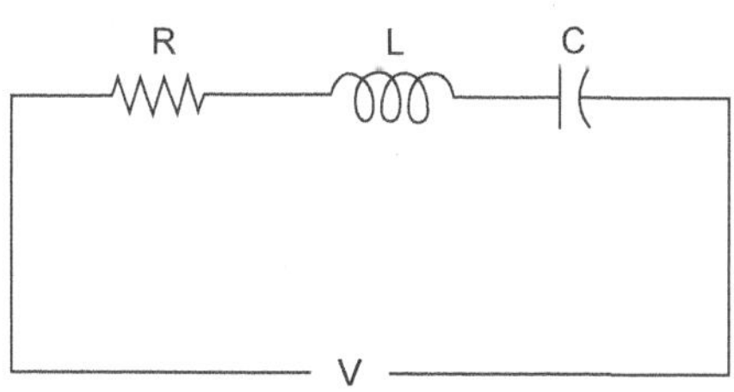

Fig. 1.14

Table 1.5: Analogies between Mechanical and Electrical Oscillators

Mechanical Oscillators	Electrical Oscillator
1. It has mechanical reactance represented by X_m and mechanical resistance by R_m.	It has equivalent inductive reactance X_L and electrical resistance as R.

Mechanical Oscillators	Electrical Oscillator
2. It has a total mechanical impedence $Z_m = R_m + jX_m$.	Its total electrical impedence $Z = R + jX_L$.
3. Its power factor (mechanical) is given by the ratio of R_m/Z_m.	Electrical power factor $\cos\phi = R/Z$.
4. Displacement in mechanical oscillators is represented by x and velocity is given by = $\frac{dx}{dt}$	The equivalent displacement is the charge q and the velocity is represented by the current, given by $i = \frac{dq}{dt}$
5. Time period of a mechanical oscillator $T = 1/f = 2\pi\left(\sqrt{m/k}\right)$.	In this case, time period of electrical oscillator is $T = 1/f = 2\pi\sqrt{LC}$.
6. The driving agent in case of mechanical oscillator is the force given by = $m\frac{d^2x}{dt^2}$	The diriving agent in this case is induced emf = $L\frac{dI}{dt} = L\frac{d^2q}{dt^2}$
7. The restoring agent in this oscillator is the elastic force $E = k.x$.	The restoring force in this case is the voltage across the capacitor given by $v = q/C$.
8. The mass of mechanical oscillator is given by m.	The equivalence of m is inductance L.
9. It has springiness.	The equivalence of springiness is capacitance.
10. It has mechanical compliance $C_m = 1/k$	It is equivalent to capacitance C.

1.11 ACOUSTICAL, MECHANICAL AND ELECTRICAL ANALOGIES—NAME AND SYMBOLS

All acoustical systems can be converted into equivalent mechanical and electrical systems. (Table 1.6)

Table 1.6: Acoustic, Mechanical and Electrical, Analogies

S. No.	Acoustical	Mechanical	Electrical
1.	Inertness m	Mass  m	Inductance 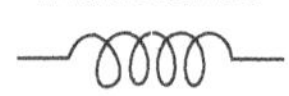
2.	Compliance  C	Compliance  $C_m = 1/k$	Capacitance 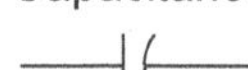 C

S. No.	Acoustical	Mechanical	Electrical
3.	Resistance 	Resistance 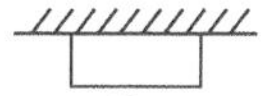 R_m	Resistance 

Note that:

(*i*) Motion of the fluid (*e.g.*, air) in an acoustical system can be taken as equivalent to the flow of current in an electrical circuit.

(*ii*) The difference of acoustic pressures in an acoustic system is just analogous to the difference in electrical pressures (potential difference) across an electrical circuit.

(*iii*) The acoustical inertness is analogous to inductance (L), acoustical resistance is analogous to the electrical resistance (R) and acoustical compliance is analogous to the capacitance (C). As a whole, acoustic impedence is analogous to impedence of the electrical circuit.

1.12 INTERFERENCE (SUPERPOSITION) OF WAVES

When two or more waves reach at a point at the same time, the redistribution of energy takes place. This phenomenon is known as *superposition* of waves.

When two or more travelling waves are moving through a medium, the resultant displacement at any point is the algebraic sum of the displacements due to individual waves.

1.12.1 Types of Interference

(*i*) When the two waves arriving at a point are in the same phase, the resultant is a sum of two individual disturbances (waves) and a wave of larger amplitude results. This is known as *Constructive Interference*. [See Fig. 1.15 (*a*)]

(*ii*) When two waves arriving at a point are in the opposite phase, the resultant disturbance (wave) is the difference of the two individual disturbances and a wave of smaller amplitude results. This is known as *destructive interference*. [See Fig. 1.15 (*b*)]

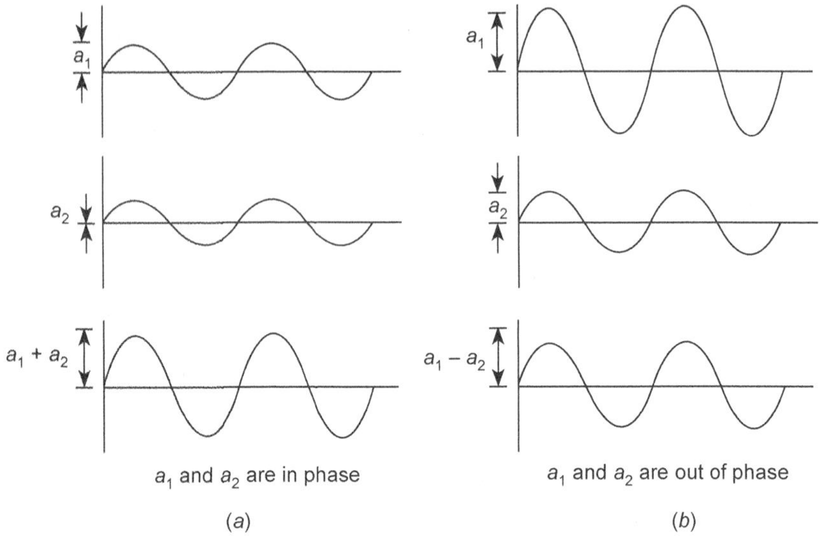

Fig. 1.15

Important

(*a*) Two waves of same frequency moving in the same direction result in *interference*.

(*b*) Two waves of slightly different frequencies moving in the same direction give rise to *beats*.

1.12.2 Application of Interference

The phenomenon of superposition (interference and production of beats) is very much used in broadcasting, transmission and reception of A.V. (audio video) signals.

1.12.3 Demonstration of Interference

The phenomenon of *interference* can be shown with two loud speakers.

(*i*) In Fig. 1.16 (*a*), two identical loud speakers have been connected separately with a same amplifier. Place the two speakers close to each other.

Now disconnect one speaker. Hear the sound coming from second speaker. A sound of certain intensity will be heard.

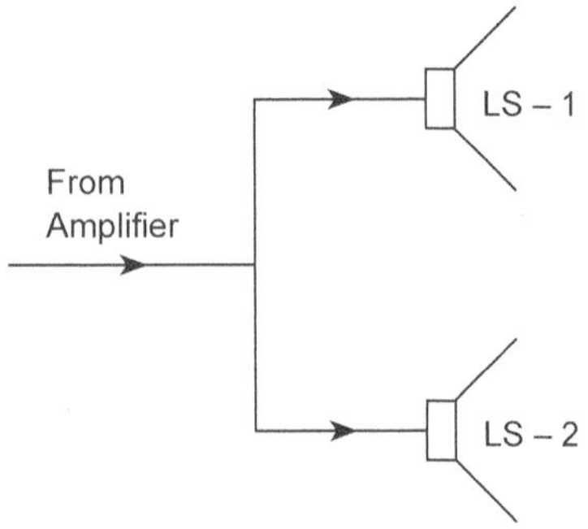

Fig. 1.16 (*a*)

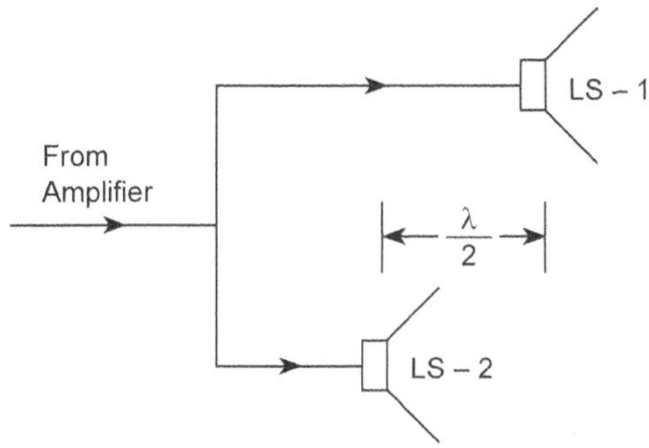

Fig. 1.16 (*b*)

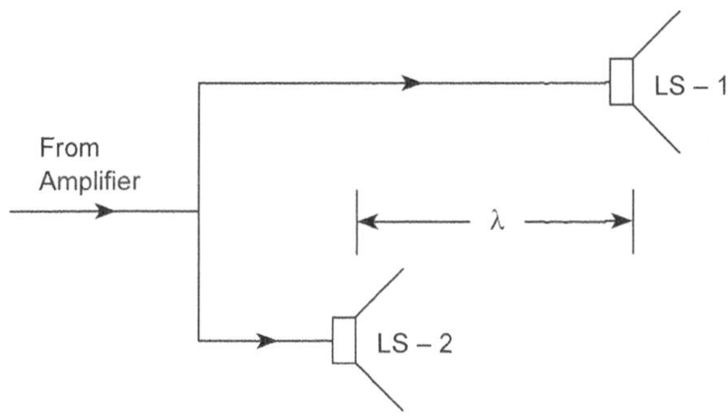

Fig. 1.16 (*c*)

Now connect the second loud speaker and hear the sound coming from both. A loud sound will be heard, it will be nearly double of the previous sound.

It is a case of *constructive interference*, as the sound waves from two speakers are in phase.

(*ii*) Now move one loud speaker backward by a distance $\lambda/2$, (i.e. half the bandwidth), a very low sound will be heard. It is because the waves coming from two speakers are out of phase. This is a case of *destructive interference*. (Fig. 1.16 *b*)

(*iii*) Now move the speaker further backward so that its distance from first speaker becomes λ. Again a loud sound will be heard. This is a case of *constructive interference*. We come to the following conclusion: (Fig. 1.16 *c*)

(*a*) When distance between two speakers is an integral multiple of λ, Constructive interference takes place.

(*b*) When the distance between the two speakers is an odd multiple of $\lambda/2$, destructive interference takes place.

1.13 BEATS

The periodic rise and fall in the loudness of sound, caused by the superposition of two sound waves of slightly different frequencies is known as *beats*.

When two sound waves of slightly different frequencies travelling along the same straight line in the same direction are superimposed on each other, there is alternately a rise and fall in the loudness of the resultant sound. This gives rise to beats.

One beat is formed by one loud and one faint sound. The frequency of beats is equal to the difference in frequencies of two sound waves.

The beats are heard only when difference in frequencies of two sound waves is not more than 12 as *persistence of human hearing* is 1/12 of a second.

If A is the amplitude of each the wave having different frequencies (difference is not more than 12), the maximum amplitude will be $A + A = 2A$ and minimum amplitude will be $A - A = 0$. It means, at one time maximum sound will be heard and soon it will be silent. So one set of "maximum sound and no sound" is known as beat.

If the frequencies of the sound waves are f_1 and f_2, the frequency of beats will be $= f_1 - f_2$.

Example. If two tuning forks of frequencies 262 and 278 are sounded together, the number of beats formed = 278 – 262 = 16.

1.14 AUDIO FREQUENCY (AF)

This is the range of frequency, that a human ear can respond. This is in between 20 Hz to 20 kHz.

The range of frequency, a human being can produce is in between 100 Hz to 10 kHz.

Table 1.7 Frequency range of different regions.

S. No.	Region	Wavelength (m)	Frequency (kHz)
1.	X-rays	7×10^{-11}	3×10^{16}
2.	Ultra violet rays	1×10^{-8}	3×10^{12}
3.	Light	4×10^{-7}	7×10^{11}
4.	Infrared rays	7×10^{-7}	4×10^{11}
5.	Radio Waves		
	(*i*) Ultra High Frequency (UHF)	1×10^{-1}	3×10^{6}
	(*ii*) Very high		

S. No.	Region	Wavelength (m)	Frequency (kHz)
	Frequency (VHF)	1	3×10^5
	(*iii*) High Frequency (HF)	10	3×10^4
	(*iv*) Medium Frequency (MF)	10^2	3×10^3
	(*v*) Low Frequency (LF)	10^3	3×10^2
	(*vi*) Very Low Frequency (VIF)	10^4	3×10^1

1.15 REFLECTION OF SOUND

When the sound waves after striking a surface are sent back, the phenomenon is called *reflection of sound.* The reflection of sound is similar to the reflection of light.

The practical examples of reflection of sound are: echo, thunder, reverberation, etc.

1.15.1 Laws of Reflection of Sound

The laws of reflection of sound are same as that of reflection of light:

(*i*) The angle of incidence is equal to the angle of reflection

$$\angle i = \angle r.$$

(*ii*) The incident ray, the normal and the reflected ray—all three lie in the same plane.

The reflection of sound is shown in Fig. 1.17 and it can be easily proved that the "sound image" A lies as far behind the surface as A is in front of the surface (as in case of image in a plane mirror).

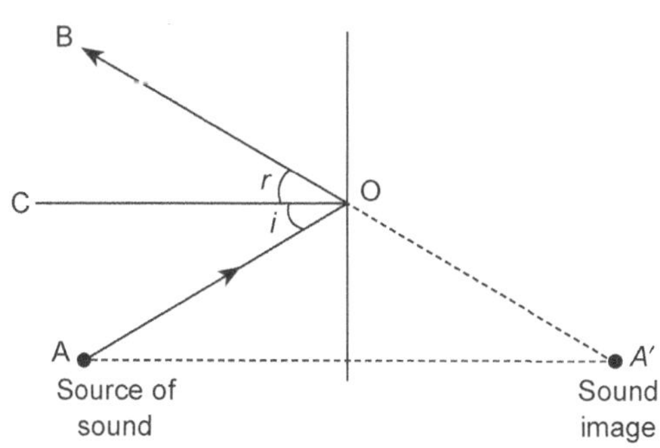

Fig. 1.17

The practical example of the *sound image* is that when an engine passes by a long wall of a building, its whisling can be heard behind the wall.

The phenomenon of reflection of sound is used to find depth of water by hearing echo of the sound. The laws of reflection of sound and that of light are same. The wave length of sound is 105 times that of light hence, the lenses and mirrors used for getting sound image are enormously large (e.g. long wall) as compared to that of optical lenses and mirrors.

1.15.2 Applications of sound reflection

(*i*) Concave surfaces concentrate the sound waves after reflection hence these surfaces are used as sound reflectors. See Fig. 1.18 (*a*).

(*ii*) Convex surfaces "diffuse" sound waves after reflection. These surfaces, hence are used in broadcasting studios to diffuse sound in the whole studio. See Fig. 1.18 (*b*).

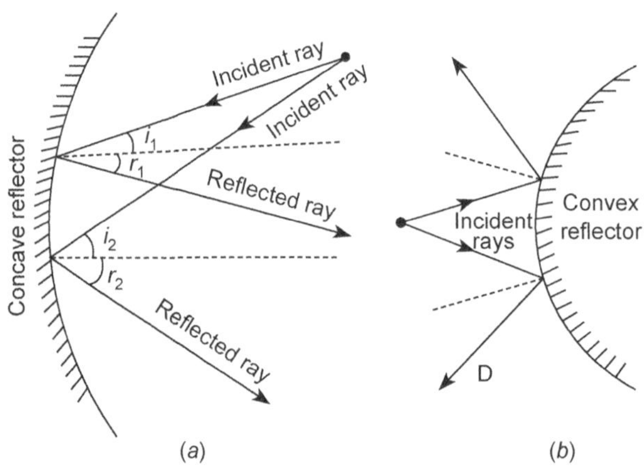

Fig. 1.18

1.16 REFRACTION OF SOUND

When a sound wave enters from one medium into another, it bends (like light rays), the phenomenon is known as refraction of sound. This is similar to the refraction of light.

When sound waves pass from one medium to another there is a change in the velocity of waves, as a result they are bent according to the laws of refraction of light. In other words, velocity of sound in air is 340 m/s and in carbondioxide gas (CO_2) is 265 m/s, hence when a sound wave enters from air into CO_2, it will bend towards normal. Similarly when sound wave travels from air into iron (velocity of sound in iron is 5000 m/s), the wave will go away from the normal.

The Fig. 1.19 shows the phenomenon of refraction of sound where, *AO* is the incident sound wave and *OB* is the refracted ray.

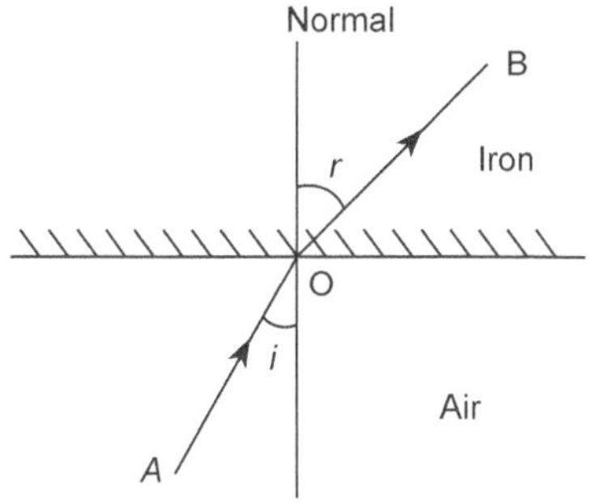

Fig. 1.19

The sound travels from air to iron, hence it is bent away the normal. That is angle of incidence μ_i is less than angle of refraction μ_r.

If the velocity of sound in air is v_1 and in iron v_2, then according to the laws of refraction

$$\frac{v_1}{v_2} = \frac{\sin \angle i}{\sin \angle r} \quad (\angle r > \angle i)$$

The phenomenon also depends upon the density of medium and its temperature.

1.17 DIFFRACTION (SCATTERING) OF SOUND (FIG. 1.20)

Whenever there is an obstacle in the way of sound waves, a change in the direction of wave takes place. This phenomenon is known as diffraction (or scattering) of sound.

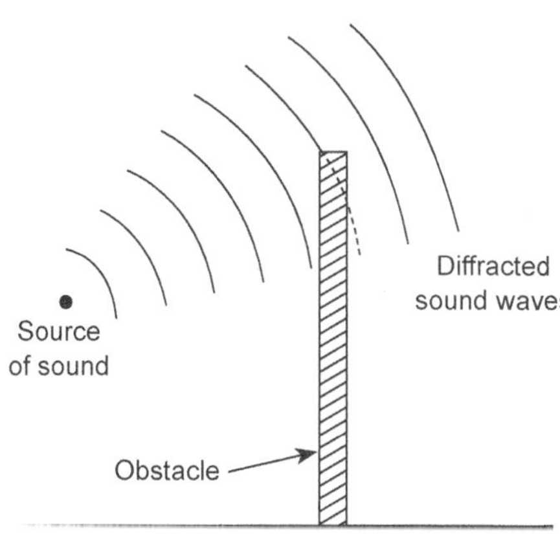

Fig. 1.20

Due to this effect of sound, we can hear a person speaking at the other side of a wall.

When a person speaks from one side of a wall, some of his sound reflects back from the wall but some crosses the wall and reaches other side of the wall. Which is called diffracted or scattered sound.

These scattered (or escaped) waves are said to be forming an image (or shadow) of the sound on the other side. The phenomenon is also affected by the wave length of the sound and size of the obstacle.

(*i*) A small obstacle does not produce any hinderance in the path of sound waves. It does not cause any "shadow".

(*ii*) An obstacle, which is much larger than the wave length of the sound waves forms a shadow *i.e.,* diffraction of sound occurs.

This knowledge is helpful in designing *acoustics of buildings*.

1.18 CHARACTERISTICS OF SOUND

Important characteristics of sound are under:

1. *Pitch.* The sensation of shrillness produced by a sound is known as its pitch. The "Cry" of a child has a high pitch, whereas soft voice of a lady has low pitch. Greater the frequency, higher is the pitch. The sound produced by a mosquito has a high pitch, the roaring of a lion has a low pitch.

   The unit of pitch is **Mel.** A pure tone of 1 kHz frequency has a pitch of 1000 mels. But frequency and pitch does not vary linearly. See Fig. 1.21. By pitch, we may compare two or more sounds. Pitch is affected by sound pressure. At increased sound pressure, pitch of a low frequency tone decreases, but at high frequency, tone increases.

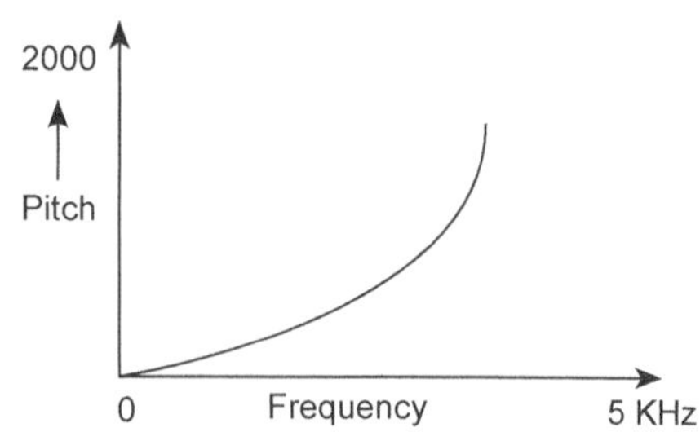

Fig. 1.21

2. *Timbre.* By this characteristic of sound, we can recognise two equal loud tones having same pitch but coming from different directions. Timbre, like pitch is also a complex characteristic and a function of intensity and frequency of sound.
3. *Harmonic structure.* Every sound wave contains a fundamental frequency and its harmonics. The fundamental frequency decides the pitch, whereas harmonics decide timbre of the sound.

4. *Loudness or Intensity.* The intensity of a sound is known as its loudness. Greater the loudness (*i.e.* intensity), greater is the effect on our ears.

The unit of intensity of sound is Bel. The bel is a logarithmic unit. More practical unit of intensity is decibel (db) which is one tenth of a bel.

$$1 \text{ Bel} = 10 \text{ db}$$

and
$$1 \text{ dB} = 10 \log_{10} \frac{P_2}{P_1}$$

Where P_2 and P_1 are output and input powers respectively.

The db has also a practical significance. The human ears do not distinguish different frequencies linearly, but they do it logarithmically. As our ears has a logarithmic response, the unit db completely matches with this quality of our ears, For example, if loudness of sound is made 100 times, the effect on our ears will not be 100 times but only double (log of 100 = 2). If it would not happen, our ears may get damaged.

The minimum intensity of sound, which a human ear can respond is equal to the energy of 10^{-12} watts per sq. meter and the maximum intensity of sound that our ear can hear without pain is 1 watt per sq. meter.

The table 1.8 gives intensity (loudness) of some sources of sound in dB.

Table 1.8: Loudness of sources

S. No.	Source of Sound	Loudness in dB (Intensity)
1.	Whispering	18
2.	Buzzing	42
3.	Normal talking	52
4.	Traffic (Heavy)	78
5.	Roaring of lion	100
6.	Thunder	110
7.	Siren	125

Intensity or loudness is

(*i*) directly proportional to the square of amplitude.

(*ii*) inversely proportional to the square of distance from the source.

(*iii*) directly proportional to the area of vibrating body (source).

Note: Intensity is also known by the amount of sound energy passing per unit area per second. It can also be measured in watts per second per cm^2 (or in Phons).

1.19 ULTRASONICS

The frequency range to which our ear can respond is known as "Audio Frequency". The AF range is from 20 Hz to 20 kHz.

The 20 Hz is known as "lower limit of audibility" and the 20 kHz as "upper limit of audibility".

Our ears neither respond a frequency below 20 Hz nor a frequency above 20 kHz.

Sound waves of frequency more than 20 kHz are known as "Ultrasonics".

Dogs can hear ultrasonics upto 50 kHz i.e. they can respond to the frequencies above A.F. The bats can produce ultrasonic frequencies and find their path by detecting the echoes after reflection from obstacles.

The frequencies below 20 Hz are known as "infrasonics". Both ultrasonics as well as infrasonics come under Inaudible Frequency range for a human being.

1.19.1 Production of Ultrasonics

Ultrasonics can be produced by

1. Magnetostriction Oscillators
2. Crystal Oscillator

1. *Magnetostriction Oscillators* (Fig. 1.22). It is based on the principle that "when a rod of a magnetic material is magnetised, it undergoes a change in its length".

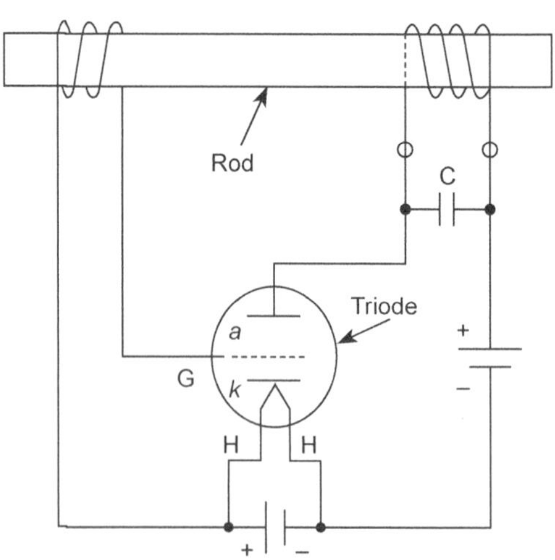

Fig. 1.22

This change in length (+ ve or - ve) has a frequency twice that of the frequency of applied magnetic field.

A rod of magnetic material (say Nickel) has two coils Q_1 and Q_2 wound on it (See Fig. 1.12). A current is passed in the coils through a triode. As a result, the rod is magnetised and demagnetised repeatedy depending upon the frequency of the supply. The length of the rod changes accordingly and its free end begins to vibrate producing ultrasonics.

2. *Crystal Oscillator* (Fig. 1.13). This oscillator works on the principle of "Piezo effect". Which says that "when HF voltages are applied across the two opposite faces of a crystal (of Quartz, Rochelle salt etc.) the crystal starts producing ultrasonics.

A thin quartz crystal is placed between two metallic plates— this forms a capacitor with crystal as dielectric. The arrangement is connected across an AC supply. (See Fig. 1.23)

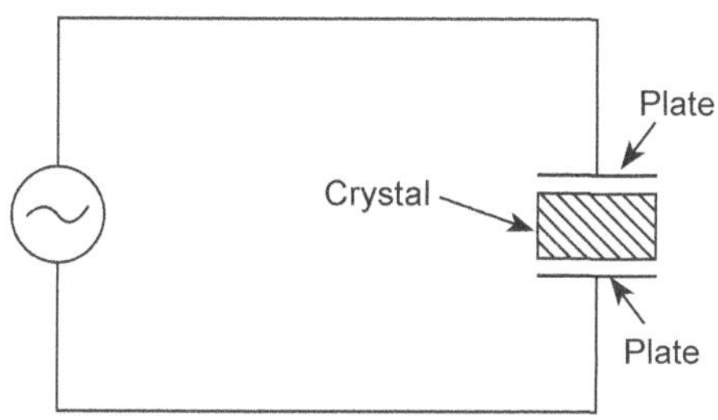

Fig. 1.23

Due to piezo effect, the crystal starts vibrating at very high frequency, thus producing ultrasonics. If the size of the crystal is selected such that the natural frequency of the crystal is equal to the frequency of the supply, "Resonance" occurs and amplitude becomes very large with this, oscillations of about 560 kHz can be produced.

The frequency that can be produced is given by

$$f = \frac{1660}{t} \text{ kHz}$$

where t is the thickness of the crystal in millimeters.

Ultrasonic frequencies are required in radio/TV broadcasting, transmission and reception.

1.20 TRANSDUCERS

A transducer is a device/equipment that converts an energy from its one form to the other. For example, microphone is a device that converts sound energy into electrical energy. Loudspeaker is another device which converts electrical energy into sound energy. Hence microphones and loudspeakers are transducers (Fig. 1.24). Transducers play an important role in audio and video systems.

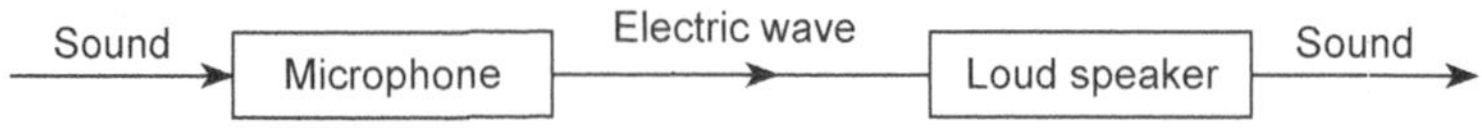

Fig. 1.24

1.21 HUMAN EAR (FIG. 1.25)

Our ear is also a transducer. When sound waves strike its membrane (diaphragm), it converts them into sensations (electrical pulses). These pulses reach our brain.

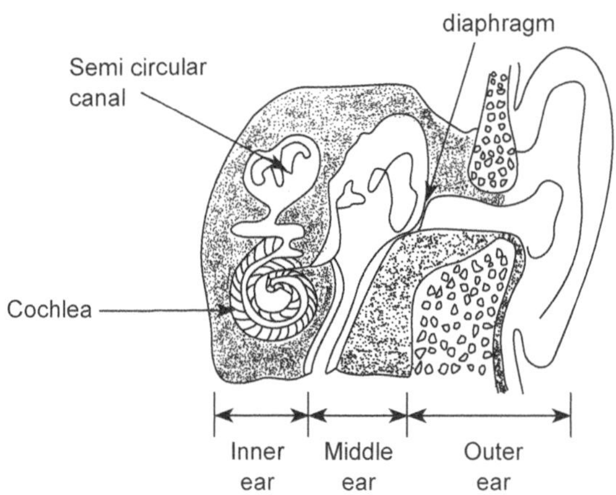

Fig. 1.25

The brain also in turn acts as a transducer and converts these pulses into the original sound, which we hear. Just as a similarity our ear acts as microphone, whereas the brain acts as a "loudspeaker".

Our ear consists of three main parts:

1. Outer ear
2. Middle ear
3. Inner ear.

1. *Outer ear*. The outer ear has a special shape, which increases its surface area and it can "catch" signals of large surroundings effectively.

2. *Middle ear.* The most important part of middle ear is the "ear drum" or diaphragm (membrane). The sound waves reach the membrane and produce oscillations in it. The membrane starts oscillating with a frequency and amplitude according to the pressure of sound striking it.
3. *Inner ear.* The oscillations reach the inner ear which is connected to our brain. These oscillations are converted into original sounds.

1.21.1 Salient Features of Human Ear (Fig. 1.26)

1. The human ear is very sensitive to sound. It can even respond to sounds having intensity as low as 0.1 Pico watt per m^2.
2. The movement of the diaphragm may be as minimum as one micron $= 10^{-6}$ m.
3. The response of the ear is not linear but logarithmic. In other words, loudness of sound hearld by our ears is not according to the intensity of sound but according to the "log of intensity of sound".

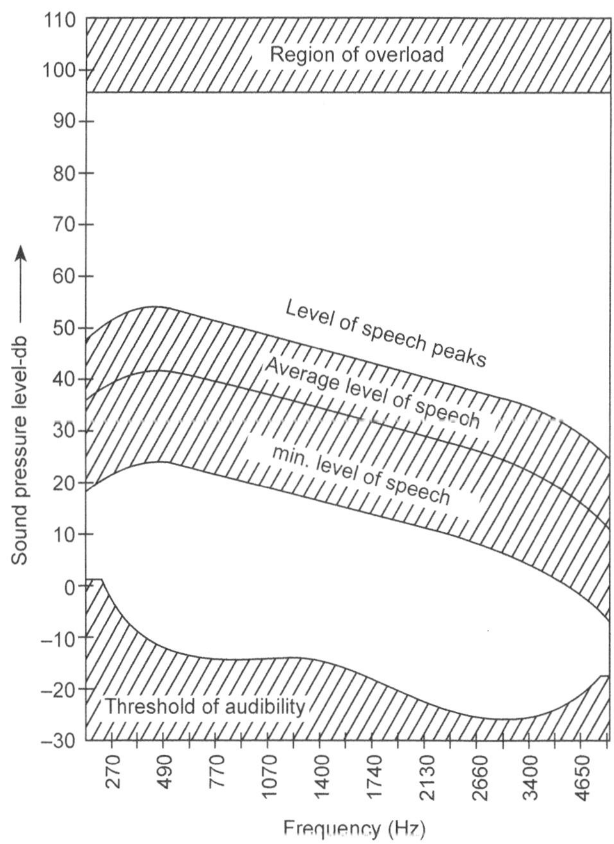

Fig. 1.26

4. Our ear can not detect a difference of 1 dB between the two sounds.
5. Our ear can do sum and difference of two sounds. It can also find the direction of a sound even, if it is weaker.
6. Our ear is more sensitive to higher frequencies and less sensitive to lower frequencies.
7. Our ear is most sensitive for frequencies of 3 kHz to 5 kHz at all ages. It decreases with age. Children's ears are sensitive upto 20 kHz.
8. Ears can detect a difference of 3 kHz in frequency and 1 db in intensity.
9. If we listen a weaker and a louder sound simultaneously, the former is supressed by the later. Movever the ear finds the direction of a sound which is received first, however weaker it may be.
10. Ear can respond sound pressures between 10^{-4} microbars to 10^{+4} microbars.
11. Ear can respond Audio frequency range (20 Hz to 20 kHz) and it can recognise a particular frequency sound out of many of different frequencies.
12. The minimum sound signal a person can respond is called his *Threshold of hearing*. This value varies from man to man and age to age.
13. Our ear judges "intervals" (ratio of frequencies f_1/f_2) and not the actual difference between 2 frequencies. For example, a change from 200 Hz to 1000 Hz will be recognised equally by our ears as a change from 50 Hz to 250 Hz as the ratio (1:5) is same (the difference may be different).

1.21.2 Frequency response curve for human ears (Fig. 1.27)

As mentioned earlier, the human ears respond to the whole audio frequency range (20 Hz to 20 kHz). The Fig. 1.27 shows a curve between sound intensity (in Phons) and frequency (in Hz), to which human ear can respond. At about 140 phons, we start feeling pains in the ear and this is known as *Threshold of pains*, though above 110 phons our ears feel discomfort. This intensity is called "*Threshold of discomfort*". But habitual (constant) hearing a sound above 160 phons can cause severe damage to our ear drums leading to total or partial deafness.

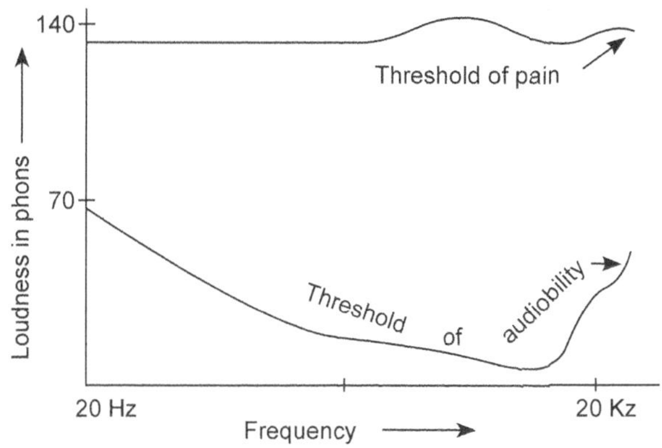

Fig. 1.27

1.22 TUNING CIRCUITS

A tuning circuit is an important component of all audio-video systems. It is basically a L-C circuit which can produce a wanted frequency. They are called tuning circuit as they are used to tune a particular station or channel in AV systems.

They may be

(*a*) *Series LC Circuit* [Fig. 1.28 (*a*)]. In which an inductance (L) and a capacitance (C) are connected in series.

(*b*) *Parallel LC Circuit* [Fig. 1.28 (b)]. In which, the inductance (L) and capacitance (C) are connected in parallel.

An a.c. supply is given to the combination.

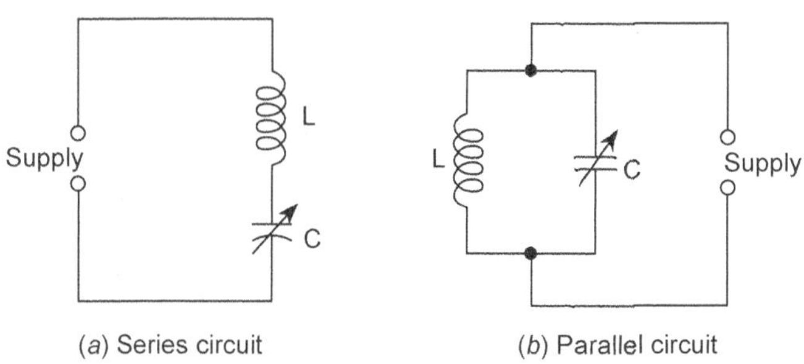

Fig. 1.28

At resonance, the value of "Inductive Reactance" becomes equal to the "Capacitive Reactance" of the circuit.

i.e., Inductive Reactance = Capacitance Reactance

or, $$2\pi f_r L = \frac{1}{2\pi f_r C} \qquad [f_r = \text{Resonant frequency}]$$

or, $$f_r = \frac{1}{2\pi\sqrt{LC}}$$

In tuning circuits, generally L is kept fixed and C as varied. By varying capacitance, a resonance is established, and a particular frequency fr is produced, known as *Resonant Frequency*. The circuit, therefore is also known, as *resonant circuit.*

For example, the frequency of Vividh Bharti radio station is 1350 kHz. By operating the radio knob, we change the value of C so that resonance is established with this frequency. Now this frequency is entertained by the radio and all other frequencies are earthed. The Fig. 1.29 shows tuning circuit of a radio receiver.

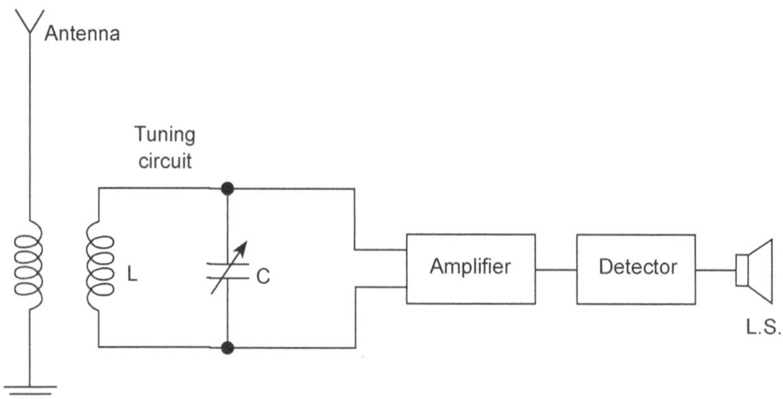

Fig. 1.29

1.23 FREQUENCY RESPONSE

It shows the "response" of an equipment (amplifier or microphone etc.) for a particular range of frequencies. The "Voltage gain" of a source depends upon the frequency. The curve between "Voltage gain" and "signal frequency" is known as "Frequency response curve".

The Fig. 1.30 shows frequency response curve of an amplifier. We can see that gain increases as the frequency increases till it becomes maximum at certain frequency Fr called "Resonant Frequency". Now if the frequency of signal is increased beyond f_r, the gain of the amplifier will decrease. In the Fig. 1.30, F_l is lower cut off frequency and F_u is upper cut off frequency,

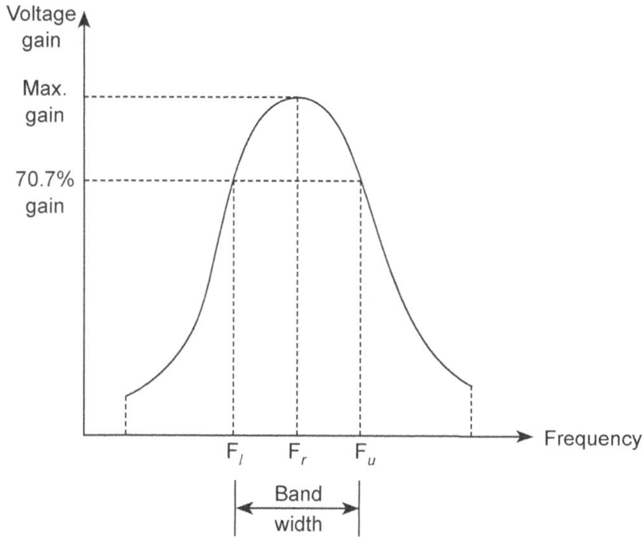

Fig. 1.30

The performance of all audio-video systems depends upon their frequency response.

1.24 BAND WIDTH (B.W.)

The range of frequency over which the gain of a system does not fall below 70.7% (of max. gain) is called as Band width.

The value 70.7% has a typical significance. It has been seen that if the gain of a system (say a loudspeaker) falls to 70.7%, our ear cannot detect this fall. But if the gain falls below this value, our ears will feel it. This is the reason that it is taken to be 70.7%.

In the Fig. 1.30, $F_l - F_u$ is the "Band width", during this range of frequency, the gain of the system does not falls below 70.7%, that is, remains constant.

For distortionless output, the range of frequency must be within B.W. of the system. In other words, the system will work "effectively" during this range.

If we calculate in db

$$\text{Fall in gain} = 20 \log_{10} \frac{100}{70.7} \text{ db} = 3 \text{ db}$$

Hence band width may also be defined as the range of frequency at the limits of which, its gain falls by 3 db from maximum gain.

Band Width For

(*i*) Telephone speech : 300 to 3500 Hz

(*ii*) Normal programs : 75 to 8000 Hz

(*iii*) HiFi program : 50 to 1500 Hz.

1.25 GAIN

The gain of a system is generally expressed in number. For example, if we say that gain of a system is 5, it simply means that its output will be "5 times of the input".

1.25.1 Bel and Decibel

For audio and video systems generally 'bel' (or decibel) is used as a unit for gain. It is a convenient unit and calculations become easier.

Bel is a logarithmic unit. It has its own practical significance. The response of ears in also logarithmic. In simple words, loudness of sound heard by our ears is not according to the intensity of sound but according to the log of intensity of sound.

For example, if the loudness of a loudspeaker is increased 1000 times, our ears will have only 3 times effect and not one thousand times, as the log of 1000 = 3.

Hence this unit tallies with the "natural response of our ears".

"The log of the "gain" to the base of 10 is called as Bel gain".

$$\text{Power gain (P.G.)} = \log_{10} \frac{P_{out}}{P_{in}} \text{ bel.}$$

$$1 \text{ bel} = 10 \text{ decibel}$$

$$\therefore \qquad \text{P.G.} = 10 \log_{10} \frac{P_{out}}{P_{in}} \text{ db} \qquad ...(1)$$

On the same lines

$$\text{Voltage gain} = 20 \log_{10} \frac{V_{out}}{V_{in}} \text{ db} \qquad ...(2)$$

and

$$\text{Current gain} = 20 \log_{10} \frac{I_{out}}{I_{in}} \text{ db} \qquad ...(3)$$

1.26 IMPEDENCE MATCHING

"If impedence of the source and impedence of the load are equal, maximum power is obtained at the load, and the limit of the maximum power which can be obtained is only 50%".

This is known as "maximum power transfer theorem".

At this stage, the load is said to be "matched" with the source and this is known as "impedence matching".

In audio and video systems, this is very important. Generally, we have amplifier as "source" and "loudspeaker" connected to the amplifier acts as load. To obtain maximum power transfer from "amplifier to speaker", it is necessary that impedence of amplifier is same as that of the "loudspeaker". In other words, "Impedence matching" between the two is essential. In such cases efficiency is sacrificed for more output. The arrangement is known as *Public Address* (*PA*) system which is a most practical example of impedence matching.

1.27 SONE

This term is used to determine increase in loudness produced by

1kHz, Remember 40 Phone sound is equal to 1 Sone.

1.28 INTELLIGIBILITY

This factor determines the frequency range of sound required for satisfactory transmission of speech. It is also called *Articulation*.

1.29 SOUND SYNTHESIS

This is the process of producing artificial sounds. We can artificially produce sound of any musical instrument or even of an orchestra. For this, audio wave forms are generated electrically and mixed by electronic tuner. So synthesised sound is a complex form produced by electronic circuits.

1.30 TYPES OF SOUNDS

Sound is a form of energy. Its generation, transmission and recaption takes place through vibrations.

Almost all sounds are of three types: (Fig. 1.31)

1. Speech
2. Music
3. Noise.

1.30.1 Speech

The speech is the common sound we speak from our mouth. This sound is generated in our vocal tract and radiated from our lips. The speech characteristic of different persons is different however, their sound pressure lies in the range of 65 dB to 80 dB, when spoken in a normal tone. The speech becomes directional at high frequencies, in other words, it can then be hearld in a particular direction only.

We can generate a frequency between 100 Hz to 10,000 Hz by mouth. The Fig. 1.31 shows energy distribution versus frequency. It is clear that at high frequency, there is loss of energy. The production of speech is described in terms of intelligibility which means how a speech can be recognised and understood by a listener. This will depend upon the energy of sound spoken, hearing power of the listener and the noise in the surrounding.

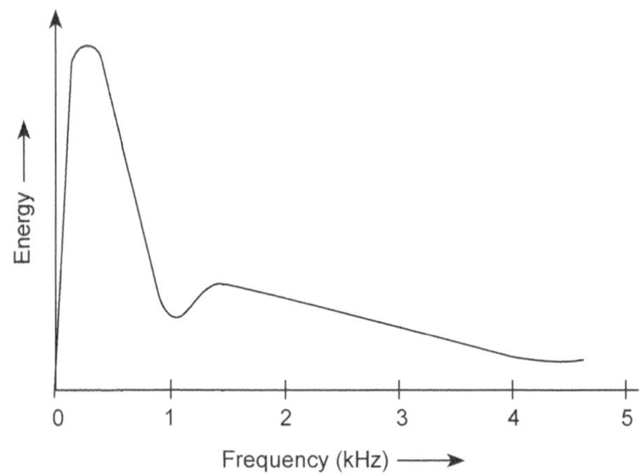

Fig. 1.31

1.30.2 Music

The sound, which gives a pleasant sensation to the listener is called as musical sound. The musical tones have simple harmonic structure with a regular wave form. It consists of fundamental frequency and its harmonics. The music tones are produced by vibrating bodies such as strings, membranes, and air columns. Different musical instruments play over different range of fundamental frequency (See Table 1.9).

Table 1.9: Frequency range of musical Instruments

Musical instruments	Frequency range (Hz)
1. Xylophone	150—3000
2. Pipe organs	20—8000

Musical instruments	Frequency range (Hz)
3. Harmonica	200—1000
4. Saxophone	50—300
5. Flute	300—2000
6. Trumpet	200—900
7. Piano	25—4000
8. Banjo	130—800
9. Cello	150—1000
10. Violin	200—3000

Most of the musical instruments are "directional" *i.e.,* they have their maximum effect in a particular direction. This knowledge helps us to arrange an orchestra.

Musical sound is different to that of a speech in the following ways:

(*a*) Musical sound contains different fundamental frequencies, whereas the speech contains one fundamental at a time.

(*b*) Musical sound has a wider frequency band as compared to the speech which has a very narrow band.

(*c*) Music is more sensitive to distortions as compared to the speech.

(*d*) Musical tones have Simple Harmonic Motion (SHM) with regular amplitude waves. A musical tone with one frequency can be represented by Fig. 1.32 (*a*).

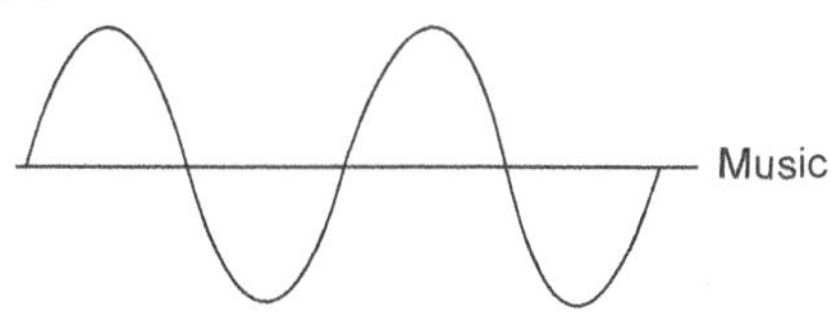

Fig. 1.32 (a)

1.30.3 Noise

Noise is that, which gives unpleasant (irritating) sensations to ears. They are irregular amplitude waves [Fig. 1.32(b)]. Noise causes fatigue and pains in ears and hence the listener avoids it.

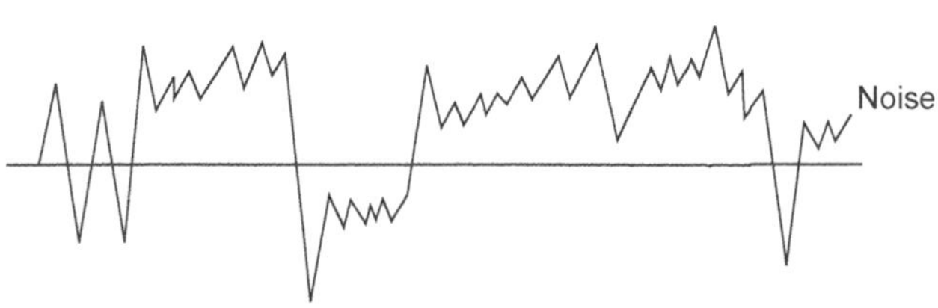

Fig. 1.32 (b)

(a) Classification of Noise

Noises can be classified as:

1. *Thermal noise.* This noise is generated in resistors due to thermal agitation of electrons. The amount of noise depends upon the value of resistor and its operating temperatures.
2. *Electrical noise or humming.* This noise is generated due to power line frequencies (50 Hz) in the audio device. The power frequencies contain harmonics (*i.e.,* 100 Hz, 200 Hz etc.) which causes noise. A good filter circuitary can reduce this noise.
3. *Cross talk.* This noise enters in an audio equipment due to presence of other circuits in the vicinity, which cause inductive or capacitive coupling with the audio circuits. This noise can also be reduced by proper filtering.
4. *Generated noise.* Thermal and electrical noises come under this class. This noise is generated in amplifiers.

   When signal enters an amplifier, this noise is produced or generated due to various components (resistors, capacitors, transistors etc. of the amplifier). This noise is added to the signal (See Fig. 1.33).

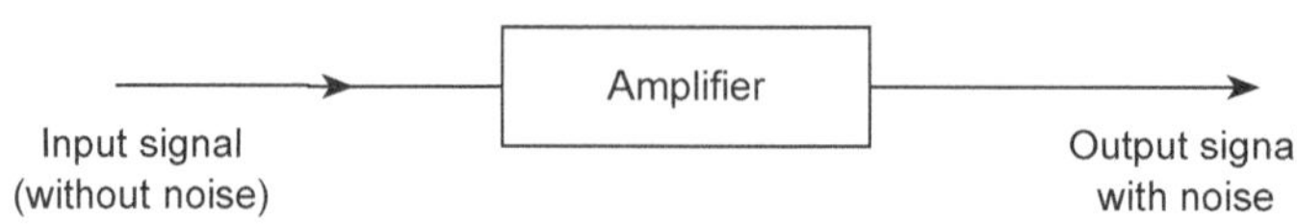

Fig. 1.33

   Generated noise further may be of the following types:

   (*a*) *Johnson nose.* This noise is produced when a signal passes through a resistor. The atoms of the resistor remain in continuous vibration and produce noise, which is added up in the signal.

   (*b*) *White noise.* The vibrations produced by thermal effects within a resistor cover a wide frequency range and therefore the noise generated consists of a wide spectrum of frequencies. This wide band noise is sometimes called *white noise.*

   (*c*) *Pink noise.* Pink noise is a signal, whose noise power per unit frequency interval is inversely proportional to frequency itself over a specified range. It is white noise passed through filter.

   (*d*) *Shot noise.* The internally generated noise in semi conductors and vacuum tubes on account of random movement of charges is called *shot noise.*

5. *Conducted noise.* The power supply to the amplifiers may itself be a source of noise as it contains ripples, spikes etc. This noise is called *conducted noise.*

6. *Radiated noise.* There may be electrical or magnetic fields in the surrounding. Unwanted signal may radiate in to the audio device and produce noise. Such noise may be called *Radiated noise.*

   The Fig. 1.34 shows sources of radiated noise.

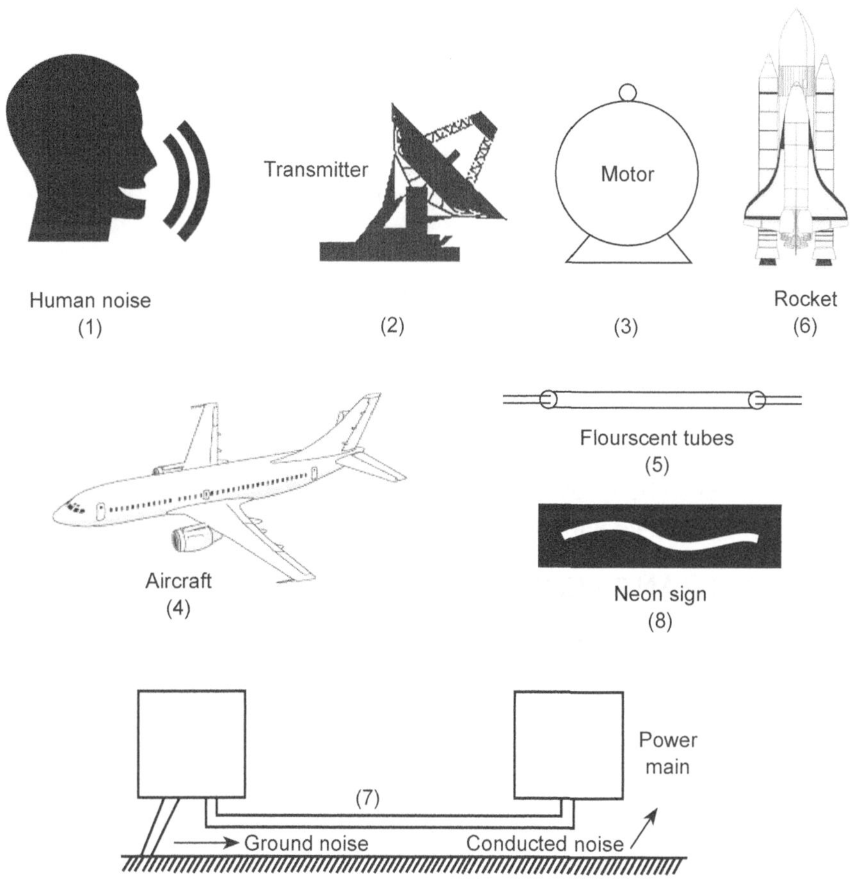

Fig. 1.34

7. *Environmental noise.* The noise present in the surrounding is called environmental noise. It may be

   (*i*) *Transient noise.* Which occur suddenly like thunder or explosion near by.

   (*ii*) General noise or Ambient noise. The noise present all the time *e.g.* class room, auditorium.

(b) Harmful Effects of Noise

Working continuously in a noisy environment has the following ill effects on human beings:

(*i*) It produces hypertension/high B.P.

(*ii*) It causes mental fatigue and lowers efficiency

(*iii*) It may retard mental growth.

(*iv*) It may cause harm to our hearing system.

(*v*) It may cause nervous breakdown.

(c) Methods of Reducing Noise

All types of noises may be broadly classified as:

(*a*) Electrical noise

(*b*) Environmental noise.

(*i*) The *electrical noise* may be reduced by using regulated power supply, use of synchronous and servo motors in drives (like recorders), regular cleaning of recorder heads, and by using techniques (*e.g.* Dolby method).

(*ii*) The *environmental* noise may be reduced by using sound absorbers/insulators in the room. In general, room should be provided with thick curtains.

1.31 IMPORTANT TERMS

1. *Signal to Noise ratio (SNR).* The ratio of desired signal to the unwanted noise is called S/N ratio. It may be expressed as

$$S.N.R. = \frac{S}{N} = \frac{\text{Signal Power}}{\text{Noise Power}} = \frac{\text{Signal expressed in (volt)}^2}{\text{Noise expressed in (volt)}^2}$$

[Power $\propto$ Volt^2]

A large SN ratio is always desired in audio video systems. This can be done by decreasing noise level by some means. If SN ratio is small (that is more noise), it may be a matter of concern.

2. *Noise figure.* The noise figure (NF) is the degradation of the SNR which may be attributed to the amplifier. It may be defined as ratio of total available output noise power (P_1) to the part of output power resulting from thermal noise or source resistance (P_2).

Thus $\quad\quad \text{NF} = P_1/P_2$

3. *Octave or Interval.* It is often convenient to divide the audio frequency spectrum into bands. The Octave is a widely used division of frequency spectrum. An octave has an 1 : 2 ratio between lower and upper frequencies. An octave band or "Interval" includes the end points and every frequency in between. The center frequency of an octave band is not the mathematical average of the upper and lower cut off frequencies, but is approximately $\sqrt{2}$ times the lower frequency.

   As explained above, the ratio of 1 : 2 is called an "octave". For example one octave of 200 Hz is 400 Hz (200 : 400 = 1:2) and two octaves of 200 Hz is 800 (400 + 400) Hz and so on.

   Mathematically number of octaves (n) between two frequencies f_1 and f_2 is given by

$$n = \log_2 (f_2/f_1).$$

4. *Harmonics.* Harmonics are the integer multiples of fundamental frequency *e.g.,* with respect to 50 Hz, its second harmonic will be = 100 Hz (2 × 50), third harmonic 150 Hz (3 × 50) and 7th harmonic = 350 (7 × 50) and so on.

   Remember 105 Hz is not an harmonic of 50 Hz as it is not integer multiple.

5. *Overtones.* All frequencies higher than fundamental are known as overtones. In this way, harmonics also come under overtones.

   The frequencies lower than fundamental are known as undertones.

1.32 ARCHITECTURAL ACOUSTICS

Architectural acoustics is the branch of engineering which deals with the production of sound effects in buildings, cinema halls, auditorium etc. The behavior of sound in enclosed spaces have been of interest to designers since long.

After Mr. sabine found a simple formula for calculating *reverberation time* for decay of sound in a room, the work in this line started on scientific grounds.

Later on, it was found that decay time of sound can be controlled by using absorbant materials on the walls of the room. After this, search started for materials with high coefficient of absorption *i.e.,* which effectively can absorb the sound incident upon.

1.32.1 Reverberation

The persistence of sound even after the original sound is ceased to operate is known as reverberation. Remember that reverberation is little different to that of an *echo*.

Reverberation time. The time taken by the sound to die completely is called reverberation time.

Actually sound is reflected back by the things placed in the room and bounces inside, till its energy is converted into heat.

Reverberation time may be defined as the time taken by the sound to decay by 60 dB from its original level (see Fig. 1.35). At point *O*, the sound source is switched ON and sound intensity rises upto point *A*, and then it becomes steady. At point *B*, the source of sound is switched off and the sound starts decaying. At point *C*, it falls down by 60 dB. Reverberation time is represented by *BC*.

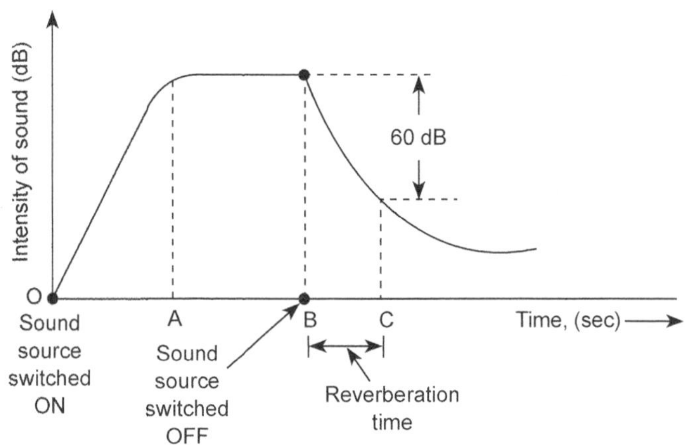

Fig. 1.35

Sabine Formula for Calculating Reverberation Time

The sabine's formula for calculations of reverberation time is given by

$$t = \frac{0.161 \times V}{a} \text{ seconds}$$

where, a is the total absorption area (of sound) of the room in m^2 and V is the volume of the room in m^3.

By absoption area, we mean the total area of the surfaces in the room absorbing sound *i.e.,* walls, ceiling, floor, curtains, furniture, audience etc.

Table 1.10: Reverberation time (t) of few places

S. No.	Place	t in seconds
1.	Drawing room	0.5
2.	T.V. Studio	0.8
3.	Concert hall	2.2
4.	Class room	0.5
5.	Temple	2.4

There is little difference between reverberation and echo. In case of echo, there is a time gap (about 50 ms) between the original sound and its echo. If the time gap is less than 50 ms, it is reverberation. The distinct echo produces irritation and hence should be totally eliminated from the room.

Reverberation has continuity of sound. A small reverberation is pleasing, therefore desirable in the design of acoustics. Total elimination of reverberation makes the sound dull and lifeless, whereas small reverberation keeps the naturality of sound.

1.32.2 Growth of Sound in a room

If a source of sound is operating in a room, the absorption of sound by the walls, surfaces and by the medium of the room prevents the sound intensity from becoming large. In small rooms the absorption in the medium is negligible. If absorption power of surfaces is small and growth of sound is low, when a sound source is started, reflections at the wall produce a sound energy distribution which becomes uniform with time. If the source of sound is pure tone of only one frequency, *standing waves* may be set up in the room. If the source produces a band of frequencies, the interference effect in a small room may be neglected.

1.32.3 Decay of Sound in a room

The intensity or sound pressure level in a room decreases with time at a constant decay rate given by

$$D = \frac{1.087\, a.c.}{V} \text{ db/sec}$$

where a is the total absorbent surface area, c the velocity of sound in air and V is the volume of the room.

As the reverberation time t is the time required for the level of the sound in the room to decay by 60 db.

i.e.,

$$t = \frac{60}{D} = \frac{60}{1.087 \times ac / V} = \frac{55.2\ V}{ac}$$

$$\left[\because D = \frac{1.087\ ac}{V}\right]$$

Now putting velocity of sound in air $c = 343$ m/s, we get

$$t = \frac{55.2\ V}{a.343} = \frac{0.161\ V}{a}$$

The volume to be taken in cubic meters and a in square meters.

Sabine formulae brought revolution in acoustics, since reverberation time can be calculated at once if volume and absorbant surface area of the room are known. By surface area we mean, the total absorption of sound, that can be easily improved by using absorptive materials on the walls. In other words, the reverberation time t can be easily controlled.

1.32.4 Absorption of Sound

When sound waves fall on a surface, the surface is set into vibrations and the damping force dissipates the sound energy into heat. Therefore the main cause of absorption of sound by materials are their porosity and vibrations.

When sound waves stike a surface, the sound is scattered in all directions. This phenomenon is a source of trouble in cinema halls, auditoriums and damages the quality of acoustics of the buildings.

The amount of absorption of sound differs from material to material and medium to medium.

"The ratio of sound energy absorbed by a surface to the total sound energy falling on it is called its *coefficient of absorption or absorbing power*".

The coefficient's value is between 0 to 1.

Table 1.11: Coefficient of absorption of few materials

S. No.	Materials	Coefficient of absorption
1.	Marble	0.01
2.	Concrete, Plaster	0.02
3.	Glass	0.03
4.	Wood	0.05
5.	Carpet, cork	0.25
6.	Furniture, Asbestos	0.30
7.	Human body	0.50
8.	Open window	0.98

S. No.	Materials	Coefficient of absorption
9.	Brick wall	0.30
10.	Wooden door	0.05
11.	Curtains	0.18
12.	Cushions	0.21
13.	Ceiling, Floor	0.05

Note. The coefficient of absorption usually increases with increase of frequency.

1.32.5 Factors in Architectural Design

(*a*) (*i*) The sound should be sufficiently loud and uniform throughout the room or hall.

(*ii*) There should be no echo.

(*iii*) The sound should die soon, after it is ceased to operate *i.e.*, reverberation time should be less.

(*iv*) Sound absorbers should be used.

(*v*) Human body is a good absorber. This is the reason that a hall full of audience has better sound effect than when it is empty.

(*vi*) The ceiling should be covered with some good sound absorbing material.

(*vii*) The hall should be free from resonance.

Echoes are produced in a room, with ceiling not sufficiently high. Due to repeated reflection of sound, echo remains in the room for sometime, even after the original sound has ceased.

As we have mentioned earlier, persistence of our ear is l/12th of a second. If another sound comes to our ear within this time, we can not distinguish it. It can be easily calculated that minimum distance for an echo to be heard should be 15 meters from the listener.

(*b*) The following considerations are also to be kept in mind, while design of acoustics, in addition to given above:

1. Reverberation time should be controlled. This may be done by controlling repeated reflection of the sound. For this purpose, good absorbing materials should be used. Open windows serve as perfect absorbers. The walls, ceiling should be covered with card boards, asbestos, heavy curtains etc. Audience is also a good absorber as told already.

2. Balconies, domes and such like structures should be avoided as they produce echoes and focussing effect of sound, with the same reasons, curved surfaces should be also avoided.
3. Acoustic treatment materials should be used to control absorption in the hall.

   The absorbing and acoustic treatment materials may be of following types:

   (*i*) *Porous materials* such as curtains in which sound penetrates and the material has no tendency to produce vibrations. "Perforated panels" is another example for such materials. Ceiling is the most appropriate surface for application of these materials. Due to this reason the false ceilings made of good absorbing materials have become popular.

   (*ii*) *Non-porous materials.* Under this class of absorbant materials wood panels, plastic boards and acoustic tiles are very much used.

1.32.6 Resonance Effect of Sound in a room

When a sound wave passes through an air column, at a certain relationship between wavelength of the wave and length of the air column "resonance" occurs. The same resonance is produced in the air column in a closed pipe. A long rectangular room (length is much more than the breadth) will also resonate due to the same reason given above. The resonance effect will depend upon the relation between length of the room and wavelength of sound.

On the same lines it can be explained that resonance will occur in sound devices like microphones, loudspeakers, recorders as they have mechanical parts capable of vibrating. They have a natural resonant frequency depending upon the mass of their vibrating parts. Similarly absorbing and reflecting surfaces will produce resonance effect when sound will penetrate through them.

SUMMARY

1. The "Audio Video Systems" deal with the electronic consumer devices.
2. The Audio Engineering is the branch of engineering which deals with the study of audio (sound) energy.
3. Sound is a form of energy.
4. Audio frequency range is between 20 Hz to 20 KHz.

5. Like light, the sound also has property of reflection, refraction and diffraction.
6. Transducer converts one form of energy into another form. The human body has many transducers like eye, ear, brain etc.
7. The decibel is a logarithmic unit to measure gain, noise, pressure etc,
8. The persistence of sound, even after the original sound ceases is called reverberation.
9. Human body is a good sound absorber.
10. The tuning circuit is an important component of all audio video systems.
11. A resonator is a device that can produce resonance. An organ pipe and other musical instruments acts as resonators.

SELF TEST QUESTIONS

1. What do you understand by audio-video systems?
2. Define Audio Engineering.
3. On what factors, velocity of sound depends? Write down the relation between frequency and wavelength.
4. Write short notes:

   Superposition of sound, Beats, Absorption of sound, Reverberation, Ultrasonics, Transducers, Tuning circuit, Selectivity, Frequency response, Band-width, Decibel, Impedence maching and PA system.
5. Explain reflection, refraction and diffraction of sound.
6. Sketch a human ear. Explain its mechanism.
7. Why woman's speech is difficult to be interpret?
8. What is the unit of sound pressure?
9. How the laws of optical reflection, refraction and diffraction are applied to sound.
10. Define signal/noise ratio. Give frequency response of human ears.
11. Explain the following terms: Interval, octave, Harmonics, Overtones, Timbre, Threshold of audibility and of pain, Resonance in a room.
12. Give various type of noises and sources of noise.
13. Explain reverberation of sound.
14. Enlist the important considerations while designing acoustics for auditoriums.
15. What are the various acoustic treatment materials?

16. What is the role of absorbing materials in acoustics?
17. Why an auditorium full of audience has a better sound effect?
18. Write a brief essay on noise, reverberation and acoustics of building.
19. Define architectural acoustics.
20. Explain sabine's formula for rise and decay of sound in a room.
21. Differentiate between reverberation and echo.

OBJECTIVE TYPE QUESTIONS

Fill in the blanks:

1. Sound is a ________
2. Sound is a ________ wave motion.
3. Velocity of sound in air is ________
4. Intensity of sound at threshold of pain is ________
5. Sone is a unit of ________
6. Unit of pitch is ________
7. Sound travels ________ than light.
8. With increase in temperature, velocity of sound ________
9. Absorption of sound at low frequencies is ________
10. High A.F. contains ________ energy.
11. The A.F. range is ________
12. Energy of a sound wave is ________ at high frequency.
13. Unit of loudness is________
14. Frequency for telephone speech is ________
15. With respect to 100 Hz, the third harmonic is ________
16. Examples of transducers are ________
17. Velocity of radiowaves is ________
18. Music-noise ratio should be ________ than one.
19. If frequency increases, wave length of a wave ________
20. A little reverberation is ________
21. Acoustics is the science of designing auditorium for creating best ________
22. Without reverberation, sound will become ________
23. Curved surfaces in buildings are ________

24. The absorption coefficient of women is ________ than men.
25. Sound effects are ________ in empty room as compared to full of audience.
26. Reverberation time in empty hall is ________ than full of audience.
27. Absorption of sound is more at ________ frequencies.

ANSWERS

1. Energy	2. longitudinal	3. 340 m/s
4. 12 W	5. loudness	6. mel
7. slowly	8. increases	9. small
10. less	11. 20 Hz—20 kHz	12. high
13. phon	14. 500 Hz—3500 Hz	15. 300 Hz
16. microphones and loudspeakers		17. 3×10^8 m/s
18. more	19. decreases	20. desirable
21. sound effects	22. dull	23. undesirable
24. more	25. poor	26. more
27. high.		

❑❑❑

2

Audio and Video Amplifiers

The amplifiers are used to amplify voltage/power in audio video devices.

The amplifiers used to amplify audio frequencies are called Audio Frequency amplifiers or simply audio amplifiers.

The amplifiers used to amplify video frequencies are called video frequency amplifiers on simply video amplifiers.

2.1 AMPLIFIERS

The amplification is carried out in stages. Accordingly, Amplifiers are of two types: (Fig. 2.1)

1. Voltage amplifiers   2. Power amplifiers.

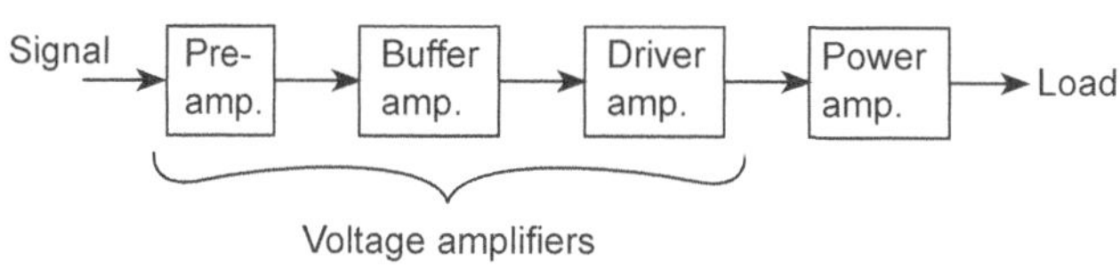

Fig. 2.1

Voltage amplifiers raise the level of voltage in stages called Pre-amplifier, Buffer amplifier and Driver amplifier. Finally a power amplifier is used, which raises the power and transfers: to the load through proper impedance matching.

2.2 CLASSIFICATION OF AMPLIFIERS

2.2.1 Audio Amplifiers

These amplify audio signals. Important terms are defined as

1. *Gain.* The ratio of voltage/power of output signal to the Input signal is called Gain of the amplifier. It is given in decibel (dB).

It may be,

(*a*) Voltage gain = $20\log\left[\dfrac{\text{Output voltage}}{\text{Input voltage}}\right]$

(*b*) Power gain = $10\log\left[\dfrac{\text{Output power}}{\text{Input power}}\right]$

2. *Bandwidth* (B.W.). Bandwidth of an amplifier is the range of frequencies, for which it gives maximum (and constant) output. Naturally B.W. of audio amplifiers should be 20 Hz—20 kHz.
3. *Distortion.* The various distortions occurring in audio amplifiers are distortion in amplitude, frequency and phase.
4. *Impedance.* Impedance of the power amplifier is more important, as it is to be connected with load through an impedance matching device.
5. *Sensitivity.* This is the minimum input signal which will give rated output of the amplifier. Lowest is the signal value in the input, better will be the sensitivity of the amplifier.
6. *Noise level.* The noise level of amplifiers should be low in other words, signal/noise ratio should be high.
7. *Slew rate.* The response of an amplifier (per second) to a changing signal is called its *slew rate*.
8. *Collector efficiency of amplifier.* The ratio of a.c. output power to the zero signal d.c. power supplied to a power amplifier is called its *collector efficiency*. This should be high for a good amplifier.
9. *Power dissipation capacity.* The ability to dissipate heat is known as power dissipation capacity of the amplifier. If heat is not dissipated, the amplifier may suffer a *thermal run away*.

2.2.2 Voltage Amplifiers

A voltage amplifier generally used is a R-C coupled amplifier.

The resistors R_1, R_2 and R_E form the biasing for the transistor, and help in stabilisation of operating point of the transistor. The capacitor C_{in} allows only ac signal to flow through the transistor. The capacitor C_E maintains maximum voltage at the output and the capacitor C_c is used to couple the different stages of the amplification. A two stage voltage amplifier is shown in the Fig. 2.2.

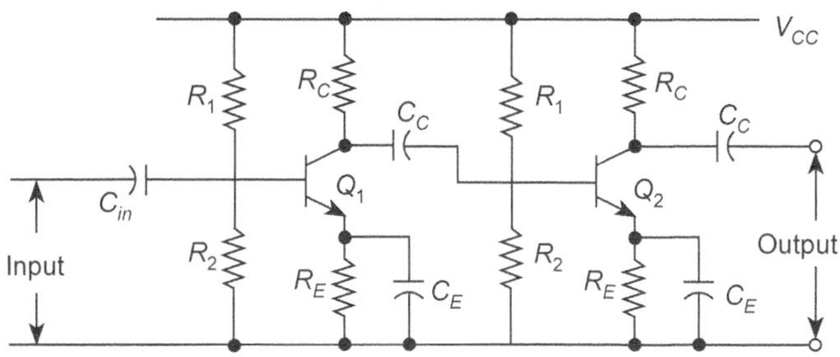

Fig. 2.2

It is called a RC coupled amplifier, as the different stages ore coupled by RC coupling.

2.2.3 Power Amplifiers

Power amplifiers may be

(*i*) *Class A power amplifiers (R-C coupled amplifiers).* In these amplifiers, collector current flows all the time during the full cycle of the signal. They are just voltage amplifiers used at final stage of power amplification. Their circuit is same as shown in Fig. 2.2 except that the output is connected to a loud speaker with a matching device.

(*ii*) *Class B power amplifier/Push pull amplifier.* In these amplifiers, collector current flows only during positive cycle of the input signal. Generally a push-pull amplifier in class B operation is used for power amplification. In it, two transistors are placed back to back. The output transformer has a centre taped primary and a loudspeaker is connected across its secondary. See Fig. 2.3

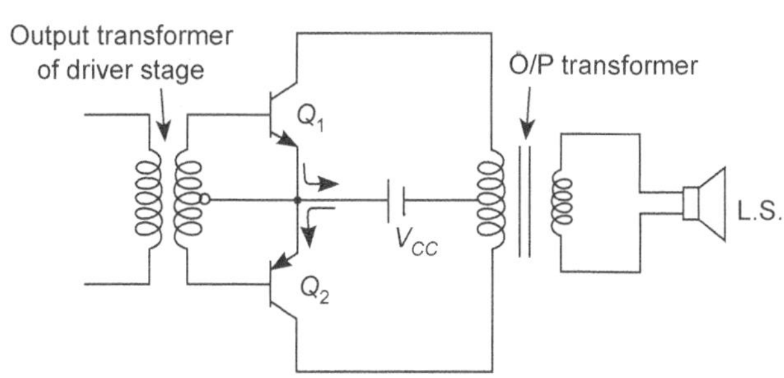

Fig. 2.3

(*iii*) Class C amplifiers. In these amplifiers, the collector current flows for less than a half cycle of the signal. However, these have a L.C. circuit and used to amplify a selected narrow band of frequencies. see Fig 2.4.

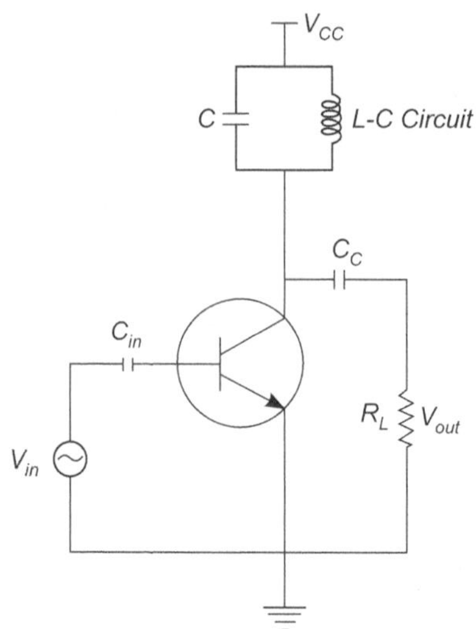

Fig. 2.4

2.3 NEGATIVE FEEDBACK (F.B.) IN AMPLIFIERS

The process of sending a fraction of output back to the input of a device is called "Feedback". The feedback may be negative or positive. In amplifiers negative feedback is employed i.e. the feedback energy is in opposition to the input signal. The advantages of negative feedback in amplifiers are:

1. The gain of the amplifier becomes stable.
2. The non linear distortion is reduced.
3. The frequency response is improved.
4. The gain becomes independent of transistor's parameters, in other words replacement of transistor will have no effect on the gain. Moreover, the gain also becomes independent of the temperature variations.

Note.

(*i*) The overall gain of an amplifier is reduced with negative feedback even then all amplifiers are provided with negative feedback to get the above advantages.

(*ii*) If F.B. fraction = 2%, it means 2% of the output is feedback into the input circuit of the amplifier.

2.4 CONTROLS IN AUDIO AMPLIFIERS

The following controls are provided in audio amplifiers:

(*i*) *Microphone Gain Control.* It is provided to control output of the individual microphone. For this, a potentiometer (P_1, P_2) is provided in the output of the microphone (Fig. 2.5).

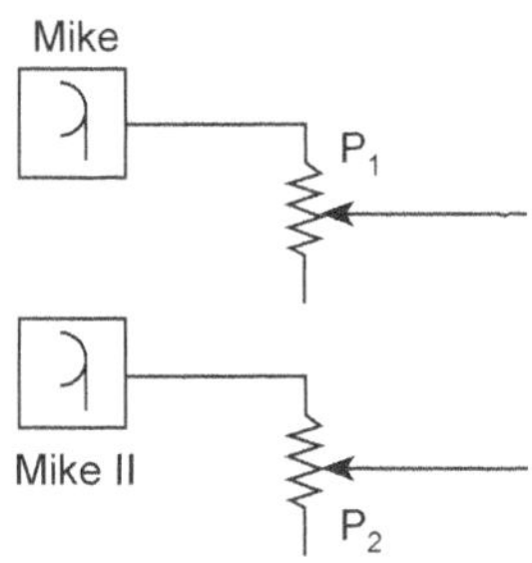

Fig. 2.5

(*ii*) *Volume Control.* It provides control on the overall gain on the volume of the amplifier. For this also, a potentiometer is provided in the collector circuit of the transistor. The potentiometer may be linear as well as logarithmic. But as our ears have logarithmic characteristics, logarithmic potentiometer shall give better results. (Fig. 2.6)

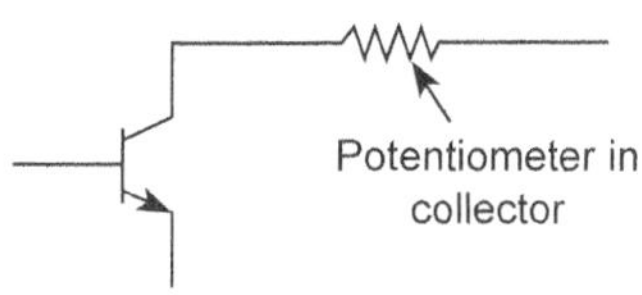

Fig. 2.6

(*iii*) *Tone (Bass and Treble) Control.* As we know depth of the sound is given by low frequencies (called bass) and sharpness is given by high frequencies (called treble). Some people want to boost *bass* and other want to boost *treble*. Therefore amplifiers are provided with both controls:

(*i*) Bass control

(*ii*) Treble control.

As we know, a capacitor has infinite impedance for low frequencies. Keeping this in view, capacitors (*c*) and potentiometers (P) are_ provided in the amplifier circuit as shown in Fig. 2.7. In this way, bass and treble controls check the amount of low and high frequencies respectively going to the next stage of the amplification.

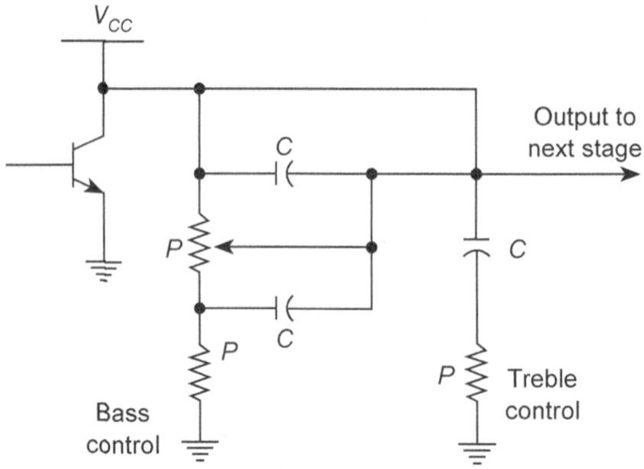

Fig. 2.7

2.5 VIDEO AMPLIFIERS

As already mentioned, few amplifiers reproduce frequencies far beyond the audible range. This extended response becomes of great importance in video (also called wideband) amplifiers. In order of merit, the wideband amplifier must be capable of responding from a few hertz to several million hertz of frequencies. The Fig. 2.8 shows frequencies response of a typical video amplifier.

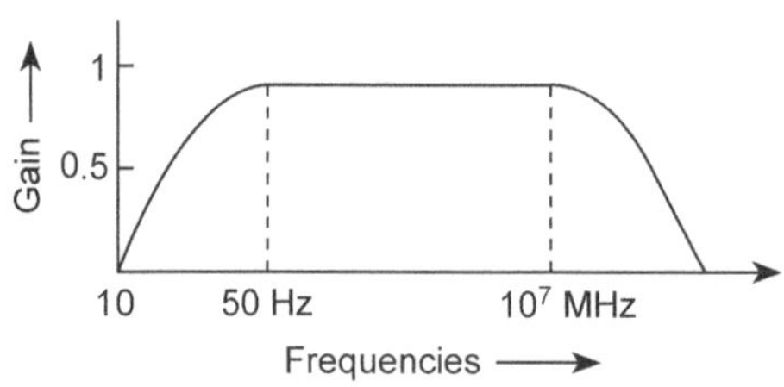

Fig. 2.8

The video amplifiers were first developed to handle the picture (video) signals of television, which require almost a flat response from 50 Hz to 5 MHz. At present, video amplifiers are used in variety of applications besides TV, such as radar systems and other devices employing pulse time signals.

An ideal video or (wideband) amplifier should meet the following requirements:

(*i*) It should amplify uniformly from a few Hertz upto hundreds of Mega Hz.

(*ii*) It should amplify a signal with minimum distortion.

(*iii*) It should amplify linearly,

The wideband amplifiers are also capable of amplifying pulses without distortion and therefore, sometimes called as *pulse amplifiers*.

Here we should remind the reader that a pulse is an abruptly changing voltage or current which may or may not repeat itself. The simplest non repetitive pulse is a stepped voltage or current shown in Fig. 2.9, which can be obtained by connecting a voltmeter across a battery through a switch and then suddenly closing the switch. The voltmeter will read zero upto a time when the switch is closed where upon, the voltage would suddenly rise to its maximum value and will stay there.

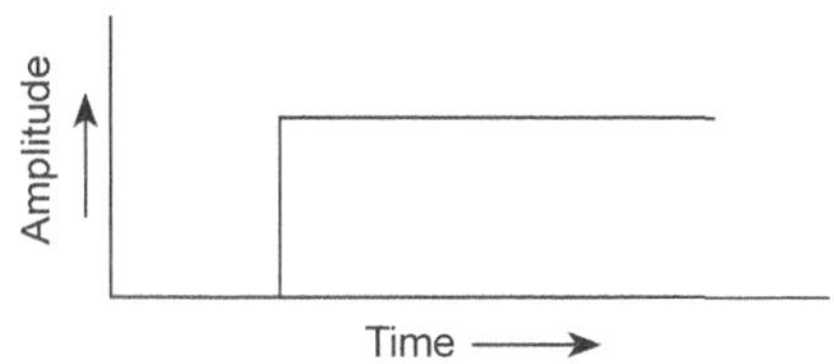

Fig. 2.9

2.5.1 Classification of video amplifier

They may be (a) uncompensated and (b) compensated.

(a) Uncompensated Video Amplifier (Fig. 2.10)

An uncompensated video or wideband amplifier is basically a RC coupled amplifier. It may use various amplifying devices such as electron tubes (Pentode), transistor, FET etc. Hence the analysis of a video amplifier is more or less same as for RC coupled amplifiers.

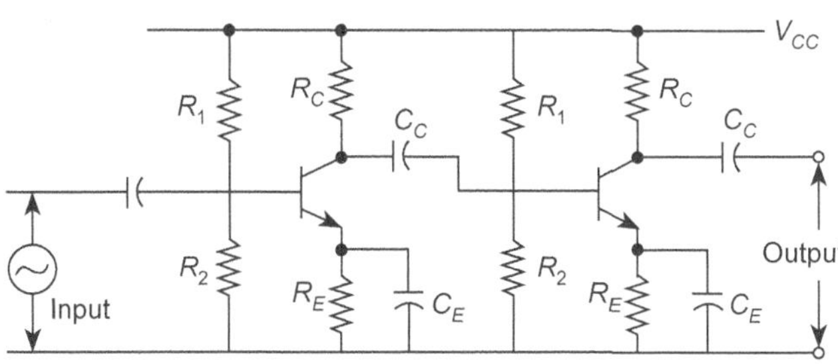

Fig. 2.10

(b) Compensated Video Amplifier

As mentioned above, a conventional R-C coupled amplifier is called as an uncompensated video amplifier.

Now it is important to increase the bandwidth without sacrificing gain. The band width may be increased by providing a high frequency compensation to the conventional (uncompensated) R.C. coupled amplifier. This is done by connecting:

(*i*) A suitable value of inductor (*ii*) is connected in series with the collector resistance RC. This is called as *shunt compensation* (Fig. 2.11 *a*).

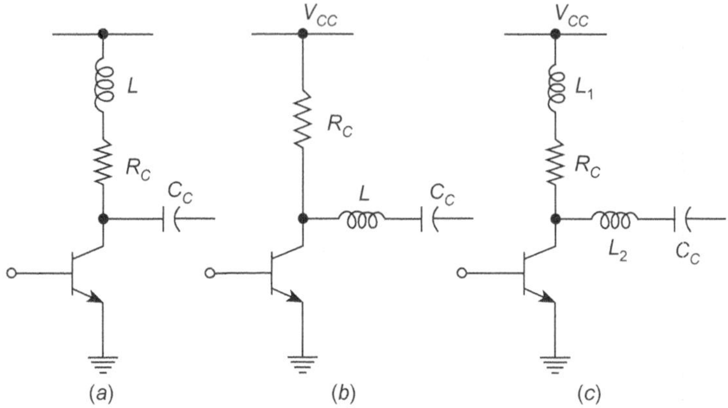

Fig. 2.11

(*ii*) An inductor in series with the coupling capacitor. This is called as *series compensation.* (Fig. 2.11 b)

(*iii*) Sometimes both the above compensations are provided simultaneously. This is called shunt series compensation. (Fig. 2.11 c)

The arrangement will give a video amplifier with a large band width.

(c) Shunt Compensated Wideband Amplifier.

The Fig. 2.12 shows a two stage common emitter wide band amplifier using *shunt compensation.* As shown, inductance L is added in series with collector resistance R_C.

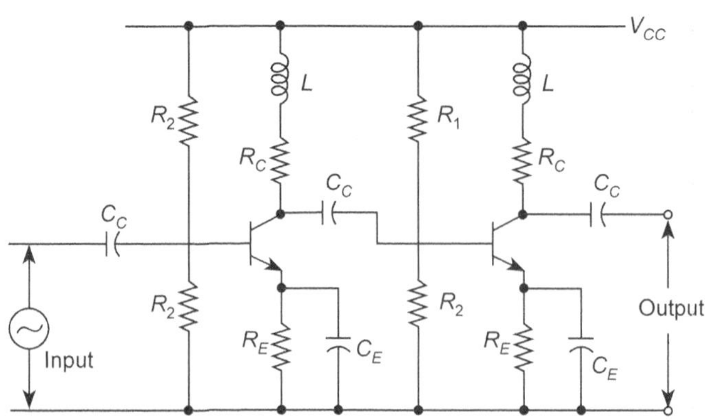

Fig. 2.12

2.5.2 Frequency response of Wideband (Video) amplifiers

The Fig. 2.13 shows frequency curves for uncompensated and compensated video amplifiers. It can be seen that bandwidth is increased by providing compensation. In practice, it is possible to extend the bandwidth by a factor of 2 to 3.

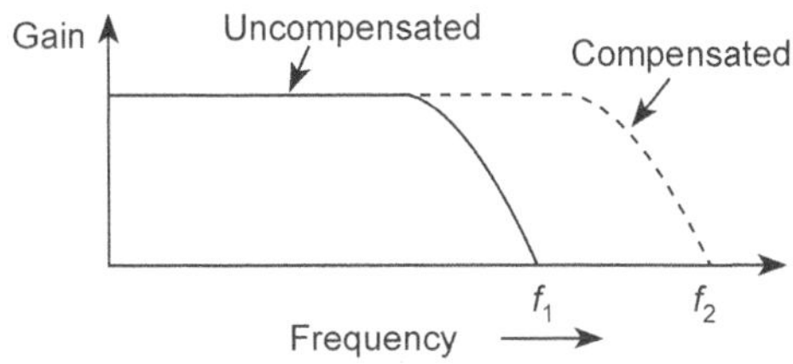

Fig. 2.13

SUMMARY

1. The amplifier may be a voltage amplifier or a power amplifier. It may be an audio amplifier or video amplifier.
2. The amplification is done in stages. The first few stages are voltage amplifiers and the last stages are of power amplifiers with which load (such as loudspeakers) is connected.
3. The power amplifiers are: class A, class B and class C amplifiers.
4. The amplifiers are provided with negative feedback, which reduces the gain (but makes the gain stable), reduces distortion and improves the frequency response.
5. A Video amplifier may be a compensated or uncompensated type.

SELF TEST QUESTIONS

1. What are audio amplifiers?
2. Differentiate between audio and video amplifiers.
3. What are voltage and power amplifiers.
4. Explain the terms—gain, Bandwidth, Sensitivity, Slew rate, Amplifier's collector efficiency.
5. Sketch a R.C. amplifier and explain function of each component.
6. What are class A, B and C power amplifiers?
7. Sketch a class B push pull amplifier.
8. What are the advantages and disadvantages of providing negative feedback in amplifiers?

9. Explain gain control and volume control of audio amplifiers.
10. What do you know by Bass and Treble controls?
11. What are video amplifiers? How are they different from audio amplifiers?
12. Why video amplifiers are also called 'wideband' or 'pulse amplifiers'?
13. Sketch a frequency response curve for a wide band amplifier.
14. Explain shunt and series compensations provided in video amplifiers. What is the purpose of providing these compensations?
15. Sketch a shunt compensated video amplifier.

OBJECTIVE TYPE QUESTIONS

Fill in the Blanks:

1. A power amplifier has ________ impedance.
2. The final amplification stage is ________ stage.
3. Resonant circuits are used in ________ amplifiers.
4. A.F. amplifier is an ________ amplifier.
5. Audio amplifiers are ________ amplifiers.
6. Band width of an Audio amplifier is ________
7. Bass boosts ________ frequencies.
8. Treble boosts ________ frequencies.
9. Negative feedback ________ the gain.
10. The band width of a video amplifier is in ________ MHz.
11. Video amplifiers are called wideband amplifier due to their ________.
12. The compensations provided in video amplifiers are ________
13. Compensations are provided to increase ________
14. A video amplifier is a R.C. coupled amplifier with ________

ANSWERS

1. low	2. power	3. tuned
4. untuned	5. class A	6. 20 Hz—20 kHz
7. low	8. high	9. decreases
10. several	11. wide bandwidth	12. shunt and series
13. Bandwidth	14. compensation.	

❑❑❑

3

Sinusoidal Oscillators

The oscillators are used in audio video devices to generate required frequencies for various purposes.

It has been established that if positive feedback is provided, an amplifier may be converted into an oscillator, *i.e.* it can produce oscillations. In other words, it can generate A.C. supply of any desired frequency.

3.1 OSCILLATORS

The use of positive feedback, which results in a closed loop gain greater than 1, and the gain in the same phase as the input, the amplifier will perform as an oscillator. In other words, output of the feedback circuit should be *in phase* with the input signal to create oscillations. An oscillatory circuit will then provide a varying output.

Accordingly, there are two types of oscillators:

(*i*) If the output voltage varies sinusoidally, the circuit is called *sinusoidal oscillator*, which produces sinusoidal output.

(*ii*) If the output voltage varies quickly to one voltage level and later drops quickly to another voltage level, the circuit is then called *non-sinusoidal/* square wave/pulse oscillator or more commonly a *multivibrator*.

Here we will discuss about sinusoidal oscillators.

Example 3.1: *The amplifier has a gain of 100. What should be the percentage of positive feedback for sustained (undamped) oscillations?*

Solution: Amplifier gain = 100

Feedback for sustained/undamped oscillations:

$$m = \frac{1}{100} = 0.01$$

percentage of feedback = 1 % **Ans.**

The gain of this oscillator should never exceed 100 to get sustained oscillations.

3.2 APPLICATIONS OF OSCILLATORS

An oscillator is required for many applications, as it can generate any frequency starting from audio frequency (20 Hz to 20 kHz) up to video frequency (in MHz). The device is used in radio transmission to produce high frequency waves (carrier signal) for modulation purposes. The device is also used (as signal generator) for testing stereo, radio, TV receivers, etc. and also for high frequency heating in industry and for several other purposes.

3.3 ESSENTIALS OF AN OSCILLATOR

The essential components of an oscillator (Fig. 3.1) are:

1. Oscillatory circuit
2. Amplifier
3. Feedback circuit

1. Oscillatory (Tanked) Circuit

This is an *LC* parallel circuit, which generates damped oscillations of a particular frequency given by its resonant (tank) frequency, *i.e.*, $f_r = 1/2\pi\sqrt{LC}$. By varying the value of C (or L), any frequency can be generated.

2. Amplifier

The amplifier amplifies these oscillations.

3. Positive Feedback Circuit

The feedback circuit supplies a part of output of the active device to the tank circuit in correct phase (i.e. positive feedback) to meet the losses in order to produce undamped oscillations.

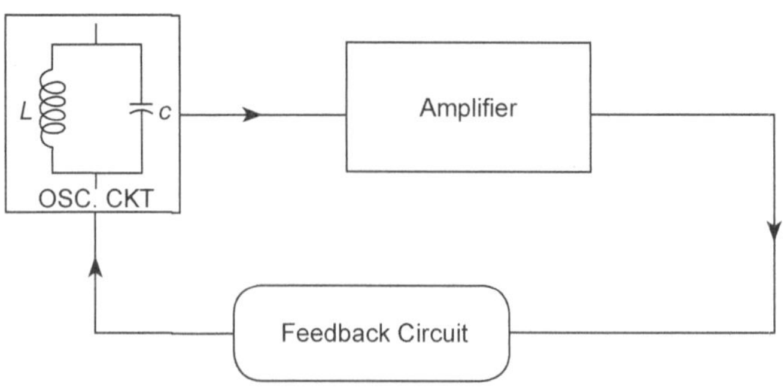

Fig. 3.1

3.4 CLASSIFICATION OF SINUSOIDAL OSCILLATORS

The oscillators may be classified in the following ways:

3.4.1 According to the Frequency Generated

The oscillators may be:

1. Audio frequency oscillators'. These are popularly known as signal generators. They can generate frequencies from 20 Hz to 20 kHz.
2. Radio frequency oscillators: They can generate frequencies from 30 kHz to 30 MHz.
3. Video frequency (microwave) oscillators: They can generate frequencies above 30 MHz.

3.4.2 According to the Circuitary used

The oscillators may be of the following types:

(a) LC tuned oscillators

They may be subclassified as:

(*a*) Tuned collector oscillator
(*b*) Tuned base/Armstrong oscillator
(*c*) Hartley oscillator
(*d*) Colpitt oscillator
(*e*) Clapp oscillator
(*f*) Armstrong oscillator
(*g*) Crystal oscillators

(b) RC oscillators

They may be subclassified as

(*a*) RC phase shift oscillator
(*b*) RC Wein bridge oscillator
(*c*) RC twin

We shall study important LC tuned oscillators:

3.5 LC TUNED OSCILLATORS

These oscillators use LC tuned circuit. Following LC oscillators are discussed.

1. Tuned collector oscillator
2. Tuned base oscillator
3. Hartley oscillator
4. Colpitt oscillator
5. Crystal oscillator

Note: An amplifier, provided with a positive feedback acts as an oscillator. If does not require any other input.

3.5.1 Tuned Collector Oscillator

The circuit contains an LC parallel (tuned) circuit in the collector of the transistor (or in the plate of the vacuum tube). Therefore, it is called 'tuned collector oscillator' (Fig. 3.2)

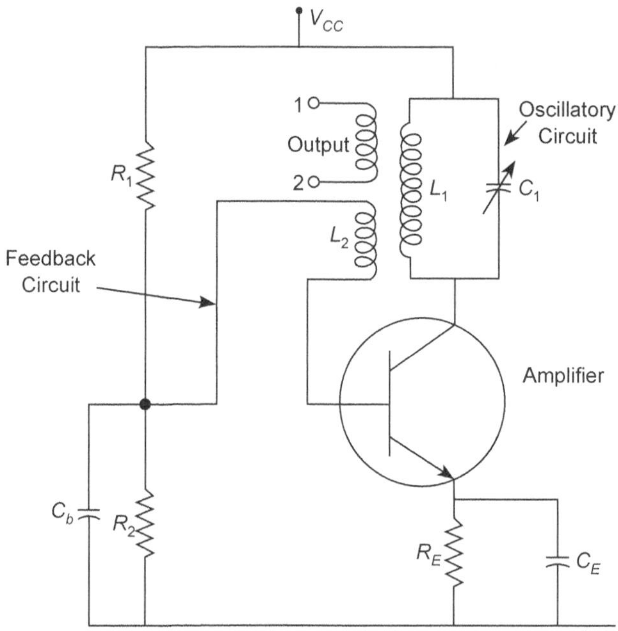

Fig. 3.2

The $L_1 - C_1$ in the collector forms the 'tuned circuit.' By varying the value of C_1 (or L_1) any frequency given by $f_r = 1/2\,\pi\,\sqrt{LC}$ can be generated. The coil L_2 in the base circuit is magnetically coupled to L_1.

When the circuit is made ON, the collector current charges the capacitor C_1, the oscillations are set up given by the expression $f = 1/2\,\pi\sqrt{LC}$. Due to mutual induction, the voltage (oscillations across L_1) are transferred to L_2 (L_1 and L_2 act like primary and secondary of a transformer.). The voltage thus obtained across L_2 is given at the input of the transistor, which is amplified by the transistor and output is obtained at the collector; thus, overcoming the losses occurring in the tank circuit.

The feedback is positive, i.e. it is in phase with the input and can be illustrated as follows: A phase shift of 180° is introduced by the transformer formed by L_1 and L_2 (There is a phase shift of 180° between outputs of primary and secondary of a transformer.). A further 180° shift is produced by the transistor due to the phase reversal phenomenon of the transistor. Thus, a total phase shift of 360° is obtained. In other words, the feedback is positive and is in the same phase as the input.

Note that capacitor C_b is connected in the base of the transistor across R_2, which provides a low reactance path to the oscillations. If this is not connected across R_2, the voltage obtained across L_2 would not go directly into the base of the transistor and some of the voltage will drop across R_2.

The output of the oscillator can be obtained across terminals 1 and 2, *i.e.,* from the tank circuit using an inductive coupling. However, the output can be taken from any part of the circuit, as oscillations are present everywhere.

The Parallel resonant circuits are preferred to series resonant circuits, due to low power losses in the former.

3.5.2 Tuned Base Oscillator

The Figure 3.3 shows a tuned base oscillator. The circuit is also called as Armstrong oscillator. In this, the tuned circuit (LC) is connected at the base of the transistor.

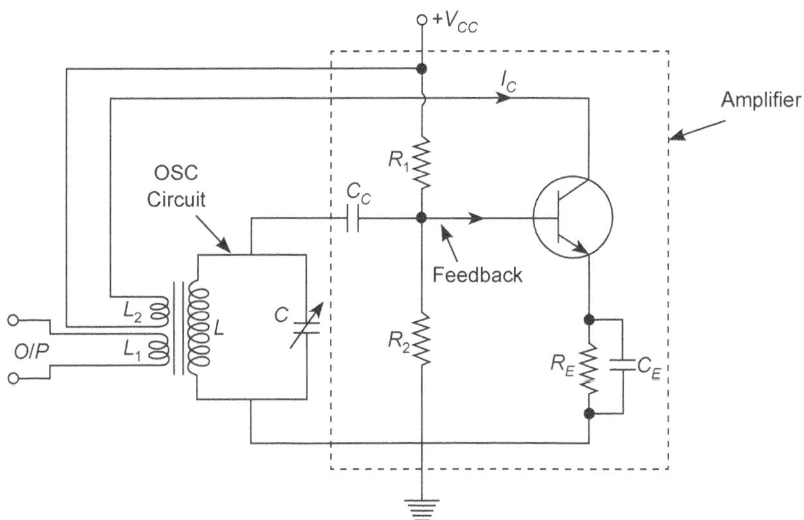

Fig. 3.3

The value of L and C determines the frequency of oscillations. The transistor provides a phase shift of 180° and another phase shift of 180° is provided by the transformer; thus total shift obtained is 360°. This is an essential condition for sustained oscillations. The frequency of oscillations is given by

$$f = \frac{1}{2\pi\sqrt{L/C}}$$

Operation:

When V_{CC} is switched ON, the rising collector current creates a varying magnetic field around L_2. This field links with L and induces a voltage in the tuned circuit and oscillations are built up. The voltage developed across C (due to its charging) drives the transistor to cut off. This reduces the gain of the transistor amplifier and the condition for sustained oscillations is satisfied. The output may be obtained across L_1 as shown.

Demerits/Drawback:

In this oscillator, the tank circuit gets loaded due to the low input resistance of the transistor; that comes in shunt. This reduces Q(Quality factor) of the tank circuit, therefore, the frequency of oscillations differs from frequency of the tuned circuit. Due to this reason, FET or vacuum triode, which have high input resistance are preferred in place of transistor.

3.5.3 Hartley Oscillator

The Hartley oscillator, known by the name of its inventor is shown in Fig. 3.4. The tuned circuit consists of L_1, L_2 and C. The L_1 and L_2 in series and are wound on the same core and therefore, some mutual inductance (M) exists between them.

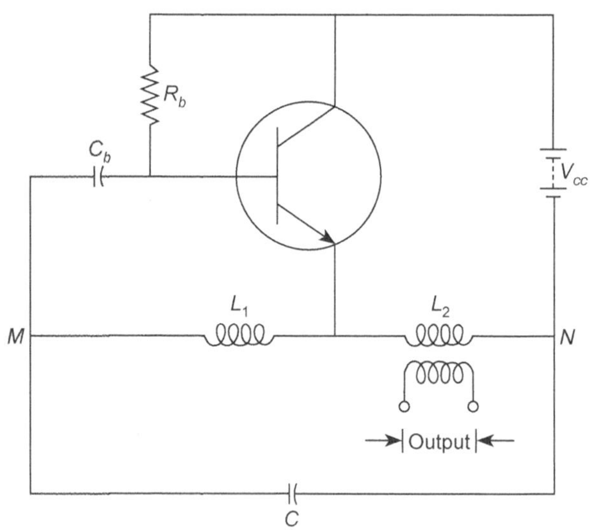

Fig. 3.4

The frequency of oscillations of this oscillator is given by the expression

$$f_r = \frac{1}{2\pi\sqrt{L_T.C}}$$

where $L_T = L_1 + L_2 + 2M$ (where M is the mutual inductance between L_1 and L_2)
The biasing is usually provided by base resistor (R_b). The capacitor C_b connected in the base circuit of the transistor provides low reactance path to the oscillations.

Operation:

The operation of the oscillator is same as of the tuned collector oscillator. The collector current charges capacitor C which later on discharges through L_1, and L_2; setting oscillations of the frequency of the expression metioned above. The oscillations across L_1 are given to the base circuit which are amplified by the transistor. The output of the transistor supplies for the losses occurring in the oscillatory circuit, thus the circuit produces undamped oscillations.

The energy supplied to the oscillatory circuit is in correct phase, because the points M and N of the transformer (formed by $L_1 - L_2$) are 180° out of phase; a further 180° phase shift is provided by the transistor. Thus, the input and output are 360° out of phase, in other words, they are of the same phase.

The Hartley oscillators are widely used in radio receivers for the following reasons:

1. They have a very wide range of frequencies
2. It is easy to tune them to the particular frequency

3.5.4 Colpitt Oscillator (Fig. 3.5)

The Colpitt oscillator is a slightly modified form of Hartley oscillator. Its construction and operation is more or less the same

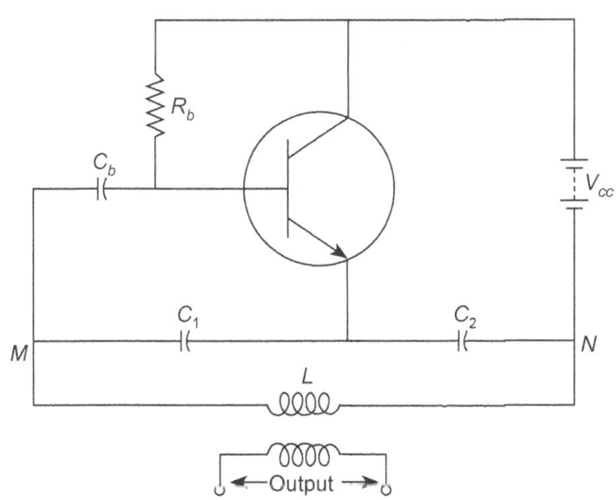

Fig. 3.5

The difference lies only in the oscillatory circuit. In this oscillator, two capacitors (C_1, C_2) are connected in series across an inductor (L). The arrangement enables the oscillator to generate frequency in MHz.

The frequency of the circuit is given by

$$f_r = \frac{1}{2\pi\sqrt{LC_T}}$$

where $$C_T = \frac{C_1C_2}{C_1+C_2} \qquad (C_1 \text{ and } C_2 \text{ are in series})$$

Operation:

When supply to the oscillator is switched on, the capacitors C_1 and C_2 are charged; later on; they are dischargd through L. The oscillations across C_1 are given to the input of the transistor, which are amplified and supply for the losses occurring in the oscillatory circuit. In this way, undamped oscillations are obtained.

The points M and N are 180° out of phase. The additional 180° phase shift is obtained through the transistor, hence the output is 360° out of phase with the input. In other words, the feedback is in phase with the input, thus condition for undamped oscillations is satisfied.

Example 3.2: *A tuned collector oscillator used in a radio receiver has L = 56 μH and C = 250 pF. Calculate the frequency of oscillations generated by the oscillator.*

Solution: $$L = 56\ \mu\text{H} = 56 \times 10^{-6}\ \text{H},$$

$$C = 250\ \text{pF} = 250 \times 10^{-12}\ \text{F}$$

Now $$f_r = \frac{1}{2\pi\sqrt{LC}}$$

Putting the values, $$f_r = \frac{1}{2\pi\sqrt{56\times10^{-6}\times250\times10^{-12}}}$$

$$= 1336\ \text{kHz}\ \textbf{Ans.}$$

Example 3.3: *Calculate the frequency of oscillations generated by a Hartley oscillator if L_1 = 100 μH, L_2 = 2 mH, M = 30 μH and C = 30pF.*

Solution: $$L_1 = 100\ \mu\text{H}$$

$$L_2 = 2\ m\text{H} = 2 \times 10^3\ \mu\text{H}$$

$$M = 30\ \mu\text{H}$$

$$C = 30\ \text{pF} = 30 \times 10^{-12}\ F$$

Now, $L_T = L_1 + L_2 + 2\,M$

$= 100 + (2 \times 10^3) + (2 \times 30)$

$= 2160\ \mu H = 2160 \times 10^{-6}\ H$

And $$f = \frac{1}{2\pi\sqrt{L_T.C}}$$

$$= \frac{1}{2\pi\sqrt{2160\times10^{-6}\times30\times10^{-12}}} = 625\ \text{kHz}\ \textbf{Ans.}$$

Example 3.4: *Find the operating frequency of a Colpitt oscillator if $C_1 = 20\,pF$, $C_2 = 70\,pF$ and $L = 20\,\mu H$.*

Solution:

Here $L = 20\ \mu H = 20 \times 10^{-6}\ H$

$$C_T = \frac{C_1C_2}{C_1+C_2} = \frac{20\times70}{20+70} = 15.5\ \text{pF}$$

$= 15.5 \times 10^{-12}\ F$

Now, $$f = \frac{1}{2\pi\sqrt{LC_T}}$$

$$f = \frac{1}{2\pi\sqrt{20\times10^{-6}\times15.5+10^{-12}}}$$

$= 9\ \text{MHz}$ **Ans.**

Note:

A Colpitt oscillator can operate at MHz frequencies.

3.5.5 Crystal (X-Tal) Oscillator

Figure 3.6 shows an x-tal oscillator. It may be a tuned collector type oscillator (as shown) or a Hartley or Colpitt type oscillator.

A tank circuit (L_1 C_1) is placed in the collector and a crystal mounted between two plates is connnected to the base of the transistor. The coil L_1 is inductively coupled to L_2. The natural frequency of L_1C_1 circuit is made equal to the natural frequency of the crystal to create resonance.

Operation

When the circuit is switched on, C_1 is charged, later on it discharges through L_1. The oscillations (A.C. voltage) produced across L_1 are transferred to L_2 through *transformer action*. The positive feedback (A phase difference of 180°

is caused by transformer and further 180° by transistor action, thus the output is in phase with the input, (*i.e.,* positive feedback) is fed into the input of the transistor thus producing oscillations. Later on, the entire circuit vibrates at the *natural frequency* of the crystal.

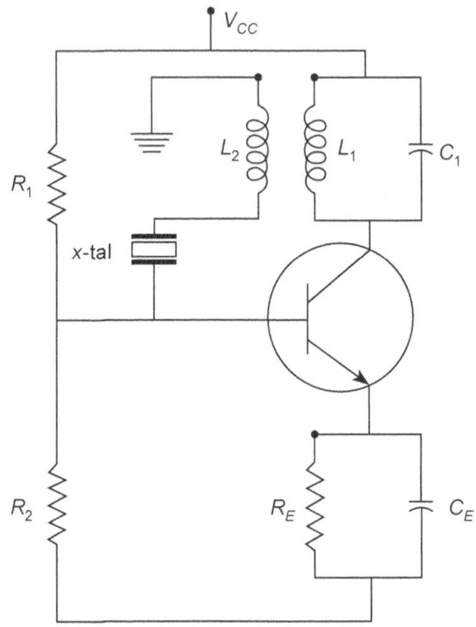

Fig. 3.6

A crystal oscillator can generate a frequency between 5 Hz and 25 MHz successfully and with high reliability. Moreover, it has a high value of Q (Quality) factor (up to 10,000). Hence the oscillator is frequently used in audio video devices.

However, the device needs careful handling as the crystal is fragile and may break, Also the frequency of the circuit cannot be varied appreciably.

Note: A crystal has two resonant circuits: series circuit and parallel circuit. A crystal oscillator therefore, can generate two frequencies: series frequency and parallel frequency. Both frequencies are almost same.

Example 3.5: *An x-tal has L = 3H, C = 0.05 pF and* C_m *= 10 pF. Calculate the series and parallel resonant frequencies for the x-tal.*

Solution: The Fig. 3.7 shows equivalent circuit a crystal. It has two resonant circuits.

(*i*) *Series circuit:* RLC series circuit

(*ii*) Parallel circuit: RLC is parallel with mounting capacitance C_n. So a crystal has two (series and parallel) resonant frequencies.

$$L = 3\text{H}$$

$$C = 0.05 \text{ pF} = 0.05 \times 10^{-12} \text{ F} \quad (\text{Assume R} = 0)$$

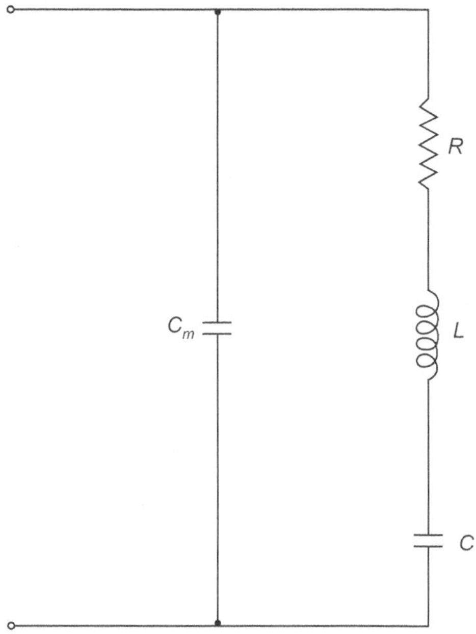

Fig. 3.7

(*i*) Series resonant frequency (RLC series circuit)

$$f_s = \frac{1}{2\pi\sqrt{L.C}}$$

$$= \frac{1}{2\pi\sqrt{3\times 0.05\times 10^{-12}}} = 411 \text{ kHz} \quad \textbf{Ans.}$$

(*ii*) Parallel resonant frequency (RLC | | C_m)

$$C = 0.05 \text{ pF}$$

$$C_m = 10 \text{ pF}$$

$$\therefore \qquad C_T = \frac{C.C_m}{C+C_m} = \frac{(10 \text{ pF})(0.05 \text{ pF})}{10 \text{ pF} + 0.05 \text{ pF}}$$

$$= 0.05 \text{ pF} = 0.05 \times 10^{-12} \text{ F}$$

$$f_p = \frac{1}{2\pi\sqrt{LC_T}}$$

$$= \frac{1}{2\pi\sqrt{(3H)(0.05\times 10^{-12}\text{F})}} = 412 \text{ KHz} \quad \textbf{Ans.}$$

Note that the series and parallel frequencies are more or less same and if this crystal is to be used as an oscillator, the frequency of oscillations must be in between 411 kHz and 412 kHz.

SUMMARY

1. Sinusoidal oscillators generate sinusoidal oscillations (a.c. voltage)
2. Oscillators use positive feedback and a tuned circuit.
3. Tunned LC oscillators are more popular

SELF TEST QUESTIONS

1. What is an oscillator? Explain how positive feedback can generate oscillations.
2. What are sinusoidal and non-sinusoidal oscillators?
3. What are the applications of oscillators?
4. Alternator and oscillator do the same job. Can they be replaced for each other?
5. What are damped and undamped oscillations?
6. How does an *LC* parallel circuit produce sinusoidal oscillations?
7. Give complete classification of oscillators.
8. What are essentials of *LC* oscillators?
9. Explain working and construction of a tuned collector oscillator.
10. Draw a neat circuit for Colpitt oscillator and explain its working.
11. Write a note on Heartly oscillator.
12. Explain the working principle of a crystal oscillator. Why is it preferred to *LC* and *RC* oscillators?

OBJECTIVE TYPE QUESTIONS

Q.I. Fill in the blanks

1. For an amplifier to generate oscillations the loop gain should be equal to ________.
2. Positive feedback improves ________ of an amplifier as \yell as produces ________.

3. The movement of electrons in a resistor generates a voltage up to ________ GHz frequency.
4. An alternator can generate a voltage only of ________ Hz.
5. A transformer as well as a ________ generate a phase shift of ________ between input and output.
6. A piezoelectric crystal has a series as well as parallel ________ circuits.
7. In an *LC* circuit, there is an exchange of magnetic and ________ energies between *L* and *C* respectively.
8. The frequency stability is of RC oscillators is ________ than LC oscillators.

❑❑❑

4

Microphones, Loudspeakers and P.A. System

The audio system process audio/sound signals: The examples of audio systems/ equipment are: Radio, Microphone, Loudspeakers, sound recorders (Tape or disc) players, audio CD, Hi Fi System, Stereo etc.

4.1 ELECTRO ACOUSTIC TRANSDUCERS

The devices which convert sound into electrical energy and vice versa are known as Electro Acoustic Transducers *e.g.,*

(*i*) *Microphone (or Mike).* It is that, which converts sound energy into electrical energy.

(*ii*) *Loudspeaker (or Speaker).* It is that, which converters electrical energy into sound.

4.1.1 Use of Electro Acoustic Transducers

Electro acoustic transducers are widely used in Audio Video systems. In radio communication, the person speaks before a mike, which converts his sound into electrical energy. This electrical energy after modulation and amplification is transmitted into space. Any radio receiver can pick up "(catch)" this signal. After passing through various stages, this signal is passed through loud speaker finally, which converts this electrical signal again into the original sound, which we can hear. See Fig. 4.1

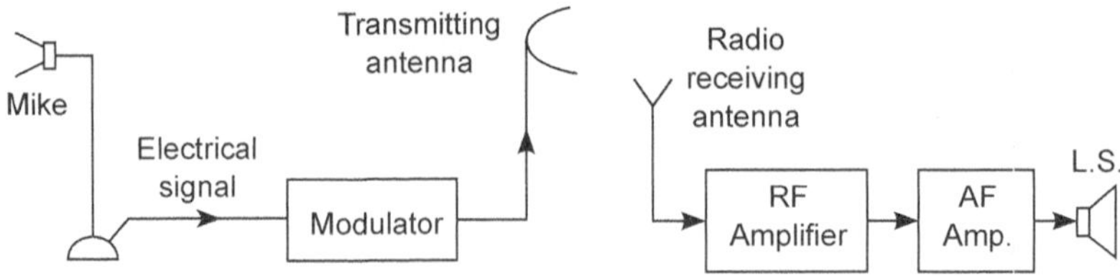

Fig. 4.1

These two devices are also very much used in Public Address (P.A.) system. A public address system is a system to address a large gathering like a political or religious meeting. The person speaks before mike, which converts his speech into electrical signal. This electrical signal is amplified (may or may not be modulated) and fed to loud speaker(s) which converts the electrical signal back into the original signal. In this way, the speech of the person can be sent to large distance and more number of people. See Fig. 4.2

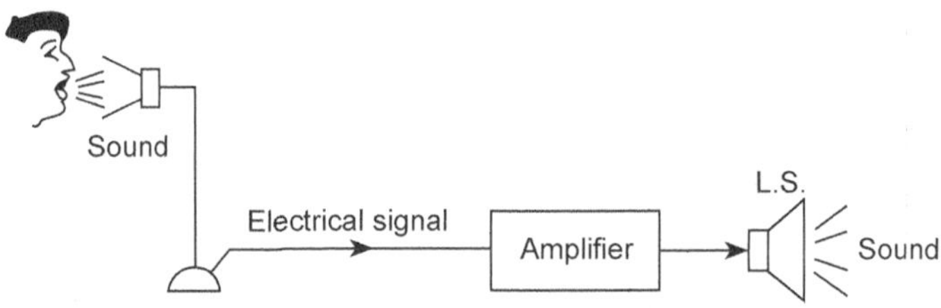

Fig. 4.2

4.2 MICROPHONES (MIKES)

Microphones, in addition to radio transmission and P.A. systems are also used in telephone communications, sound recording and broadcasting and scientific measurements etc.

Each microphone has a "diaphragm" which is a very sensitive part. This intercepts the sound waves and vibrates, thus converting sound into electrical pulses.

(*i*) When sound strikes on only one surface of the diaphragm, the mike is said to be "Pressure type microphone".

(*ii*) When sound strikes on both surfaces of the diaphragm, the mike is said to be "Pressure gradient" microphone.

4.2.1 Characteristics of Microphones

The important characteristics of microphones are:

1. *Sensitivity.* Sensitivity of a microphone is its ability to pick up the weakest sound. This can be measured by the voltage obtained at output of the mike at a given sound pressure. It is usually expressed in milli volts per microbar or in decibels (db) referred to 1 volt.

   The common sound pressure used for measurement is 94 dB (10 μ bar). Hence sensitivity is measured as:

   (*i*) Open circuit voltage at 0 dB = 1 volt/μ bar.

   (*ii*) Max. power output at 0 dB = 1 mW/10 μ bar.

2. *Signal/Noise Ratio (SNR).* This is defined as a ratio of the output with noise to the output in absence of noise at a sound pressure of 1 μ bar.

   It is to be kept in mind that some noise is produced in the mike itself in its transformer and in the resistance of its circuitary.

   ∴ Signal Noise Ratio,

   $$\text{S.N.R.} = \frac{\text{output in the presence of noise}}{\text{output when the noise is not present}}$$

   e.g., If the output of a microphone (at a sound pressure of 1 μ bar) is 1000 mV. In absence of noise the output is 100 mV. The S.N.R. of the mike will be

   $$= 20\log\left(\frac{1000}{100}\right) = 2\log(10)$$

   $$= 20 \times 1 = 20 \text{ dB.}$$

3. *Directivity.* By directivity, we mean the sensitivity of a microphone towards sounds coming from different directions. The directivity of mikes is described by *polar diagrams*. The polar diagram shows output of a mike at different angles of incidence of sound on its diaphragm. According to this characteristic, the mikes can be classified as:

   (*a*) *Uni-directional.* Some mikes can pick up sound from one direction only. The person must stand in front of the microphone when speaking e.g. moving coil microphones are unidirectional. The shape of polar diagram of uni directional mike is of the shape of the heart as shown in Fig. 4.3.

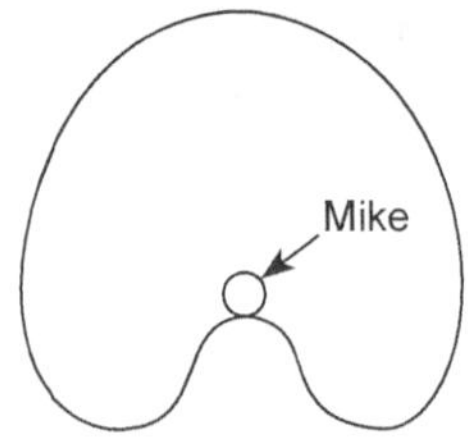

Fig. 4.3

In these mikes, diaphragm is left open at the front. Such an arrangement produces a uni-directional response of the mike towards sound.

(*b*) *Bi-directional.* These mikes can pick up sound from two directions *i.e.,* from front as well as from the back. A *ribbon microphone* is an example.

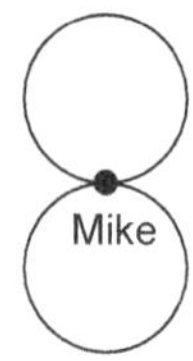

Fig. 4.4

While using these microphones, the source (S) of the sound are to be positioned carefully. To obtain bi directional (two directional) response, the diaphragm is kept open on both sides, thus it operates on the principle of *pressure gradient* on the two surfaces of the diaphragm.

Its polar diagram is of the shape of the digit Eight (8). See Fig. 4.4

A velocity microphone is another example which gives "eight" directivity.

(*c*) *Non-directional or Omni directional or all directional.* These mikes can pick up sound from all directions. In other words they are without any direction. These mikes pick up sound from all sides with same intensity. The arrangement of source of sound is no concern. Crystal microphone is one example.

Fig. 4.5

The polar diagram of these mikes is a simple circle with mike at its centre. See Fig. 4.5

The performance of this mike is well at low and at mid frequency of the sound signal, but at higher frequencies, it produces a "shadowing effect" and does not pick sound from sides and back with equal intensity. To reduce the shadowing effect at H.F., the size of the mike may be reduced.

These mikes have their diaphragm open at front but completely closed at back side.

All pressure microphones have this type of polar diagram.

Note:

(*i*) Sharp direction can be obtained by combination of multiple microphones.

(*ii*) A cardioid (Heart) polar diagram can also be obtained by connecting a bi-directional mike in series with a non-directional mike.

4. *Output impedence of microphones.* A microphone has a certain output impedence. It is given in ohms. It helps to provide suitable impedence matching between microphone and the transformer for maximum transfer of power to amplifier and then to the loud speaker. (Fig. 4.6)

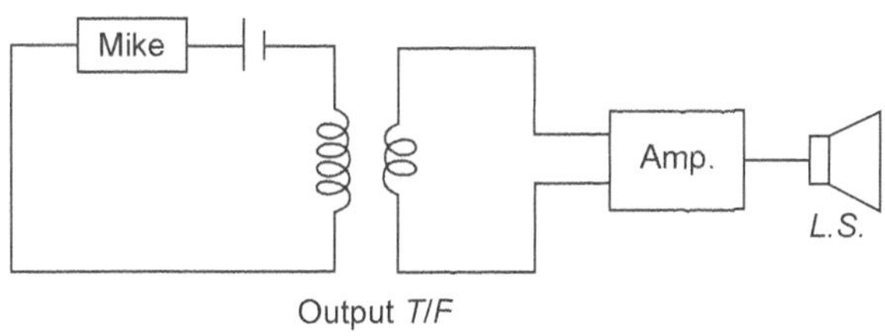

Fig. 4.6

Few mikes have very low output impedence. Such mikes have a built in transformer for impedence matching with the line.

Note:

(*i*) Crystal and condenser mikes have high impedence and need an emitter follower circuit for impedence matching.

(*ii*) The Ribbon and moving coil mikes have low impedence and need an ordinary transformer for impedence matching.

5. *Frequency response:* How a microphone responds to various audio frequencies will be known by its frequency response. In other words, the output voltage of a mike in response to the different sound frequencies falling on its diaphragm is called its *frequency response*. Naturally, it will be a graph obtained between output voltage of the mike versus incident frequencies. A mike with flat curve between 50 Hz to 16000 Hz will be considered as Hi-Fi mike. However for normal, a mike with a band width of 70 Hz to 8000 Hz may be acceptable.

Frequency response of few microphones are shown in Fig. 4.7.

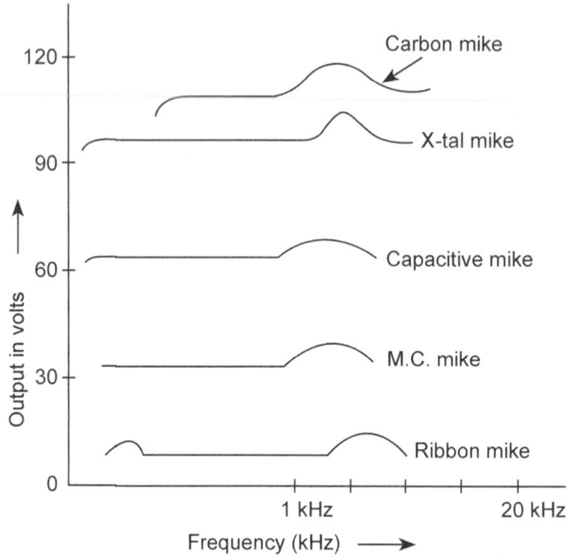

Fig. 4.7

6. *Distortion*

The various distortions produced by microphones are:

(*i*) *Frequency distortion.* Described above.

(*ii*) *Amplitude distortion.* A mike also produces amplitude distortion which results in the production of harmonics, not present in the original sound. This distortion should not be more than 5% for a normal quality mike and not more than 1% for a hi-fi mike.

(*iii*) *Phase distortion.* The mike produces phase distortion between different components of a sound input.

> *A good microphone should have*
>
> (*i*) High sensitivity
>
> (*ii*) High S/N ratio
>
> (*iii*) Directivity according to the application and requirement
>
> (*iv*) Output impedence, which can be easily matched
>
> (*v*) A good frequency response giving a flat curve for wide range of audio frequency.
>
> (*vi*) Very low distortion upto 5%.

4.2.2 Classification Of Mikes

Mikes can be classified as:

1. *Pressure Microphones.* Examples are:
   (*i*) Carbon Microphone
   (*ii*) Capacitive or Condenser Micrphone
   (*iii*) Moving Coil Microphone
   (*iv*) Crystal Microphone.
2. *Pressure gradient microphones.* Example is:
   (*i*) Ribbon/velocity microphone.
3. Other special microphones are
   (*i*) Radio/wireless microphones
   (*ii*) Lavalier or mobile microphones
   (*iii*) Noise cancelling microphones
   (*iv*) Electret microphones
   (*v*) Tie clip microphone.

These will be discussed one by one.

1. Pressure Microphones

The working of these microphones is based on the sound pressure incident upon the diaphragm (membrane) of the microphone.

Important pressure microphones are described below:

(a) Carbon Microphone—Construction (Fig. 4.8 a)

In this microphone there is a carbon block and also a diaphragm of carbon. The space between the two contains "carbon granules".

When a person speaks before it, the diaphragm begins to vibrate depending upon the sound pressure on it. By the inward and outward movement of the diaphragm, the granules are 'compressed' and 'rarified' and their resistance changes accordingly.

With change of resistance of the granules the primary of the transformer carries a varying current and thus the secondary current. The output of transformer is given to amplifier and then to the speaker.

The connection of the microphone with amplifier and speaker are shown in Fig (*b*).

Merits

1. This mike is cheaper and durable.
2. It is mechanically rigid.
3. It is very cheap and small in size.
4. It is very sensitive and can pick up even weak signals.

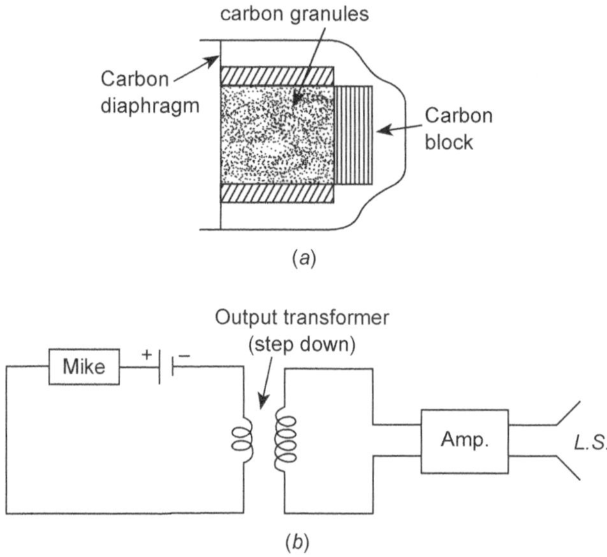

Fig. 4.8

Demerits

1. Due to local heating of granules, it sometimes produces noise.
2. It has poor frequency response hence not suitable for h_i-f_i work.
3. It is affected by heat and moisture.

Characteristics

(*i*) Sensitivity of carbon mike is high. Its value is about 100 mV.

(*ii*) The signal/noise ratio is poor. The mike creates self noise due to change of resistance of carbon granules.

(*iii*) It has a bandwidth from 150 Hz to 5 KHz, hence not suitable for hi-fi work.

(*iv*) Distortion is about 10% due to variation is resistance and sticking of carbon granules.

(*v*) It is omni-directional *i.e.* it can pick up signal from all directions.

(*vi*) Output impedence is less upto 400 μ.

Working of a Carbon Mike

When a person speaks before the mike due to the sound pressure, the diaphragm moves and changes the resistance of the carbon granules. This produces a change in the current flowing from battery. The change in the current depends upon the sound pressure incident on the mike. This change of current is amplified by the step up transformer and then by amplifier. When carbon granules release back, their resistance increases and current decreases.

Applications

(*i*) Due to rigidness, high output, low cost, carbon mikes are very much used in telephone and radio communication systems.

(*ii*) As mentioned earlier, it is not suitable for hi-fi applications due to more distortion, low bandwidth and more self-noise.

(b) Capacitive or Condenser Microphone

It basically consists of a air (parallel plate) capacitor of which the diaphragm makes one plate and other plate is a *back plate*. The dielectric between the two is air. The diaphragm is movable depending upon sound pressure. The back plate is fixed and has grooves. [Fig. 4.9 (*a*)].

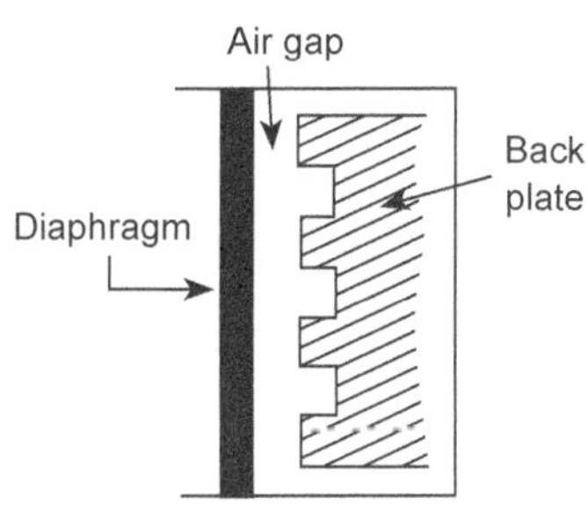

(*a*)

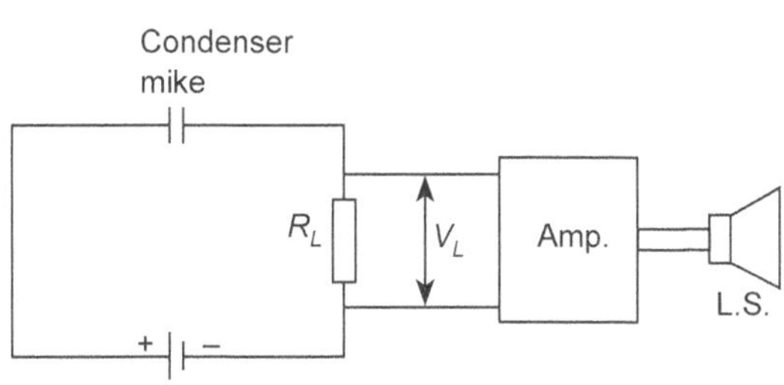

(*b*)

Fig. 4.9

When sound waves strike its diaphragm, it starts vibrations depending upon the intensity of sound pressure. With the inward and outward motion of the diaphragm, the distance of air gap changes and thus capacity of the parallel plate capacitor also changes accordingly.

The potential drop (p.d.) across output resistance R_L varies with capacity of the microphone. Which is fed to amplifier and then to the speaker.

The principle of capacitive microphone is based on the fact that when capacity of a capacitor changes and if charge on it remains constant, the voltage changes accordingly.

We have $$V = \frac{Q}{C}$$

Where V is the voltage, Q is charge and C is capacitance of the capacitor. It is clear that on changing C, Voltage (V) will change, provided Q is kept constant.

The connections of capacitor microphone is shown in Fig. (*b*).

Merits

1. It has almost a constant frequency response.
2. It can be used for hi-fi quality work.
3. It is almost free from noise.

Demerits

1. Its sensitivity is very low.
2. The amplifier needs to be kept very near to the mike.
3. The mike is to be protected from dust and dirt.
4. It is also affected by heat and moisture.
5. It needs a separate dc bias, hence costly.
6. It is a delicate mike, hence needs careful handling.

Characteristics

(*i*) The output is very low and therefore a amplifier is inbuilt in the mike.

(*ii*) They have a high signal-noise ratio.

(*iii*) They have excellent frequency response and have a flat bandwidth between 20 Hz to 12 kHz.

(*iv*) They have very low distortion (about 2%).

(*v*) They are omni-directional.

(*vi*) They have high input impedence of about 100 μ.

(*vii*) Sensitivity is about 50 dB.

Working of a Capacitive Mike

The diaphragm acts as one plate and the fixed *back plate* acts as second plate. The two plates are 25 μm apart and form a parallel plate capacitor of about 40 pF. A 200 V dc is applied as a bias between the two plates.

When sound waves strike its diaphragm, it moves. During compression the distance between plates decreases, it increases the capacitance. During rarefaction, the distance increases, it decreases the capacitance. This change in capacitance gives a varying voltage across the two plates depending upon the sound pressure.

Applications

(*i*) Due to excellent frequency response, a capacitive mike is used as a standard mike to calibrate other mikes and testing of loud speakers.

(*ii*) Due to above reason, It is used in *sound level meters* which can measure sound level of source.

(*iii*) It is used in hi-fi recording work.

(*iv*) It is useful for accurate laboratory measurements.

(*v*) It is also used for theate and studio work.

(*vi*) Due to more cost, they are not used for ordinary work.

It is to be kept in mind that a capacitor microphone forms a parallel plate capacitor, and the formulae for its capacitance is given by:

$$C = \frac{K.A.\varepsilon_0}{d}$$

where A = area of each plate

d = Distance between the plates

ε_0 = Absolute permittivity

$= 9.85 \times 10^{-12}$ Farads/m

K = di electric constant.

As the area of plate cannot be varied, the capacitance of the microphone is varied by varying the space between plates.

(c) Moving Coil or Dynamic Microphone (Fig. 4.10)

In this microphone, a coil having large number of turns moves between the magnetic poles of a "Pot magnet".

When sound waves strike on the diaphragm, it starts moving forward and backward in accordance to the sound pressure on it. As the coil is fixed with the diaphragm, the coil also starts moving accordingly. There is a relative motion between coil and the magnet, hence an emf is induced in the coil (according to the Faraday's laws of electro magnetic induction).

Construction and Working

The magnet is POT type as a result, a uniform magnetic field is produced. The diaphragm is of non magnetic material. A protective cover is used to save diaphragm and coil assembly against mishandling.

As mentioned earlier when sound strikes the diaphragm, it moves and hence the coil moves in the magnetic field. Due to change of flux linking, an emf (e) is induced in the coil given by the following formulae:

$$e = N.\frac{d\phi}{dt}.$$

where, N = No. of turns in the coil.

$$\frac{d\phi}{dt} = \text{Rate of change of flux.}$$

As a result of this emf, a current is produced which later on is amplified.

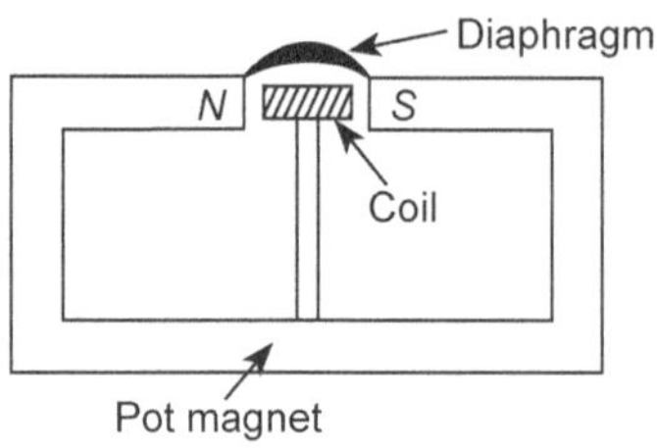

Fig. 4.10

This induced emf is fed to the amplifier which in turn is coupled with the loudspeaker. The connections are shown in Fig. 4.11

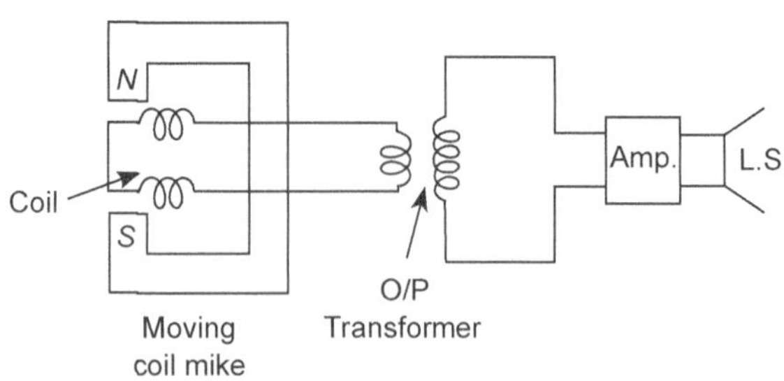

Fig. 4.11

Merits

1. Its impedence is low in between 25 to 40 ohms.
2. It is not effected much by atmospheric changes and hence it is more durable.
3. It is a moderate mike, suitable for public adress systems.
4. The output of this mike is about 50 db.
5. They are unidirectional.

Demerits

1. This microphone is to be kept very near to the source of sound.
2. The magnet is large and is to be protected in a heavy case.

Characteristics

(*i*) Sensitivity of the mike is about 35 μV, which is stepped up to 100 μV by a step up transformer.

(*ii*) S/N ratio is about 35 dB.

(*iii*) It gives a flat response from 50 Hz to 8 kHz.

(*iv*) Its distortion is about 4%.

(*v*) The mike is uni-directional.

(*vi*) The output impedence is about 30 μ hence a step up transformer is built within the microphone to increase impedence for matching purposes. However modern mikes use many turns of thin wire which makes their resistance 200 μ and then an *in built* transformer is not needed.

Applications

(*i*) The mike is uneffected by atmospheric problems, hence used in public-address system.

(*ii*) It is also used for broadcasting work.

(*iii*) If this mike is connected in series with a ribbon mike, a very good directivity pattern (Heart or cardioid) is obtained. This makes the system very suitable to be used for theatres. This assembly is available in the market.

(*iv*) It is omni direction, hence it is specially useful for conference halls. Better results are obtained, when it is suspended from ceiling.

(*v*) The mike is similar in construction to a moving coil (dynamic) loud speaker. The same unit can be used as microphone as well as a loud speaker and it is often customary in Intercom systems.

(*vi*) In tape recorders, condenser mike is *built in*, whereas the external microphone used is the moving coil type. (See Fig. 4.12)

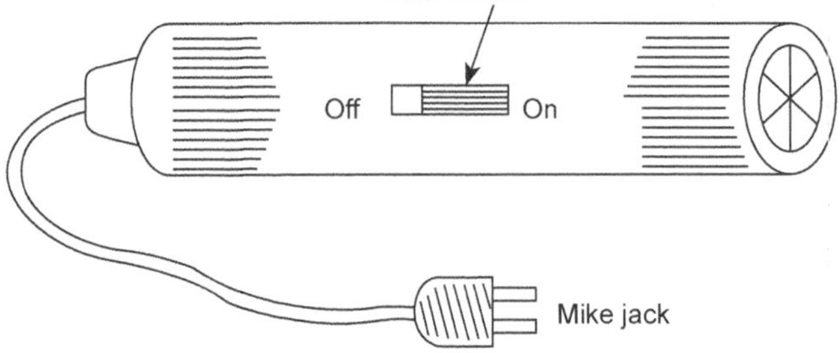

Fig. 4.12

(d) Crystal Microphone

The crystal microphone works on the principle of *Piezo-electric* effect. According to this, crystals like quartz, when applied pressure starts producing "vibrations" or "oscillations".

Crystal microphone makes use of crystals of roschelle salt, lihium sulphate, ammonium dihydrogen phosphate (ADP) etc. (See Fig. 4.13) as these materials give better results.

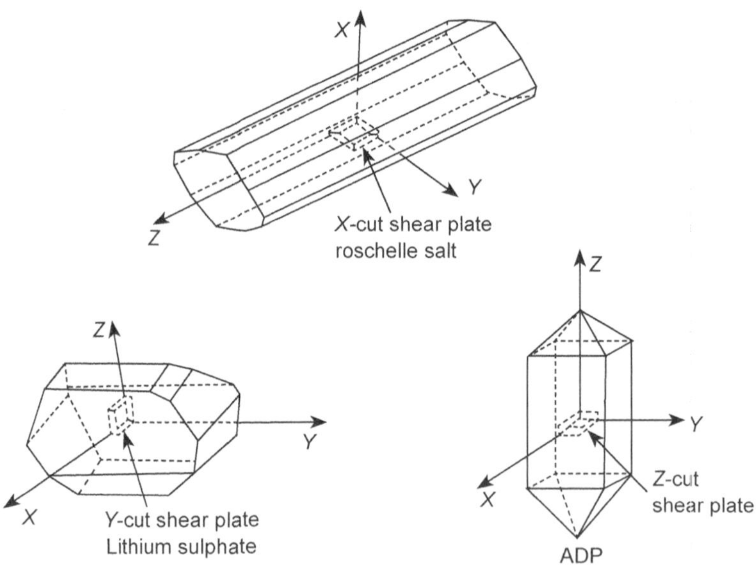

Fig. 4.13

The mike has a diaphragm, when sound waves strike it, it starts vibrating. The frequency of vibrations and their amplitude depends upon the pressure of sound incident on it. This diaphragm in turn exerts pressure on the crystal.

As a result we obtain a varying electrical output which is amplified and fed to the speaker.

Construction and Working (Fig. 4.14)

We know that few crystals like quartz, rochelle salt, ceramic and touramaline etc. show Piezo effect, thus when opposite faces of these crystals are subjected to a pressure, a potential difference is produced between the faces. Out of the above materials Ceramic and Quartz are widely used for microphones, as they show good Piezo effect and also they can work even in unfavourable atmospheric conditions.

Two thin slices of quartz, cut along suitable axis are placed with air spaces between them. An aluminium diaphragm is also attached with the crystal such that with sound pressure when diaphragm moves, it exerts a mechanical pressure on the faces of the crystal.

Usually, a crystal is formed by joining two crystal slices of opposite polarity. Hence when sound waves strike the diaphragm, it pushes one crystal in tension and other in compression and opposite charges are developed on their surfaces. The emf so produced is fed to the amplifier.

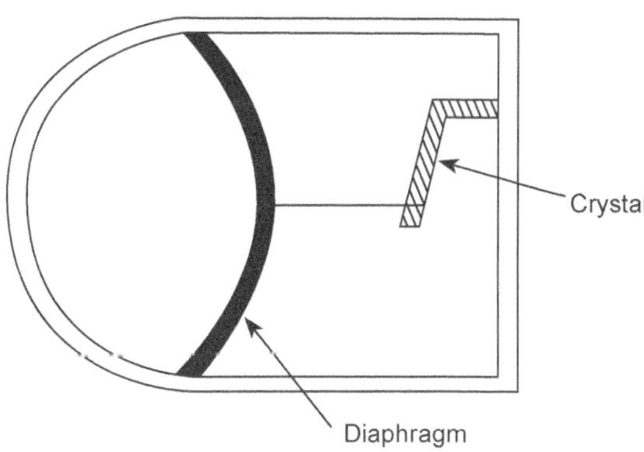

Fig. 4.14

Merits

1. This microphone has a very high impedence such that it does not need any output transformer and the microphone can directly be connected with the amplifier coupled to the speaker.
2. The output is about 50 db.
3. Its frequency response is in between 30 Hz to 10 KHz.
4. It is no-directional.

Demerits

1. The microphones cannot bear high temperature. Their use is recommended for below 100°C.
2. They cannot bear direct sunlight.

Characteristics

(*i*) The mike has a good sensitivity of 50 mV.

(*ii*) Its S/N ratio is quite high, a typical value is 45 dB.

(*iii*) It responds to a bandwidth of 30 Hz – 10 KHz.

(*iv*) Distortion is very less, about 2%.

(*v*) It is omni (i.e. all) directional.

(*vi*) Output impedence is about 1 M .

(*vii*) The person can stand close to the microphone.

(*viii*) The mike does not need any external biasing.

(*ix*) It should not be exposed to direct sun light.

(*x*) Now a days, a new crystal has been developed which has replaced the conventional quartz. This crystal is known as Ammonium Dihydrogen Phosphate (ADP). [See Fig. 4.13 (*c*)] It has better characteristics than the other conventional crystals.

(*xi*) Due to high impedence, long length of cable should be avoided. Coaxial cables are better to be used.

Applications

(*i*) The mikes are extensively used in PA system due to their robust construction and cheapness.

(*ii*) They are used in sound level meters.

(*iii*) Sometimes they are also used in hearing aids.

7. Pressure Gradient Microphones (Ribbon Velocity Microphone)

Construction

The ribbon velocity microphone has a thin ribbon shaped aluminium foil suspended between the two poles of a magnet. The ribbon is free to move. When sound waves strike, the foil moves according to the velocity of sound waves. The net effect is proportional to the difference in the sound pressures on the two sides of the ribbon foil. [See Fig. 4.15 (*a*) and (*b*)]

Working (Fig. 4.15 c)

When the aluminium ribbon placed in the field of a permanent magnet moves due to sound pressure on it, due to change in the flux linking with the ribbon, an emf is induced in the ribbon. This emf is fed to amplifier and then to the loudspeaker.

The mechanical force working on the ribbon is proportional to the difference of sound pressures acting on its both surfaces. Hence the mike is called pressure gradient microphone. Further the emf induced is dependent on the velocity of the ribbon, hence it is also known as a "velocity mike".

The ribbon is a single thin conductor, the emf induced is very small, hence a transformer is used to step up the voltage before feeding it to the amplifier. The transformer steps up the induced voltage to about 50 times.

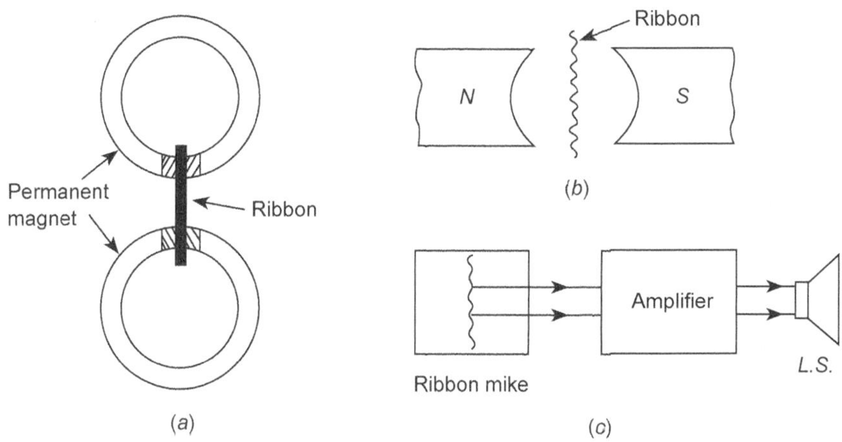

Fig. 4.15

Merits

1. The output of the microphone is about 60 dB.
2. The mike is bi-directional. For best results, the person to speak should stand atleast 1.5 feet away from the mike.

Demerits

1. The mike should be protected from the high velocity winds.
2. Its sensitivity is less than moving coil mike.

Characteristics

(*i*) These mikes have a very good frequency response for low as well as for high frequencies. It has a bandwidth of 15 Hz to 13 KHz. This is due to the very small size (few microns) of the ribbon.

(*ii*) These mikes have *eight shape* directivity. These mikes now, are available such that their one part is *pressure operated* and other part is *pressure gradient* operated. Such mikes give a very good "hyper cardiod" directivity pattern.

(*iii*) Their impedence is very small about 0.25 ohm, which is to be raised to 200 ohm for impedence matching with the line.

(*iv*) Their sensitivity as mentioned above is very less. It is 3 μV or 100 dB. With the help of transformer, the voltage may be raised to 100 μV.

(*v*) Signal/Noise ratio is higher than the moving coil mike. Its value is about 45 dB. This mike therefore is more suitable for hi-fi work.

(*vi*) Their distortion is low about 2%.

(*vii*) Ribbon is a very delicate part of the ribbon mike and hence needs careful handling.

(*viii*) Ribbon is sensitive even to breathing, hence the person to speak should never stand less than 1.5 feet from the mike. This is very important for these mikes.

(*ix*) It does not need any external bias.

Applications

(*i*) The ribbon mike is suitable for broadcasting purposes.

(*ii*) It is very suitable for dramas and theatre work.

(*iii*) It is not suitable for close talking as the source of sound should be kept minimum 1.5 feet away from the mike.

(*iv*) It is not suitable for outdoor work as the wind can vibrate the diapragm and this will produce unwanted noise within the mike.

Table 4.1 Microphones—a comparison

S.No.	Particulars	Carbon mike	Capacitive mike	Moving coil (Dynoamic) mike	Crystal mike	Ribbon velocity mike
1.	Sensitivity	100 mV	4 mV	35 μV	50 mV	3 μV
2.	Frequency response	150 Hz—5Hz	20 Hz—12 kHz	30 Hz—10 kHz	30 Hz—10 kHz	5 Hz—13 kHz
3.	Distortion	10%	2 %	4 %	2 %	2%
4.	Directivity	Omni directional	Omni directional	Uni directional	Omni directional	Bi directional
5.	Impedence	400 Ω	100 mΩ	40 Ω	1 M	0.3 Ω
6.	External Biasing	Required	Not required	Required	Not required	Not required
7.	Distance of the person speaking before mike	Close (3 inch)	1.5 feet	(6 inch)	6 inch	1.5 ft.
8.	Cost	Cheapest	High	Medium	Cheap	High
9.	Self noise	Highest	Lowest	High	Lower	Low
10.	Remarks	Used in telephone and portable recorders	Used in recording and calibration	Used in P.A. system and Broadcasting	Used in Recording and mobile commuications	Used in Drama, Broadcasting and orchestra.

3. Special Microphones

(a) Radio/wireless Microphone

It is used as a VHF frequency modulated transmitter such that no cable connection is needed from mike to amplifier. These mikes are also used in films, broadcasting, drama stage etc. It is a mobile microphone.

(b) Lavalier

It is a small moving coil type microphone, which can be hanged around neck for mobile purposes.

(c) Noise cancelling (or Differential) microphone.

This mike is suitable to be used in noisy place like a factory, sports ground, fairs etc. The mike cancels the surrounding noise. This mike is kept touching the lips by the person.

(d) Electret microphone.

It is a condenser microphone without requiring an external bias thus making it cheaper. This also makes it suitable for field work. It has a metal diaphragm and back plate is made of polytetra fluro ethane (PTFE or Teflon). This is negatively charged during manufacturing. This charge is retained by it for many years of its use. The *–ve* charge of the back plate induces an equal and positive charge on the diaphragm. The two opposite charges with a gap in between form a condenser. See Fig. 4.16.

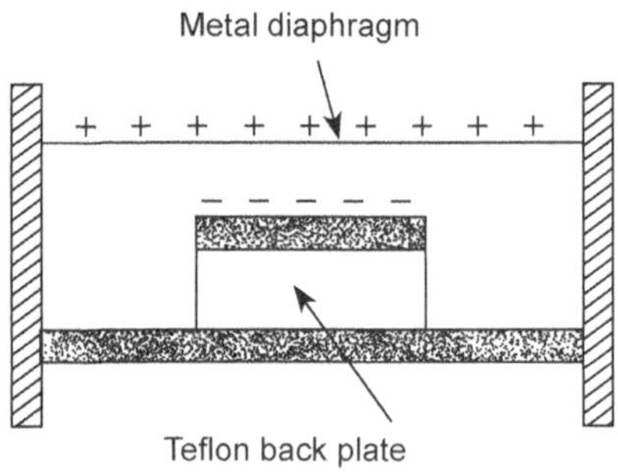

Fig. 4.16

This mike is used in P.A. system and also in wireless micro-phones.

(e) Tie clip microphone.

It is an *electret* microphone which can be clipped on a tie, shirt button, collar etc.

Example 4.1: *The output voltage of a microphone is 5 μV find sensitivity.*

Solution: Sensitivity $= 20 \log\left(\frac{1}{V_0}\right)$

$$= 20\log_{10}\left(\frac{1}{5\times10^{-6}}\right)\left(5\ \mu\text{V} = 5\times10^{-6}\ \text{V}\right)$$

$$= 106.02 \text{ dB} \quad \textbf{Ans.}$$

Example 4.2: *The microphone with a sound signal gives an output voltage of 30 mV and without sound signal, the output is 5 mV. Find the Sound Noise ratio.*

Solution: SN ratio $= 20 \log \frac{30\times10^{-3}}{5\times10^{-3}} = 15.55.$ **Ans.**

Example 4.3: *The output resistance of a microphone is 15 Ω. It is desired to feed the output to a line of impedance of 240 Ω. Find turn ratio of the transformer so that maximum power is delivered to the line.*

Solution: For maximum power, impedence matching is to the provided between source and the load.

$$\frac{Z_2}{Z_1} = \left(\frac{N_2}{N_1}\right)^2 = \frac{240}{15} = 16$$

$$\frac{N_2}{N_1} = \sqrt{16} = 4$$

It is a step up transformer. The turn ratio = 4 : 1 **Ans.**

Example 4.4: *A directional microphone gives an output voltage of 0.5 mV. When an omni directional microphone is placed in the same position, the output is 0.5 μV, for the same sound pressure. Find directivity of the microphone.*

Solution: Directivity $= \frac{\text{Actual output in the direction of max. output}}{\text{Output in that direction for omini diectional mike}}$

$$= \frac{0.5\times10^{-3}}{0.5\times10^{-6}} = 1000$$

or $20 \log (1000) = 60$ dB. **Ans.**

Example 4.5. *A moving coil microphone has a flux density of 0.02 coulomb/ m^2, length of coil is 1 cm and velocity due to sound pressure is 0.1 cm/ Sec. Find emf induced.*

Solution: The emf, $e = \text{B}.l.\text{V}$

$$= 0.02 \times (1 \times 10^{-2}) \times (0.1 \times 10^{-2}) = 2 \text{ mV.} \quad \textbf{Ans.}$$

4.3 LOUD SPEAKERS

The audio chain in audio video systems starts from microphone and ends at loud speaker. For example in a TV, the audio signal is delivered to the amplifier which boosts the signal to make it powerful enough to do real physical work. This amplified signal is given to the terminals of loud speaker—a device that converts electrical signals received into original sound waves. This job, the speaker does by causing the air particles to vibrate synchronously with the electrical signal.

In loudspeaker, two different magnetic fields interact each other to cause mechanical vibrations which produce sound waves. One of the two fields is constant and is a permanent magnetic field and the other magnetic field is caused by the electrical signals which is to be converted into sound waves. Hence the mechanical vibrations and therefore the sound waves produced are proportional to the electrical signals.

Definition

A loudspeaker is an electro-acoustic transducer which converts electrical input obtained from microphone (after amplified in the amplifier) into the acoustic (sound) output of same frequency as of the input. In this way, it is "reverse" of the microphone.

The diaphragm of the speaker is made to vibrate—this in turn sets the molecules of the surrounding medium into motion, such that the sound can be heard at a distance.

4.3.1 Characteristics of Loudspeakers

The following characteristics of loud speakers determine their performance.

(*i*) *Efficiency.* It is defined as the ratio of output sound energy obtained from a loud speaker to the amount of electrical energy as input from the mike to the loud speaker. Naturally the output power will depend upon the impedence matching between the source (loud speaker's coil) and the load (air to be disturbed by the speaker). The efficiency of horn type loud speakers is about 10% and that of domestic (cone) type loud speaker is only 1%.

(*ii*) *Sensitivity.* It is the input signal which can provide an output sound pressure of 1 microbar at a distance of 1 m from the loud speaker.

(*iii*) *Signal/Noise ratio.* The loud speaker creates "self noise" as its mechanical parts also produce vibrations due to their natural frequency. The S/N ratio of loud speakers is equal to

$$\text{SNR} = \frac{\text{Signal output}}{\text{Signal output without noise}}$$

(*iv*) *Frequency response.* This is the range of audio frequency to which a loud speaker responds. Ideally, a loud speaker's frequency response should be for the entire range of audio frequency i.e., from 20 Hz to 20 kHz, but practically, this is affected by many factors e.g., mass of the vibrating components of the loud speaker etc. Practical loud speakers have a response from 100 Hz to 16 kHz. See Fig. 4.17.

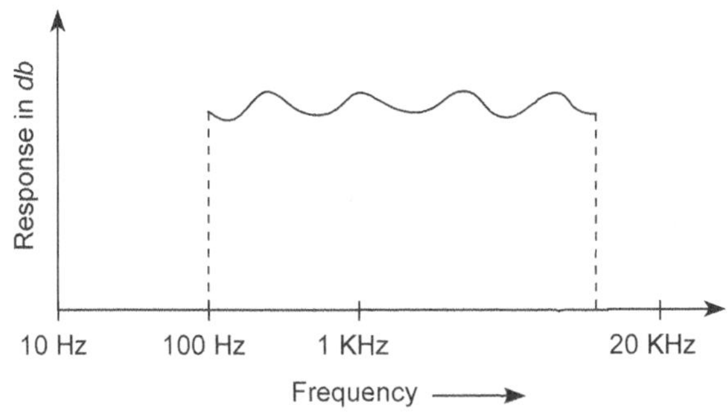

Fig. 4.17

(*v*) *Distortion.* The output of the loud speaker may not be exact replica of the sound input at the mike. This is called distortion. This may be an amplitude, frequency or phase-distortion.

Frequency distortion is more important in loud speakers as they produce second and third harmomics of frequencies. See Fig. 4.18.

Truly speaking, any of the distortions is not desirable in loudspeakers.

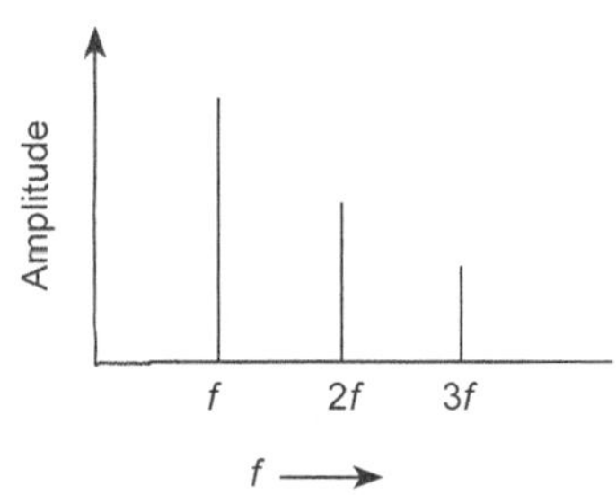

Fig. 4.18

(*vi*) *Directivity.* Loud speakers are always directional *i.e.,* they produce maximum intensity in a particular direction. Ratio of sound intensity in the direction of maximum intensity to the sound intensity that would be available if the loud speaker would have been non directional, is called its "Directivity".

(*vii*) *Impedence.* The input impedence (of speaker's coil) is given in ohms. It helps in its matching with the amplifier e.g., If a loud speaker has been marked as 16 Ω, the output impedence of the amplifier should be 16 Ω in order to get maximum transfer of power from amplifier to the loudspeaker. (Fig. 4.19).

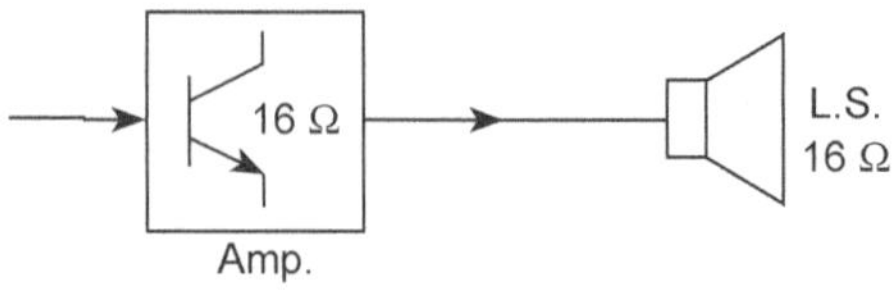

Fig. 4.19

At low frequencies its impedence is higher and at high frequencies, its impedence is low. Usually loud speaker is available in 4, 8 and 16 ohms. Knowledge of impedence is also helpful, when loud speakers and connected in series or parallel.

(*viii*) *Power Handling Capacity.* Power in watts, which can be handled by a loudspeaker without distortion is known as its power handling capacity. If a loudspeaker is marked as 1 W, it means maximum power the loud speaker can handle is 1 W. If it is asked to handle power more than 1 W, it may damage the loudspeaker.

4.3.2 Classification of Loudspeakers

A loud speaker has a vibrating surface which radiates sound into the surrounding air. The loud speaker encloses an arrangement which sets the vibrating surface in motion. In addition, the loud speaker has an *enclosure* to couple sound energy with the surroundings.

Accordingly, loud speakers are classified into:

1. *Direct radiation (or cone) type.* Which has a vibrating surface (the diaphragm) directly coupled with air. They are compact in size, cheaper and have a frequency response for a wide audio band.

   But, they have low output, narrow directivity and poor response at higher frequencies.

2. *Indirect radiation (or Horn) type.* The indirect radiation types loud speaker does not radiate sound power directly into the surroundings, but the sound is first delivered to an air column enclosed in a fixed non vibrating "horn" and from there, it is thrown into the surroundings.

Table 4.2: Comparison between the two types of speakers

S. No.	Direct radiating (or cone type) speakers	Indirect radiating (or horn type) speakers
1.	It works on motor principle.	It also works on motor principle.
2.	The cone is of made of paper.	There is no paper cone. A diaphragm is used.
3.	The paper radiates sound directly into the air.	The diaphragm radiates sound in air through a horn.
4.	There is no matching.	Some impedence matching exists between diaphragm (Horn) and the air.
5.	Its efficiency is 10%.	Its efficiency is 40%.
6.	It has a frequency response from 500 Hz to 5 kHz.	Its frequency response is from 20 Hz to 12 kHz.
7.	Distortion is 10%.	Distortion is 2%.
8.	It can handle power upto 80 W.	It can handle power upto 400 W.
9.	Its input impedence is 200 Ω.	Its input impedence is 16 Ω.
10.	It is used in Radio, TV, VCR etc.	It is generally used in P.A. system.

1. Direct Radiating (Cone) Speakers

There are of two types (1) Moving coil with permanent magnet and (2) moving coil with electromagnet type.

(*i*) **Moving Coil (Permanent magnet) or Dynamic Cone Speaker:** It is the most popular loudspeaker, because of its simplicity in construction, precision, easy to coordinate with other equipment and freedom from electrical troubles. It is widely used in all kinds of domestic reproducing systems.

The other major factor of its wide use is its "own power house". It requires no other power other than the signal to make it operational. Its power house is its permanent magnet. It does not lose its magnetism for the whole of its life.

(*a*) **Construction.** The major parts of a dynamic loud speaker are:

(*i*) Voice coil

(*ii*) Permanent magnet

(*iii*) Spider

(*iv*) Diaphragm (cone)

The Fig. 4.20(a) shows the assembled speaker and (b) shows the dismantled speaker its various components.

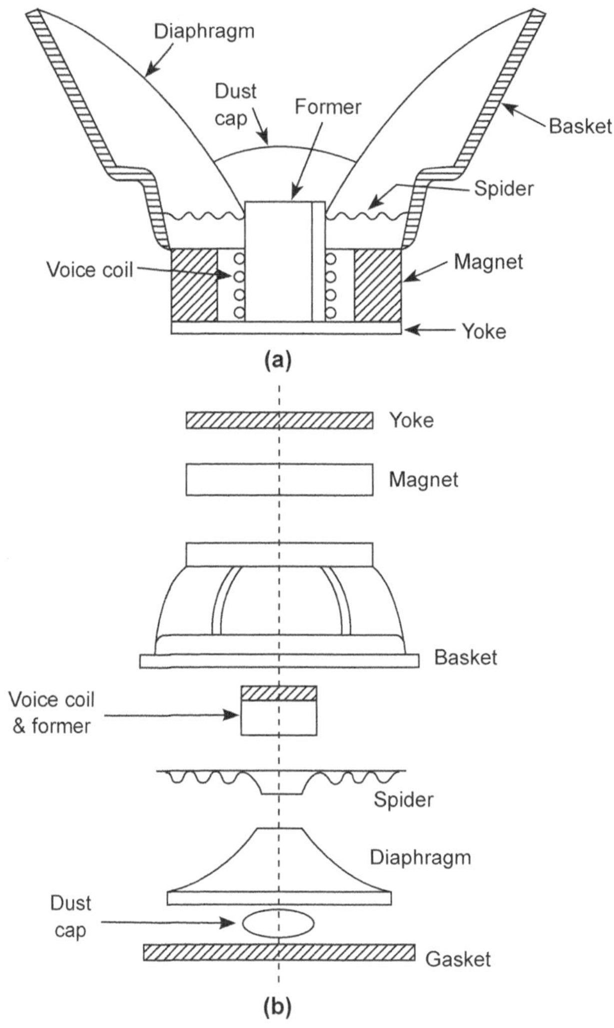

Fig. 4.20

(*i*) *Voice coil.* The voice coil is the only thing in a speaker, that carries the electrical signal. As the name indicates, the voice coil is the part of the loud speaker which "talks", when energised by the signal (or speech) currents coming from the amplifier.

(*ii*) *The permanent magnet.* The magnet is pot type and made of Alnico (A1 = 10%, Ni = 2% and cobalt 10%).

The magnet is the heart of the loud speaker. The current flowing through the voice coil sets up a magnetic field around the coil. The interaction between this field and the magnetic field produced by the permanent magnet sets up a torque on the voice coil, which forces the coil to move. Thus the working of the loud speaker depends upon the "motor action".

Due to the interaction of the two magnetic fields—one stationary field of magnet and other movable field of the voice coil, the voice coil will move in up or down direction. The direction of the voice coil field will be determined by the direction of the signal currents through the coil. The direction of motion of the voice coil is generally parallel to the length of the gap, in which it is balanced.

(*iii*) *Spider.* The spider is a device, which balances the voice coil in the air gap. It holds the coil centred in the gap hence it is also called "centering spider". The coil, while in motion vibrates in and out of the gap. The spider does not allow any lateral movement of the coil, at the same time it does not obstruct the movement of the cone. The coil also should not strike the metal walls of the gap. In this way, spider is an important role to play in the proper functioning of the loudspeaker.

(*iv*) *Diaphragm (Cone).* The voice coil is attached to the diaphragm, which actually puts the surrounding air into motion. The cone is made of special paper or of impregnated cloth. The cone moves back and forth alongwith the voice coil depending upon the strength of the signal currents flowing through the coil. The cone causes the surrounding air to vibrate exactly in the same manner as the signal, this creates sound waves in the air.

(*v*) *Basket.* All the components are housed on a sheet metal body called Basket.

(b) Impedance of the Dynamic Speaker

This loud speaker is "current operated", as it is actuated by the signal currents through the voice coil. For being current sensitive, the device should have minimum impedance. Therefore this speaker has an impedance range of 2, 4, 8 and 16 ohms. The 8 ohm loud speaker being the most popular. Further the impedence is not constant, it varies significantly with the frequency. The greatest variation is near "resonant frequency" (Fig. 4.21).

Explanation for the curve (Fig. 4.21) of loud speaker (impedence vs frequency) may be given as under.

When the voice coil moves in the magnetic gap, it acts like a "Generator" and produces an emf. This emf in fact may be considered as "Back emf" of the generator. As we know the emf generated is in opposition to the applied voltage (Lenz's law), current in the voice coil will be lesser hence loud speaker impedence will be higher

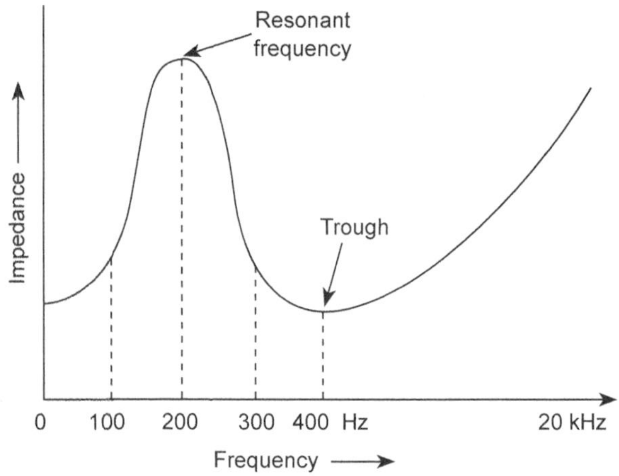

Fig. 4.21

(*i*) At resonant frequency, back emf becomes minimum, hence impedence is maximum.

(*ii*) After the trough area, as the frequency (f) of the voltage applied to the voice coil increases, the inductive reactance (X_L) (*i.e.*, impedence) of the voice coil increases at uniform rate ($X_L \propto f$) as shown.

(*c*) Efficiency of Moving coil dynamic Speakers

The efficiency of a moving coil dynamic speaker is less than 5%. The reason of low efficiency of speakers is that the electrical energy fed to the loud speaker is not converted directly into sound but it has following stages:

(*i*) First of all, electrical energy is converted into magnetic energy.

(*ii*) This magnetic energy is then converted into mechanical energy.

(*iii*) Now this mechanical energy is converted finally into sound energy. (Fig. 4.22)

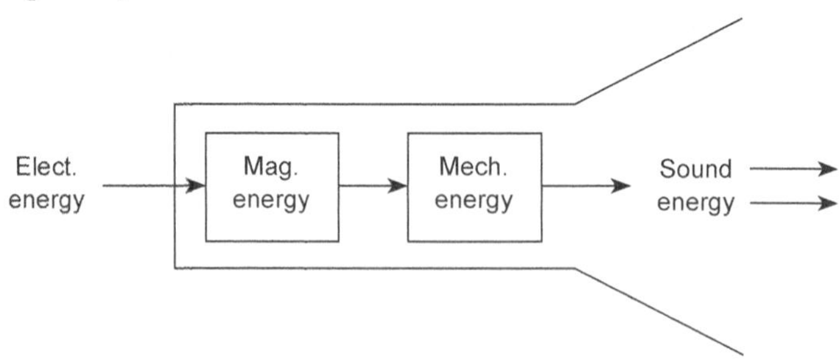

Fig. 4.22

Further, during the stages, the following losses also take place.

(*i*) When electric energy flows through the coil, power loss takes place in the coil, which will depend upon the resistance of the coil.

(*ii*) During conversion into magnetic energy, a loss will occur again, which will depend upon flux linkage between voice coil and the flux in the gap.

(*iii*) During conversion from mechanical energy to the sound, losses will occur in the mechanical drive system of the loud speaker.

(d) Working

As mentioned earlier the, dynamic loud speaker works on the motor principle. When the signal current passes through the voice coil placed in the magnetic field of a permanent magnet, the coil experiences a torque and moves. This makes the paper cone to vibrate, which produces sound oscillations in the air.

The force in Newton (Nw) acting on the coil may be given as:

$$F = BIl \sin \theta \; Nw,$$

where, B = Flux density of the magnetic field.

I = audio currents in the coil

I = length of the coil

θ = angle between coil and the field.

The Fig. 4.23 shows an equivalent circuit of a cone speaker. In this figure:

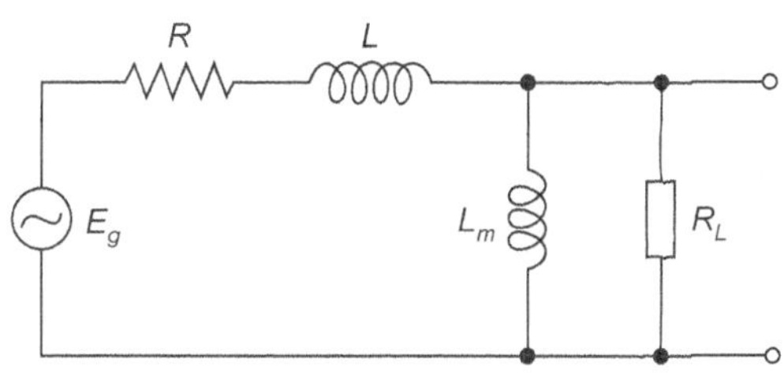

Fig. 4.23

E_g = Electrical input.

R = Resistance of voice coil.

L = Inductance of voice coil.

L_m = Mechanical inductance (mass) of coil and the cone assembly.

R_L = Load on the speaker (Resistance of the air replaced by the speaker).

(ii) **Moving coil with electromagnet (Electrodynamic) cone speaker:** The moving coil speakers have low wattage therefore moving coil type speakers with an electromagnet (in place of permanent magnets) are used. In these speakers, power handling capacity may be upto few hundred watts.

Their construction is similar to dynamic type with a difference that these have an electomagnet in place of the permanent magnet.

The magnet here is also pot type and has *E* shape, the South pole being at the centre surrounded by two North poles. The voice coil wound on an aluminium former is placed in the annular gap as shown in Fig. 4.24.

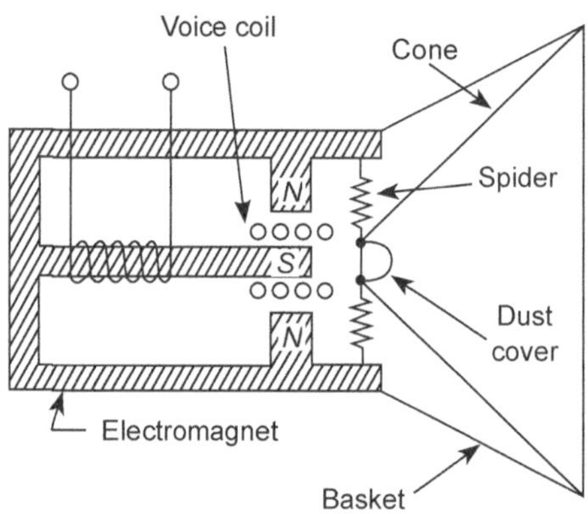

Fig. 4.24

Working of electrodynamic speaker

The working is also similar to that of the *dynamic speaker*. The audio signal obtained from amplifier is made to flow through the voice coil and produces a varying magnetic field around the coil according to the signal.

The interaction of magnetic fields of the coil and of the electromagnet produces a torque in the coil which starts vibrating. These vibrations are transmitted to the cone, which produces vibrations in the surrounding air.

Electrodynamic speaker has high power handling capacity and has better frequency response. On the other hand, external supply is needed for the electromagnet and also it is heavier and costlier than the dynamic speaker.

Characteristics of Cone Speakers

(*i*) Efficiency, as explained already is less than 5%. In addition to the reasons of low efficiency described above, there is one more reason that there is complete mismatch between impedance of the speaker and of the surrounding air.

(*ii*) The SN ratio is about 28 dB.

(*iii*) It gives better frequency response at mid frequencies. The typical value is from 200 Hz to 6 kHz. See Fig. 4.25.

(*iv*) Distortion of the loud speaker is about 12%.

(*v*) Due to baffles (enclosures) used, the speaker has best directivity at the front.

(*vi*) The loud speakers are available in 4, 6, 8 and 16 ohms.

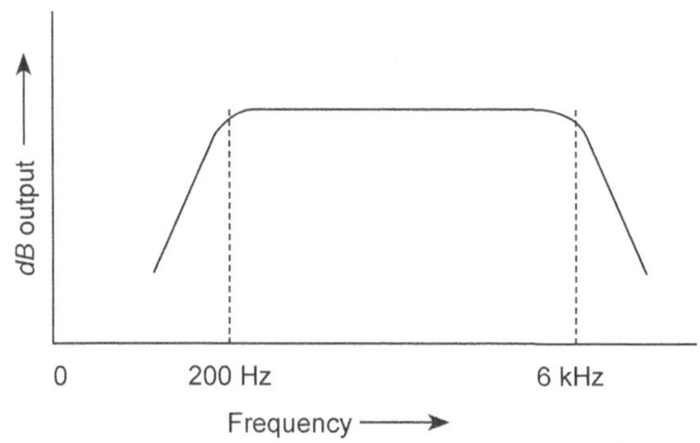

Fig. 4.25

(*vii*) Maximum power handling capacity of these speakers is 25 W.

(c) More about cone speakers

(i) **Baffles/Enclosures/Cabinets:** In a loud speaker, air is present on both sides i.e. at front as well as at the back. There is compression and rarefaction of air columns during the operation of the speaker. The sound waves produced at front and back are out of phase. The sound waves from the back come and meet the sound waves at the front and cancel each other, as they are out of phase. This is more serious problem at low frequencies. To solve this problem, it is necessary to increase path of the waves by using a barrier. This barrier is called a **Baffle**.

Thus a baffle is a structure, that increases the length of sound path and stops interference of anti-phase sounds coming from different sides of the cone.

The baffles are made of good sound insulating materials. Wood or preferably celotax is used for this purpose. The speaker should be mounted on the baffle firmly, so that they do not produce rattling sound.

It can be shown that for 50 Hz, the size of "baffle" required will be atleast 3 meters, therefore "enclosures" have been designed, which can replace baffles. Thus there is a little difference between baffles and enclosures. The other name for baffle is **Cabinet**.

The Fig. 4.26 (*a*) shows interference of sound waves and (*b*) shows, how length of sound path has been increased by using baffles.

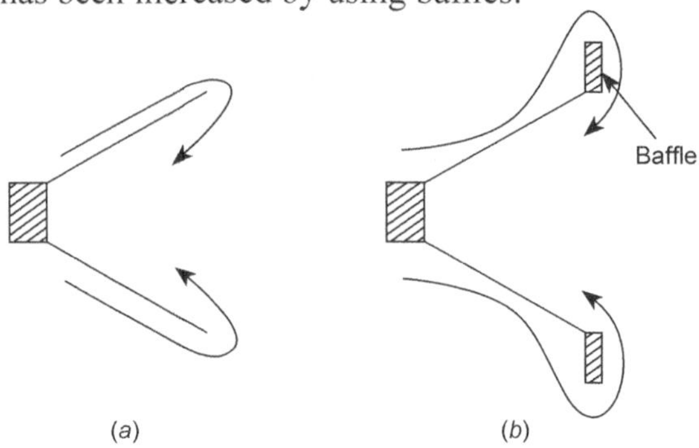

Fig. 4.26

Types of Baffles/Enclosures Cabinets

The important types of baffles used are:

1. *Finite Baffles (Fig. 4.27).* The *cabinets* used in radio, TV receivers are "finite baffles". They are not so effective in prevention of interference of front and back sounds.

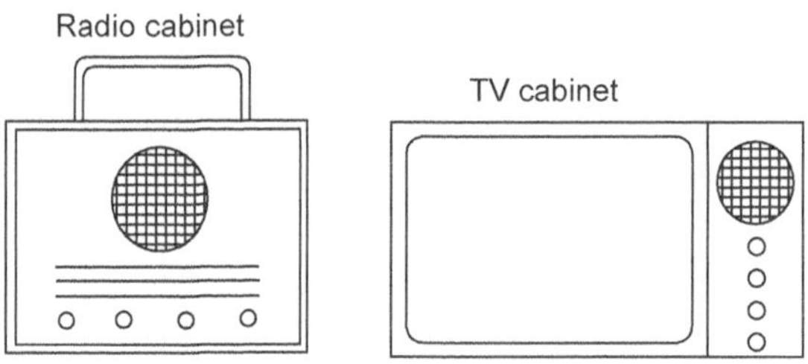

Fig. 4.27

2. *Infinite baffles.* Theoretically the size of baffle should be equal to half the wavelength of the sound, hence, for lowest frequency, we will need an infinite baffle, which will not be practical.

3. *Folded baffles (Fig. 4.28 a).* A baffle bent at the edges of the one speaker is called a folded baffle. This is a practical shape of an infinite baffle.

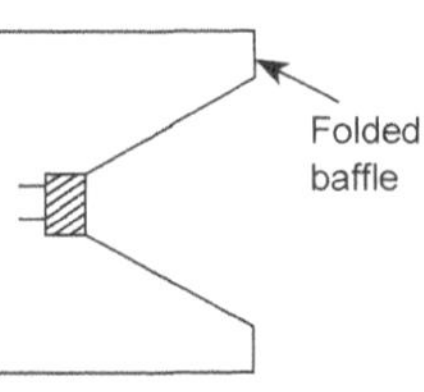

Fig. 4.28 (a)

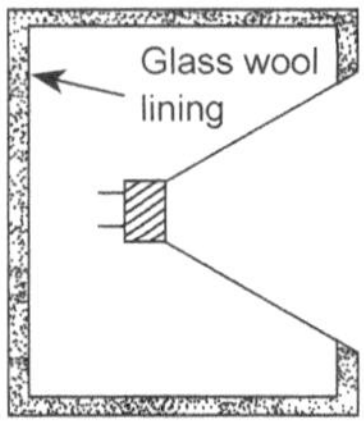

Fig. 4.28 (b)

4. *Enclosure baffle (Fig. 4.28 b).* If the folded baffle is closed at the back. It will give an *enclosure baffle*.

   The stereos and decks use these type of baffles. Their loud speakers are enclosed in air tight plywood boxes. These boxes may have a lining of some absorbent material like glass wool, cork etc. to absorb sound waves reflected from inside walls of the enclosure.

5. *Reflex (Vented) baffle [Fig. 4.29 (a), (b)].* In this baffle, a cut is made in the baffle at the front. This cut is known as "Port" and it is so designed that, the rear and front sounds become *in phase*, thus one strengthens the other. This may also be lined internally by a sound absorbent material.

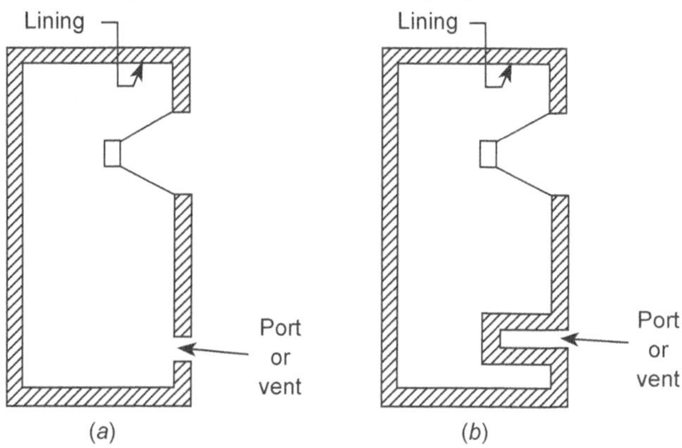

Fig. 4.29

6. *Acoustic labyrinth Baffle (See Fig. 4.30).* For Hi-Fi work, where atleast two speakers (called woofer and tweeter) are enclosed in the same cabinet, the cabinet is given the shape of a acoustic labyrinth. For this the cabinet is divided into partitions. This arrangement also helps in strengthening the front and back sound waves.

   This is a modified form of reflex baffle. For very low frequencies, this arrangement gives very good result. Lining of absorbent material may be provided inside the walls.

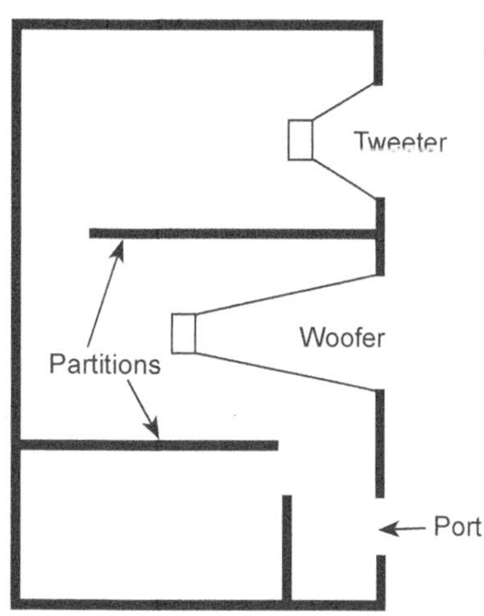

Fig. 4.30

(ii) Multiway Speaker System for Cone Speakers

As we know the audio frequency (AF) range is from 20 Hz to 20 kHz. A single loud speaker cannot give satisfactory response for the whole range. As there is an interference between back and front sound waves and the same cannot be totally eliminated by using baffles, therefore the total AF range has been divided into 3 parts.

(*i*) 20 Hz to 1 kHz—Low Frequencies

(*ii*) 1 kHz to 5 kHz—Medium Frequencies

(*iii*) 5 kHz and onwards—High Frequencies

The speakers which cover low frequencies are called *woofers*, the speakers which cover mid frequencies are called middle range speakers or *squawker* and those covering high audio frequencies as called as *Tweeters*. The woofer has small "angle" (θ), where as mid range and tweeters have larger angles.

The woofers and tweeters can be separately mounted in the same baffle, as shown in Fig. 4.31 (*a*).

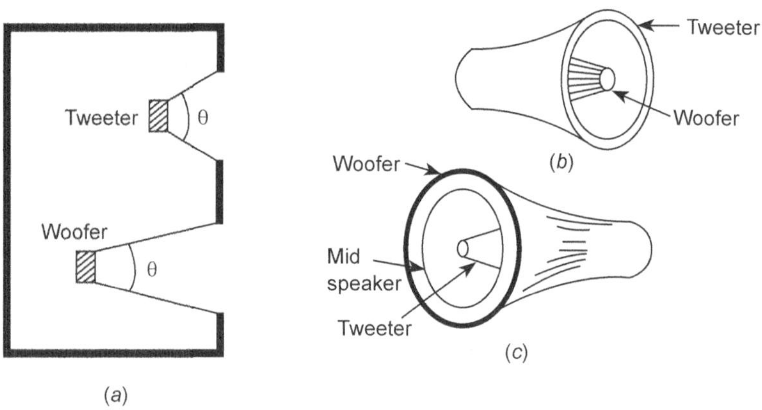

Fig. 4.31

Some speakers have two diaphragms. The larger diaphragm works as woofer and smaller as a tweeter. They are called Double cone or *co-axial* speakers [Fig. (*b*)]. These have better frequency response.

The Fig. (*c*) shows a *tri-axial* speaker unit in which three speakers are mounted in a single unit, thus the whole *AF* range is covered by a single speaker.

(iii) Cross Over Network for Cone Speakers

The output from the amplifier should be divided to the loud speakers such that each speaker gets its own frequency range. In other words, in a multispeaker system, where woofers, tweeters and squawkers are used, the woofer should get only low frequencies, tweeters should get high frequencies and squawker should be fed with only mid frequencies. This "cross over" of frequencies is performed by *cross over networks* (or circuits). If this is not done properly, the speakers will be over heated and it will reduce efficiency of the system.

The networks may be:

1. *Two way network (See Fig. 4.32).* This network is used when only tweeter and woofer are there. The *L-C* circuit is used to distribute the frequencies into low and high ranges. The inductor (*L*) checks high frequency from entering into woofer and capacitor (*C*) does not allow low frequencies to enter the tweeter. The Fig. 4.32 (*b*) shows response curve for the two way not work.

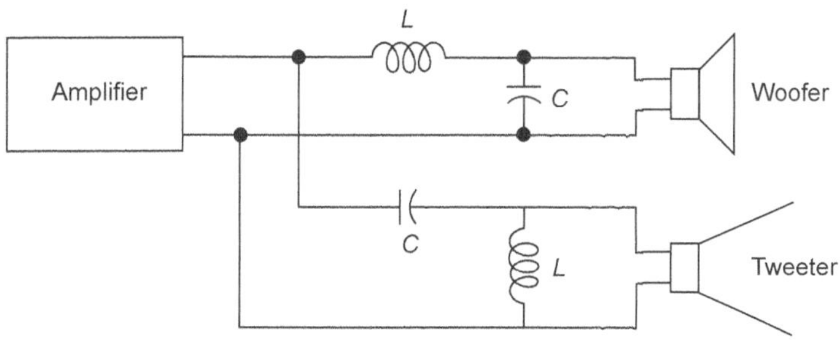

Fig. 4.32 (a)

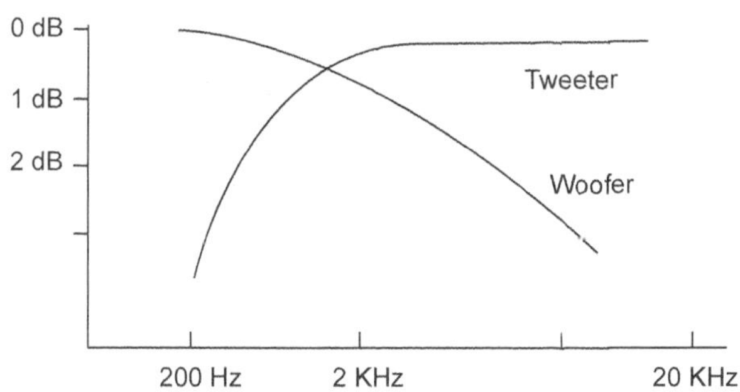

Fig. 4.32 (b)

2. *Three way network (See Fig. 4.33).* This circuit is used, where all the three speakers—Tweeter, woofer and squawker are connected in the circuit.

   The capacitor *C* prevents low and mid frequencies to enter tweeter. The inductors L_1 and L_2 prevent low and high frequencies to enter the squawker and the inductor L_3 prevents low and high frequencies to enter the woofer.

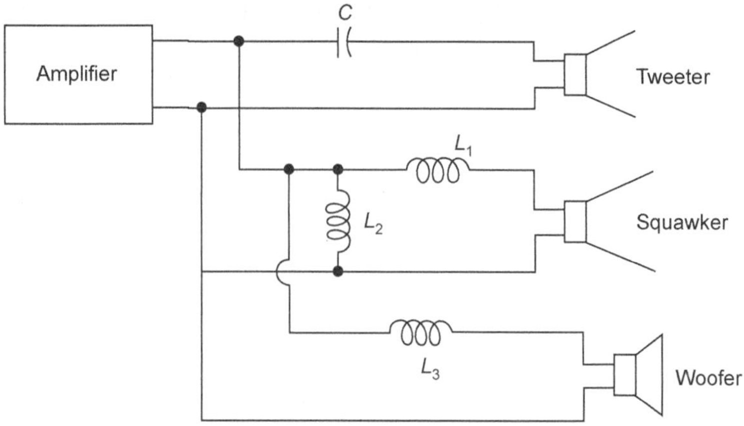

Fig. 4.33

Cross over frequency. The cross over frequency is the point where output curves of speakers intersect each other in frequency response diagram.

For a three way cross over network, the frequency range to the cross over point are:

Woofer—15 Hz to 600 Hz

Squawker—600 Hz to 4 kHz

Tweeter—4 kHz to 16 kHz

For two way cross over network, this range is

woofer—15 Hz to 1 kHz

Tweeter—1 kHz to 16 kHz

Example 4.6: *Design a cross-over network to give 12 dB per octave attenuation for weeter and woofer for a critical frequency (f_c) 1 KHz.*

The loudspeaker resistance is 8 Ω.

Solution: The values given are

$$f_c = 1 \text{ KHz} = 1000 \text{ Hz}$$

$$R_L = 8\ \Omega.$$

Now,

$$\boxed{\text{Capacitance, } C = \frac{1}{2\sqrt{2}\pi f_C R_L}}$$

$$C = \frac{1}{2\sqrt{2}\times 3.14\times 1000\times 8} = 14.2\ \mu\text{F. } \textbf{Ans.}$$

and

$$\boxed{\text{Inductance, } L = \frac{R_L.\sqrt{2}}{2\pi f_C}}$$

$$L = \frac{8\sqrt{2}}{2\times 3.14\times 1000} = 1.8 \text{ mH}$$

The cross-over network is shown (Fig. 4.34).

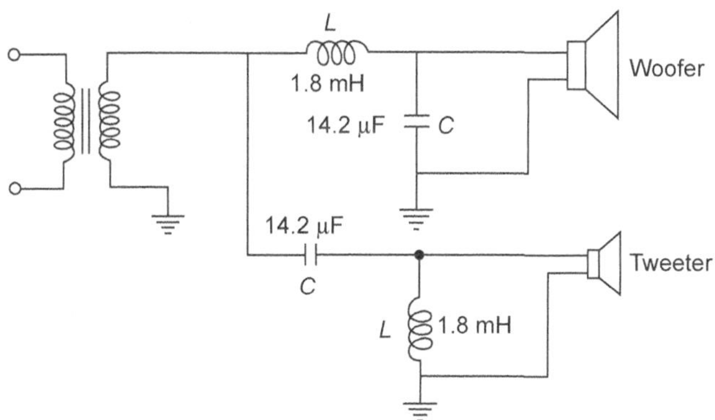

Fig. 4.34

Example 4.7: *A cone type speaker gives an output of 2 W equivalent sound. The resistance of the voice coil is 12 ohms and voltage input is 21V. Find input power and efficiency of the speaker.*

Solution: Input power $= \frac{V^2}{R} = \frac{(21)^2}{12}$

$= 36.75$ W. Ans.

$$\text{Efficiency} = \frac{\text{output power}}{\text{Input power}}$$

$$= \frac{2}{36.75} = 5.44\% \quad \textbf{Ans.}$$

Example 4.8: A power amplifier has an output resistance of 45000 ohm. It is connected to a stepdown transformer having $N_1/N_2 = 10$. What should be the resistance of the loud speaker connected at the secondary.

Solution: For impedence matching

$$\frac{Z_2}{Z_1} = \left[\frac{N_2}{N_1}\right]^2$$

$$Z_2 = Z_1\left(\frac{N_1}{N_1}\right)^2 = 4500\left(\frac{1}{10}\right)^2 = 45\ \Omega$$

(iv) Cone Speakers Shapes

In shapes, speakers may be round, rectangular etc. Their size varies from 2″ to 10″. The selection of the size depends upon the cabinet of the radio/TV. (See Fig. 4.35)

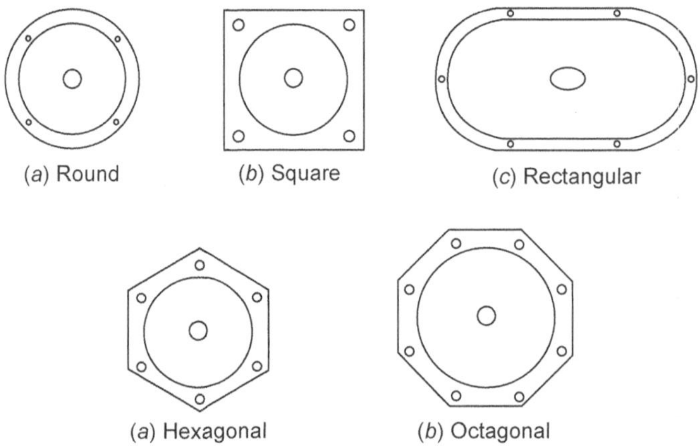

Fig. 4.35

(v) Headphones

Headphones are basically loud speakers enclosed in a small cover. Their construction is also more or less same. They have a "voice coil" suspended in the magnetic field of a permanent magnet and a diaphragm to vibrate.

Advantages

(*i*) By using headphone, we can listen a program without disturbing others in the vicinity.

(*ii*) We are also not disturbed by the noise of the surrounding.

(*iii*) We can hear sound at the desired loudness.

(*iv*) By headphone we can hear stererophonic sounds also. This is not possible by an earphone.

(*v*) A headphone needs a very small amplifier to give a satisfactory frequency response.

(*vi*) We need a fraction of a watt to drive a headphone.

2. Indirect Radiating (or Horn) Type Speakers

There is no major difference between cone and horn type speakers, except that the cone has been replaced by a horn to get more audio output, as a horn provides better impedance matching with the surrounding air.

There is also a very little difference in working of the two. The cone speaker radiates sound directly into air but in horn type, the sound is radiated into air through the horn. The horn does this job acoustically which has been done by the cone mechanically. Moreover horn acts as a sound transformer. All this give it a better efficiency. The horn type speakers can give upto 40% efficiency whereas, cone type speaker has an efficiency of 5%.

(a) Horn

A horn may be defined as a tube tapered such that, its diameter increases from a small value at one end called *throat* to a large value at another end called *mouth* (Fig. 4.36). Due to its special shape, it can work as sound transformer. The shape of the horn is major factor, which determines its performance.

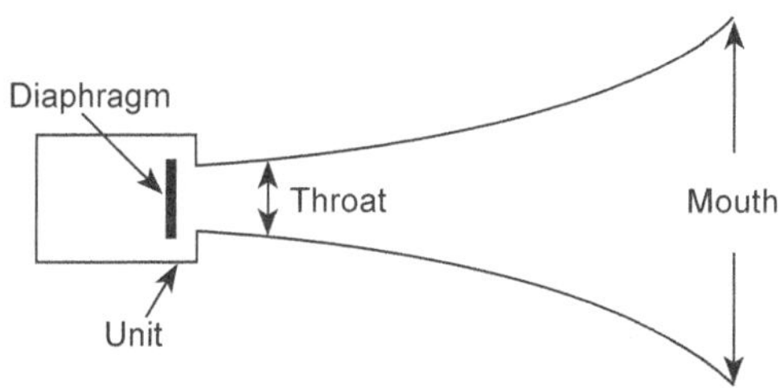

Fig. 4.36

The various shapes used for horn speakers (as they are popularly called) are: (See Fig. 4.37)

(*i*) Parabolic

(*ii*) Exponential

(*iii*) Hyperbolic

(*iv*) Pyramid

(*v*) Conical.

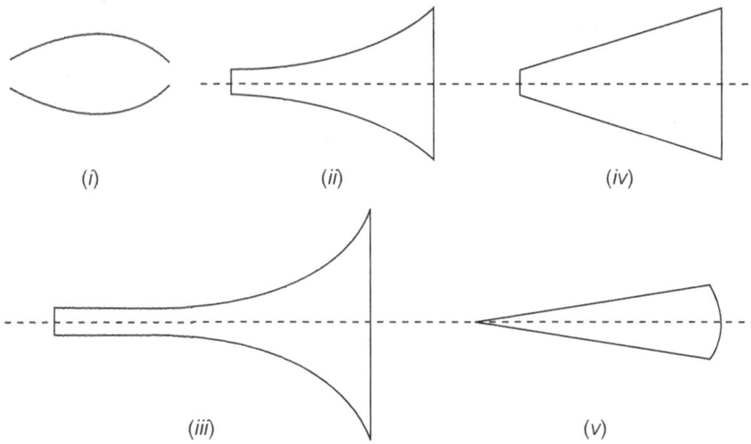

Fig. 4.37

The most commonly used shape is exponential, in which the change in cross section area from throat to mouth is proportional to the length of the speaker and increases logarithmically.

"Small throat" horns are used in the frequency range of 100 Hz to 6000 Hz and "large throat" horns are used for low frequencies. Every horn has a frequency below which it does not radiate sound at all. This is called its "cut off frequency".

Note:

(*i*) Cut off frequency of a horn speaker can be found by following equation:

$$f_0 = \frac{V.a}{2.\pi.V_{air}} \quad \text{(Fig. 4.38)}$$

where f_0 = Cut off frequency

a = Cross section area of throat = $\pi/4\ d^2$

V_{air} = Volume of air.

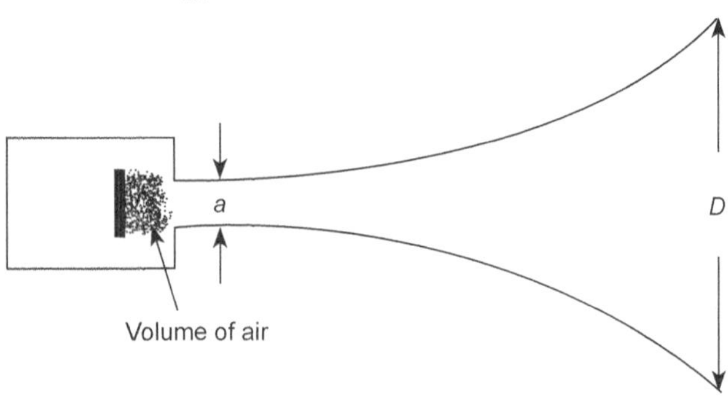

Fig. 4.38

(*ii*) The lowest frequency for horn speaker can be found by following equation:

$$f_{\text{low}} = \frac{169}{D}$$

where f_{low} = lowest frequency in Hz

D = Diameter of mouth.

(*iii*) Special loud speakers are used for hi-fi systems. They have a "cone" as well as a "horn". The mouth of cone is kept in a direction opposite to the audience, where as mouth of the horn is kept towards the audience. This arrangement gives a very good frequency response.

(*iv*) *Frequency response of horn speakers*. In *exponential* horns, the sound waves have medium expansion, in hyperbolic speakers expansion of waves is very slow and in Pyramid (and conical) speaker, the expansion of the waves is fast. The frequency response for these three have been shown in Fig. 4.39.

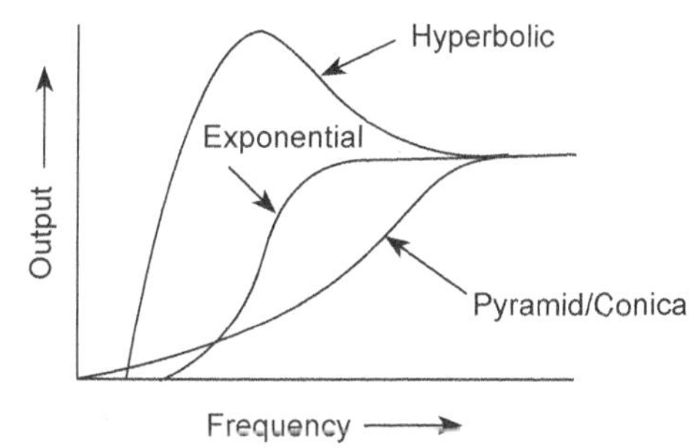

Fig. 4.39

(b) Types of Horn Speakers

As mentioned already, there is no major difference between cone and horn speakers. If the cone is replaced by a horn, it will become a horn loudspeaker. As horn speakers give large output than the cone speakers, the size of the unit is also larger. It is important to mention that they are classified depending upon the principle involved in the working of their unit. Important types of horn speakers are:

(*i*) Moving coil or dynamic speakers

(*ii*) Moving coil, or Electrodynamic speakers
(*iii*) Armature speakers
(*iv*) Induction speakers
(*v*) Capacitive (condenser) speaker.

(*i*) **Moving Coil dynamic speaker (with permanent magnet):** The speaker is very popular due to its versatile properties and is widely used. They have a permanent magnet and do not require any external source of supply.

It consists of a coil moving in the field of a permanent magnet. Thus basically it is an electric motor. With the coil, a diaphragm is connected which is also movable.

The output current from the amplifier is given to the coil which is called "Voice coil". As the current carrying voice coil is suspended in a magnetic field, it starts rotating (motor rule). This sets the diaphragm (Horn) into vibrations. The diaphragm (*i.e.* Horn) puts the surrounding in to motion and sound can be heard at a distance. The Fig. 4.40 shows this speaker.

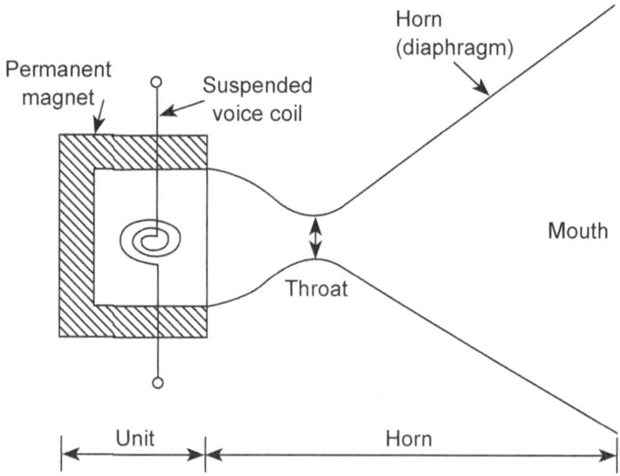

Fig. 4.40

The "Horn" which acts as a diaphragm is a type of sound transformer. The sound while travelling from unit to the horn becomes louder due to the transformer action of the horn. Actually it establishes a type of "impedence matching" between driver and the driven air and thus output power is improved. The horn's shape may be—conical, exponential or flat. A conical is that in which cross-section of the horn is increased in proportional to its length, whereas in exponential horn, the same is an exponent of the length.

The ratio of diameters of throat to mouth is 1 : 7 which means due to narrow throat, the pressure of sound is very high at the throat (about 7 times) than that at mouth. As the sound moves from "throat to mouth"— it goes on expanding thus its pressure goes on decreasing and the sound waves are spread over to a large surroundings. This is the reason that the sound can be heard at a large distance.

However the speakers may be with the unit resistance of 2, 8 or 16 ohm and Wattage of 10, 20 and 80 W etc.

Efficiency

The efficiency of a speaker is the ratio of the useful acoustic power radiated to the electrical power supplied. In case of these speakers, this efficiency is very low and maximum upto 35%.

(*ii*) **Moving coil Electrodynamic Speaker (with temporary magnet):** Moving coil Electro dynamic speaker contains an electromagnet and hence needs a separate source of supply. It has a three pole magnet. The central limb contains a coil which is to be energised by a separate d.c. source. (Fig. 4.41)

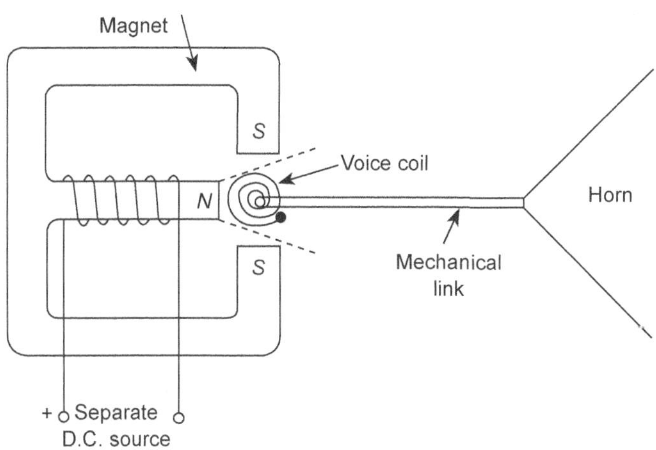

Fig. 4.41

There is a voice coil which is connected with output of the amplifier. With the same principle, the voice coil starts rotating; the torque acting on the coil is in accordance to the current following through the voice coil.

With oscillations of the voice coil, the diaphragm starts oscillating due to its mechanical coupling with the voice coil.

(*iii*) **Armature Speaker:** In this speaker, an armature is made to rotate between the poles of a permanent magnet as shown in Fig. 4.42.

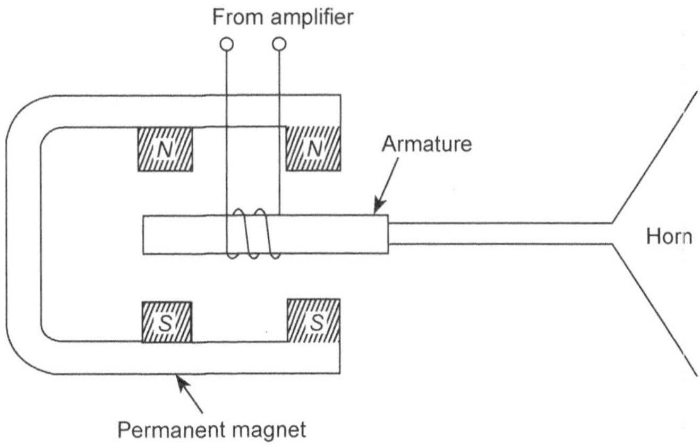

Fig. 4.42

On the armature, a coil is wound which is connected with the output of the amplifier. When signal current obtained from amplifier flows through the coil of the speaker, it produces its own separate magnetic field. An intersection between this magnetic field and the magnetic field of the permanent magnet, a torque is resulted which tends the armature to rotate. The armature being mechanically linked with the horn, makes horn to oscillate accordingly.

For obtaining a strong magnetic field, the air gap between the armature and the poles should be as small as possible. But this has got limitations. If the gap is made very small, due to very strong magnetic fields, the armature will rotate with a tremendous high speed even striking the poles and *rattling sound* will be produced.

(*iv*) **Induction Speakers:** These speakers work on the principle of electromagnetic induction. Which says "that emf (hence current and torque) is produced whenever there is a change in the flux linking with a coil".

The speaker has a diaphragm which is placed between two sets of co-centric coils. The direct current is supplied to the coils in opposite direction. The signal current also flows through the coils. (Fig. 4.43)

With the principle of induction, a torque acts on the diaphragm and it starts vibrating.

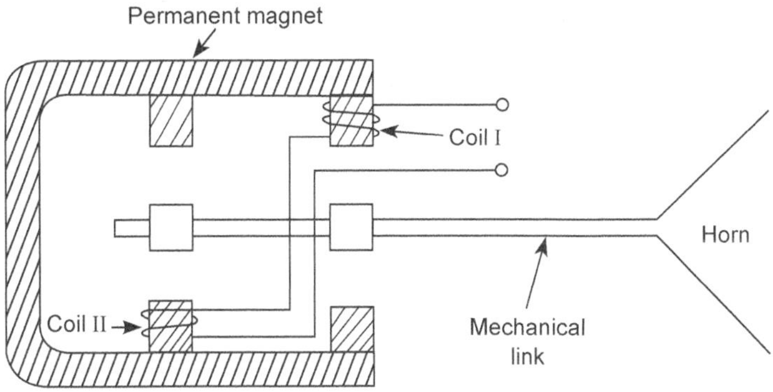

Fig. 4.43

Due to permanent magnet, it does not need any battery supply and therefore very much used with a cone in portable radio receivers.

(*v*) **Capacitive or Condenser Speakers:** This speaker contains two plates separated by some dielectric, thus forming a capacitor or condenser. The stationary plate is of copper or aluminium and generally has perforations to prevent compression of air between the plates. The movable plate or diaphragm is made of some light material. The dielectric between the plates is thin and flexible with high dielectric constant and of high breakdown voltage. (See Fig. 4.44)

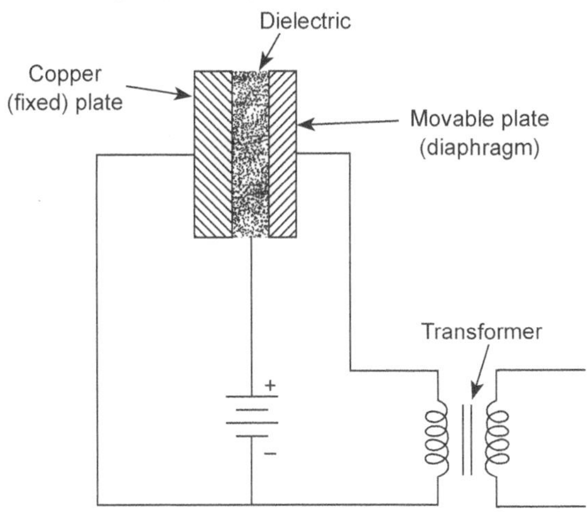

Fig. 4.44

When the plates are given voltage, an electric field is built up. On this, the field of the signal is superimposed. This causes attraction and repulsion between the plates as a result the movable plate (diaphragm) starts vibrating corresponding to the audio signal.

4.4 P.A. SYSTEM

The P.A. system stands for "Public Address System". This is an electro-acoustic system to send an audio signal to a distance.

It is a common experience that intensity of sound goes on decreasing with increase in distance. If a sound is to be transmitted to a distance (e.g. a public meeting), the sound should be suitably amplified so that it may reach to every corner and a person sitting there may listen comfortably. For this amplifiers are used.

The electronic circuit, which amplifies the sound and processes the same so that it may reach to the audience is called a P.A. system.

This system is used in public speeches, sportsmeet, cultural functions, auditoriums and also at railway stations and aerodromes for giving necessary information to the passengers.

4.4.1 Components of a P.A. System (Fig. 4.45)

The important components of a P.A. system are:

1. *Microphone.* The person speaks before a microphone, the function of which is to convert the sound (speech) of the person in to electrical pulses; as it is easy to transmit a sound in the form of electrical energy. The selection of suitable microphone and its positioning in the P.A. system is an important factor. The mike used in P.A. system is generally dynamic (moving coil) type or sometime ribbon velocity type.
2. *Mixer.* The output of microphone is fed to a mixer circuit which isolates different frequencies from each other.

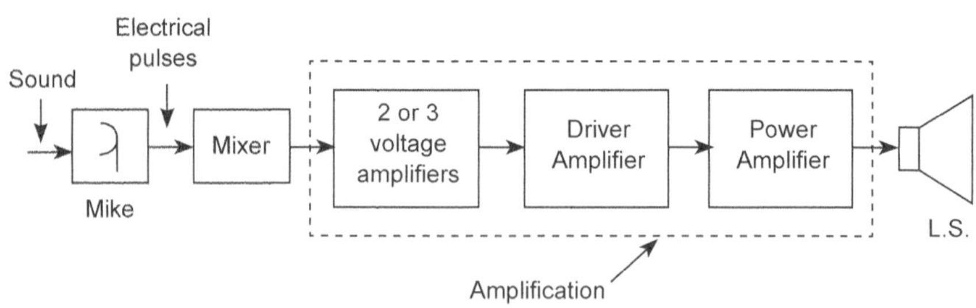

Fig. 4.45

3. *Amplifier.* The output of the mixer stage is now amplified. Generally amplification is done in following stages:

   (*i*) *Voltage amplification.* The signal obtained from the mike have extremely low voltage (2 mV) therefore the voltage level of weak signal is raised by two or more voltage amplifiers. Generally RC coupled amplifiers are used for voltage amplification.

(*ii*) *Driver stage.* The output of the last voltage amplifier is given to a driver amplifier. It supplies necessary power at its output stage. The driver amplifier is a class A transformer coupled power amplifier. Here concentration is given on the maximum power gain.

(*iii*) *Output power stage.* The output of the driver amplifier is given to the output power amplifier. It is the final stage which delivers power to the loud speaker. This stage necessarily uses a class *B* push pull amplifier. Concentration is given on the maximum power output to the speakers. The amplifier used for P.A. system has a rating ranging from 20 W to 130 W.

4. *Loudspeaker.* A loud speaker is the last stage of P.A. system. It converts electrical signal into the original sound produced at the mike, The output of the power amplifier is connected to the speaker through a matching device, which is generally a transformer. Impedance matching delivers maximum power from the amplifier to the speaker. In some designs a separate impedance matching device is not needed.

   The loud speakers used for a P.A. system are horn type loud speakers.

   For maximum power transfer, a series parallel combination of the speakers should be arranged.

4.4.2 Characteristics of P.A. System

1. There should not be any feed back of power from loudspeaker to the microphone. This results in a whistling sound from the speaker.
2. It is better to use many small loud speakers than a single large speaker. It provides a better distribution of sound intensity. As audience is the best sound absorber, the speakers should be directed towards the audience and not towards walls. Moreover, the number and wattage capacity of speakers should be sufficient to handle the power obtained from amplifiers.
3. The reverberation problem can also be solved by using a no. of small speakers rather than a single large speaker. The speakers should be placed at suitable points in the auditorium. The reverberation reduces intelligibility of sound in a closed room (auditorium). The loud speakers should throw more power in that portion of the auditorium which has more echo effects.

4. The noisy places like market, mob, sportsmeets, the P.A. system should boost high frequencies more than the low frequencies. The level of amplification should also be kept high, The "Noise cancelling microphones" can also be used.
5. The microphones used for a P.A. system should have a *heart shape* directivity pattern. In dramas, stage performance a column of unidirection microphones should be installed.
6. The loud speakers should be so positioned that sound appears coming from the original source. This maintains intelligibility of the sound.
7. Sometimes, the speaker stands very near to the microphone. This makes the system overloaded. For this, a sound level stabiliser (sound limiter) is used in P.A. systems, which keeps the output level of sound at a constant value irrespective of the input level.
8. For better results, stereophonic P.A. systems should be used. It gives high fidelity (hi-fi) sound effects. Such P.A. systems employ hi-fi components. However, in hi-fi P.A. systems, speaker columns containing tweeters, woofers and mid frequency speakers are used.
9. The microphone leads should be of short length as possible so that it may not pick up hum from AC mains. There should also be a proper matching between impedance of the mike and impedance of amplifier.

The characteristics of P.A. system may be summarised as under.

1. It should send sound equally in all directions.
2. The system should not pick up sounds from the surroundings.
3. It should take care of "reverberation" *i.e.,* the sound should immediately die when source of sound is removed.
4. Position of mikes and loudspeakers should be such that "faithfulness" is maintained.
5. For better results sterophonic PA system may be used.
6. The mike should not pick up hum.

4.4.3 Positioning of Microphones in P.A. System

1. The microphones used in P.A. system should be unidirectional *i.e.,* they should pick up sound from the front only. These mikes have a *cardioid* type directivity.
2. The mike should not pick up the neighbouring noise or any sound from the loud speakers.

3. As mentioned earlier, the microphone should possess *sound limiter* which will keep sound level constant irrespective of input and thus will compensate the drawbacks of the person standing very near before the mike.

4. At stage performances, a portable microphone will be best; as the actors are to speak dialogues while moving on the stage. The other alternative is to install a column of unidirectional microphones. See Fig. 4.46 (*a*). Sometimes microphones are hung from the ceiling. The Fig.(*b*) shows its directivity pattern.

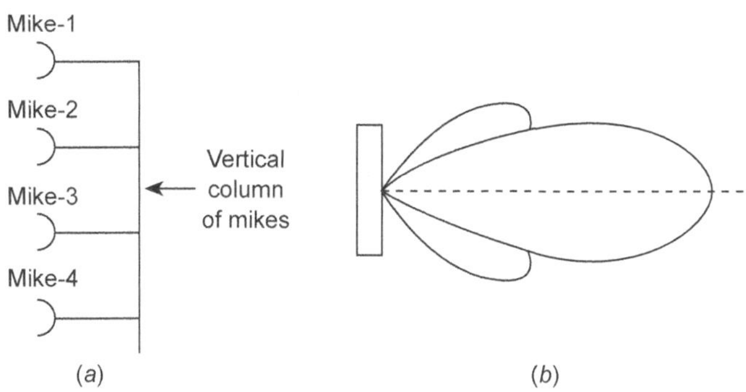

Fig. 4.46

5. In classroom, where a teacher is to deliver lecture before a class, a wireless microphone or a hand held portable mike will be best. This will fecititate the teacher to face the students as well as the black board.

6. At railway stations, air ports and other noisy places, noise cancelling microphones should be used.

4.4.4 Positioning of Loud Speakers

1. Generally, a horn type loud speaker or a column of loudspeakers is used. The arrangement ensures proper distribution of sound in the area and discourages power feed back to microphones.

2. Best sound effect is felt if the sound seems to be coming from the original source. Interestingly, human ears can locate the direction of sound, if it happens, the original rythem of the sound is maintained.

   It is also an interesting characteristic of human ears, that they can determine the direction of sound source from the sound first received by them. If an arrangement is made that the same sound strikes the human ears again and again (with a time delay), the natural effect of the sound is maintained. This is, what is practically done in P.A. systems.

In stereophonic P.A. system, a time delay of 25 ms is maintained between the two sounds. The second sound is produced by another loud speaker. In this arrangement, microphones are placed along the flood lights and a column of loud speakers is installed on either side (See Fig. 4.47). By this, a correct directional perception is achieved and naturality is maintained.

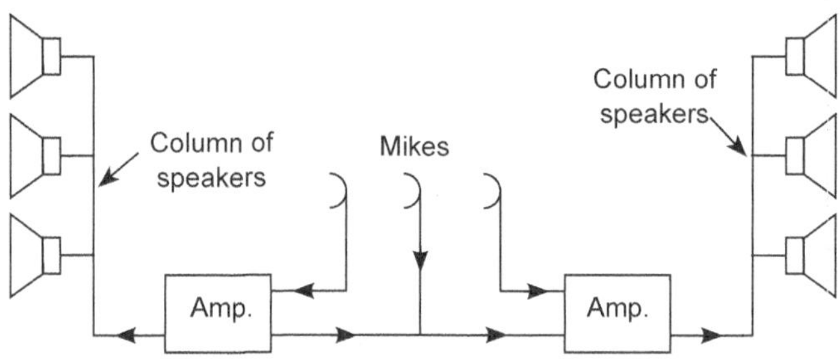

Fig. 4.47

3. (*a*) In open air theatres/auditoriums or on roadsides, the installation of loud speakers need much attention due to the effect of wind and the temperature. At such places, the area is distributed into zones and in each zone, a separate loud speaker is installed in place of one central loud speaker.

   Installation of only one central loud speaker will create following effects:

   (*i*) When there is no variation in temperature with height, the sound waves are straight and whole audience is covered [Fig. 4.48 (*a*)] *i.e.,* no problem.

   (*ii*) When the velocity of sound increases with increase in temperature, a drop in temperature with height causes the sound waves to bend upward and whole audience will not be covered [Fig. (*b*)].

   (*iii*) Similarly when temperature rises with height, the sound waves will bend downward. In this case also, whole audience will not be covered [Fig. (*c*)].

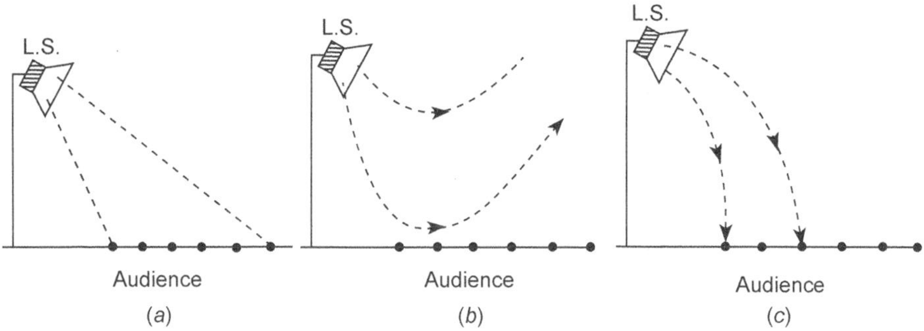

Fig. 4.48

(*b*) The other problem will be created by the echo coming from nearer loud speakers, thus an audience will hear more than one sound with a time delay; therefore the sound will lose its originality. It can be shown mathematically that minimum distance between two loud speakers radiating sound in the same direction should be 16 meters.

This problem can be solved by an arrangement shown in Fig. 4.49. The sound from different loudspeakers is delayed such that the audience feels originality of the sound.

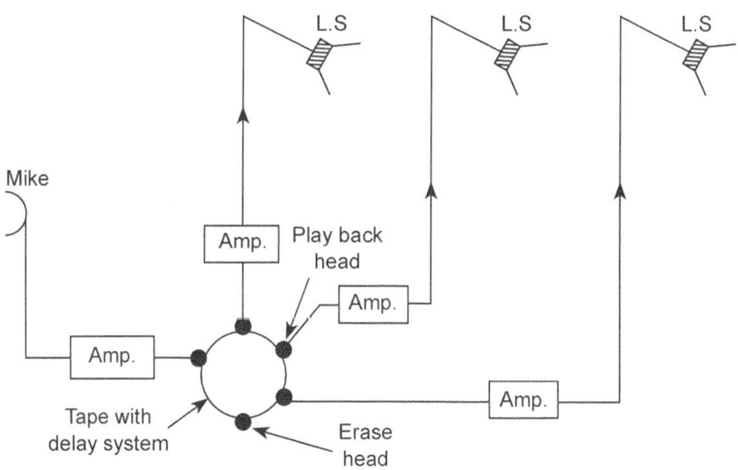

Fig. 4.49

In this arrangement, the sound obtained from the microphone is amplified and recorded on a tape. The sounds from different loudspeakers are picked up by corresponding heads and passed through the respective amplifiers. Now the recorded sound is picked by playback head with a delay, amplified and fed to the corresponding loud speakers. Before fresh recording, the previous recording is erased.

4.4.5 Frequency Response of P.A, System

Following are the factors determining frequency response of a P.A. system:

(*i*) Absorption of sound depends upon frequency. As we know the *higher sound notes* of sound suffer more attenuation as a sound travels in an auditorium, therefore, the frequency response of P.A. system should he such that, it amplifies the higher notes more before sending them towards the audience. Moreover a P.A. system used in a reverberating auditorium should attenuate the lower notes.

(*ii*) The *beaming of sound* on the beam angle of loud speaker is also an important factor. The loud speaker of a P.A. system should be so positioned that it covers the audience with a vertical angle between 20—30°.

(*iii*) Our physiology of hearing is such that at higher sound levels the lower notes are heard stronger. In an auditorium, considering the noise, the sound level of the P.A. system should be theoretically more than 64 db. Moreover the P.A. system should also attenuate the lower notes to maintain the naturality of the sound. When the noise level is very high, in that case only higher notes should be amplified in addition to attenuate the lower notes.

(*iv*) The loudness and intelligibility are also dependent on frequency. As we know that loudness of speech is determined by lower notes and intelligibility of speech is determined by its higher notes.

It has been seen that sound notes below 1 kHz, supply 95% loudness and 25% intelligibility, while the notes above 1 kHz supply only 5% loudness and 75% intelligibility. It is therefore clear that a P.A. system should attenuate lower notes.

4.4.6 Role of Mixer in P.A. System

A mixer is a very important stage of a P.A. system. As already explained, the output of different microphones is fed to the mixer, which isolates different channels from each other before feeding to the amplifier.

A mixer can be of three types:

(*a*) One, which does not use amplifiers, only uses gain control.

(*b*) Second, which has separate pre amplifier for separate channels and a common main amplifier.

(*c*) The third, which is used in practice has a common pre amplifier as shown in Fig. 4.50.

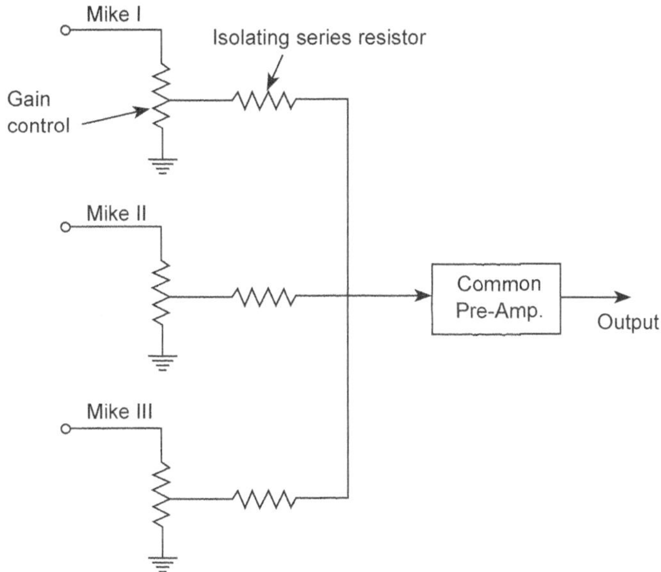

Fig. 4.50

Controls Available on a Mixer

The various controls available on a mixer of a P.A. system are;

(*i*) *Microphone Inputs.* The input of microphone into mixer should be of atleast 0.3 mV sensitivity. This can be controlled from the mixer.

(*ii*) *Tone controls.* Different tone controls are also available. We may have in a mixer, separate tone control for each microphone input.

(*iii*) *Reverberation Control.* A control is provided on the mixer to vary the amount of reverberation.

(*iv*) *Output Control.* We can also control output of a mixer. We can obtain a stereophonic as well as monophonic sound.

4.4.7 Problems in P.A. Systems

Some common problems of P.A. systems are:

(*i*) The unwanted couplings between the different stages of amplification cause a problem of instability. At audio frequencies, the magnetic coupling between loud speaker feeder and stereophonic cable results in a whisling. This overloads the amplifier, This problem can be reduced by keeping *loud speaker* and *microphone cables* apart. The problem is produced also when loud speaker feeder develops an earth leakage.

(*ii*) Radio interference is another possible problem for P.A. systems. The radio signal is picked up by input leads which is got amplified along with the original sound signal. For this, Field Effect Transistors (FET) are used at input circuits. The microphone sockets also should be kept clean so that they do not pick up radio frequencies.

(*iii*) Similarly electrical interference is also a problem with P.A. systems. The input leads may pick up an electrical signal on the switching operation of a heavy electrical load, if any being operated near the P.A. system. The same methods may be used for reducing this problem as discussed for radio interference.

(*iv*) The P.A. system should be earthed only at one point. If it is earthed at more than one point, It leads to a "hum" which is not desirable in the P.A. system.

SUMMARY

1. The device which converts sound into electrical energy and vice versa is called electro-acoustic transducer e.g., microphone and loud¬speaker.
2. The microphones convert sound into electrical energy and may be pressure type or pressure gradient type. Carbon microphones are most popularly used.
3. A loudspeaker converts the electrical energy into sound. They may be cone type or horn type. Cone types are used in radio, TV etc. and horn types in Public Adress System.
4. A Baffle is used to increase the length of sound path in cone loudspeakers.
5. The Public Adress (PA) system is an electro acoustic system to send an audio signal to a distance.
6. The components of a PA system are : microphone, amplifier and loudspeaker.

SELF TEST QUESTIONS

1. What are transducers. Give examples.
2. What do you know about sensitivity and directivity of microphones?
3. Describe

   (*i*) Carbon microphone

   (*ii*) Capacitor microphone.

4. Describe
   (*i*) Crystal microphone
   (*ii*) Moving velocity microphone.
5. Describe various characteristics of microphones.
6. What is the useful frequency range for microphones?
7. Explain the working principle and frequency response of moving coil microphones?
8. Explain sensitivity, directivity and polar diagrams in reference to moving coil mikes.
9. What are the various factors for selecting a microphone for a particular purpose?
10. What do you understand by power handling capacity of mikes?
11. Describe construction and working of a velocity ribbon microphone. Also draw its frequency response and polar response curves.
12. Explain parameters, on which quality of microphone depends. Give their values for moving coil and crystal mikes.
13. Why a ribbon microphone is called a velocity microphone?
14. Why ceramic gives good results than a rochell salt in crystal mikes.
15. Explain an electret microphone. Draw its frequency response curve and mention its applications.
16. Write short note on the following:
    (*a*) Tie clip microphone
    (*b*) Wireless microphones
    (*c*) Noise cancelling microphone.
17. Compare moving coil, crystal, ribbon and carbon microphones for their important parameters.
18. What are the characteristics of loud speakers? Why efficiency of indirect radiating speakers is higher than direct radiating speakers ?
19. Why a moving coil speaker is called a direct radiating speaker?
20. What is difference between a cone and a horn speaker?
21. With the help of a neat sketch, explain an electrodynamic horn speaker.
22. Give merits and demerits of permanent magnet and electromagnet cone speakers.

23. Mention the field of applications of cone and horn speakers.
24. What do you mean by baffle? What is their need?
25. Explain in brief important types of baffles.
26. Explain how a reflex baffle gives better intensity of sound than an infinite baffle.
27. Give practical examples of finite and infinite baffles.
28. Why inner surfaces of a baffle are lined with sound absorbers?
29. What is the frequency range of squawker speakers?
30. Why electrolytic capacitors can not be used in cross over networks ?
31. How impedance matching is achieved between amplifier and loud speaker?
32. With the help of a block diagram, explain a P.A. system.
33. Explain the various requirements of a P.A. system.
34. Discuss the positioning of microphones and loud speakers for effective functioning of a P.A. system.
35. Explain various factors which determine frequency response of a P.A. system.
36. Discuss P.A. systems problems.
37. Explain the following terms related to P.A. systems:

    Reverberation, ambient noise, noise cancelling microphones, hum, Acoustic feed back.
38. What is the function of mixer in a P.A. system. What are the various controls available over it.

OBJECTIVE TYPE QUESTIONS

Fill in the Blanks:

1. A.F. range is ________.
2. The M.C. mike works on the principle of ________.
3. The most widely used mike is ________.
4. A mike is like a ________.
5. The ________ mike is not a pressure microphone.
6. The other name of ribbon mike is ________.
7. ________ is a pressure microphone.

8. ________ is a pressure gradient microphone.
9. ________ microphone has a bi-directional directivity.
10. ________ microphone works on the principle of Piezo electric effect.
11. Roshell salt is not used in crystal microphones, because ________.
12. ________ mike has best frequency response.
13. ________ mike has an inbuilt amplifier.
14. ________ mike needs an external d.c. simply.
15. The ________ mike does not need an external d.c. supply.
16. In telephone, a ________ microphone is used.
17. A ________ mike has highest sensitivity but poorest frequency response.
18. ________ microphone has highest impedence.
19. The M.C. mike is also called ________.
20. M.C. mike is an example of ________ transducer.
21. The M.C. microphone is ________.
22. Microphones have inbuilt transformer to ________.
23. The impedance of carbon microphone is ________.
24. Frequency response of carbon mike is ________.
25. Change of sound pressure on diaphragm of carbon mike ________.
26. Frequency response of capacitor mikes is ________.
27. Sensitivity of microphones is determined ________.
28. The frequency response of microphones is ________.
29. P.A. system is used for________.
30. The most popular microphone used in P.A. system is ________.
31. The matching used in between amplifier and loud speaker is ________.
32. The intensity of sound for comfortable hearing is ________.
33. In P.A. system, we prefer ________ directional microphone.
34. In a noisy atmosphere ________ microphones should be used.
35. Frequency above 1 kHz gives ________ percent power.
36. A loud speaker works on the ________ principle.

37. The efficiency of direct radiating speakers is ________ than indirect radiating speakers.
38. The power handling capacity of an electro dynamic speaker is ________ than the dynamic speaker.
39. A cone type speaker is ________ radiating type speaker.
40. Baffles are used with ________.
41. Enclosure acts as an ________ baffle.
42. Woofer handles ________ frequencies.
43. Reactance of a capacitor decreases with ________ of frequency.
44. When the load resistance is less than the source resistance, a ________ transformer is used.
45. A horn can be called as a ________ transformer.

ANSWERS

1. 20 Hz—20 KHz
2. Electro magnetic Induction
3. moving coil mike
4. dynamo
5. ribbon
6. velocity microphone
7. moving coil mike
8. Ribbon velocity mike
9. Ribbon velocity
10. Crystal
11. It cannot bear high temperature
12. Carbon
13. Capacitive
14. Carbon
15. Electret
16. Carbon
17. Carbon
18. Capacitive
19. Dynamic mike
20. Active
21. Omni directional
22. Boost the signal voltage
23. 50—400 n
24. 400 Hz to 2500 Hz
25. Changes pressure on granules
26. 50 Hz—10 KHz
27. by measuring output voltageat a given sound pressure
28. 40 Hz to 16 KHz
29. Mike as well as speaker
30. Moving coil microphone
31. Transformer matching

32. 0.1 mW
33. Uni
34. Noise cancelling
35. 5
36. motor
37. less
38. more
39. direct
40. speakers
41. infinite
42. low
43. increase
44. step down
45. sound.

❑❑❑

5

Audio Recording, Hi-Fi and Stereo

In this chapter, we will discuss about Audio Recording, Hi-Fi and Stereo.

Audio recorders perform the following three basic functions:

(*i*) *Recording.* They record a particular information and can store the same for longer period.

(*ii*) *Playing.* They can play or reproduce the recorded information whenever they are required to do so.

(*iii*) *Erasing.* Before recording a new signal or information, they can erase out any previous recording, if any on the tape.

5.1 AUDIO (SOUND) RECORDING

The various recording methods are following:

1. Disc Recording
2. Magnetic Recording
3. Optical Recording

5.1.1 Disc Recording

In disc recording, a needle known as "Stylus" moves on a disc and makes a "track" on the disc. The stylus is connected with a small motor.

The arrangement has been shown in the Fig. 5.1. Basically an armature is kept in between two poles of a permanent magnet. The stylus is attached to the armature. The current of the signal to be recorded is made to flow through the armature. According to the motor rule, armature experiences a torque in accordance to the strength of the signal- The stylus moves in accordance to the torque. The disc in turn is rotated by a "Turn table" at a designed speed. As a result "Signal track" is formed on the disc.

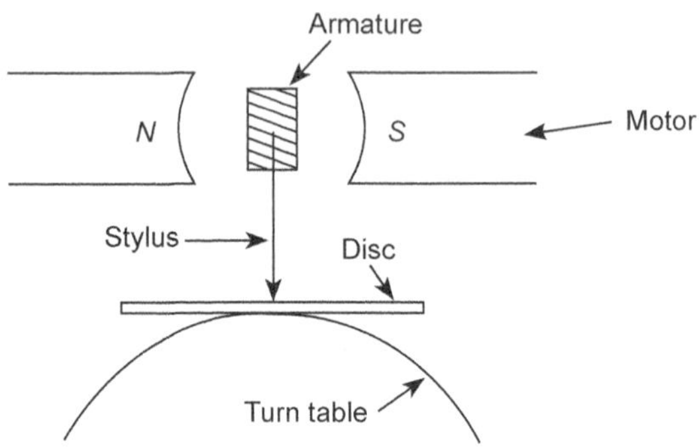

Fig. 5.1

(i) The Disc

The disc is made of wax or lacquer. It has grooves, which are uniformly spaced in its normal condition. During recording, the stylus moves from edge towards centre of the disc (which rotates at a designed speed). The grooves on the signal take a shape as shown in Fig. 5.2 which is according to the audio signal to be recorded.

Fig. 5.2

Modern discs are made of vinyle resins (and not of wax) and are more flexible. Stereo recording is also possible on these discs i.e. two audio channels can be recorded in a single groove. The Fig. 5.3 (*a*) shows stylus motion for stereo recording and Fig. (*b*) shows dimensions of a L.P. disc in millimeters.

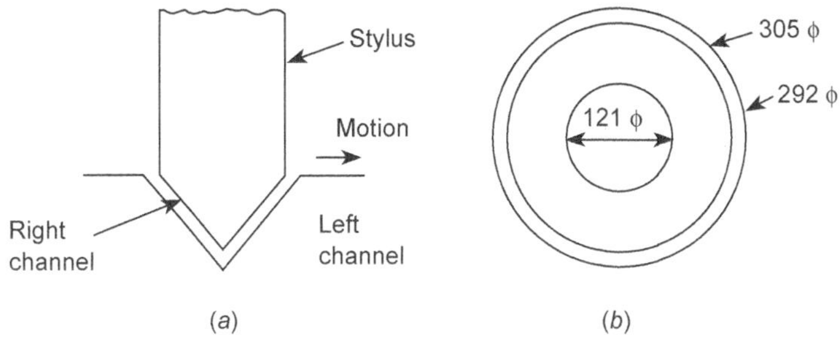

Fig. 5.3

Now a days, the discs are made by coating a layer of cellulose, nitrate lacquer on a base of Aluminium (Fig. 5.4 shows two views of the disc). The coating is made on both sides of the base. The lacquer (cellulose nitrate with resins and oils) is soft to be cut by diamond stylus and hard enough to give sufficient number of play backs.

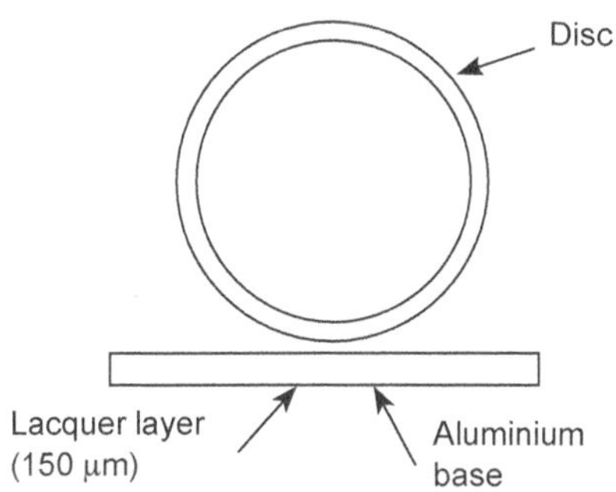

Fig. 5.4

(ii) Principle of Disc Recording & Reproduction

(a) The important elements of a disc recording are (See Fig. 5.5):

1. *Microphone.* The microphone converts sound waves into electrical waves. A microphone with cardiode (heart) pattern should be used for recording.

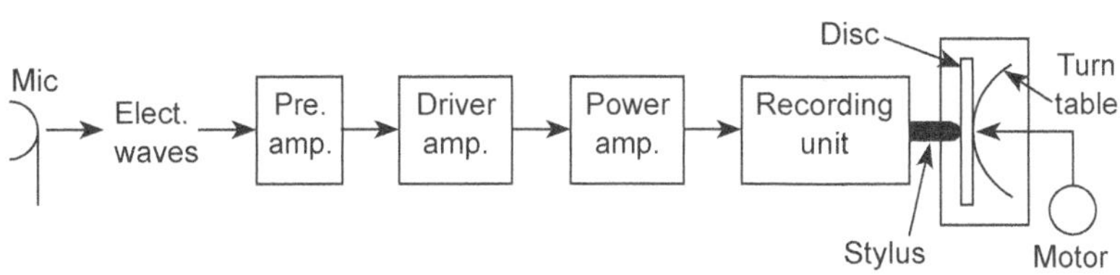

Fig. 5.5

2. *Preamplifier.* The weak output of mike is given to preamplifier to raise its level to the required value.

3. *Driver amplifier.* The signal is further amplified here for sufficient voltage.

4. *Power amplifier.* Here, we get power amplification of the signal.
5. *Recording unit.* The power output from power amplifier is used to drive the recording unit. It converts the audio signal into mechanical vibrations.

   (*i*) *Disc.* It is of wax or lacquer with grooves.

   (*ii*) *Turn table.* The turn table carries the disc over it and rotates the disc in a particular direction during recording.

   (*iii*) *Motor.* A motor is used to drive the turn table and disc. Generally, a synchronous motor is used, which gives a constant speed.

(b) Disc Reproduction

The principle of disc reproduction is just reverse of the recording process.

For reproduction, a "*cartridge*" is used which changes the vibrations of playback stylus into electrical signals. These electrical signals are amplified and given to a loudspeaker, which converts it into original sound.

Block diagram of reproduction system can be drawn as shown in Fig. 5.6. The important elements are:

1. *Motor (M).* Usually, a fractional horse power synchronous motor rotates the turn table at a calculated speed.

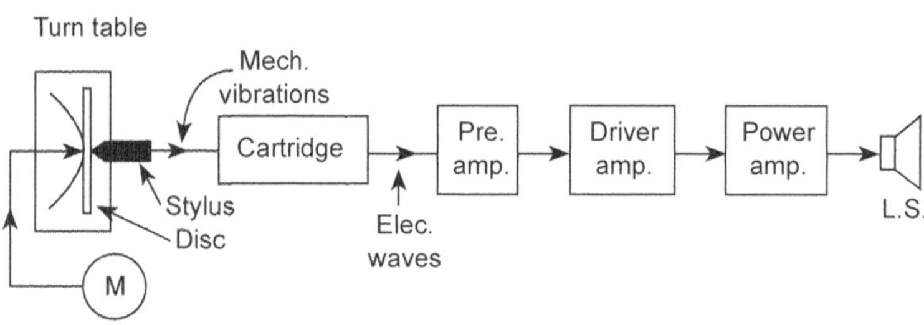

Fig. 5.6

- *Turn table.* The turn table carries the disc.
- *Disc.* It is the disc on which the signal to be reproduced has been recorded.
- *Stylus.* It is the reproducing needle, which moves in the grooves of the disc and its vibrations are fed to the cartridge.

2. *Cartridge or Pickup mike.* The stylus moves in the recorded grooves. These mechanical movements (vibrations) of the stylus are fed to a cartridge. The cartridge converts these mechanical vibrations into electrical waves.

3. *Pre amplifier.* It pre amplifies the weak signal output of the cartridge.
4. *Driver amplifier.* It further amplifies voltage of the signal to sufficient value for getting maximum power output from the next stage.
5. *Power amplifier (P.A.).* This is the final stage of amplification. It gives maximum output of power.
6. *Loudspeaker (L.S.).* This converts the electrical signal into original sound. There is transformer impedence matching between power amplifier and the loudspeaker.

(iii) Disc Grooves

The disc grooves may be:

1. Coarse grooves
2. Micro grooves

1. *Coarse grooves.* (See Fig. 5.7) Approximately 50 grooves are cut per cm of the disc. The speed of disc is 80 RPM. These discs are *monophonic* type. The details of these grooves and disc are given below:
   (*i*) groove angle = 87°
   (*ii*) groove depth = 75 um (3 mil)

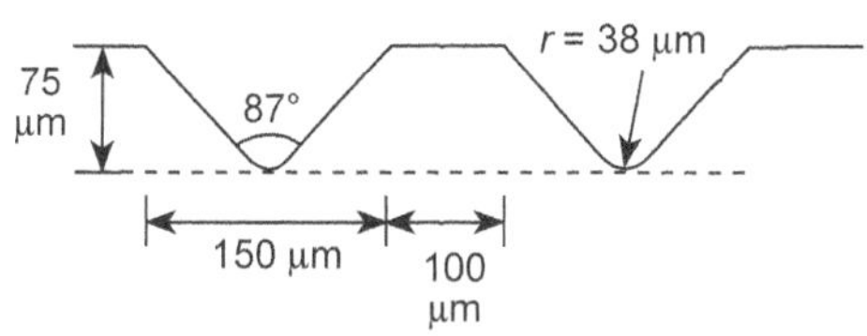

Fig. 5.7

   (*iii*) groove width = 150 µm (6 mil)
   (*iv*) Wall depth = 100 µm (4 mil)
   (*v*) Groove bottom radius = 38 µm (1.8 mil)
   (*vi*) Dia of disc = 30 cm
   (*vii*) Speed of disc = 78 RPM
   (*viii*) Radius of stylus tip = 63 µm.

2. *Micro grooves.* (Fig. 5.8) As the name indicates, micro grooves are much smaller then the coarse grooves. Approximately 100—120 grooves per cm can be cut. They are different for monophonic and stereophonic recordings. The speed of the disc is 35 RPM to 45 RPM. The details of these grooves and disc are:

(*i*) groove angle = 85°

(*ii*) groove depth = 25 μm (1 mil)

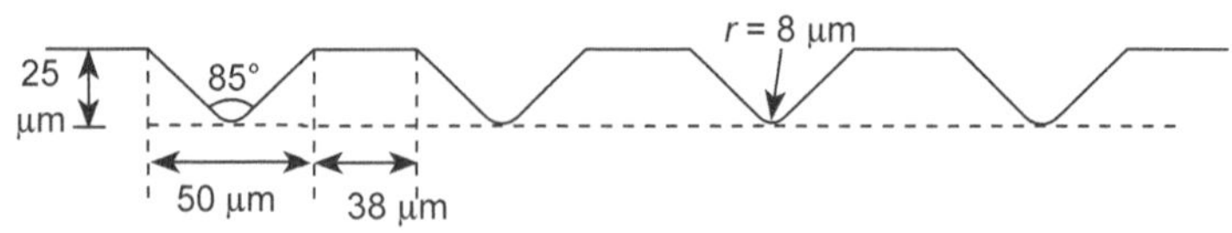

Fig. 5.8

(*iii*) groove width = 50 μm (2 mil)

(*iv*) Wall depth = 38 μm (1.8 mil)

(*v*) Groove bottom radius = 0.8 μm (0.3 mil)

(*vi*) Diameter of disc = 20 cm to 30 cm

(*vii*) Speed of disc = 35 to 45 RPM

(*viii*) Radius of stylus tip = 23 μm.

Micro grooves are further classified as:

(*a*) *Hill and Dale grooves.* (Fig. 5.9) These grooves were used in earlier days. These grooves were moving up and down, only depth of the groove was changing whereas, the gap between grooves remained constant. The width of all the grooves was about 150 μm (constant).

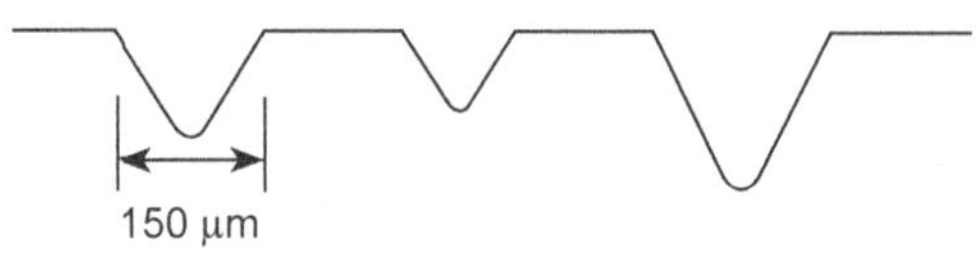

Fig. 5.9

(*b*) *Lateral grooves.* (Fig. 5.10) Later on with development of technology, lateral grooves replaced the *hill and dale* grooves. In these, the stylus is moving laterally on the disc. The depth remained constant but width between grooves changes. These grooves give better results than the earlier.

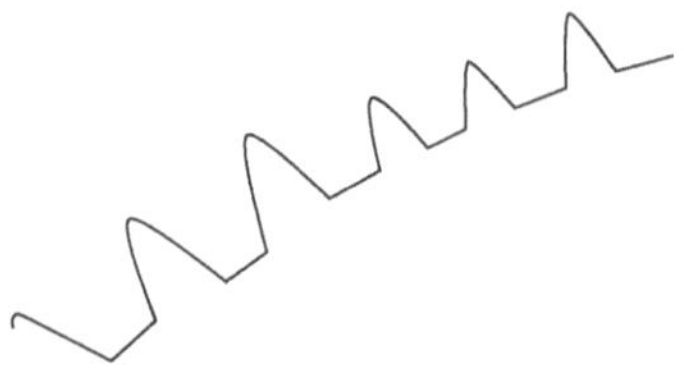

Fig. 5.10

(*c*) *Sterophonic grooves.* These grooves are cut while recording two channels simultaneously. In this system, two channels are recorded on two walls of the groove inclined at 45° to the horizontal. When currents in the two channels are equal, the stylus moves in the groove {Fig. 5.11 (*a*)], when the currents are unequal, the stylus will move on one of walls of the groove [Fig. (*b*) and (*c*)] depending in which channel, the current is more.

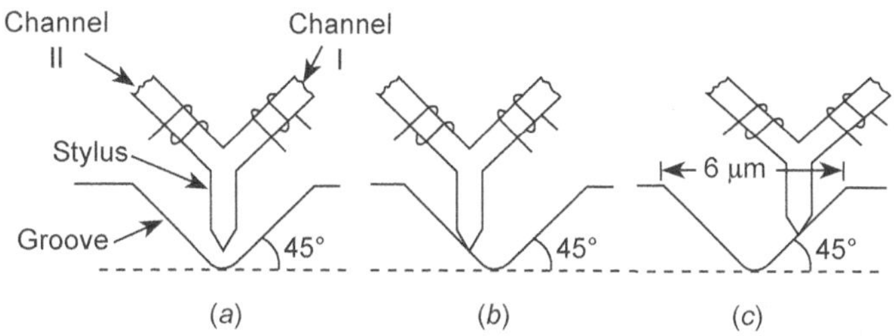

Fig. 5.11

(iv) Stylus Cutter

It is a chisel shaped tool which cuts grooves in the disc. The materials used for its construction are diamond, saphire and fused aluminium oxide but mostly diamond is preferred which gives best results. The stylus is mounted on a long screw and moves from edge of the disc towards its centre. The cutter moves radially while the disc has rotatary motion.

When the cutter moves and cuts the disc, it produces "swarf" which should be removed immediately, for this purpose, a suction device is also provided in the cutter.

The Fig. 5.12 (*a*), (*b*) and (*c*) show construction of stylus.

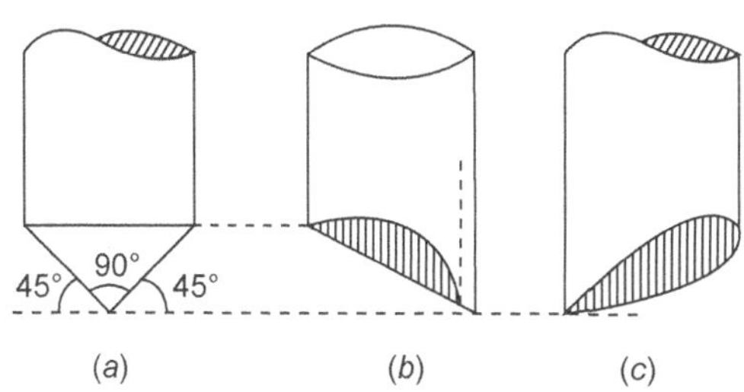

Fig. 5.12

For smooth cutting, the stylus is heated by current. If grooves cut are not smooth, when the playback needle tracks these grooves, a hissing noise is produced.

For *monophonic* recording the tip radius is 25 μm (1 mil) and for stereophonic recording it is 12.5 μm (0.5 mil). The sides of the cutter are at an angle of 90° to each other and at 45° to the vertical. For smooth cutting the stylus is moved on the disc at an angle of 20° with the vertical. A stylus has a life of about 25 cutting hours which further depends on the working conditions.

1. Stereophonic Disc Recording Head. (See Fig. 5.13)

The stereophonic disc recording head contains two moving coils one for each channel. Each unit contains a feedback coil also. It makes use of negative feedback. The system provides a motor action, the stylus is attached to its armature.

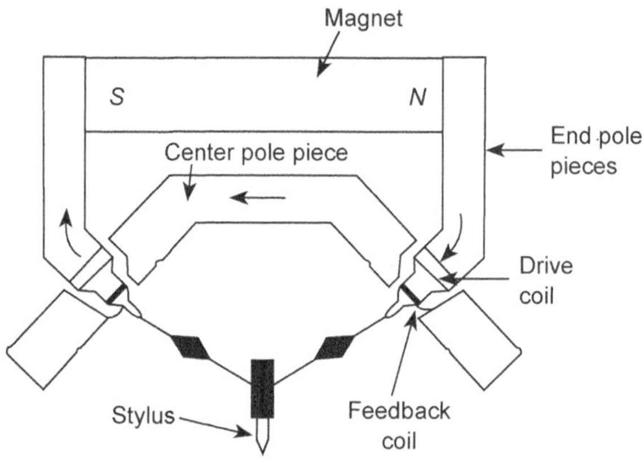

Fig. 5.13

2. Characteristics of Stylus

The desired characteristics of cutter stylus are:

(*i*) It should be light in weight, so that it may move at a higher speed.

(*ii*) The speed of turntable which rotates the disc should be constant. This is the reason that a constant speed motor (like synchronous motor) is used to rotate the turn table.

(*iii*) The stylus should follow the path of sound signal, to be recorded.

(*iv*) It should perform faithful recording.

(v) Types of playback stylus

The playback stylus tips are made of osmium, sapphire or diamond. The osmium is the cheapest material and is used in low quality gramophones. The saphire is a better material than osmium having moderate cost and is commonly used. The diamond is the hardest material and having longest life. This is used for hi-fi work. The other materials used for making stylus tip are *steel* and *tungsten carbide*.

According to the construction, playback stylus may be:

(*i*) *Conical or Chisel stylus.* [Fig. (5.14 (*a*)] Which are cheapest, sharp and cause much wear and tear, therefore not used for playback. However, as mentioned earlier, these are more suitable for *recording*.

(*ii*) *Round or spherical stylus.* [Fig. (*b*)] This also causes much wear and tear of the grooves and hence not used for replay.

(*iii*) *The elliptical stylus.* [Fig. (*c*)] This is the most widely used stylus. This causes minimum wear and tear and also minimum distortion.

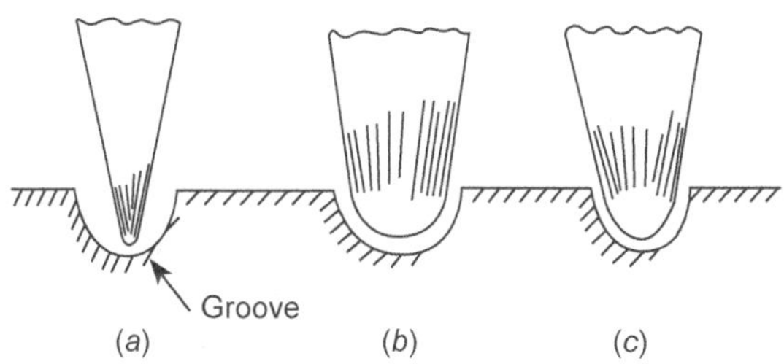

Fig. 5.14

Note: (*i*) The useful life of a diamond stylus is 200 hours and of saphire stylus is 2000 hours.

(*ii*) The radius of its tip is 65 |am for coarse grooves and 14 μm for micro grooves.

(vi) Cartridges (or Pickup Mikes) for reproduction

As mentioned earlier cartridges work as a microphone in reproduction system. They pickup mechanical vibrations from playback stylus and convert them into original audio signal.

These are of the following types:

1. *Moving coil type or dynamic cartridges.* [See Fig. 5.15 (*a*)] Like microphones, they may be also permanent magnet type or electromagnet type.

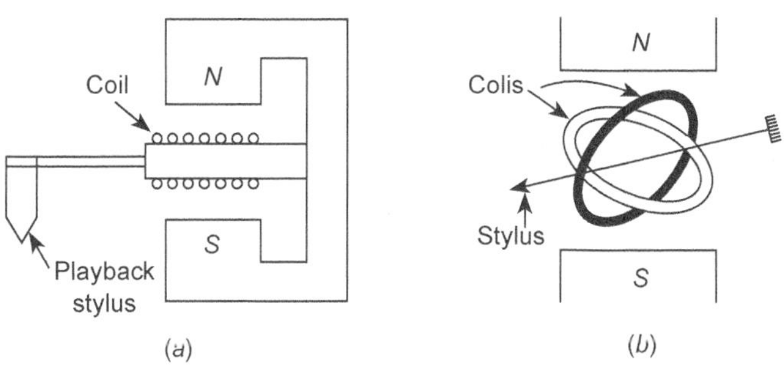

Fig. 5.15

They consist of a coil which can move in a magnetic field. The replay stylus or needle is attached to the magnet. When the stylus moves on the grooves, it generates mechanical vibrations. This moves the coil in the magnetic field and an emf generated in the coil due to generator action. The amount of emf induced will depend upon the amount of flux linkage with the moving coil (Faraday laws). The Fig. (*b*) shows another type of the moving coil pick up. It contains two coils coupled to the stylus.

Their output voltage is very small (in micro volts) and hence they need a step up transformer to boost the voltage. However, they have a very good frequency responses,

2. *Magnetic cartridge.* (Fig. 5.16) This is reverse of the moving coil type cartridge. It has a fixed coil and moving magnet. The stylus is attached with the moving magnet, which moves with movement of the stylus and an emf is induced in the coil. It has a better voltage output of about 8 mV. This is a widely used cartridge in record players.

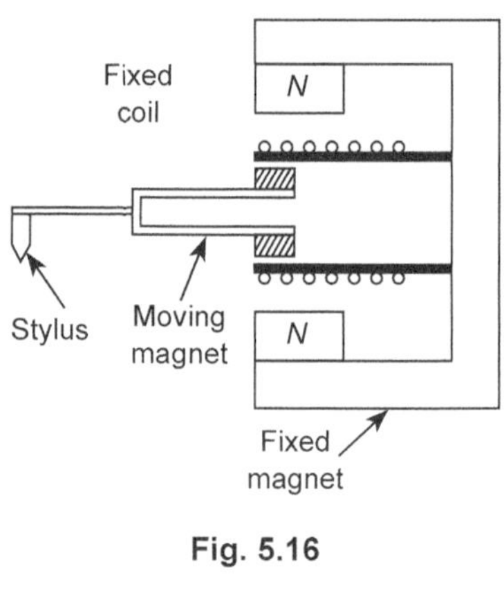

Fig. 5.16

3. *Crystal cartridges.* (Fig. 5.17) As we already know that when a crystal is subjected to a pressure, it exhibits *piezo electric effect i.e.,* an emf is generated in the crystal. Generally a rochelle salt crystal is used. The arrangement consists of two wafers of crystals cemented together with a metal foil. One end of the crystal is clamped and other end is connected to the stylus.

   They have a satisfactory frequency response. But their performance depends upon many factors like temperature and methods of their formation, cutting and bending, therefore sometimes they do not give good results.

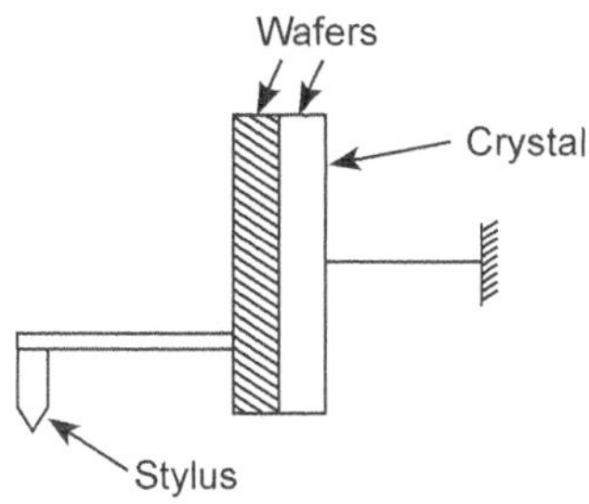

Fig. 5.17

4. *Capacitor cartridge.* (Fig. 5.18) It consists of a parallel plate capacitor (*C*) with one plate fixed and other movable. The stylus is connected with the movable plate. When the stylus moves in the grooves, the moving plate of the capacitor also moves, this changes the distance (*d*) between the plates and hence the value of the capacitance. But the charge on the capacitor remains constant, the potential across the capacitor goes on changing with the movement of the stylus. This output voltage can be taken across a resistor (*R*) as shown.

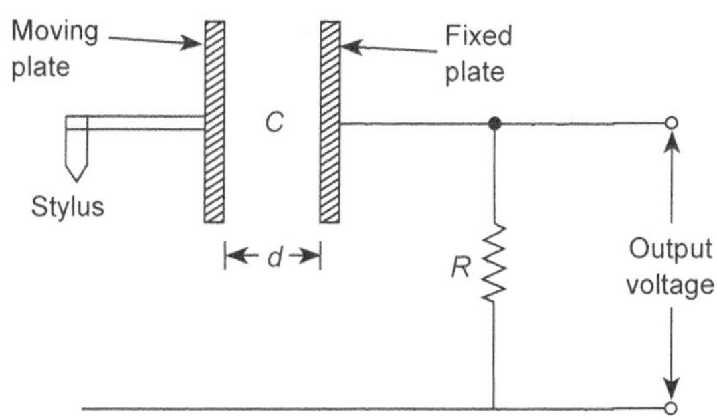

Fig. 5.18

It has a good frequency response but output voltage is low which needs pre amplification.

(vii) Equalisation

The technique of improving signal noise (S/N) ratio and accommodating signals in the limited space (of the disc and the tape) is known as Equalisation. For this during recording high frequency signals are boosted (raised) and the low frequency signals are attenuated (lowered). As the amplitude of sound at low frequencies is more, which will occupy too much space on the disc therefore their attenuation is necessary.

Further it is also desired that recording between 1 kHz—2 kHz should be done at constant velocity. We know that amplitude of sound increases with decrease of frequency and vice-versa, hence the amplitude of sound has to be reduced proportionately as the frequency decreases to keep the recording constant below frequencies of 1 kHz. Conversely above 2 kHz, the amplitude should be increased accordingly. The arrangement will give a constant recording between 1kHz and 2 kHz as desired.

During replay, the process is reversed. That is the frequencies below 1 kHz are boosted and above 2 kHz are attenuated.

The Fig. 5.19 shows curves for recording and replay equalisation process.

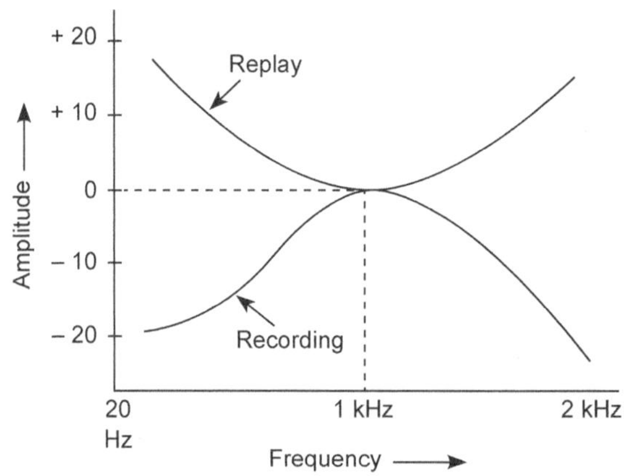

Fig. 5.19

It is important here to note that magnetic pick up cartridges 4 need equalisation circuit but the crystal pick up cartridges do not need this circuitary as they have a constant amplitude response.

(viii) Important Terms

1. *Signal/Noise ratio.* This is the ratio of power carried by signal to that of noise in a replay system. It is measured in decibels.
2. *Cross talk.* It is unwanted "cross" of audio energy from one channel to the other. It is also expressed in dB.
3. *Vow and Flutter.* When the speed of turn table suffers variations in the nominal speed, this is known as Vow and Flutter. Vow is caused by slipping of the disc on turn table and irregular rotation of motor causes a Flutter. This changes the pitch of the replayed signal.

   The phenomenon is a form of *frequency modulation* and may also be caused, if the record is eccentric.

Note: The disc recording is now almost absolete.

5.1.2 Magnetic Recording

This is the widely used system for recording audio signals.

The basic principle involved in magnetic recording is the phenomenon of electro-magnetic induction. The audio and video tapes are coated with a magnetic material. The signal (audio or video) currents induce magnetism in the tape. The track is in the form of magnetic path on the tape.

(a) Basic Knowledge of Magnetism

Before understanding magnetic recording, we should revive our basic knowledge about magnets and magnetism.

(*i*) **BH Curve:** The curve between magnetic intensity or magneto motive force H and flux density (B) is known as magnetic or B-H curve.

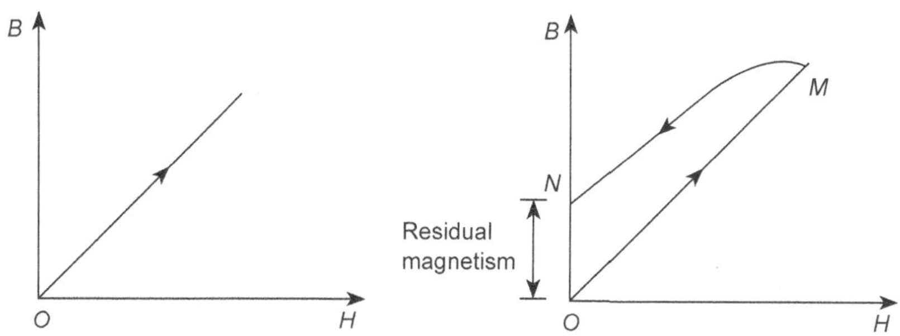

Fig. 5.20

When H is applied on a magnetic material, B is produced which is proportional to H. But when H is removed, B is not zero, but still persists in the material, which is known as "Residual Magnetism". (See Fig. 5.20). For one complete cycle of AC supply connected to the material [Fig. 5.21 (*a*)], we can get a complete "Hysterisis curve". [See Fig. 5.21 (*b*)]

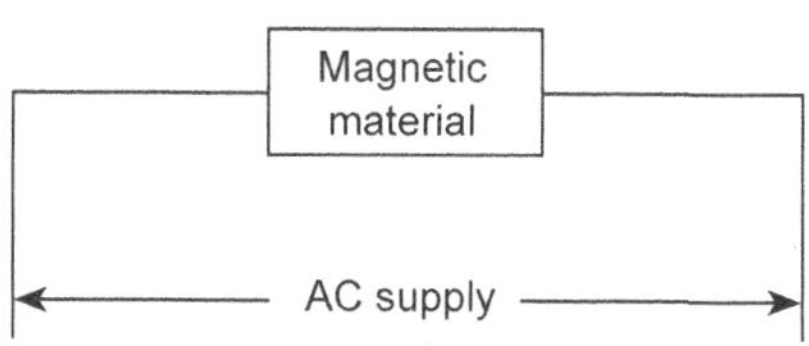

Fig. 5.21 (a)

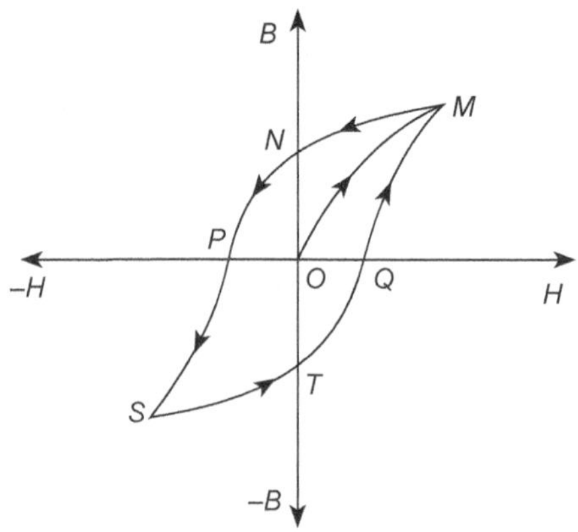

Fig. 5.21 (b)

This principle is used in signal "recording" as well as "Erasing". During recording, the tape passes through "Hysterisis" curve and during erasing the hysterisis is first "saturated" and then is made to "vanish" by reversing the applied magnetic field at high frequency.

(*ii*) **Retentivity:** By this term we mean the duration or time for which a magnetic material can "retain" residual magnetism in it. For some materials, period of retentivity may be few "seconds" and for others it may be "years".

This property is being used to prepare small loudspeakers/ microphones and audio video equipment. These devices are equipped with permanent magnets and do not need external supply. The iron has low and alloys like "Tungsten steel" and "Cobalt steel" have higher retentivity.

(iii) Electromagnets: The electromagnets are temporary magnets made of soft magnetic material (iron, cobalt etc.) wound with a coil. When they are connected with supply, they become magnet and when supply is cut off, they cease to be a magnet.

(b) Magnetic Heads

The magnetic heads play an important role in magnetic systems.

Magnetic heads are electro-magnets with an air gap. In shape they may be round or rectangular having one or more air gaps.

Accordingly, they possess one or two sets of north-south (N-S) poles. Through the air gap, useful flux is made to fall on the tape. [See 5.22 (*a*) to (*e*)]

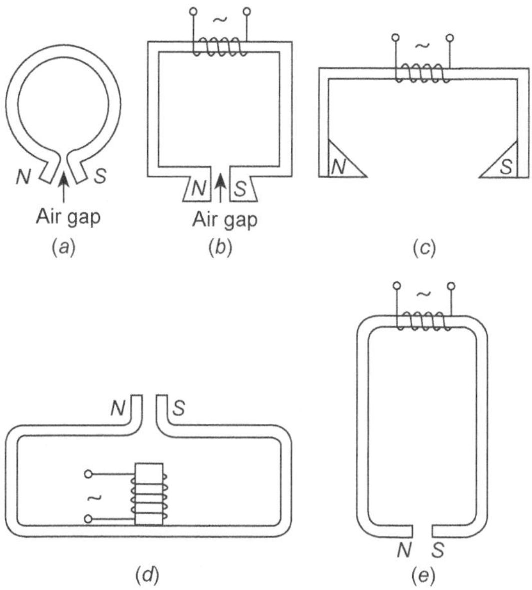

Fig. 5.22

Three types of heads are used:

(*i*) *Record head.* This is used for recording audio/video signals.

(*ii*) Play head. This is used for replay or "reproduce" an already recorded audio/video signal.

(*iii*) *Erase head.* These heads are used to erase out a recorded signal before recording a new signal on the same tape.

Sometime a single head may be used for all the three purposes.

The core of the heads is made of some soft magnetic alloys like ferrites, permalloy, mumetal etc.

When current is passed through the coil of the head tips, it forms N and S poles on its tips, according to the direction of the current. On switching off the supply, the head does not remain as magnet, and the poles also vanish from its tips.

(c) Magnetic Tape Recording of Audio Signals: Principle

For magnetic recording, the audio signal is given to the microphone which converts it into electrical pulses. This electrical signal is amplified and the amplified current is sent through the audio head coil, which magnetises the head. The magnetic strength of the audio head goes on changing in accordance with intensity of the audio signal being produced at the microphone.

Now, the magnetised audio head is moved on the audio tape. The audio tape is covered with a layer of ferromagnetic material like ferric oxide. As a result tape is also magnetised, the magnetic strength goes on varying according to the signal. By magnetic induction "magnetic track" of sound is formed on the tape which contains audio signal. (See Fig. 5.23)

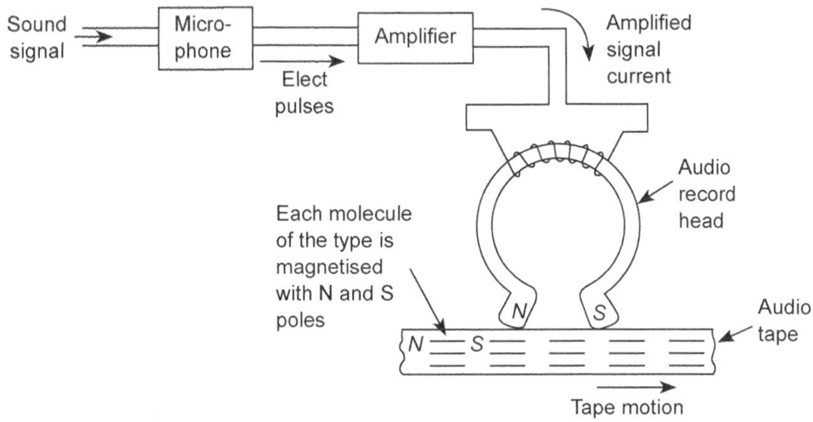

Fig. 5.23

The voltage obtained at the output of amplifier should not have any distortion otherwise during replay, the same will come into play. The core of the head is made of soft alloy, hence it does not have any residual magnetism, therefore it will not give any distortion. However, "demagnetisers" may be used to demagnetise the head immediately after it has done its job.

Almost, a reverse method is used for replaying a recorded signal. (See Fig. 5.24)

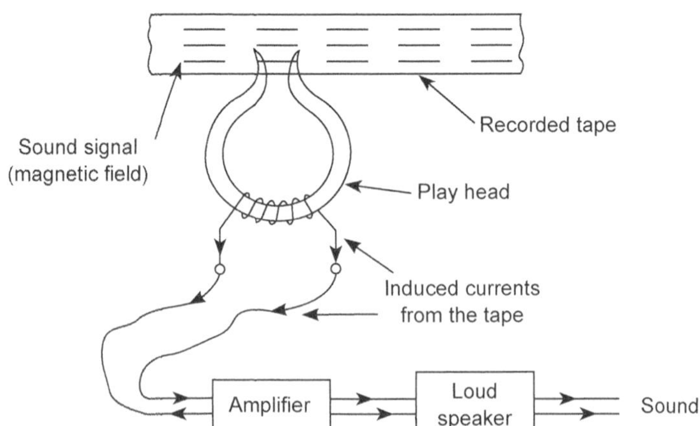

Fig. 5.24

For magnetic recording, generally the following methods are used:

(*i*) Longitudinal Magnetic Recording

In this method, the tape is moved longitudinally to the recording head. The north and south, both poles are kept on the same side of the tape.

The method is simple and gives satisfactory results. This is commercial and a very popular method of recording. (See Fig. 5.25)

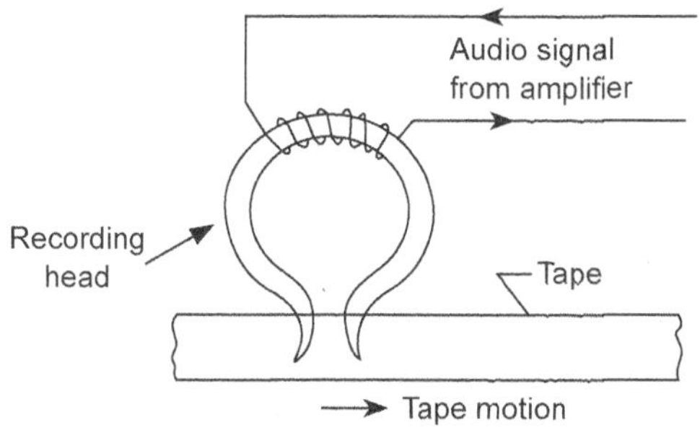

Fig. 5.25

(ii) Transversal Magnetic Recording

In this recording, the poles of the record head are kept transversal to the tape. In this case the record head needs special construction and also it is a complicated method. Results are also not so satisfactory as in the first case, hence it is not used at commercial level. (See Fig. 5.26)

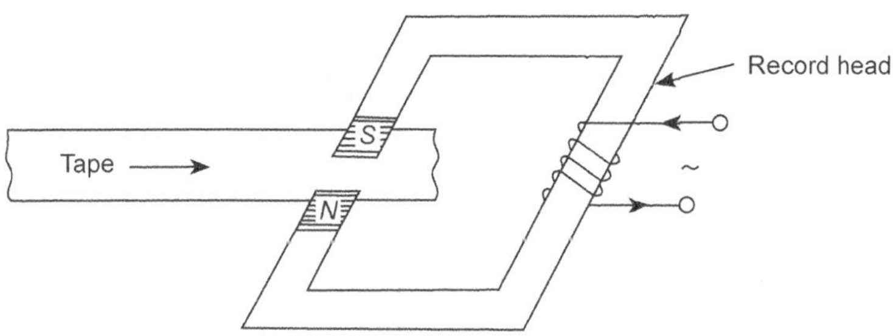

Fig. 5.26

(iii) Perpendicular Magnetic Recording

In this method, the tape moves normal (at perpendicular) to the poles of the record head— one pole on each side.

This is also a complicated method as movement of the tape cannot be controlled effectively. This method is also not adopted commercially. (See Fig. 5.27)

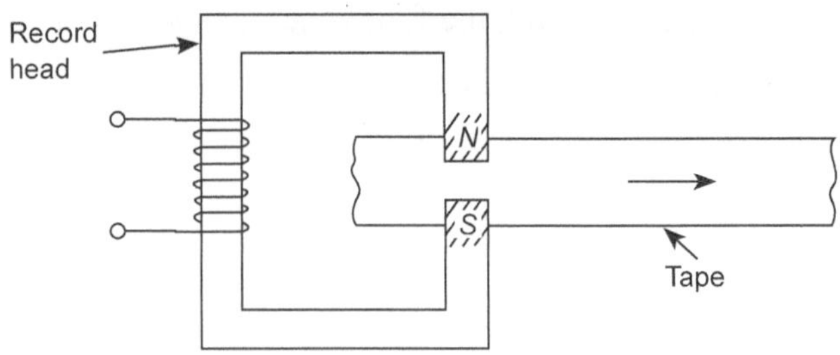

Fig. 5.27

(d) Reproduction of Magnetic Recording

For reproduction of the audio signal "playhead" is moved on the tape. This activates the playhead. The recorded magnetic path (sound track) induces currents in the playhead coil (by electromagnetic induction). These induced currents are amplified and given to speaker which converts the currents into the original sound.

(e) Erasing

The following methods are used to erase/wipe off the previous recording from the tape.

(i) Permanent Magnet Erasing: In this method, the tape is passed first through a strong "forward" and then through "reverse" magnetic fields. This pulls the previous recorded magnetic field first to "saturation" and then to "neutral" state. This method is not good and does not give satisfactory results. A small previous magnetic field is still left which creates noise and distortion during reproduction.

(ii) HF Erasing: In this method, the eraser head has a wider air gap and it works at high frequency. As the tape is passed over the wider air gap, the magnetic field recorded on the tape is reversed at a rapid speed. As a result, it turns into neutral state as in the first method.

The Fig. 5.28 (*a*) shows a HF eraser head with a gap wider than in the record head.

The Fig. (*b*) shows how progressively smaller hysterisis loops are transversed as the tape passes out of the alternating field till the saturated condition at (1) decays to the neural state at (6).

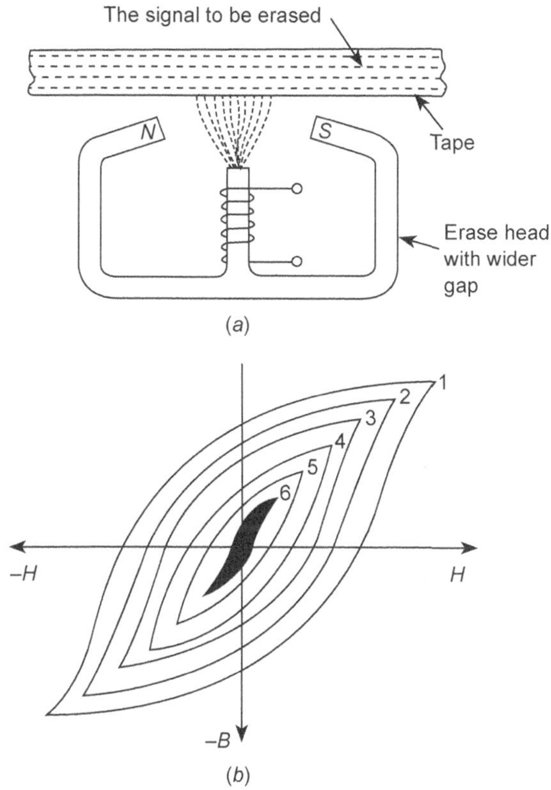

Fig. 5.28

(iii) Magnetic Bulk Erasing: This is a rapid method of erasing. The eraser head is moved on each side of the tape and the magnetic fields on the tape are quickly turned into the neutral state. This method has less distortion and noise as compared to the ordinary eraser head method.

Note. The recording and play heads both are almost identical. The pole shoes and gap in both the heads are such that magnetic field decreases where the tape emerges from it. The eddy currents losses in both the tapes are less and both have high resonant frequency and maximum possible impedance.

The recording head can be used as playback head and vice versa.

(f) Important Parameters of recording

(*a*) Wavelength

The length of tape magnetised by one cycle of the sound signal is called recorded wavelength. It is expressed in λ and given by

$$\text{Wavelength } (\lambda) = \frac{\text{Speed of tape in cm/s}}{\text{Frequency of sound signal in Hz}}$$

$$\lambda = \frac{v}{f} \text{ or } v = f\,.\,\lambda.$$

It is clear that at high audio frequencies, the wave length will decrease.

(*b*) **Gap width**

As explained above, as the audio signal frequency increases, wavelength decreases and width of the head gap measured in terms of wavelength increases *i.e.* the same gap width becomes $\lambda/4$, $\lambda/2$ etc. It can be shown that maximum output is obtained if gap width is $\lambda/2$. The gap width is kept in the order of few microns.

The relation between speed of the tape, gap width and maximum frequency is given by:

$$\text{Maximum frequency} = \frac{\text{Speed of tape}}{2 \cdot \text{Tape width}}$$

$$f_m = \frac{v}{2 \cdot w} \text{ or } v = 2.w.f_m.$$

or

$$w = \frac{v}{2f_m}$$

(*c*) **Choice of tape speed (v)**

(*i*) If the speed is very high, program of small duration can be recorded, as distance (λ) occupied on the tape by each cycle of signal will be large, (as $\lambda = v/f$).

(*ii*) If the speed is kept very low the audio cycles will occupy very small space on the tape and at high frequency they will even overlap and hence will produce distortion.

Here, we can inform the reader, that for *long plays* (low frequencies) low speed (9.5 cm/s and 4.7 cm/s) and for short plays (high-frequencies) high speed (38 cm/s and 19 cm/s) is used. In brief, the tape must move by atleast a distance equal to the "head gap" which may be as small as 0.025 mm or even less.

(g) Biasing Techniques

We have studied hysterisis curve for magnetic materials which shows that the curve (characteristic) is non- linear. While recording and reproduction, the magnetic tape passes through the hysterisis curve. The magnetising force (H) is determined by the number of turns on the head winding, current through it, produced magnetism (B) and the induced magnetisation in the tape.

When a particle on the magnetic tape comes across the head gap, it is magnetised upto M [See Fig. 5.29]. When the particle leaves the head, though there is no magnetising force (*H*) on it, even it has some magnetism called residual magnetism ON (Fig. 5.29).

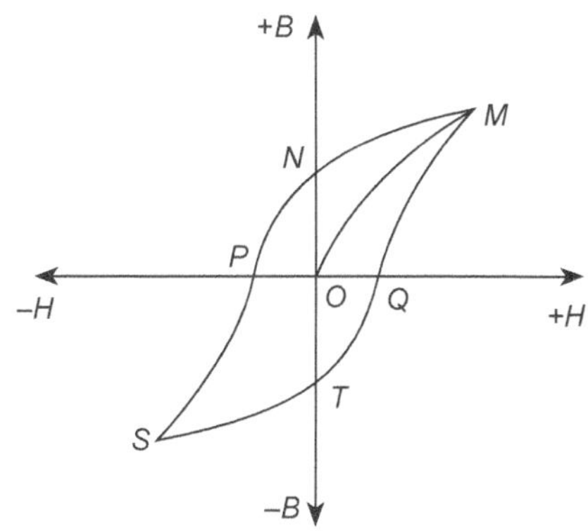

Fig. 5.29

This residual magnetism leads to distortions. Therefore, a technique is used during recording, reproduction and erasing, which makes the above operations only in the linear region of the hysterisis curve OM, which makes the operations free of distortions and noise. This explains the need of biasing, for which *biasing oscillators* are used.

Biasing is of two types:

1. D.C. Biasing
2. A.C. Biasing

1. *D.C. Biasing (Fig. 5.30).* This is done by adding a constant direct current (DC) to the audio signal to be processed and is operated in the linear region of hysterisis curve. This gives a distortion free output.

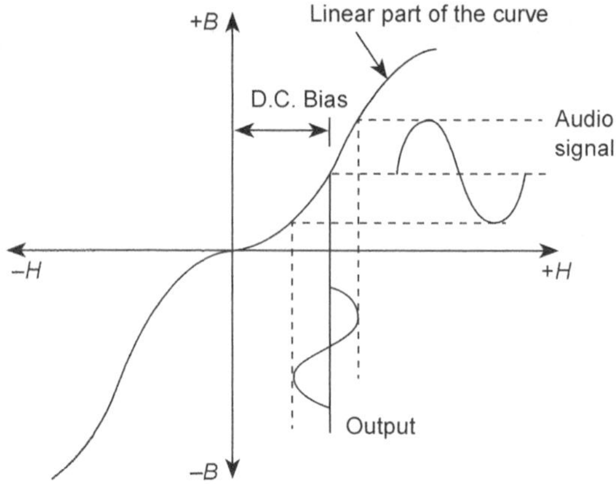

Fig. 5.30

Advantages

(*i*) It improves the quality of the reproduced signal.

(*ii*) Output is free of distortions.

Disadvantages

(*i*) Enough d.c. bias is required for getting desired results.

(*ii*) The amplitude of the signal that can be processed is small.

(*iii*) The tape remains magnetised due to d.c. bias, even there is no audio signal. This produces a continuous hissing sound during replay.

2. *A.C. Biasing (Fig. 5.31).* In this, high frequency alternating currents (A.C.) of the order of 100 kHz is added to the audio signal. The output is obtained on both sides of the magnetising curve.

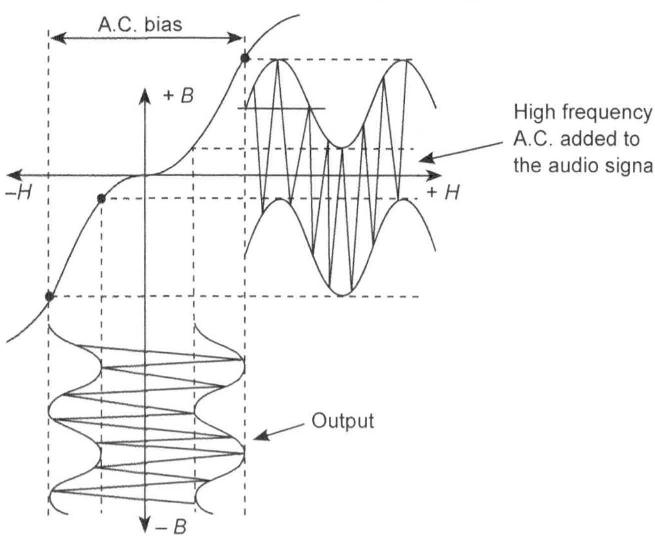

Fig. 5.31

It is also to be noted that the recording signal and the high A.C. bias are simply added without any *modulation* and the frequency of the pure sinusoidal a.c. bias voltage should be about four times the highest frequency to be recorded.

Advantages

(*i*) Higher audio signals can be recorded.

(*ii*) Noise is reduced.

Disadvantages

If A.C. bias applied is not proper:

(*i*) Distortion is increased, if the bias is low.

(*ii*) If the bias is higher, high frequencies are erased from the tape.

(h) The Magnetic Tape

In earlier days, a paper coated with red oxide of iron was used as the tape material. But now the magnetic tape has a base of myler or polyster coated with magnetic oxide like ferric oxide (Fe_2O_3) or chromium oxide (CrO_2). These oxides are retained on the tape with the help of binder which helps also in its lubrication. The backside of the base is coated with a thin layer of carbon which improves its binding characteristic. The Fig. 5.32 shows cross section of a tape.

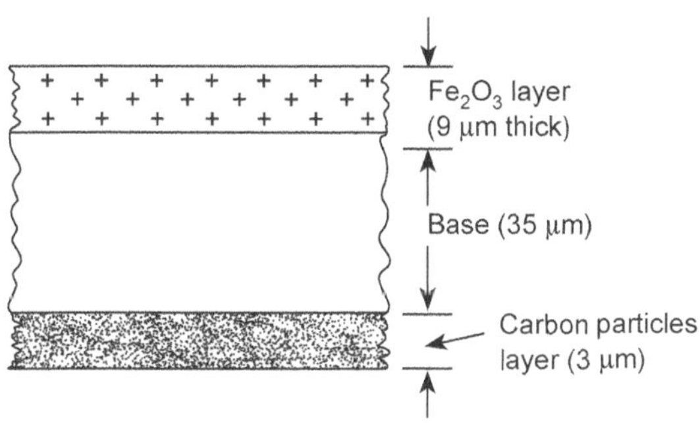

Fig. 5.32

Note:

(*i*) The tapes are available in following standard widths:

6.3 mm (1/4 inch)

12.7 mm (1/2 inch)

25.4 mm (1 inch)

50.8 mm (2 inches)

(*ii*) The 6.3 mm (1/4 inch) tape can accommodate 2 tracks each 2.5 mm (1/10 inch) wide separated by a guard band of 1.3 mm (1/20 inch). See Fig. 5.33.

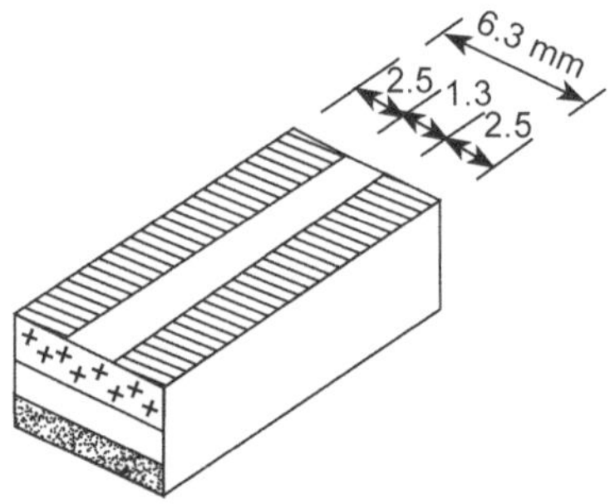

Fig. 5.33

(i) **The Magnetic Audio Cassette:** The audio tape is enclosed in a plastic cassette. The cassette protects the tape from mishandling (Fig. 5.34)

(*a*) **Important Points:**

1. The duration of the cassette is marked on it *e.g.* C—30, C—60, C—90 and so on. The C—60 means its duration is 60 minutes - 30 minutes on each side and so on.

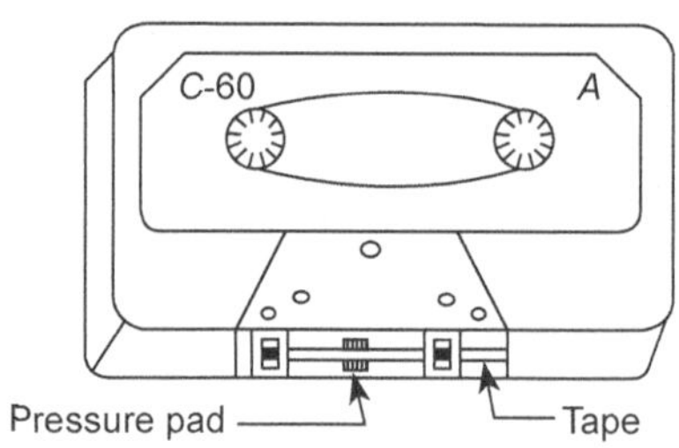

Fig. 5.34

2. The tape is 0.15″ wide and 0.002″ thick. The length will naturally depend upon its duration.

3. The speed of the tape is generally 4.6 cm per second. This is the standard speed to obtain good results.

4. The tape should not be touched with hands, it produces distortions in the replay.

5. It should be protected from external magnetic fields. An external magnetic field can erase the recording.

6. The tape should also be protected from dampness. It spoils the tape as well as the heads.

(*b*) **Repair of a broken tape**

1. The broken tape pieces should be cut slant as shown [See Fig. 5.35 (*a*)] and join the two pieces with a cello-tape.

2. Or use an adhesive to join the pieces. [Fig. (*b*)]

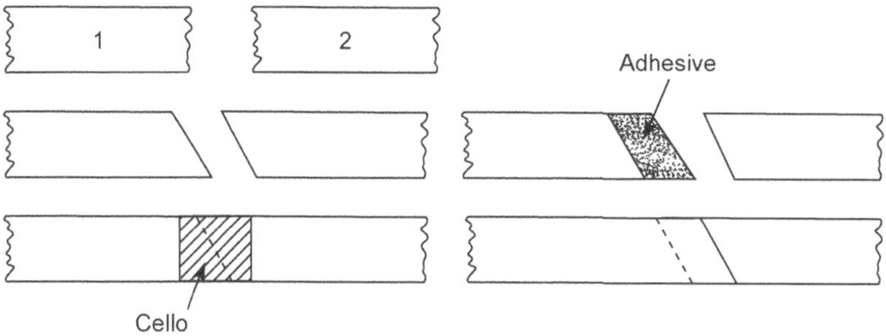

(*a*) By cello (*b*) By adhesive

Fig. 5.35

(j) Magnetic Tape Recorder

The tape recorder may have the following controls on its panel. [See Fig. 5.36 (*a*) and (*b*)]:

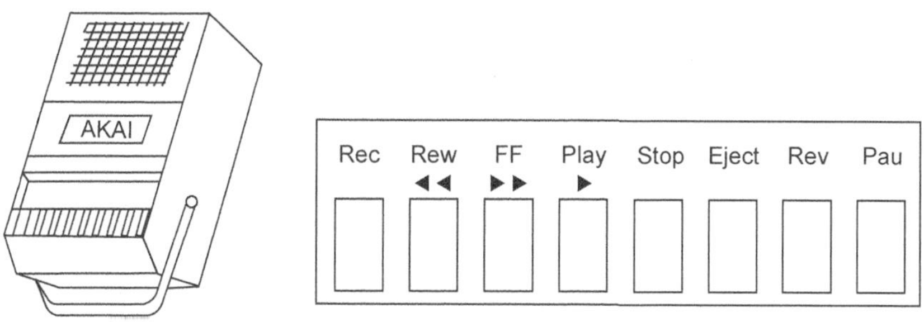

(*a*) (*b*)

Fig. 5.36

1. Record (Rec)
2. Rewind (Rew)
3. Fast Forward (FF)
4. Play or Play back
5. Stop
6. Eject
7. Pause (Pau)
8. Review (Rev)

(i) Block Diagram of a Magnetic Tape Recorder/Player (Fig. 5.37)

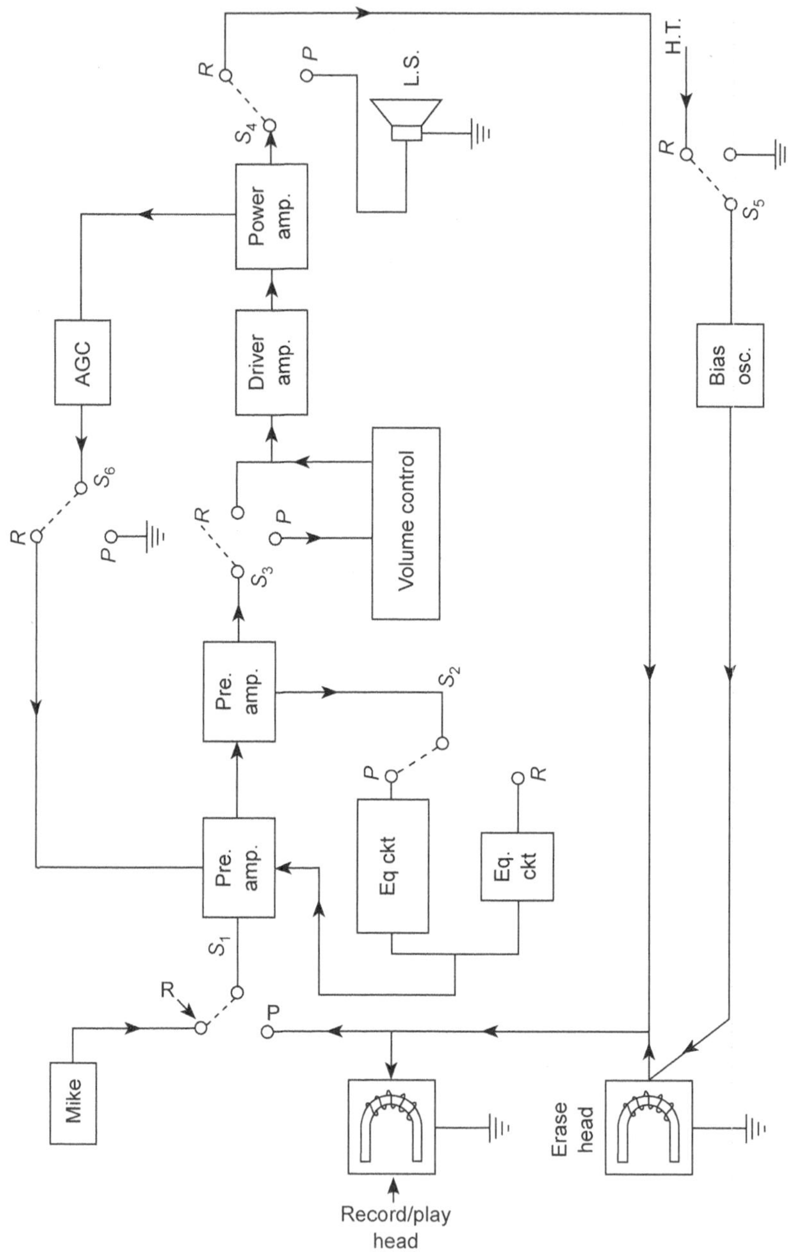

Fig. 5.37

1. *Mike.* This is an acoustical transducer, that converts sound (audio) signal into the electrical pulses. In record mode, its output is given to a pre-amplifier, which amplifies it to the desired level. In play mode of the tape recorder, it has no function to perform.

2. *Equalising circuits*. (Eq. ckt.) Low frequency signals produce weaker magnetic fields whereas higher frequency signals produce stronger magnetic fields. To get desired results, a voltage feed back is provided by equilising circuits during play.

   During recording, high frequencies obtained in the output of mike are weaker and hence they are given feedback from equilising circuits. They are two in number and a switch S_2 selects one of these circuits each during record and play (R/P).

3. *Volume Control.* The output of the last pre-amplifier is given to volume control. During play, this is connected in the circuit by S_3 otherwise during recording this is cut off and whole the audio signal may go to the amplifier.

4. *Record/Play Head.* The same head works as record as well as play head. The switch S_1 is to be put on the desired mode—Play (*P*) or *R* (record).

   In 'play', this head converts the magnetic field (audio track) of the tape into electrical pulses. The output of the play head is in micro-volts and it is to be amplified.

   In 'Record' mode, this head converts the sound signal into magnetic field which makes audio track on the tape. For a stronger magnetic field, the output is amplified through an amplifier. To reduce the hysterisis effect, a bias voltage of high frequency is produced, thus the record head is given the audio voltage obtained from microphone as well as a HF bias voltage obtained from bias oscillator.

5. *Erase head.* In construction, it is similar to R/P head. Its function is to "erase" or remove any previous recording on the tape before recording new one.

   The tape at first is passed through erase head and then through the record head. During recording, it gets "erase voltage" however, during play it is earthed.

6. *Bias Oscillator.* As explained, the R/P head also needs HF bias voltage for producing a strong sound track. This HF voltage is produced by a Bias oscillator. It is a *Hartley type oscillator* which gives an 80 kHz output. In play mode, it is earthed as it has no function to perform. [See Fig. 5.38 (*a*) and (*b*)]

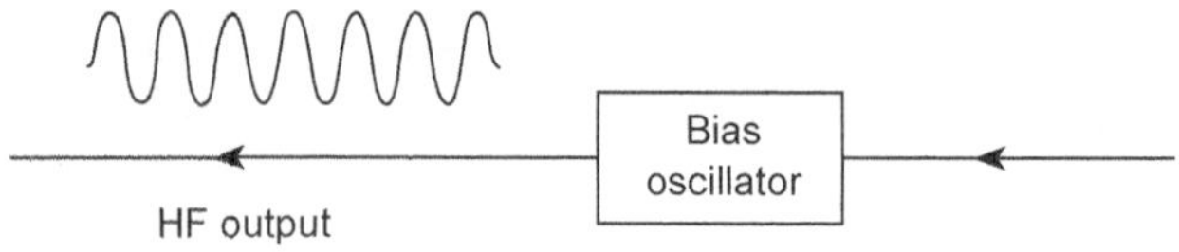

(a)

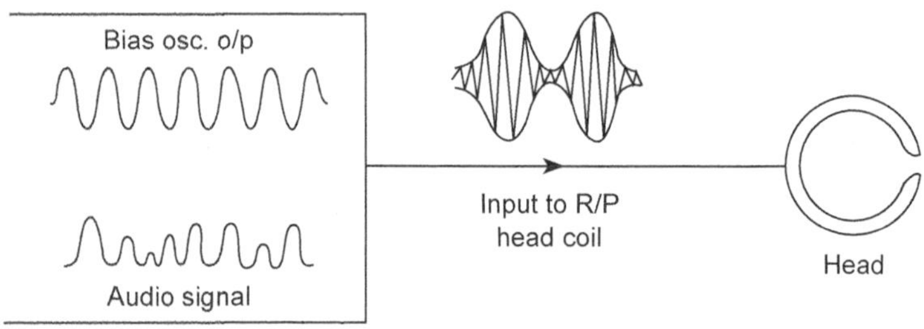

(b)

Fig. 5.38

7. *Pre-amplifiers.* There are two or three voltage amplifiers which raise the audio voltage to a desired value. They work in stages with negative feed back.

8. *Driver Amplifier.* The output of last pre-amplifier is given to driver amplifier which further amplifies the audio signal.

9. *Power Amplifier.* This is the last stage of the amplification. Generally push pull amplifier is used.

10. *Automatic Gain Control (AGC).* This is a device which controls gain of the amplifiers automatically. If the gain is increased beyond limits the amplifiers go in saturation, which will create distorted output. A part of output of power amplifier is given to AGC and then to first pre-amplifier as a reverse bias, which will reduce the gain of the amplifier and will keep it within limits.

11. *Speaker (L.S.).* This is the final block, which converts the electrical pulses into original sound.

(ii) Record and Play Modes

Now we split the block diagram into "Record mode" and "Play mode", which will clearly show which particular blocks function in each mode. The Fig. 5.39(a) shows recording mode.

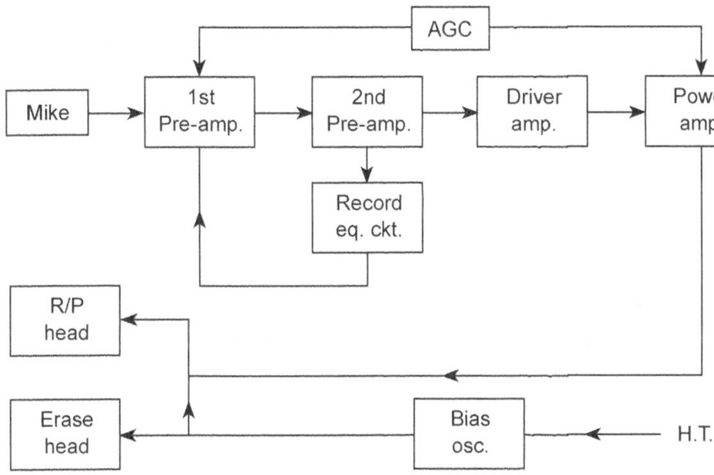

Fig. 5.39 (a)

The Fig. 5.39(b) shows play mode.

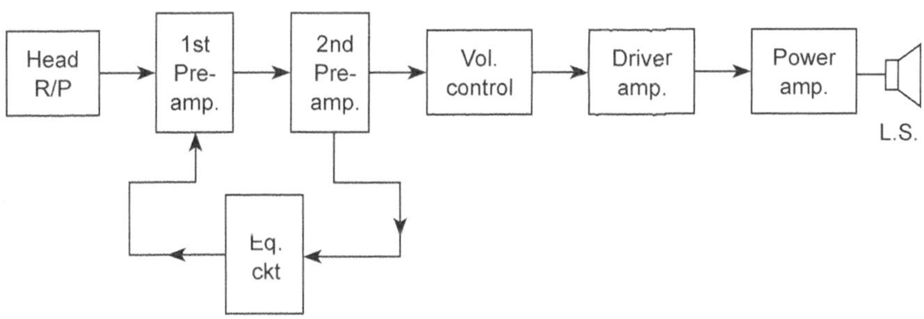

Fig. 5.39(b)

(iii) Tape Transport Mechanism (Fig. 5.40)

The tape transport mechanism is the most important part of a tape recorder. It moves the tape at a constant speed over the 3 heads (*i.e.* Erase, Record and Play heads). The tape movement at a predetermined and calculated speed is very important for proper functioning of the recorder/player.

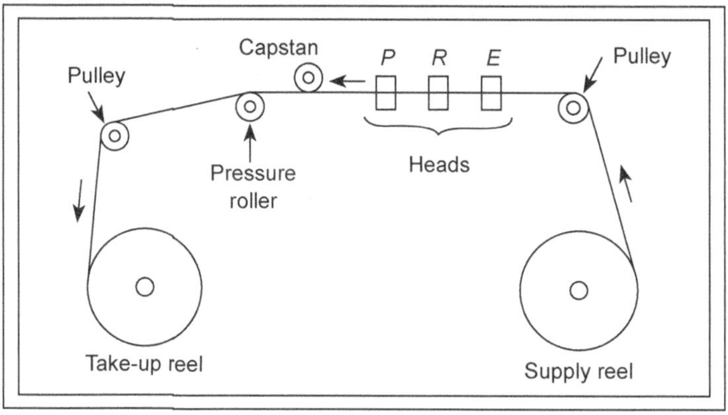

Fig. 5.40

The servo tape moves between *supply* reel and the *take up* reel and maintains continuous contact with various rollers, which help in the smooth movement of tape.

As the tape comes out of the supply reel, it passes on a pully (or idler) and as the tape passes the erase (*E*), record (*R*) and replay (*P*) heads, the motion should be completely free of any noise. After passing through the 3 heads, the tape moves through capstan and pressure (or Pinch) roller assembly. The pressure (or Pinch) presses the capstan against the tape and in turn capstan provides a constant angular velocity to the tape. In modern recorders, the capstan is driven by servo-controlled motors which operates at speeds of 19, 38 and 76 cm/s. From the capstan, the tape goes to the take up reel through a pulley.

The motor used in the tape transport mechanism is usually a synchronous motor, which provides a constant speed irrespective of the fluctuations in the supply voltage.

(*i*) Capstan-Pinch Assembly (See Fig. 5.41)

The main function of the capstan pinch assembly is to provide a constant speed to the tape. Capstan is fitted to a small flywheel which absorbs the variations in the speed, if any. The tape is pressed against the capstan through a pinch, which is nothing but a pressure wheel the tape moves in between capstan and the pinch.

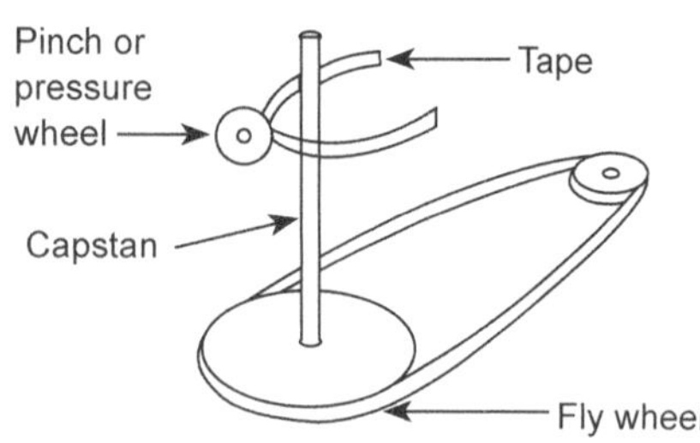

Fig. 5.41

During playback, the tape speed is also controlled by the capstanpinch assembly.

(iv) Vow and Flutter

The momentary variations in the tape speed is called Vow and Flutter. The variations in the speed at lower frequencies are known as *Vow* and the variation in speed occuring at higher frequencies are known as *flutter*. The Vow and Flutter are expressed in percentage variations of the speed from the normal.

The constant tape speed keeps vow and flutter within 0.3% which is the limit of quality tape records. Remember that narrow tapes have this problem more, hence special care is needed.

Vow is also caused by a buckled tape reel and an *off the centre* disc while the Flutter is also caused by a damaged capstan-pinch assembly or a damaged disc.

The tape transport mechanism as told earlier, should have the following controls: Play or Playback, Rewind, Fast Forward, Record, Stop, Eject, Pause, Review, Bass, Treble.

The other special controls may be

(*i*) *Auto stop.* The mechanism shuts off automatically when recording or replay reaches the end of the cassette.

(*ii*) *One touch recording (OTR).* By pressing the recording button only, the recording starts. No need to press both "Record" and "Play" buttons.

(*iii*) *Shuttle.* By pressing this button, the tape can be shuttled back and forth at a given speed.

(v) Care of Recorder Heads

As discussed, three heads are used in tape transport mechanism of tape recorder, viz, record head, reproduce (replay) head, and erase head. Practically two heads are used as the same head performs the functions of recording as well as replaying.

A head is basically a *transducer*. The *record head* converts the electrical signal into magnetic signal, the *playhead* reconverts the magnetic signal back to electrical signal, which is given to the loudspeaker to get the original sound. The *Erase head saturates* the previous magnetic field and then pulls it into neutral state at a particular frequency.

The core of these heads is made either of permalloy (Ni-Fe) or Alfesi (Al-Fe-Si) or sometimes of ferrites, and are wound with a coil of specified turns. These heads have a gap ranging from 12 μm (Record head), 6 μm (Replay head) and 100 μm (Erase head).

The life of a recorder depends upon these heads, therefore they need utmost care. Any dust partile on the heads or the tape is going to damage the head and the tape, therefore regular cleaning of heads is utmost necessary. A cleaned head establishes a good contact with the tape. Alcohol, Neptha or petrol may be used for cleaning a head. Along with the rollers, capstan, pinch etc. should also be cleaned. The heads should also be periodically "damagnetised" using a damagnetiser.

(vi) Noise in Recording and Replay

The noise in recording and replay is produced due to following effects:

1. *Contact effect.* Poor contact between tape and the heads is a source of noise. Pads are provided to give better contacts. As the magnetic circuit is completed through the gap, the variation in contact disturbs the circuit and also the process going on.

2. *Gap effect.* The length of head gap has a critical effect on the recording as well as replay process. The magnetic circuit completes only when the gap width is half of the wave length (*i.e.* $\lambda/2$) of the signal, in this case, maximum output is obtained (See Fig. 5.42). In this way, gap effect plays a major role on the performance of the recorders.

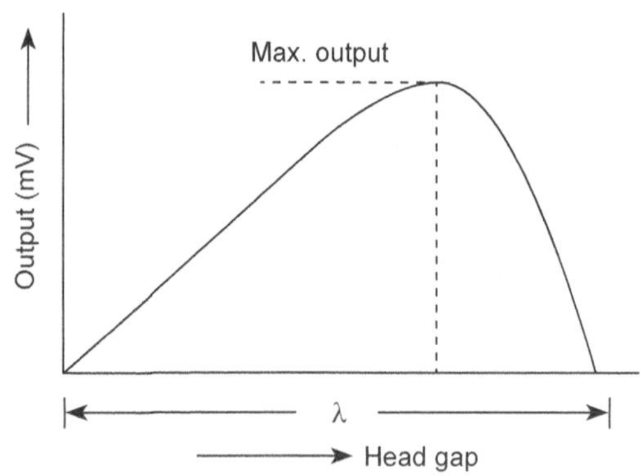

Fig. 5.42

3. *Azimuth effect.* The recording and the play heads both should be at right angles to the tape. In other words the both heads should be exactly parallel to each other. If not so, misalignment takes place. This is known as Azimuth effect. The Azimuth (Misalignment) causes losses and also noise called the *Azimuth noise*. This adjustment may be done by a device called alignment tape.

4. *Hissing noise.* A hissing noise is produced due to non- uniform tape coating. In multitrack recording, this noise is more serious.
5. *Rumble.* The frequency distortions produced by mechanical vibrations of motor or capstan are called rumble. To prevent this, vibration absorbers may be used.

Table 5.1: Comparison between disc and magnetic recording

Disc Recording	Magnetic Recording
1. Disc recording has moving parts like stylus and, therefore the system has more wear and tear.	Magnetic recording has no such moving parts and therefore it has less wear and tear.
2. The disc has a short life, as it loses its quality with time.	A tape can be played thousand times.
3. Editing and dubbing is very difficult.	Editing and dubbing is easy.
4. Immediate playback is not possible in the disc recording.	Immediate playback is possible in magnetic recording.
5. The same disc cannot be used again and again for recording.	The same tape can be used again and again by erasing the previous recording.
6. There is no modulation noise in the disc.	In tape, modulation noise is present.

Note. Disc and magnetic recordings are now almost absolete and have limited use.

5.1.3 Optical Recording

The optical recording is done on a photo film reel (not on disc or magnetic tape). The sound signal recorded can be reproduced with the help of a light beam and a photo cell (Film Projector). The sound signal is recorded in the form of a photographic image 2.5 to 5 mm wide (called optical sound) on the right hand side of the picture (video) signal next to sprocket holes of the film (See Fig. 5.43)

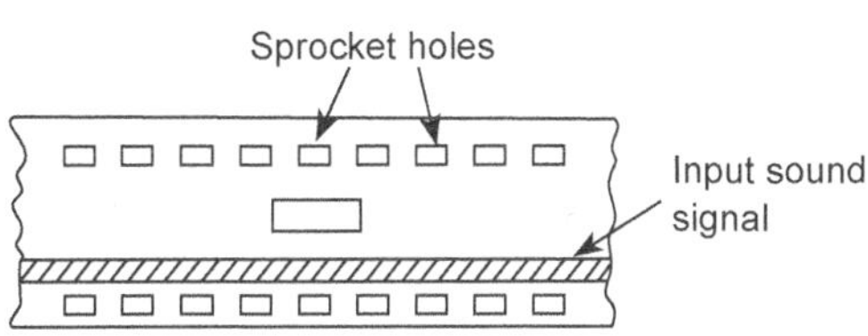

Fig. 5.43

At first, the sound and picture signals are recorded on separate strips for editing. After this, the sound is recorded on the film 0.37 mm ahead of the corresponding picture signal.

The films which are used for motion pictures (tape recording and photography) are mostly made of thermoplastic polymers both artificial (cellulose easter) and synthetic (polyethylene, polypropylene PVC, PTFE etc.). These films are made through rolling process, or the plastics are directly applied to fabrics, or through intermediate materials. Cheaper films are prepared through *extrusion process*.

The optical recording of sound is of two types:

(a) **Photographic Recording:** The first type is the sound recording done on photographic films in movie cinema. Therefore this is also known as motion picture recording. In this, the audio (sound) signal to be recorded is converted into light, whose intensity varies in accordance to the sound pressure. This is also called as *photographic film recording*.

(b) **Digital or CD Recording:** The second type is the sound recording done on compact discs (CD). This is done through LASER. This type of recording is done in binary system (0 and 1). This method maybe called as Digital recording or compact disc recording.

We shall study both types in detail.

1. Photographic Recording

(i) There are two methods:

• *Variable density method.* In this method, the weak sound signal obtained from microphone is first amplified and then fed to a special vacuum tube (lamp). The intensity of light obtained from the lamp varies according to the sound pressure to be recorded. This light is made to fall on the photographic film, which moves with a calculated speed. The image is formed on the film in the form of bars of varying density in accordance to the sound pressure.

The Fig. 5.44 (*a*) shows the method and Fig. (*b*) shows that tracks so formed.

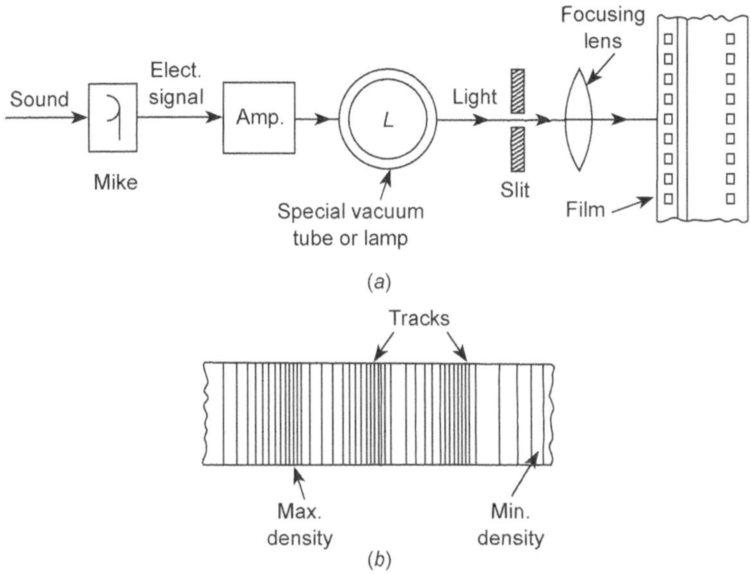

Fig. 5.44

•• *Variable area method.* In this method, sound is fed to a microphone and the weak electrical signal obtained is amplified and fed to the coil of a mirror galvanometer. The coil moves and sets the mirror into motion. A focussed light is made to fall on the mirror which deflects this light. The area of the slit opened for this light varies according to the sound pressure and also the picture formed on the film. The Fig. 5.45 (*a*) shows the method and Fig. (*b*) shows the sound track so formed.

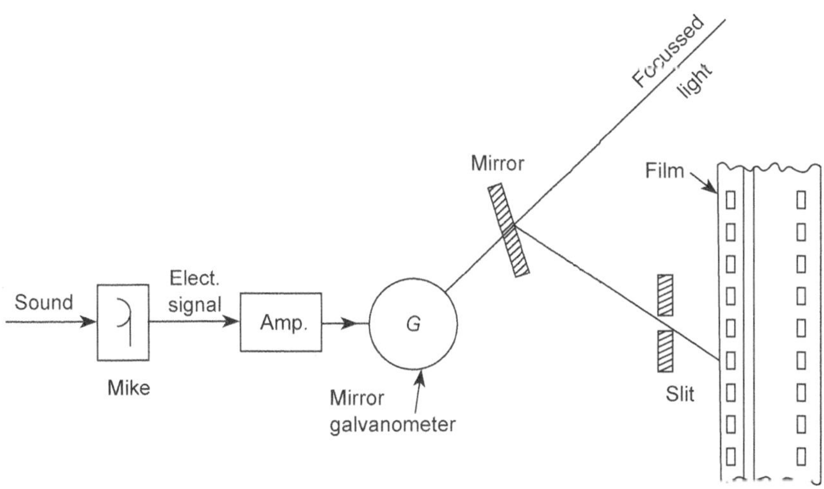

Fig. 5.45 (a)

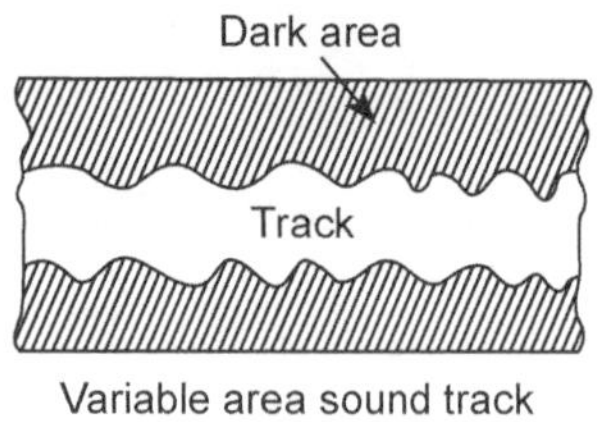

Fig. 5.45 (b)

(ii) Reproduction of photographic Signal (See Fig. 5.46)

The reproduction of the recorded sound tracks on the film by the above two methods is very simple. A focussed beam of light is made to fall on this sound track. The film is moved at a designed speed. This light now is fed to a photo cell (Film Projector), which converts it into electrical signal. This signal is amplified and passed through a loudspeaker, which reproduces the sound originally recorded.

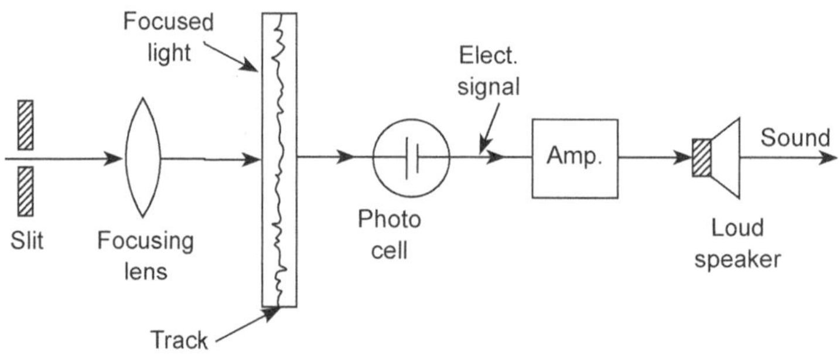

Fig. 5.46

Film Projector

The reproduction of sound image (optical sound) is done by a machine called Film Projector. Based on the principle of persistence of vision of our eyes (our eyes can retain a picture only for 1/16 seconds), the projector projects at the rate of 25 pictures per second to create an "Illusion of continuity".

The various important parts of the film projector are given below: (See Fig. 5.47)

1. *Sprocket.* They are the toothed wheels and constitute the /I transport system for the reel. A guide aligns the film.
2. *Spools.* There are two spools called 'Supply reel' and "Take up reel". The reel is unwound from one spool and is wound on the other.
3. *Optical system.* The system produces a beam of focussed light, which falls on the sound track of the film.

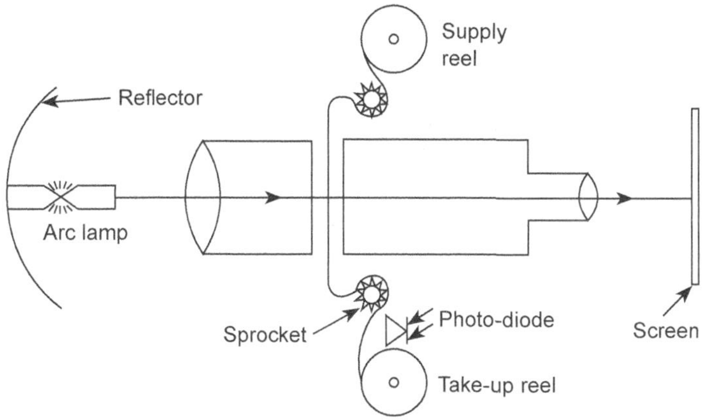

Fig. 5.47

4. *Carbon arc lamp.* A carbon arc lamp and a reflector assembly is provided which produces a bright parallel beam of light to be reflected towards optical system.
5. *Photo cell/photo diode.* The photo cell converts the input signal into electrical signal, which is further processed.

(iii) Dimensions of a Film

The Fig. 5.48 (*a*) shows a 35 mm (1.378 inch) film with monophonic sound tracks.

The Fig. (*b*) shows a 35 mm film with stereophonic sound tracks.

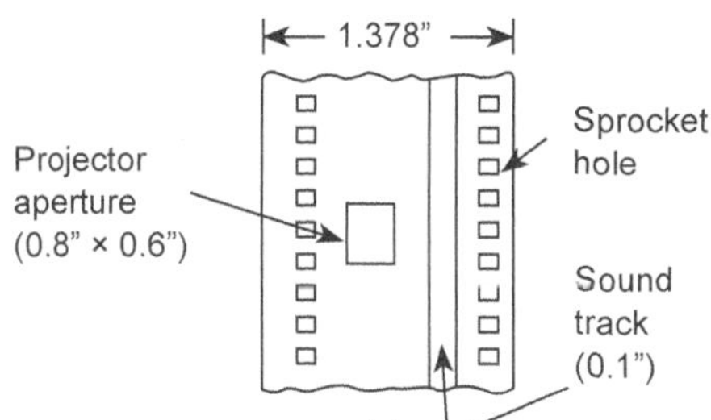

(*a*)

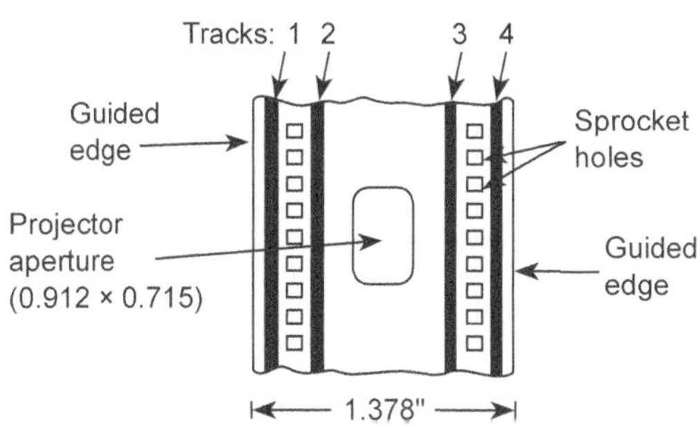

(*b*)

Fig. 5.48

Table 5.2: Comparison of photographic recording Methods

S. No.	Variable area recording	Variable density recording
1.	In this method, nearly 85% modulation can be achieved.	In this method, about 35% modulation can be achieved.
2.	It has a better S/N ratio.	It has comparatively poor S/N ratio.
3.	In overload input conditions, distortion is more.	In this case, distortion is less.
4.	In this case, the quality of the output is very much dependent on the quality of recording/ reproducing equipment.	Quality of the output is independent of the quality of recording/ reproducing equipment.
5.	The track varies in area (width) according to the modulation index but has constant density.	The track varies in density but has constant area (width).
6.	The sound track is easy to process and has a greater sound volume	Its processing is not so easy and has a comparatively smaller sound volume.
7.	The variable area tracks under adverse conditions cause distortion while reproduction.	The variable density tracks under adverse conditions give satisfactory reproduction.

(b) Digital (CD) Recording

So far we have studied disc and magnetic recording which are examples of Analog recording. Now digital recording is becoming popular which is done on compact discs (CD) and digital audio tapes (DAT). Digital recording is done in Binary code (0, 1).

Advantages of digital recording

(*i*) A high quality of recording is obtained by digital method.

(*ii*) It is free from noise as signal-noise ratio is quite high.

(*iii*) The recording can be copied for indefinite number of times.

(1) Compact Disc (CD)

A compact disc (CD) is made of polycarbonate protected by a transparent layer of lacquer. Aluminium also added in the disc to give it a reflecting property. The disc is coated with a photo resist material.

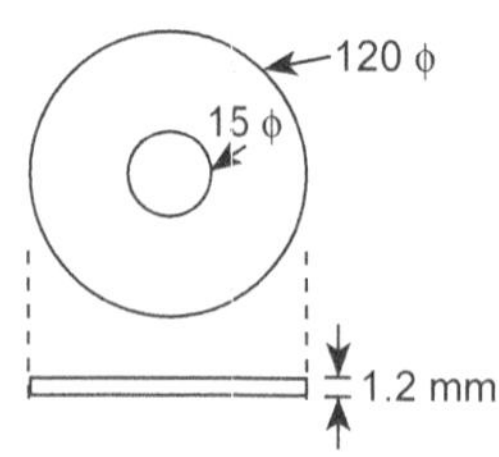

Fig. 5.49

The programs are recorded on one side of the disc. Recording is done in spiral pits going from centre towards edges of the disc. The whole disc is protected by a plastic cover.

Table 5.3: Specifications of CD (See Fig. 5.49)

S. No.	Specifications	Details
1.	Diameter	120 mm
2.	Thickness	1.2 mm
3.	Dia. of centre hole	15 mm
4.	Direction of rotation	Anti clockwise
5.	Speed	250-500 RPM
6.	Playing time	above 60-74 minutes
7.	Length of the pits	0.833 μm to 3.00 μm
8.	Depth of pit	0.12 μm to 1 μm
9.	Width of pit	0.6 μm
10.	Pitch of the pit	1.5 μm
11.	Sampling frequency	44.1 kHz
12.	Digital Quantisation	16 bits
13.	Number of levels	65, 536 per channel
14.	Bit rate	7.3 Mega bits/second
15.	Scanning rate	1.2-1.44 m/s
16.	Wave length	780 nanometer
17.	S-N ratio	90 dB
18.	Number of channels	2 or 4
19.	Channel separation	90 dB
20.	Frequency response	entire AF range (20 Hz to 20 kHz)

(2) CD Layers

The Fig. 5.50 shows different layers of a CD.

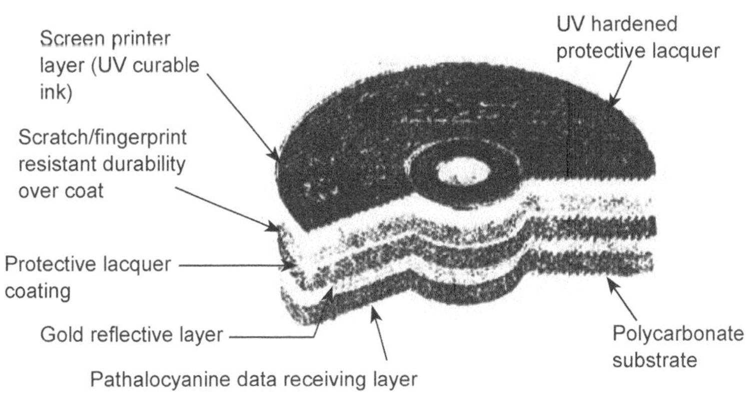

Fig. 5.50

(3) CD Tracks

The Fig. 5.51 (*a*) shows details of tracks. Fig. (*b*) shows one pit. The Fig. (*c*) shows pitting in the disc.

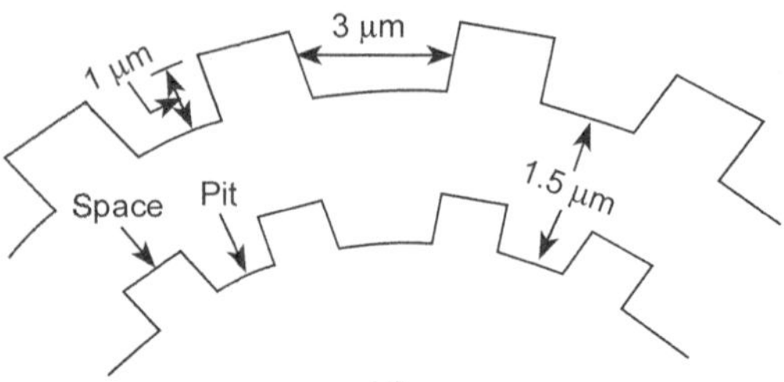

(*a*)

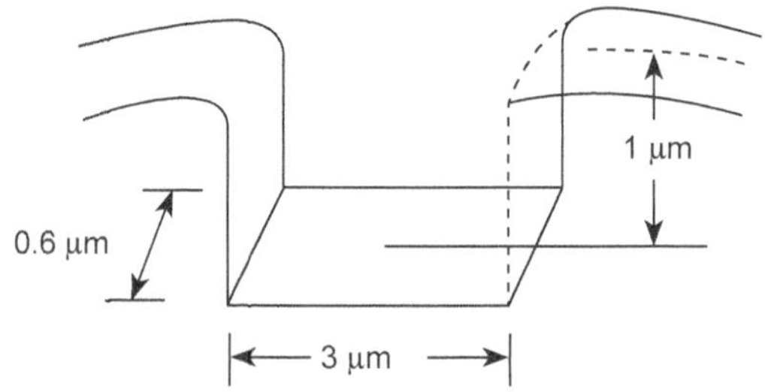

(*b*)

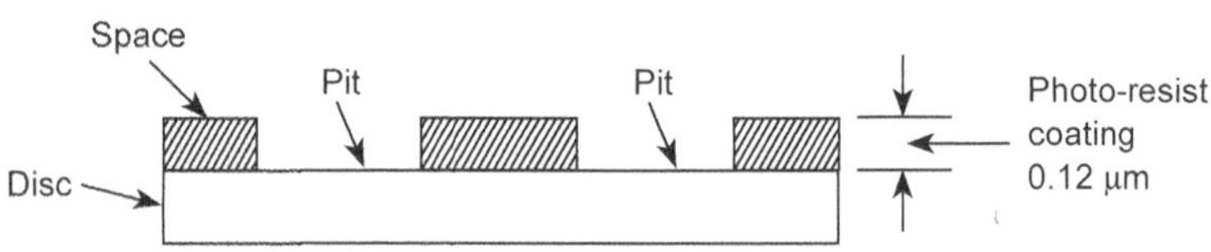

(*c*)

Fig. 5.51

(4) CD Recording (Fig. 5.52)

For recording, first a master disc is prepared. The laser beam acts as a stylus, which is modulated by digital audio signal. For this the audio signal to be recorded is obtained from the microphone, amplified and sampled @ 44.1 kHz with digital quantisation of 16 bits. Thus 7 mega bits $(44.1 \times 10^3 \times 16)$ per channel per second are used to modulate the laser. This modulated laser is made incident on the CD. When laser beam is ON, the output is 1 and when the beam is OFF, the output is 0. It forms pits on the photo resist material of the CD.

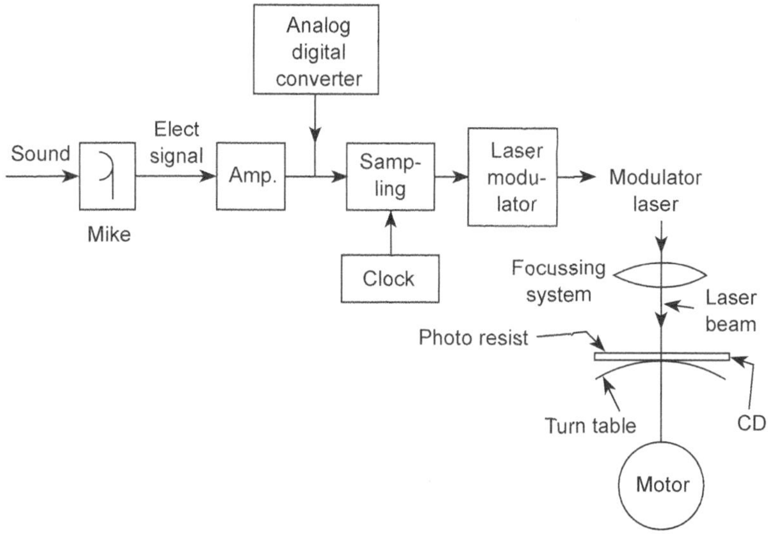

Fig. 5.52

(5) CD Playback (Fig. 5.53)

For replay a disc, a focussed laser beam of suitable wave length is made incident on the CD tracks through a mirror. A half polished mirror is used, as it allows the laser to pass through it, but does not allow it to return. The reflection of laser beam from pits represents 0 and from space represents 1. This reflected beam is again passed through a focussing system and then through a photo diode detector. The output of the photo diode is fed to a *digital to analog converter*. The analog signal so obtained is fed to a loudspeaker for reproduction.

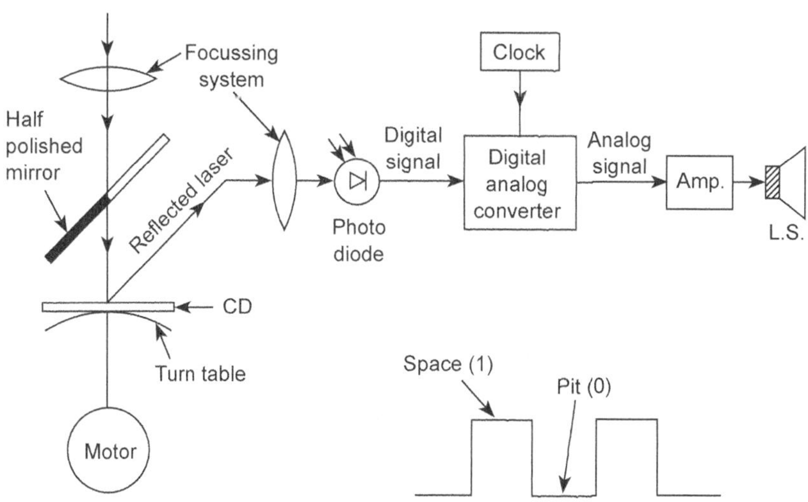

Fig. 5.53

(6) Important Notes—Digital Recording

1. The digital recording is concerned with two saturation states 0 and 1 therefore H.F. bias is not needed.
2. Erasing is also not required.
3. Digital recording finds use in computers, satellite communication, electronic games and audio video recorders.
4. There is no degradation in the quality of the digital recorded program irrespective of infinite reproductions.
5. Mostly we find one hour of stereo per side of the disc though theoretically it can be more.
6. For digital recording, digital audio tapes (DAT) are also popular. A stationary head DAT (S-DAT) produces parallel tracks and a rotary head DAT (R-DAT) produces spiral tracks on the tape. The digital recording has a band width of 3 MHz. The digital tape has much thinner magnetic oxide coating than the analog tape. An R-DAT cassette has dimensions of $10.5 \times 53 \times 75.5$ mm.
7. The recording on floppy disc is almost similar. A floppy disc is made of plastic and is coated with a magnetic material forming circular tracks. The most common floppy disc has a diameter of $5\frac{1}{4}$ and is kept inside a square plastic case. This is also known as "Floppy diskette". These discs may be single sided, or double sided (Fig. 5.54). A high density floppy can store a memory of 1.2 Mega bites.

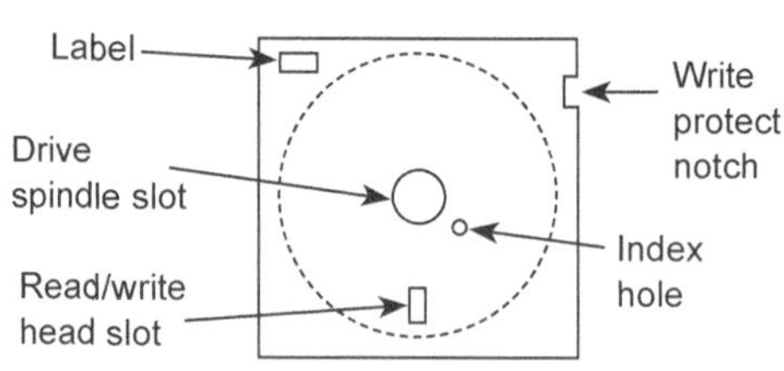

Fig. 5.54

(7) CD (Compact Disk) System

The CD audio system is similar to any audio system plus digital signal processing circuit. The CD player is a kind of LP record player that plays a pre-recorded CD, which can contain any audio signal i.e. song, speech, music, lecture etc. A piece of song/speech stored in a disk is called a *track*.

(a) **Audio CD:** An audio CD is a plastic coated round disk of 120 mm (4.25″) diameter. It is called 'compact' disk, because it is small and compact. The audio signal is recorded on one side of the disk. The CD contains about 1 hour audio program. The Fig. 5.55 shows an audio CD, all dimensions are in mm.

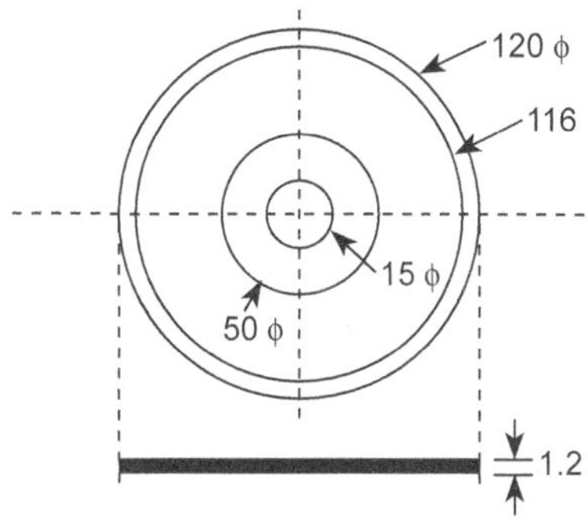

Fig. 5.55

The audio program on the CD is 'read' by laser (light amplification by stimulated emission of radiation) beam, generated by a laser diode made of aluminium-Gellium-Aresenide (Al-Ga-As) inside the CD player.

(b) **Audio CD Vs LP Record**

1. On the CD (Compact Disc) player, there is no direct contact between the reading circuitary and the disk, whereas in case of LP (Long Play) record player, the "pick up" moves on the record surface. This causes wear and tear and the life is reduced.
2. The reading circuitary in CD player is constituted of LASER system, this gives high class sound output, where as in case of a LP record, the sound output is of interior quality.
3. In CD, we can access and replay any track/part of the disk easily, whereas in case of LP, it is not easy.
4. In CD and as well as LP record, the audio signal is recorded through grooves in spiral form but in case of LP, the grooves are small and of varying depth (Fig. 5.56 (*a*)) whereas in case of CD, the signal is stored in "pits" and "flats" (Fig. *b*).

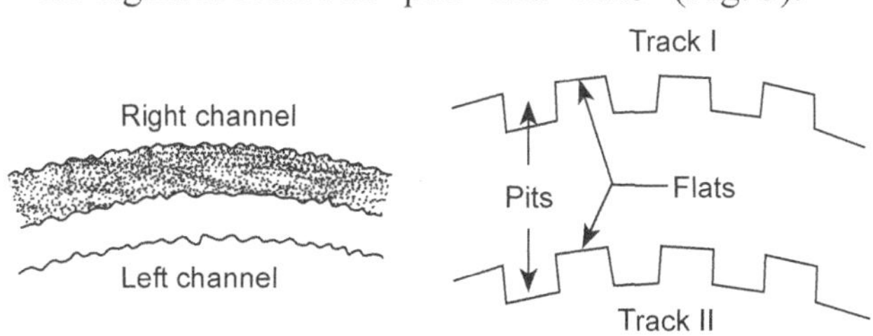

(*a*) (*b*)

Fig. 5.56

5. When a CD is played, the playing starts from centre towards outside. But in case of a LP record, the playing takes place from outside towards its centre.
6. In LP, the signal is stored is analog whereas in case of CD, the signal stored in digital.

   There are few advantages of the digital storage like: in replay, the reproduction is free from any disturbance or hissing noise, more over "cross talk" problem is also absent, and real stereo sound is obtained.

(c) Audio CD Manufacturing: The CD manufacturing is done in the following steps: (Fig. 5.57)

1. First of all a "master" disk is prepared. For this, a polished glass disk is taken and covered with 0.1 mm thick photo resist coating. A photo resist is a material whose properties are much effected with light. On this photo-resist layer, with the help of "LASER", an image of the CD to be produced is obtained. See Fig. 5.57(*i*). This is the original disk, on which audio signal is first recorded. This is also called "Resist master disk" (RMD).
2. The image is developed to create "pits" and flats on the portions where the modulated laser strikes. See Fig. (*ii*). The master disk is coated with sliver to make it electrically conductive.
3. Now over the pits, "nickel" is plotted. See. Fig. (*iii*).
4. The nickel is now peeled off. This is the "negative" of the master disk. This is also called FATHER disk. See Fig. (*iv*).
5. Now the "father" is used to produce few "positives" known as "mother" disks. Generally 10 mother disks are obtained from a single father disk [Fig. (*v*)].
6. In turn, the mother produces few negatives called "Sons". These son disks are used further to produce required number of disks. The son disk is also called a "stamper". See Fig. (*vi*).
7. Now, the pits and flats of the "son" are filled with a liquid plastic (polycarbonate). See Fig. (*vii*).
8. The plastic is now removed and the disk is coated with aluminium coating. The aluminium makes the disk reflective to "read" (replay) the signal by the CD player, when LASER strikes over it. See Fig. (*viii*).

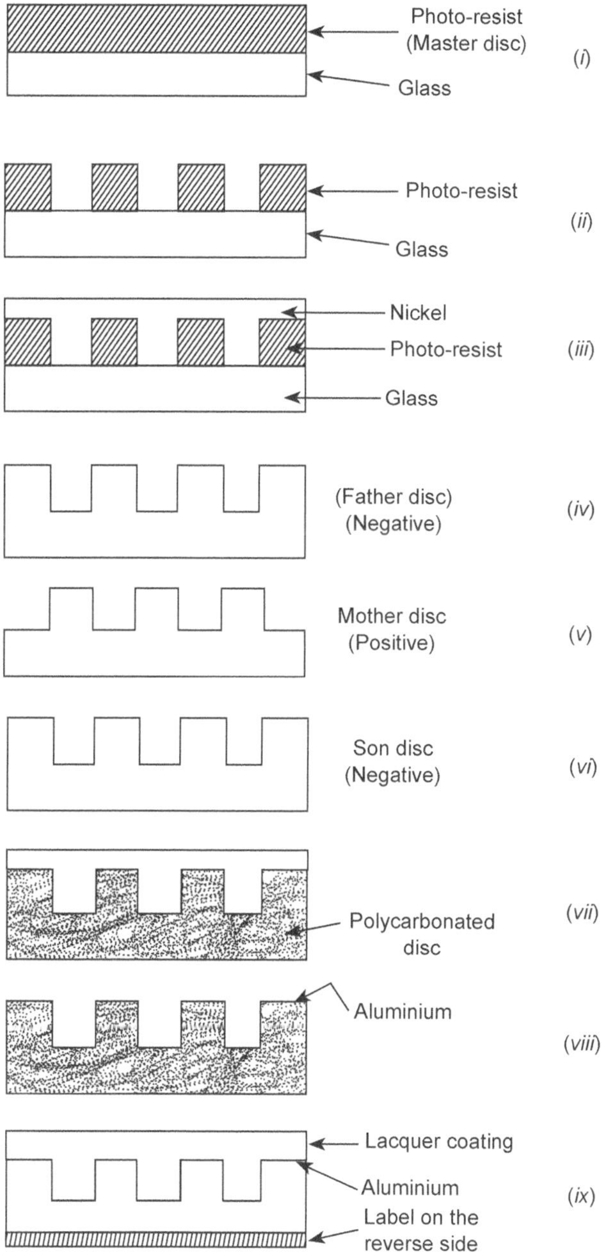

Fig. 5.57

9. Now the aluminium coating is again covered with a plastic or lacquer coating. It saves the disk against dirt, dust etc. and the disk can be washed with soap solution to keep it clean. The final CD or consumer disk is ready. Fig. (*ix*). On the reverse side, a label is provided which gives details of the contents of CD.

(8) Audio CD Player

The audio signal on a CD is recorded in digital form (0, 1). A pit represents 0 and the flat represents 1. See Fig. 5.58. The CD player converts this audio information in to analog form by a device called D/A (digital-Analog) converter. The CD player is a constant linear velocity (CLV) device, whereas the LP record player is a constant angular velocity (CAV) device. The CD contains more information on its outer area as compared to its inner area, as the diameter goes on decreasing from outer to inner side. The standard linear velocity of a CD player is 1.2 m/s, and this is maintained constant by changing the rotational speed from 500 RPM at the inner side to 250 RPM at outer side of the disk.

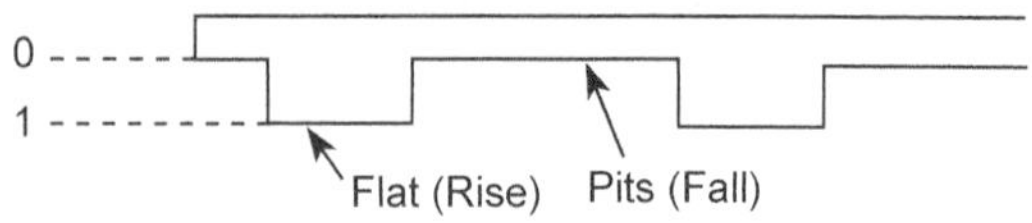

Fig. 5.58

The total track length is 5.5 km, and playing time is about 1 hour. The frequency response of a CD is for total AF range and S/N ratio is 100 dB.

(i) Block Diagram of a audio CD Player: For reproduction, LASER beam is thrown on the tracks. It is reflected from the "flats" and not from the "pits". Now the audio digital signal is produced with the help of a photo diode. Through a digital to analog converter, this signal is converted into analog and fed to a loudspeaker, which converts the same into original sound.

The Fig. 5.59 shows the block diagram explaining working of a CD player.

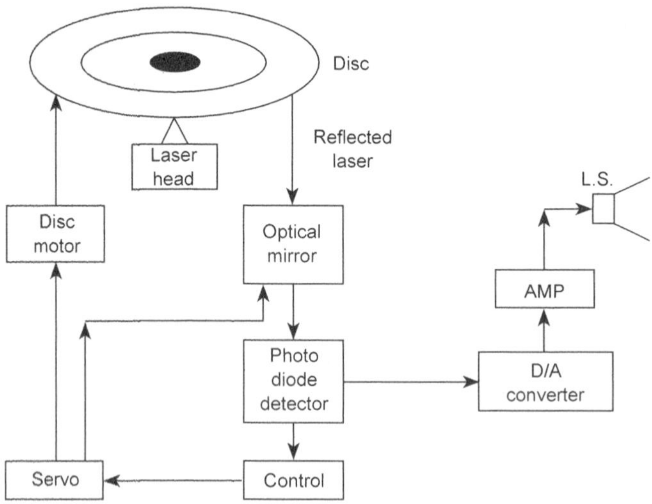

Fig. 5.59

(ii) Controls on Audio CD Player (Fig. 5.60): Following are the important controls on a CD player.

1. *Push to open.* This button is provided to open the disk compartment for placing in and removing out the disks. The disk is so placed that the "label" side is facing up.

   The compartment closes by pressing its lid down.

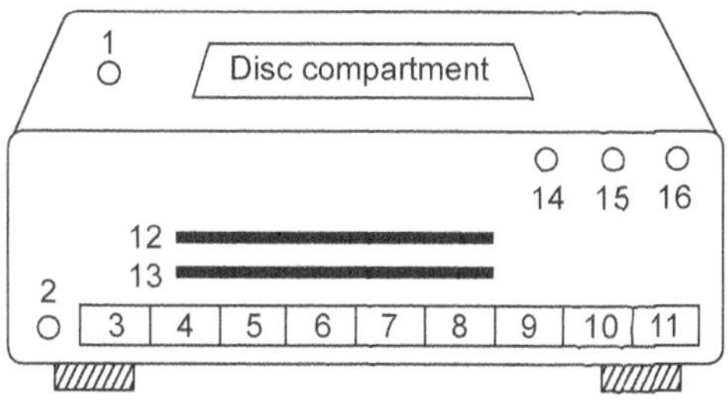

Fig. 5.60

2. *ON/OFF Switch.* This is for making the player ON/OFF. When the player becomes ON, as an indication a LED glows.
3. *Select.* This is to select a particular track (program/song).
4. *Store.* This button is used (along with the Select) to program the sequence or the "order" of the tracks to be displayed. This "selects and stores" the tracks in the memory.
5. *Cancel.* This is used (along with Select) to cancel or remove any program stored in the memory.
6. *Stop/CM.* This is used to stop the track being played or to *Clear the Memory* (CM) of the CD player.
7. *Play/Next.* With the disk placed and this button pressed, the player will start playing the tracks. This button can also be pressed to stop the "track" being played and to play the NEXT track on the disk.
8. *Rev (Reverse).* This is the button to go reverse on the disk.
9. *Fwd (Forward).* This is the button to go forward on the disk.
10. *Pause.* This is used to stop the play for a brief period. This stops the playing but the disk goes on rotating for the pause period.
11. *Repeat.* By pressing this button, the player again repeats to play the disk from the beginning.

In addition to above controls, a CD player has the following displays:

12. *Program display*
13. *Track display*
14. *Error LED.* Which glows, showing a player operating error.
15. *Pause LED.* Which glows, when "Pause" button pressed.
16. *Repeat LED.* Which glows when repeat button is pressed.

(iii) Symbols Used on CD Players: The Fig. 5.61 shows symbols used on CD players and their purpose.

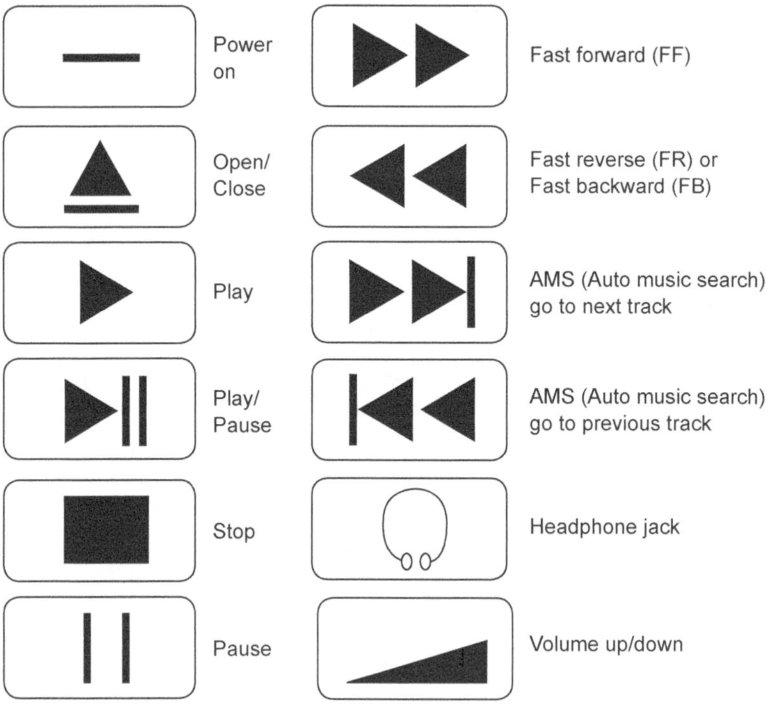

Fig. 5.61

5.2 HIGH FIDELITY (HI-FI)

High fidelity is the attempt to reproduce by electronic and electromechanical devices (and without distortion) the full range of audible sound from 20 Hz to 20 kHz. A reproducing system is called as "Hi-Fi", if it fulfills the above conditions and comes as close as possible to sounding like original. In other words, the reproduced sound should be exact replica of the recorded sound.

'Fidelity' means "Faithfulness". An ideal hi-fi system should

(*i*) be complete free from noise. The S/N ratio should be infinite.

(*ii*) It should have frequency response from 20 Hz to 20 kHz *i.e.* for complete AF band. In other words, there should be no frequency distortion.

(*iii*) There should also be no amplitude or phase distortions.

(*iv*) It should be able to detect the direction, from which a particular sound is coming.

(*v*) It should create stereophonic (3-dimensional) effects.

5.2.1 Distortions of Sound in Hi-Fi Systems

If the reproduced sound is not an exact replica of the input sound. The output is said to be distorted. The distortion is measured in percentage.

In hifi system, the following distortions are common

(*i*) Amplitude or non-linear distortion

(*ii*) Frequency distortion

(*iii*) Phase distortion

(*iv*) Transient distortion or hang over

(*v*) Spatial distortion.

(*i*) *Amplitude or non-linear distortion.* This distortion is produced by operating the amplifiers over non-linear part of its characteristics. The sound becomes rough and hoarse. This distortion takes place in power amplifiers handling very large powers.

In this, amplitude and hence the wave shape of the output is different from the input. The Fig. 5.62 shows, that output is different in amplitude from the input, as amplifier is being operated on its non-linear part of characteristic.

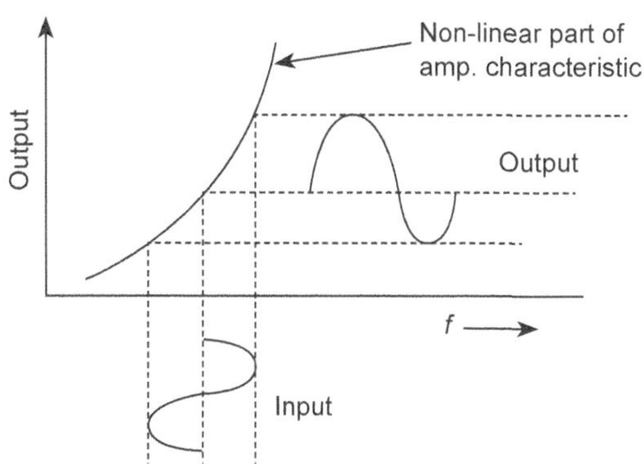

Fig. 5.62

This distortion can be reduced by working only on the linear part of the amplifier's characteristic and by using negative feedback in the amplifiers.

(*ii*) *Frequency distortion.* This distortion occurs in the system, when different frequencies of the signal are not amplified equally.

To illustrate, we take the example of a Piano. Let its fundamental frequency is 4400 Hz, it will also produce 2nd, 3rd and 4th harmonic frequencies. The fourth harmonic of piano is 17600 Hz, which is the approximate upper limit of human ear. Although the intensity of these harmonics is much less than the fundamental and varies from instrument to instrument, their presence determines "quality" and "timbre" of different instruments. A Hi-Fi system must reproduce the full range of fundamental and harmonics of all musical instruments.

The Fig. 5.63 shows frequency response of human ear. The lower curve shown as "threshold of hearing" is a measure of the minimum sound intensity which can effect our ears and hence is audible. The upper curve "threshold of pain" is the sound intensity which will produce sensation of pains to the ears. The sounds in between the two curves are audible.

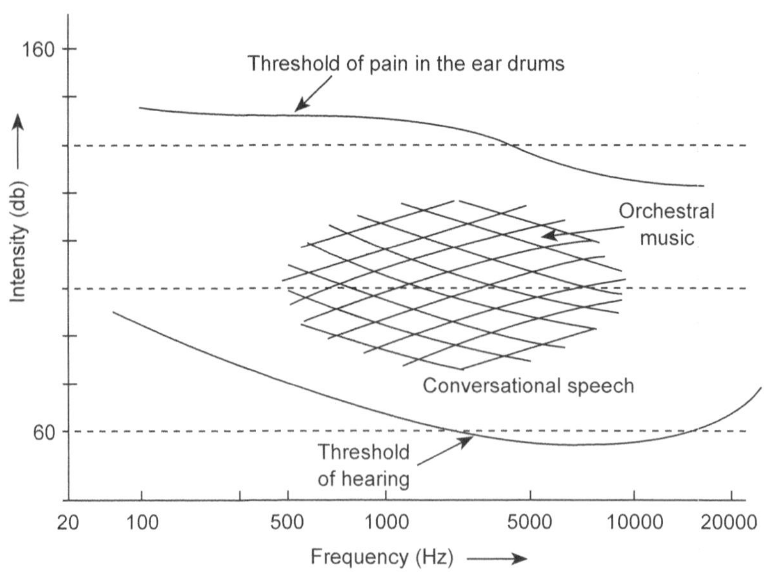

Fig. 5.63

Actually in frequency distortion, the frequency response curve does not remain flat for the output (Fig. 5.64 (*a*)) but is distorted (Fig. (*b*)). Low frequency response depends upon series capacitor and high frequency response depends upon shunt capacitor and series inductance of the system. The frequency distortion may be caused in amplifiers or moving coil microphones and also in loudspeakers due to their directivity pattern.

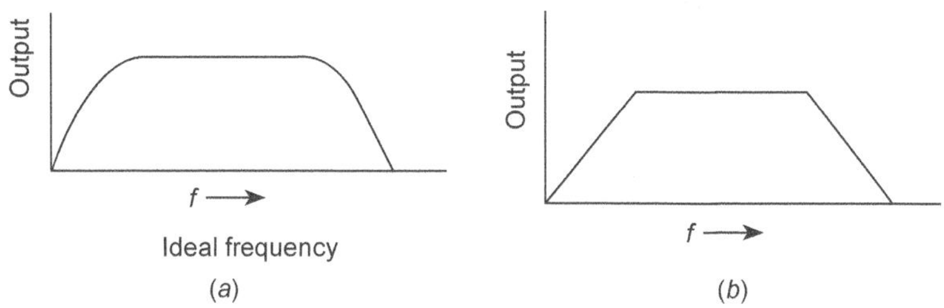

Fig. 5.64

For getting good frequency response, the value of reactance of the series capacitance and inductance should be low and value of the shunt capacitance should be high.

It may be recalled here, that the capacitance and inductance in an audio video system may be electrical or mechanical e.g., in loudspeakers, its mass acts as inductance, therefore cone of hifi speakers (tweeter) are made small.

(*iii*) *Phase distortion.* When different frequencies are reproduced in improper time w.r.t. to each other, phase distortion is said to be occurred. The reason of phase distortion is same as of the frequency distortion.

As already discussed, our ears cannot detect phase shift therefore phase distortion is not important in audio systems however, it is very important in video systems (i.e., T.V. reception). But, if the two sounds coming from two directions are completely (180°) out of phase, they will cancel each other and the listener will hear nothing. For this reason all the loudspeakers should be connected with same phase of the supply and at the same terminal. See Fig. 5.65

(*iv*) *Transient distortion or hang over.* When a system fails to follow sudden large changes in sound level, it is known as transient distortion. The transients so produced set the loudspeaker into vibrations at its natural frequency.

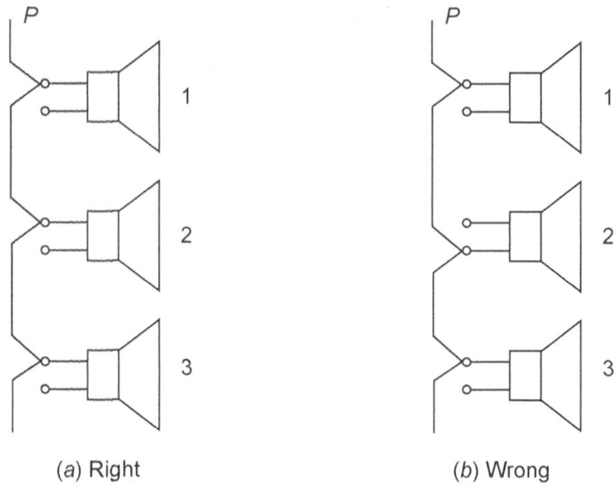

Fig. 5.65

(*v*) *Spatial distortion.* When a person gets the feeling of wrong location of a sound. This is spatial distortion. In other words, absence of stereophonic effect is known as spatial distortion.

Note:

(*i*) There is a difference between distortion and noise. A distortion is the deformation of the signal and is the internal matter, while noise comes from outside and mixes in the signal i.e. noise is an external matter.

(*ii*) All distortions can be reduced by using stereophony and providing a designed amount of negative feedback in the amplifiers.

(*iii*) Compact disc is the answer to all hifi problems.

(*iv*) For obtaining hifi output, distortion as well as noise both should be reduced.

The distortions can be reduced as discussed above. The electrical noise (humming, hissing, cross talk, vow and flutter, rumble etc.) may be reduced by using servo motors, and providing proper grounding. The environment noise can be reduced by using sound absorbers. Reverberation should also be controlled.

5.2.2 Components of Hi-Fi System

The quality of a reproduction system will depend upon many factors. The acoustics of the room and deficient (poor quality) electromechanical components (microphones, speakers) will also effect output of the system. However a Hi-Fi system at the minimum, should contain an input source such as record player, an audio amplifier and a loud-speaker. A more elaborate system as shown

(Fig 5.66) may contain more than one input source, separate pre-amplifier and a multiple speaker system at the output.

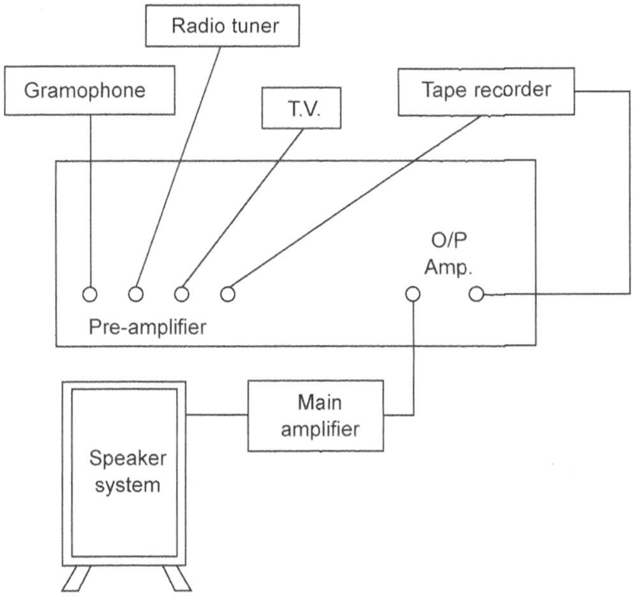

Fig. 5.66

The reason, why people prefer higher cost of separate components instead of a radiogram in a single cabinet is the better quality of each individual component—especially the amplifiers and speakers and also because of the absolute need for the speakers to be housed separately for high quality (Hi-Fi) sound reproduction.

Below a list is given of the components required for a Hi-Fi system:

1. Radio tuner
2. Record player-pick up
3. Preamplifiers
4. Main amplifier
5. Loud speakers
6. Tape-recorders
7. TV.

1. *Radio Tuner.* A ratio tuner is a radio receiver minus amplifier. It performs all the functions—tuning, demodulation etc., except the amplification. Since Hi-Fi system has its own amplifiers, hence it is not required in case of a tuner.

To be used as a hi-fi component, a tuner (AM or FM) must have high quality design features. The R.F., I.F. and detector circuit of a tuner must pass the entire audio modulation band impressed on the carrier. It should have a high selectivity *i.e.* able to select two overlapping stations. The sensitivity of a tuner should be of the order of μ-volts to pull even the weakest signal. A basic tuner is equipped with volume control, tuning, AM/FM selector and in special cases—treble and bass controls also. [See Fig. 5.67 (a)]

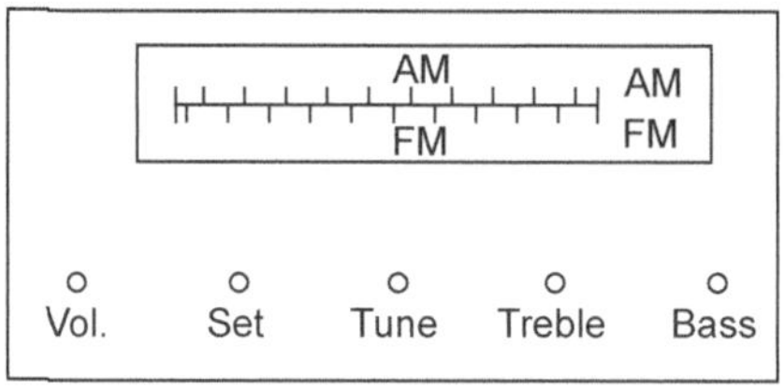

Fig. 5.67 (a)

2. *Record Player Pick ups.* The pick up, consisting of cartridge and stylus mounted at one end of the tone arm constitutes starting point of a Hi-Fi system. It should be able to faithfully translate the record "grooves" into corresponding original sound. [See Fig. 5.67 (b)]

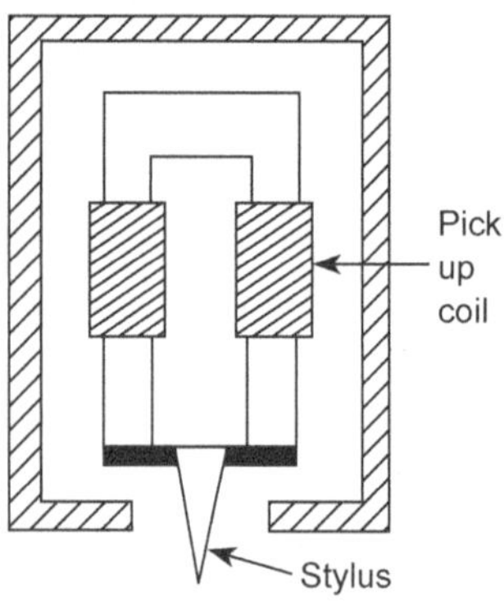

Fig. 5.67 (b)

Stereophonic pick up. A stereophonic recording on tape contain two physically separate channels of sound which are reproduced separately to create an illusion of solid sound. A stereophonic gramophone record also contains two separate channels of sound but instead of being in separate channels, both are recorded in the same groove. Thus a stereo pick up has to do a difficult task of converting it into two separate electrical signals.

The Fig. 5.68 (*a*) shows a monophonic (single channel) crystal pick up and (*b*) shows a stereophonic pick up. Two signals are amplified separately and fed to two loudspeakers respectively.

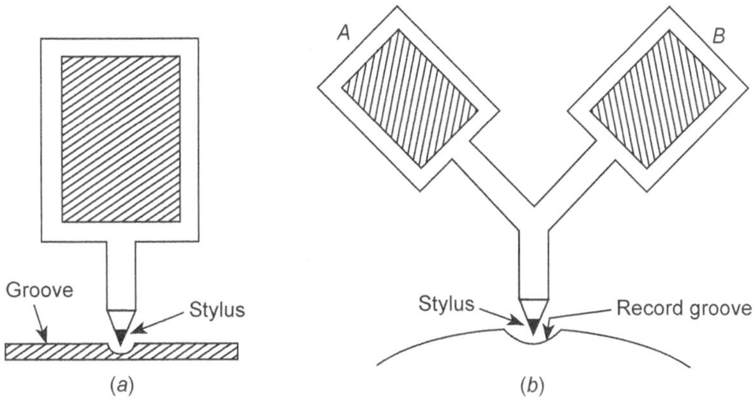

Fig. 5.68

3. *Pre-amplifiers.* The voltage output of a pick up is in millivolts, hence amplification is necessary. The additional amplification is achieved by using a pre-amplifier between pick up and the main amplifier.

   The Fig 5.69 shows a typical hi-fi pre-amplifier. The selector switch enables to select the particular audio device, whose output is to be amplified e.g. it can select mike, tuner, T.V., recorder etc.

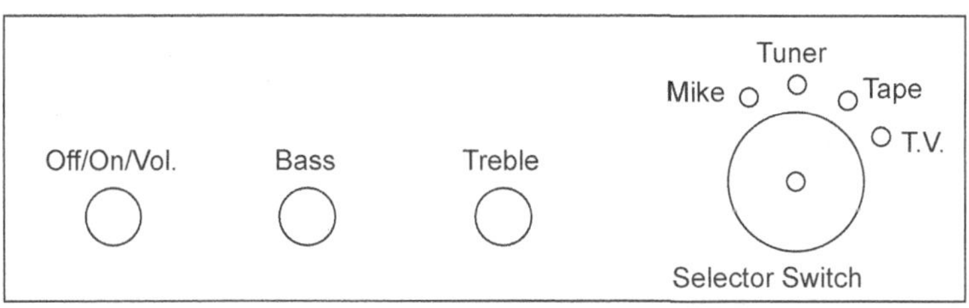

Fig. 5.69

   The ON/OFF control also serves the purpose of volume control. It also has bass/treble controls which permit boosting or reducing the low and high frequency response respectively.

4. *Main amplifier.* The basic function of main amplifier is to strengthen the output of pre-amplifier to sufficient level so as to drive the loudspeaker. A stereophonic amplifier is made up of two separate amplifiers. Both amplifiers-one for each channel are fed from same power supply—Its other component is balance control—a potentiometer which governs

the ratio of voltage amplification between two channels. The output of pre-amplifier is the input of the main amplifier. (See Fig. 5.70)

Fig. 5.70

5. *Loudspeaker.* Loudspeaker is the most important component of a hi-fi system. It is the last link of a system and converts electrical signal into the original sound signal.

   The Fig. 5.71 shows the basic moving coil (or dynamic) speaker. It carries a voice coil in the magnetic field of a permanent magnet. The coil takes signal currents from the output of the amplifier. At one end of the coil, there is "cone" made of paper or cloth which radiates sound. The spider enables forward and backward motion of the entire "cone and voice" assembly.

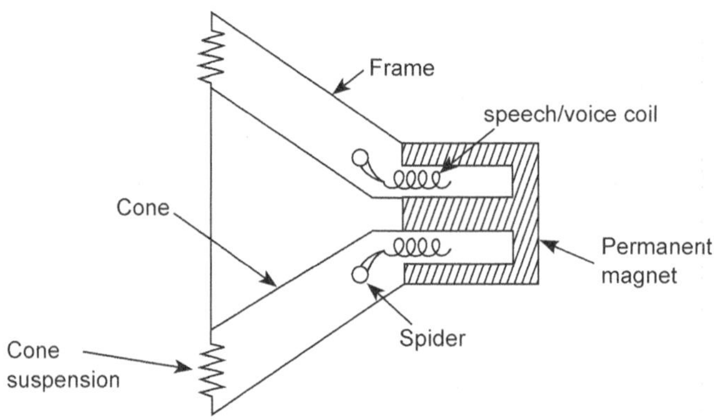

Fig. 5.71

The magnetic field produced by voice coil is at right angles to the permanent field. The interaction of the two fields make the voice and cone assembly to move forward and backward, naturally their movement will be governed by the signal strength.

(*a*) **Co-axial and Tri-axial speakers**

A single speaker cannot cover the whole A.F. range. To do this, either separate speakers for separate frequency range may be employed and then interconnected or to mount separate speakers in a single unit co-axially. The Fig. 5.72 shows a co-axial speaker unit in which larger speaker covers the L.F. *i.e.* acts as a woofer and smaller one covers the HF *i.e.*, acts as a tweeter.

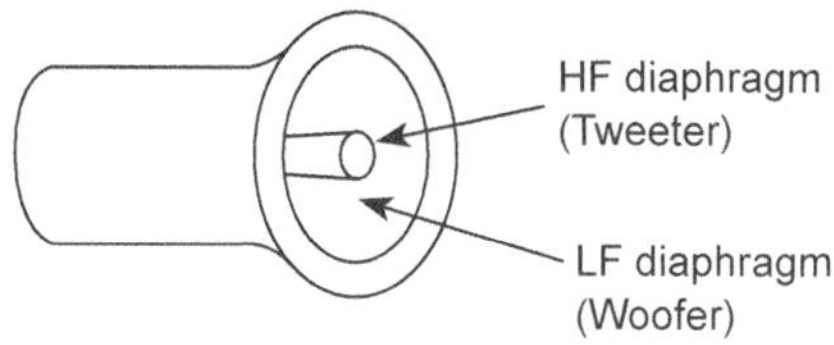

Fig. 5.72

The Fig. 5.73 shows a tri-axial speaker unit in which three speakers have been mounted in a single unit. Thus the whole audio frequency range-low, medium and high is covered by a single speaker.

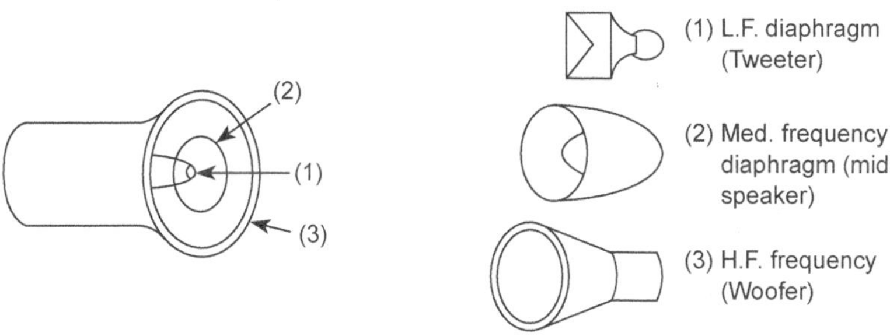

Fig. 5.73

(*b*) **Speaker systems**

To cover the A.F. range, speaker systems may be employed consisting of separate speakers connected through a network. Two systems are more popular:

1. *Two way system.* In this system, a woofer and a tweeter are connected through a high pass filter network. The network limits the frequency range for the speakers to avoid their overloading or underloading. For example woofer may be given to handle frequencies from 20 Hz to 4000 Hz and above this range may be passed to the tweeter. In this way, whole audio range may be covered. (See Fig. 5.74)

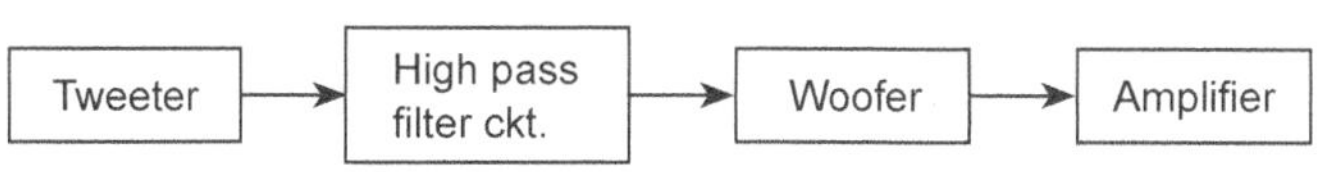

Fig. 5.74

2. *Three way system.* For better division of frequencies, three way system has been designed. The AF may be divided into low, medium and high range and can be given to woofer, medium

and tweeter respectively. This will further improve fidelity of the system. (See Fig. 5.75)

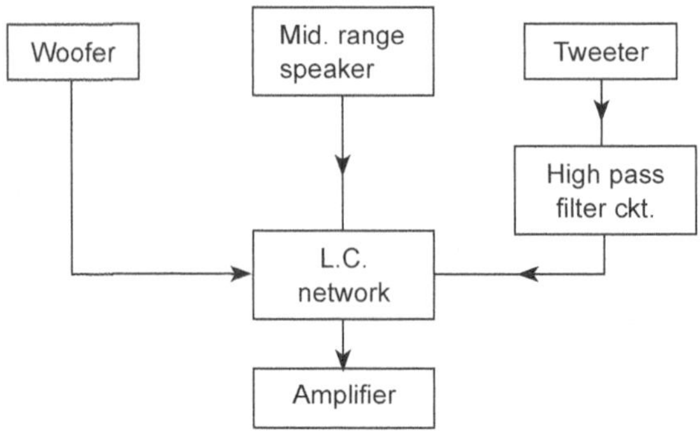

Fig. 5.75

(*c*) **Loudspeaker enclosures or Baffles**

The loudspeaker enclosure or frame also helps in maintaining quality of a loudspeaker. The importance and role of the enclosure may be experienced by holding a best hi-fi speaker (without enclosure) in hand.

What actually the enclosure (or baffle as it is called) does is that it acts as an "obstacle" around the speaker and prevents the sound waves of front and back side of the speaker cone to "meet", so that they may not cancel each other. Basically it increases the length of sound-path from front to the rear of a speaker.

The Fig. 5.76 shows the different types of baffles used for hi-fi speakers.

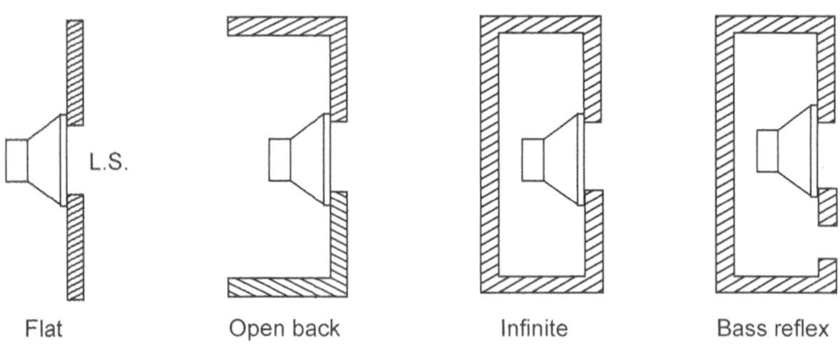

Fig. 5.76

5.3 STEREOPHONY

The stereophony is made up of two terms stereo + phone. Stereo means solid (*i.e.,* 3 dimensional) and phone means sound, *i.e.,* stereophony means "3 dimensional sound". Stereophony gives original sound effects and we can feel its *depth* as well as *direction*. The human system of hearing is also stereophonic.

5.3.1 Stereo-Phonic Reproduction

Stereo-phonic (Stereo means two) reproduction is based on the fact that we have two ears. If we close one ear for some time, we can realise the importance of two ears. One ear cannot preserve the depth of sound. The stereo requires broadcasting, recording and reproduction over two separate channels to satisfy our hearing by two ears. Stereo also satisfies our "two eared three dimensional" hearing. The mono-phonic reproduction cannot differentiate between left and right, up and down or front and back sounds.

Actually the stereo system involves separation of low and high frequencies. Human voice has low frequency, while the musical instruments have high frequency. The system records, amplifies and reproduces them on separate channels. Each channel may have one or more speakers. We can illustrate the principle of stereo through the following explanation:

Let a person is sitting in a room. The sound on left side reaches first to his left ear and after some time to the second ear. Similarly the sound from his right side will first reach to his right ear and then to his left ear. Naturally all this will create a confusion.

If, as shown in Fig. 5.77 two separate channels are created for left and right sides' sounds and more over they are separately produced, amplified and passed through separate speakers such that left ear covers sound coming from left side and right ear covers sound coming from right side, it will give better effects on ears and the quality and timbre of the sound will also be maintained.

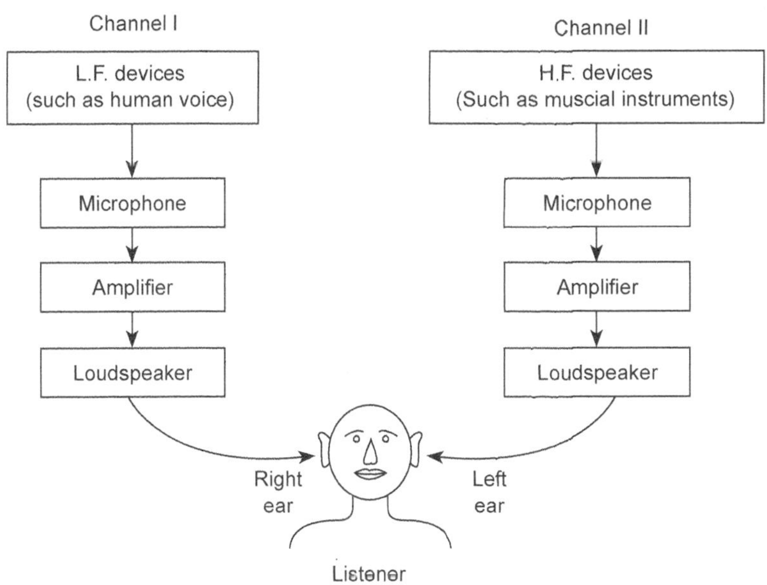

Fig. 5.77

We can also take example of a public address system. If on a stage there is only one mike and the various signals of all the sound sources are fed to the same mike, then the signal of the sources nearest to this mike will be stronger than other sources which are distant from the mike. For example an orchestra is going on the stage, the singer is singing before the mike and his group is playing different musical instruments (like flute, clarionet, harmonium etc.) at various distances from the mike. The output of the singer will be strongest and of other instruments will be weaker and weaker in accordance to their distance from the mike. The public cannot enjoy the program fully. This is the monophonic production of sound. (See Fig. 5.78)

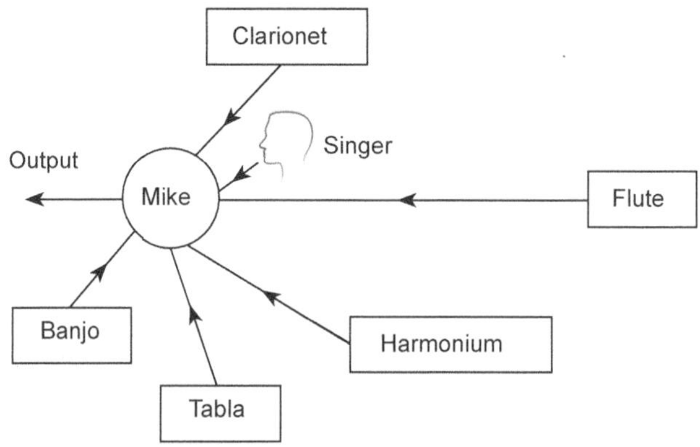

Fig. 5.78

Now if at least two microphones may be arranged and all the sound sources are divided into two separate channels as shown in Fig. 5.79 a better sound effect will be produced and the public can enjoy the actual "depth" of the program. This explains the principle of stereophonic sounds.

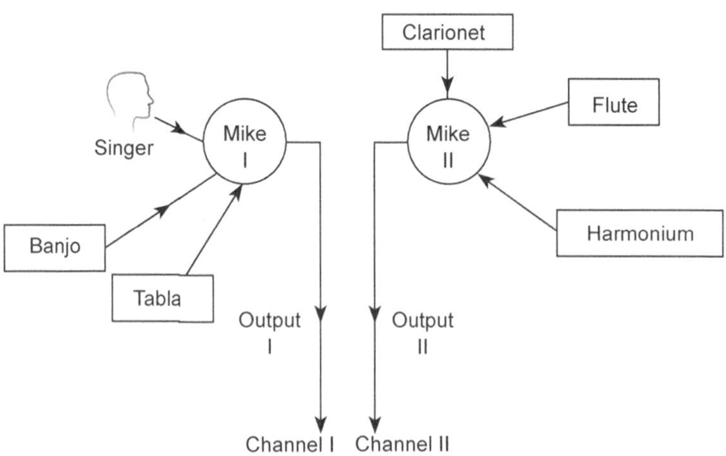

Fig. 5.79

The channels are required to be kept separate upto loudspeakers. If at any stage, the two channels are mixed together—stereophonic sounds will cease to exist. In other words, the two channels should remain separate throughout from reproduction to transmission and upto reception. The Fig. 5.80 shows the transmission system of stereophonic sounds in a block diagram.

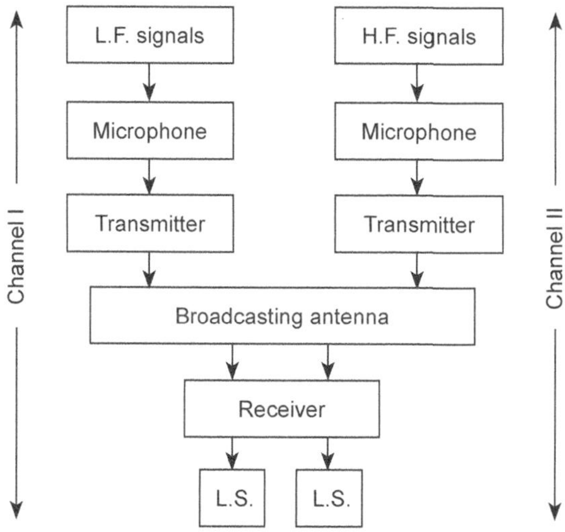

Fig. 5.80

5.3.2 Stereo Head R/P (Replay/Play)

To keep the channels separate, the amplifier of a stereo tape recorder has also two channels. The stereo head (record and play) is a "Double track" head as it is generally called. The head has two separate cores and two air gaps. It has two separate coils giving out 4 terminals. Thus a stereo head has 4 pins whereas ordinary tape recorder head has 2 pins. (See Fig. 5.81)

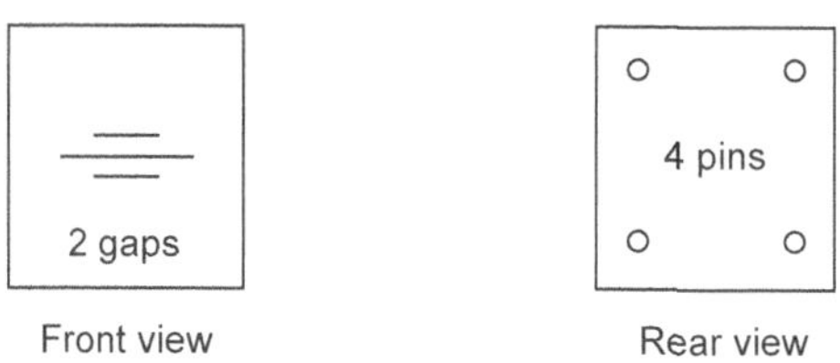

Fig. 5.81

5.3.3 Stereo Tape

The ordinary cassette tape may be used for stereo recording. As the stereo head has 2 gaps. There are 4 tracks on the tape two on one side and two on the other side. There is a guard band between two tracks and also between two channels. See Fig. 5.82

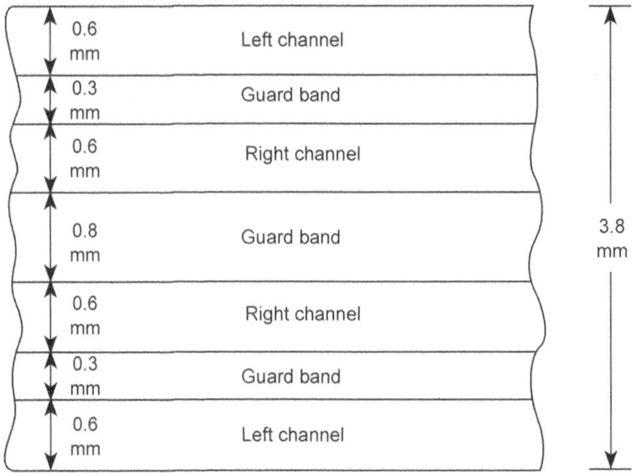

Fig. 5.82

5.3.4 Block Diagram of a Stereo (Fig. 5.83)

(*i*) *Stereo in recording mode.* The blocks working in recording mode are shown in Fig. 5.83. Each channel has separate mike. Their audio signals are separately amplified and fed to the stereo heads. As both channels get one H.T. (high tension) line, one "Bias oscillator" has been shown. To prevent mixing of the two channels, a "channels separating network" comprising of resistors and capacitors is used.

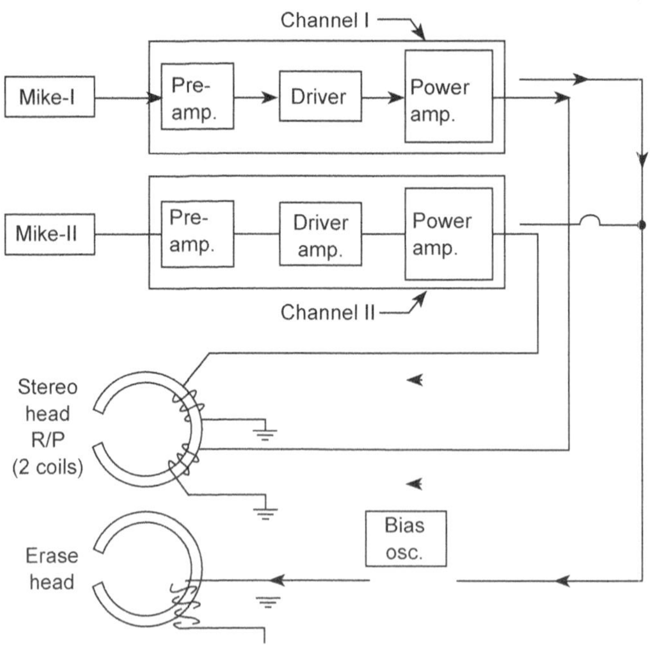

Fig. 5.83

(*ii*) *Stereo in play Mode (Fig. 5.84).* In play mode, mike, erase head and bias oscillator have no function to perform and hence thrown out of the circuit, they have not been shown. From R/P head, signals of both the channels are fed to separate amplifiers and then given to the loudspeakers separately. The separating network is used as in the record mode.

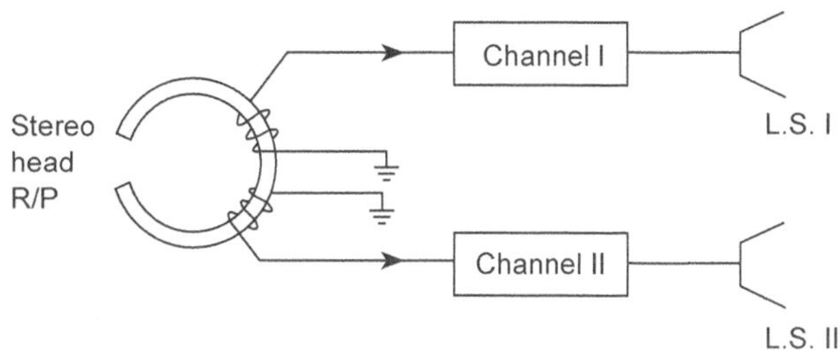

Fig. 5.84

5.3.5 Position of Speakers in Stereophony

For getting good stereo-phonic sound effects, high quality hi-fi speakers are used. They are generally Dual Cone type or Column speakers.

The position of speakers and their distance from the listeners is also an important factor. The distance between the two channel speakers should not be less than 2 meters and distance of the listener (or listeners) from the speaker should not be less than 3 meters. If the listener is equidistant from the speakers, best stereo-phonic effects can be obtained. (See Fig. 5.85)

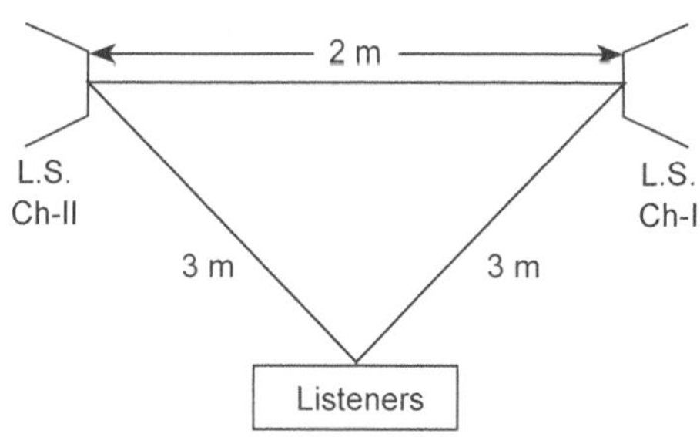

Fig. 5.85

For best stereophonic effects, the loudspeakers should be positioned in the corners of the room. Moreover these loudspeakers should be placed at 90° to each other and the area between the speakers should be non-reflecting. Curtains or sound absorbers may be provided in the room on all sides.

If the loudspeakers are concealed by using a net curtain, the stereophonic effect is further improved. See Fig. 5.86.

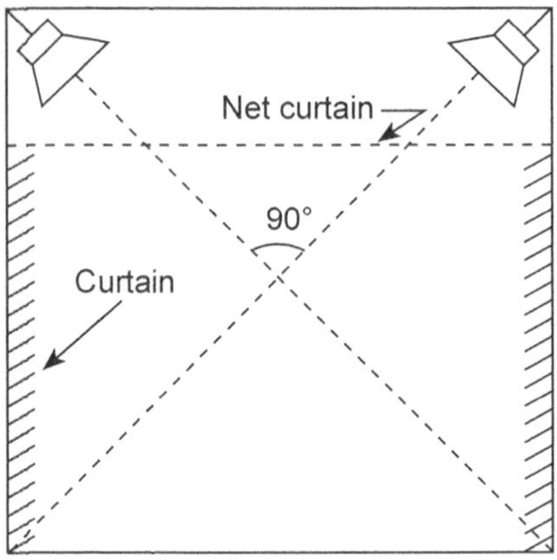

Fig. 5.86

5.3.6 Commercial Method of Stereophonic Recording (Disc-Westrex Method or 45° Method)

Stereophonic recording requires recording of the signals in synchronism to each other on a single medium. For this, stereophonic cutter stylus unit (See Fig. 5.87) is used which records two separate channels in the same groove. The two signals are amplified and fed to two loudspeakers separately during reproduction.

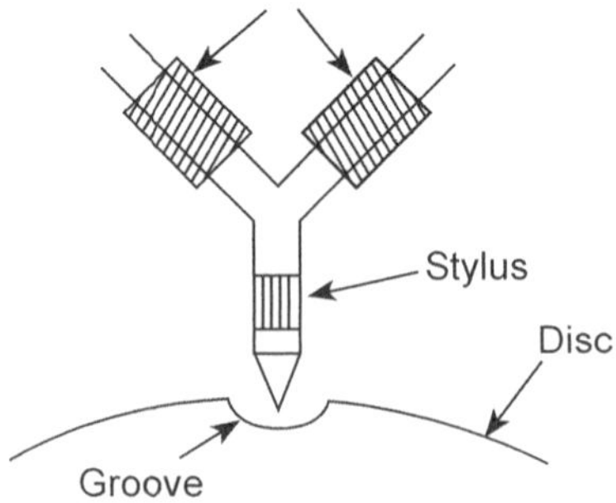

Fig. 5.87

Each channel is recorded in the direction of 45° with the horizontal, clearly the two channels will be at 90° to each other and hence will not interfere. See Fig. 5.88

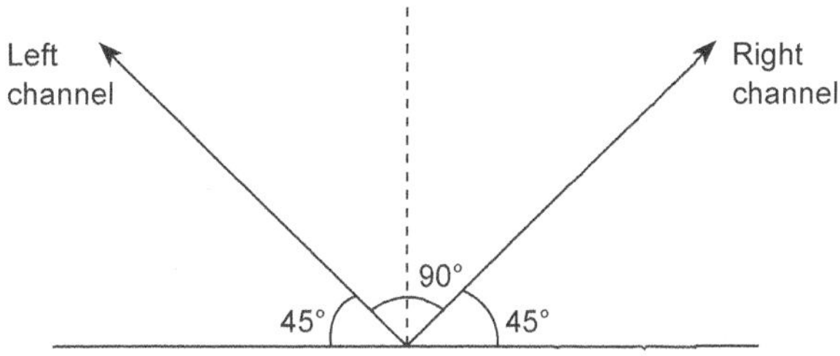

Fig. 5.88

(*i*) When the signal currents in the two coils of the cutter stylus are equal, the cutter will make vertical grooves. (The resultant of two equal forces at 90° is given by the diagonal of the square so formed). See Fig. 5.89.

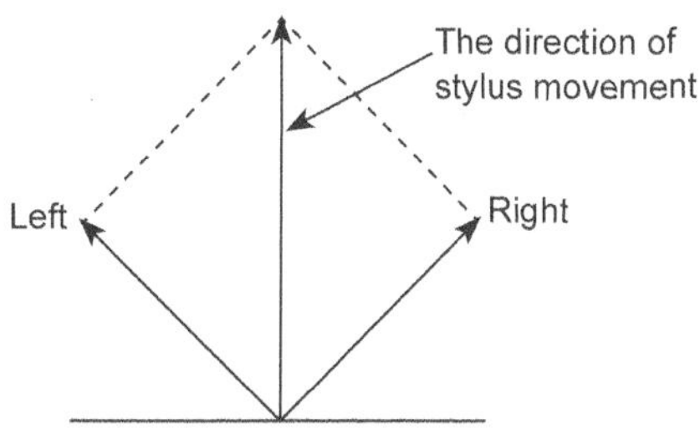

Fig. 5.89

(*ii*) If the signal currents in the two coils are unequal:

(*a*) If current in the left is more, the stylus will tilt towards left [Fig. 5.90 (*a*)].

(*b*) If the current in the right coil is more, the stylus will tilt to the right [Fig. 5.90 (*b*)].

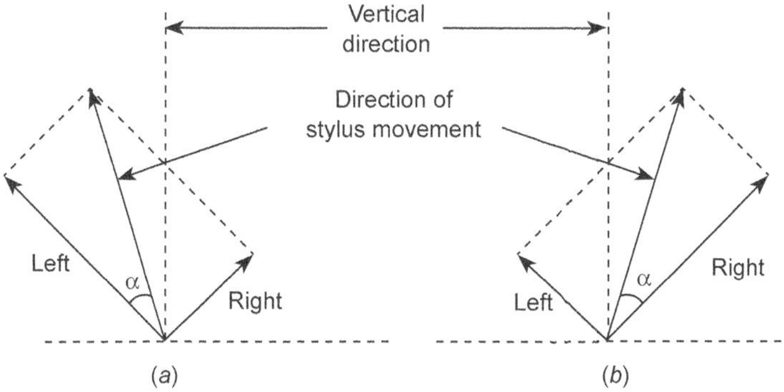

Fig. 5.90

Motion of the stylus is mostly vertical, as the both signal currents are almost equal.

SUMMARY

1. The sound recording may be done by any of the following methods: Disc recording, magnetic recording, optical recording and Digital recording.
2. In magnetic recording, electromagnets are used as record head, play head and erase head.
3. The magnetic recording is done on a tape and disc recording is done on a disc.
4. The optical recording is done on a film, which can be reproduced by a film projector.
5. Digital recording is done on compact disc, usually it is done in Binary system.
6. The data is stored in floppy, hard disc etc. These are called a data storage devices.
7. When the reproduced sound is exactly similar to the original sound, This is called a Hi Fi system.
8. The stereophonic sound is two dimensional. It preserves the original "depth" of the sound.
9. In streophony, the different frequencies are amplified separately.
10. Westrex method is a commercial method of stereophonic recording.

SELF TEST QUESTIONS

1. What are the various functions performed by an ACR?
2. Write a short note on ACR heads.
3. Describe various erasing methods.
4. What do you understand by "Bass" and "Treble" controls?
5. Draw block diagram of an ACR and explain each block in brief.
6. Explain the need of biasing. How dc and ac biasings are provided.
7. What is the material used for making magnetic tapes?
8. Explain the tape transport mechanism.
9. What is the function of capstan and pinch roller?
10. Explain Dolby system of noise reduction.

11. What are Vow, Flutter and Rumble noises? How these can be reduced?
12. Compare disk and magnetic recordings.
13. What are the 3 basic elements of magnetic recording?
14. Write a brief note on care of recorder heads.
15. Explain optical recording method.
16. Explain how sound is recorded on a film and reproduced?
17. Explain why the sound track runs 35 cm advance of the corresponding picture track?
18. Why a picture frame is projected twice by a projection?
19. Write a brief note on 'hi-fi'.
20. What are the essentials of a hi-fi system?
21. Comment "the compact disk is the answer to all hi-fi problems.
22. How is stereophony different to that of monophony?
23. Explain the stereophonic character of our ears.
24. Explain 45° method of stereophonic recording on a disk.
25. Explain with the help of a block diagram a hi-fi stereo recording and reproduction.
26. What are the hi-fi equipment? Explain positioning of loud speakers to achieve stereophonic effects.
27. What is stereo effect? How is it achieved in an audio system?
28. Differentiate between a hi-fi system and a stereo system.
29. What are the various controls provided on a stereo amplifier?

OBJECTIVE TYPE QUESTIONS

Fill in the Blanks:

1. A stylus tip is made of ________.
2. The disk speed is ________.
3. The monophonic disks have ________ grooves.
4. Length of tape magnetised by one cycle of signal is equal to its ________.
5. If tape speed is high, the duration of tape will be ________.
6. Standard tape speed is ________.
7. The magnetic tape has a coating of ________.

8. Commercially a record head and a play head is ________.
9. Erase head ________ the record head.
10. For tape transport ________ motors are used.
11. Vow and Flutter are produced due to variation in ________.
12. Rumble is produced due to ________.
13. Dolby method reduces ________.
14. Duration of a C-90 tape is ________.
15. The cutter stylus should be ________ in weight.
16. The width of a standard audio tape is ________.
17. Biasing frequency is kept ________ than the audio frequency.
18. Vow and Flutter are caused due to change in ________.
19. A monophonic head has ________ coil.
20. A stereophonic head has ________.
21. Bass control is used for boosting ________ frequencies.
22. Treble control is used for boosting ________ frequencies.
23. At lower tape speed ________ frequencies are lost.
24. At higher tape speeds ________ frequencies are lost.
25. Dolby system expands signals above ________ hertz.
26. Width of sound track on motion film is ________.
27. A photo cell converts light into ________ signal.
28. A picture frame is projected twice on the screen to reduce ________.
29. In a hi-fi system S/N ratio is more than ________.
30. Multispeaker system improves ________.
31. The CD is totally free of ________.
32. Stereophony gives a sense of ________ of sound.
33. Our hearing system is ________,
34. In stereophonic system, there should be atleast ________ channels.
35. The other name for 45° method of disk recording is ________.
36. When we cannot detect the direction of sound, This is a ________ distortion.
37. The speed of a stereo tape is ________.

38. A greater modulation percentage can be achieved in optical recording, if we use variable ________ method.
39. For getting sufficient acoustic volume in larger cinema hall, the acoustic power of ________ watt is needed.
40. The transducer used in film projector is a ________.
41. Non-uniform motion of the film will produce ________.
42. The human eye has a persistence of vision equal to ________ second.

ANSWERS

1. diamond	2. 78 RPM	3. lateral
4. wave length	5. less	6. 38 cm/s
7. Ferromagnetic material		8. same
9. preceeds	10. synchronous	11 speed of motor
12. vibrations of motor		13. Noise level
14. 90 minutes	15. light	16. 1/4 inch
17. more	18. tape speed	19. one
20. two	21. low	22. high
23. high	24. low	25. 500
26. 2.5 mm	27. electrical	28. flickering
29. 50 dB	30. fidelity	31. Vow and flutter
32. direction	33. stereophonic	34. Two
35. westrex method		36. spatial
37. 14.7 cm/s	38. Area	39. 30
40. Photo cell	41. Vow and Flutter	
42. 1/12.		

❑❑❑

6

Video Engineering VCR, VCD and DVD

In this chapter, we will discuss about Video Engineering/Video Recording, VCR, VCD and DVD.

6.1 VIDEO ENGINEERING : VIDEO RECORDING

The basic principle of recording (audio as well as video) is same. At first the audio/video signal is converted into electrical signal. This electric current is given into the coil of the recording head (See Fig. 6.1). The tape is moved at a calculated speed.

When the video signal currents flow through the coil of the head, magnetic flux is produced in the head. When the head is pressed on the tape, the flux passes to the tape through the high reluctance head gap. As the head moves with a constant speed, the video signal is converted into *magnetism* on the tape. The variations of the magnetism on the tape is just according to the strength of the changing video signal.

Later on, while reproduction, this magnetism induces a varying voltage (i.e. current) in the coil of the *replay head*, which is processed into original video signal.

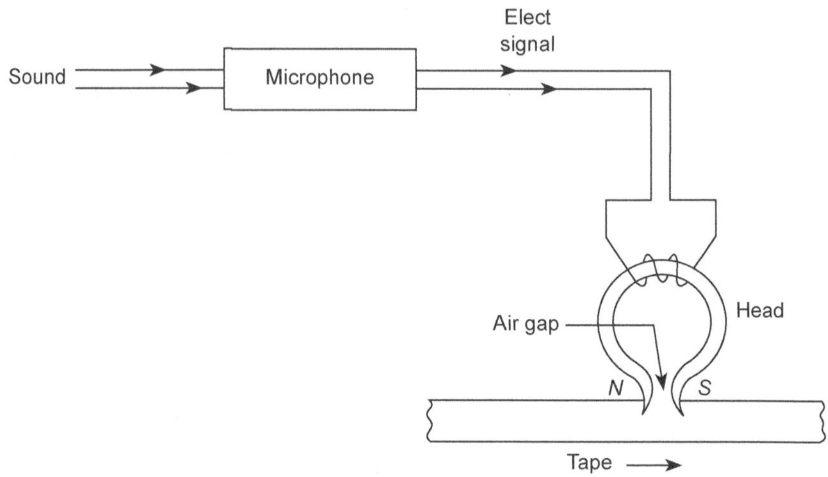

Fig. 6.1

6.1.1 Need of rotating video head

In audio recording, the head remains stationary below we explain why this system cannot work in video recording.

In case of audio recording, the speed of audio tape is fixed approximately at 4.6 cm per second (while the head does not move); that is if frequency of the signal to be recorded is assumed to be "one hertz per second", in one second, when the record head will give magnetic effect of 1 Hz, it will cover a distance of 4.6 cm on the tape.

It means in +ve cycle of 1 Hz, that is in 2.3 cm length of tape, North pole will be set up and in the next 2.3 cm, South pole will be set up. Such that length of the tape covered in one cycle is 4.6 cm, In other words, the wavelength of the recorded signal is 4.6 cm.

This can be shown by using the basic formulae

$$v = n\lambda$$

where v = Speed of the tape = 4.6 cm/sec

f = frequency = 1 Hz

λ = wavelength = ?

$$\lambda = \frac{v}{f} = \frac{4.6}{1} = 4.6 \text{ cm. (Refer Fig. 6.2)}$$

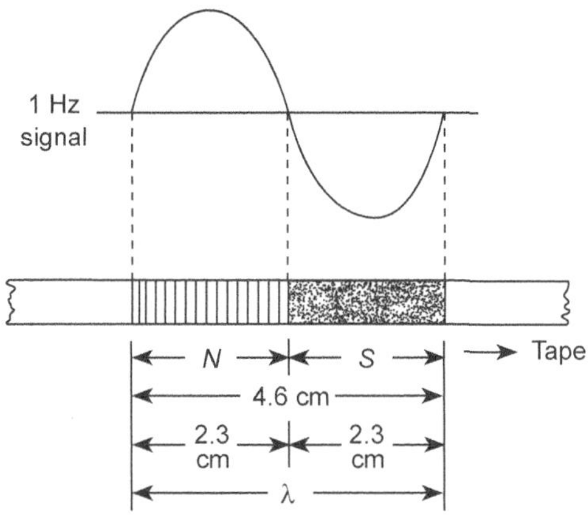

Fig. 6.2

Now, if the frequency of the recorded signal is 20 Hz and the tape speed is same *i.e.* 4.6 cm/s.

In this case, the wavelength $\lambda = \dfrac{4.6}{20} = 0.23$ cm.

This shows that by increasing the signal frequency, the wave length will be proportionately reduced (See Fig. 6.3). If the signal frequency is further increased, the wave length becomes equal to the "air gap" of the head.

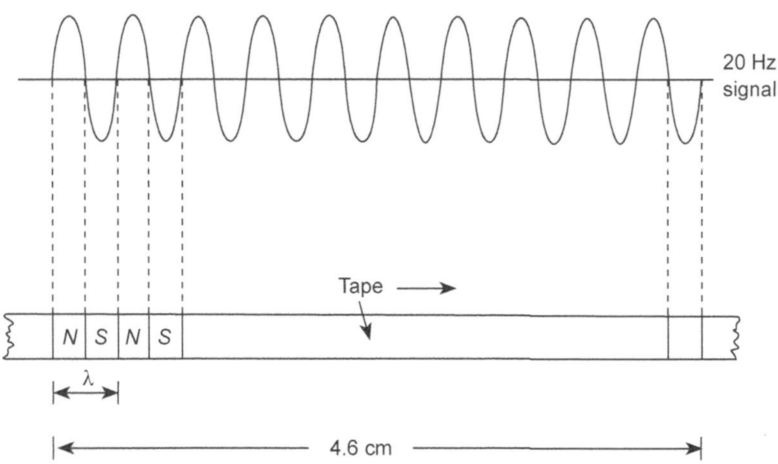

Fig. 6.3

As there is no magnetic effect in the air gap of the head, the replay of such a signal will be defective and full of distortions. Similarly, the wave length reduces further and if it becomes less than the air gap, again the replay will be defective. If one pole of the head is under the effect of N Pole of the signal cycle and the second pole of the head is under the effect of S pole of the signal cycle, in this case on replay also, original recorded signal will not be obtained.

The above discussion reveals that in video recording, wavelength of the recorded signal should be more than the air gap of the recorded head.

The air gap of the record and play head in case of audio recording is about 0.0003 cm and if the tape speed is assumed to be 4.6cm/s and assuming head gap equal to the wave length, the frequency of the recorded signal will be,

$$f = \frac{v}{\lambda}$$

$$= \frac{4.6 \text{ cm/s}}{0.0003 \text{ cm}}$$

$$= 15000 \text{ Hz}$$

$$= 15 \text{ kHz.}$$

The above calculations show that with the condition of "Head gap equal to wavelength", we can faithfully record (or play) a signal less than 15 kHz.

However, it should be mentioned here that if the wave length is "one and a half times" of the head gap, best results are obtained in recording as well as in replaying (See Fig. 6.4).

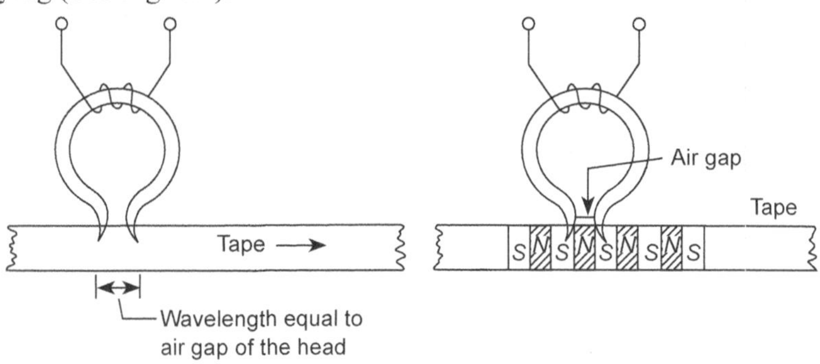

Fig. 6.4

Therefore in order to record and play higher frequency signals of video signals either head gap is to be reduced or tape speed may be increased to maintain wave length more than the head gap. Increasing the tape speed will mean that the tape (during recording and play) in a very short time would be unwound from one "spool" and will be rewound on the other spool—this will reduce the length of the tape and if we want the tape speed to be normal (4.6 cm/s), we will have to reduce the head gap.

Keeping the tape speed at 4.6 cm per second and head gap 0.0003 cm, we can record a signal with maximum frequency of 15 kHz. But video signals are recorded in *Megahertz*, hence the same cannot be used in video recording. Moreover, we cannot increase the tap speed beyond a limit, *hence for a video recording*:

(*i*) Tape speed is increased to 5 meters per second, while this is 4.6 meters per second in case of audio recording and

(*ii*) The air gap of the video head is reduced to 0.00006 cm, which is about one fifth of the audio head.

Now we will calculate the frequency of the video signal which can be recorded at these conditions,

$$f = \frac{V}{\lambda}$$

$$= \frac{5 \text{ m/s}}{0.00006 \text{ cm}}$$

$$= 8 \times 10^6 \text{ Hz}$$

$$= 8 \text{ Mega hertz}$$

In this calculation, we have taken the tape speed as 5 m/s which is not possible as at this speed, in a very short time, the tape will be unwound from one spool to the other, this will reduce the length of tape and duration of the cassette. Hence pratically we donot increase "tape speed" but increase its "Effective speed" by adopting a special technique. This effective speed of the tape is known as the "*writing speed*".

To obtain *writing speed* of the tape in video recording, we rotate the video head at a determined speed, the tape speed keeping at 4.6 cm/s. In video recording/replay along with the tape the video head also moves with the help of a small motor, this increases the "effective or relative speed" of the tape.

Table 6.1: Comparison between audio and video recording.

Audio Recording	Video Recording
1. Only tape moves at approx. speed of 4.6 cm/s. The heads remain stationary.	The heads also move along with the tape.
2. The audio recording is done at lower signal frequency (which is in kHz).	The video recording is done at higher signal frequencies (in MHz).
3. Modulation is not needed during recording.	Frequency modulation is used for recording.
4. The shape of tracks is longitudinal.	The shape of tracks is slant.

Audio Recording	Video Recording
5. No. of audio signal on the tape is only one.	No. of video signal on the tape are four (audio, luminance, colour and guard).
6. The width of tape is 1/4".	The width of tape is 1/2".
7. Highest frequency is 15 kHz.	The highest frequency is 5 MHz.
8. The head gap is 3 μm.	The head gap is 0.3 μm.

6.1.2 Video Heads

The rotating video head has a special construction. It is a drum type structure, on whose periphery there are two air gaps. During recording, the tape is kept horizontal and the head in slant direction so that, at first the tape passes over the upper part of the head and later over the lower part of head. This is the reason that we get "inclined or slanting" video tracks on the tape.

After the signal is recorded by the first head gap on the tape, and as this gap moves clockwise, the second head gap from the other side comes in contact with the tape. But during this period as the tape advances new signal is recorded on the new track formed on the tape. The rotating video head and the formation of video tracks is shown in Fig. 6.5 (*a*) and (*b*) respectively.

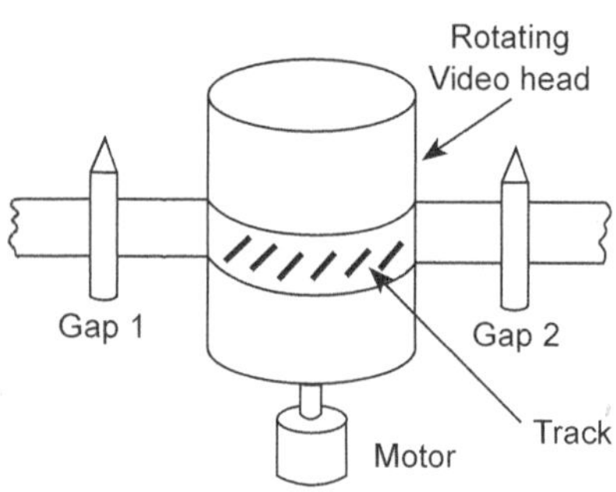

(*a*)

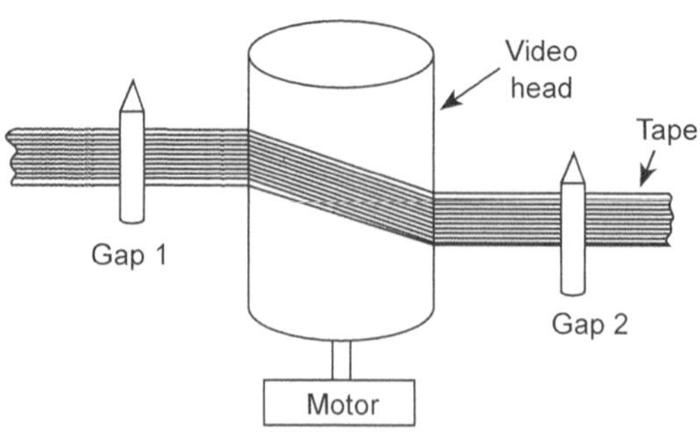

(*b*)

Fig. 6.5

The time taken by a head gap in forming a track will depend on the speed of the video head. Generally video head travels by "half cycle" in formation of one track. We should therefore rotate the video head at a speed that one head gap takes 1/50 second in forming one track. In other words we should rotate the video head at a speed of 25 cycles per second. The theory behind this is "scanning" which is a basic principle of video transmission.

6.1.3 About Video Tape

(a) Calculations on Video Tape

(*i*) *Total length of video tape.* The video signal is recorded on a tape in the form of slant video tracks. So the total length of the recorded program is not the length of tape but length of all the tracks formed on the tape. If length of one track is l and the total number of tracks is N. The length of recorded program will be $N \times l$. Now we will calculate the length of the tape in a 3 hrs cassette (See Fig. 6.6)

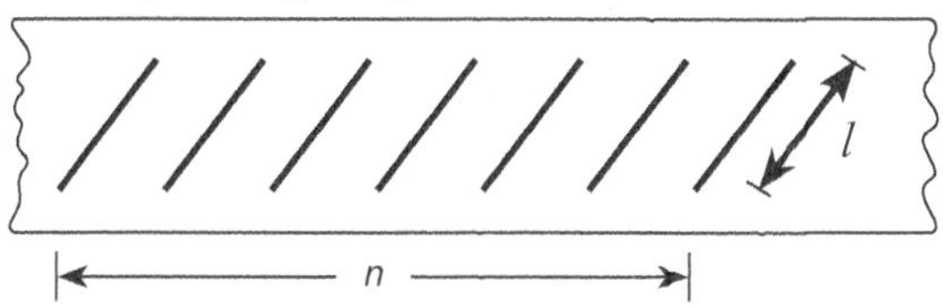

Fig. 6.6

In 1 sec, tape moves = 2.3 cm

In 1 hr, tape moves = 2.3 × 3600 cm

In 3 hrs, tape moves = 2.3 × 3600 × 3 cm

= 260 meters (approx).

(*ii*) *Number of tracks.* Keeping in mind that tape speed as 2.3 cm/s and one head gap takes 1/50 seconds to travel half periphery on the video head, we can calculate the total number of tracks on a 3 hrs cassette.

In 1 sec, tape moves = 2.3 cm

In 1/50 sec, tape moves $= 2.3 \times \frac{1}{50}$ cm = 0.048 cm

In other words, the length of 1 track will be 0.048 cm.

As we have already calculated that 3 hrs cassette contains 260 m (260 × 100 cm) tape.

Hence, Total no. of tracks $= \frac{260 \times 100}{0.048} = 541666$ approx.

(*iii*) *Length of one track.* (Fig. 6.7) As already told, video tracks are slanting on the tape. These tracks make an angle of 6° with horizontal.

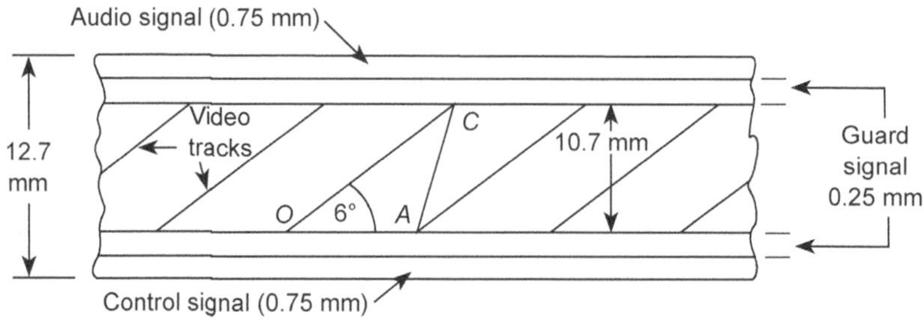

Fig. 6.7

The width of video tape is 12.7 mm and the video tracks are formed in a width 10.7 mm. The remaining part of the tape carries "audio, guard and control tracks" as shown in the Fig. 6.7.

In right triangle AOC (See Fig. 6.8)

Length of the video track = OC

Width of the video track = AB.

$$\angle AOC = 6°$$

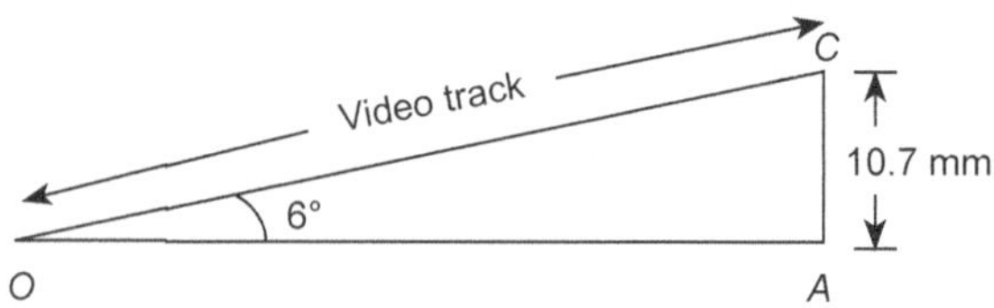

Fig. 6.8

$$\text{Now,}\sin \angle AOC = \frac{AC}{OC}$$

$$OC = AC \times \sin \times \angle AOC$$

$$= AC \times \sin 6°$$

$$= 10.7 \times 0.1045 = 10 \text{ cm (approx).}$$

(*d*) *Total length of tracks.* We have calculated already that total number of tracks on a 3 hrs tape is 541666. We have also calculated that each track is approx. of 10 cm length. Hence total length of the tracks.

$$= 541666 \times 10 \text{ cm}$$

$$= 54 \text{ kilometers.}$$

Note. (*i*) Control track controls the motor speed during replay.

(*ii*) There is a small gap between control track and video track. It does not allow the intermixing of audio and video signals during replay.

(b) Manufacturing of Video Tape: Generally, two polymers are used to prepare the tape. They are *Myler* and *Cellulose accetate*. Their tensile strength is more and are unaffected due to heat, temperature and dampness. They are made in the form of thin ribbons. The thinner tape can work on higher frequencies, i.e., thinner the tape, higher is the frequency. The video recording is done on megahertz frequencies.

The ferromagnetic materials which are used are ferric oxide and Chromium oxide. The thickness of myler is 25 μm and thickness of magnetic material layer is 10 μm. At first with the help of some adhesive, their paste is formed and is worked upon the tape in a thin layer. The process of applying layer has been shown in Fig. 6.9. The material is filled in a tank. Rollers one on other are moved through the tank and through the rollers, it reaches to the tape, where it is applied in layers.

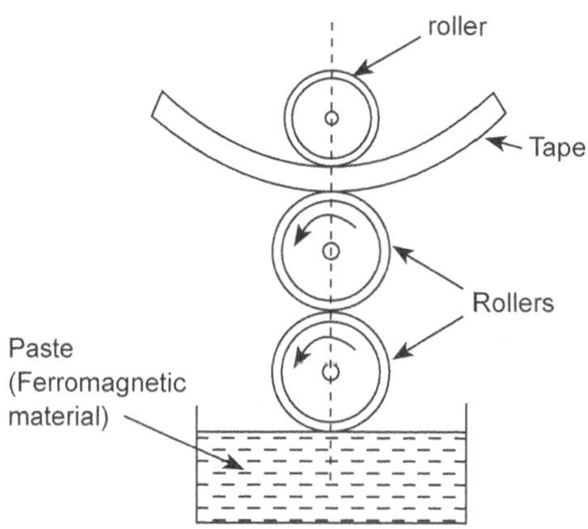

Fig. 6.9

The Ferro-magnetic "dipoles" initially are "at random" but during recording when magnetic field is applied due to signal currents, they are oriented in the direction of signal currents. See Fig. 6.10 (*a*) & (*b*).

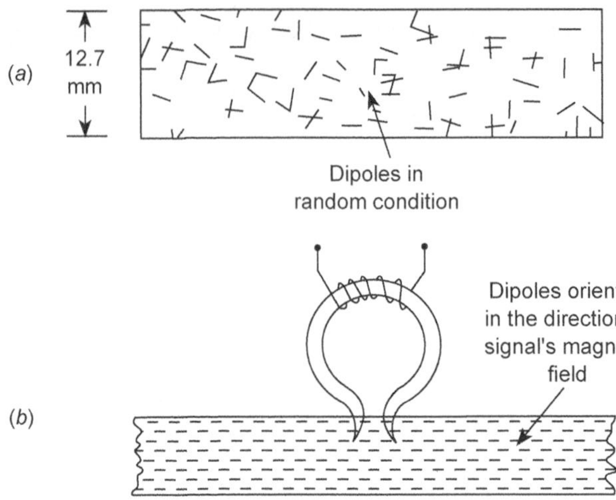

Fig. 6.10

(c) Tracks On The Video Tape: The following tracks are formed on a video tape: (Fig. 6.11)

1. Audio/horizontal track
2. Video/Picture track
3. Control track
4. Guard band or track
5. Synchronous signal band (or track).

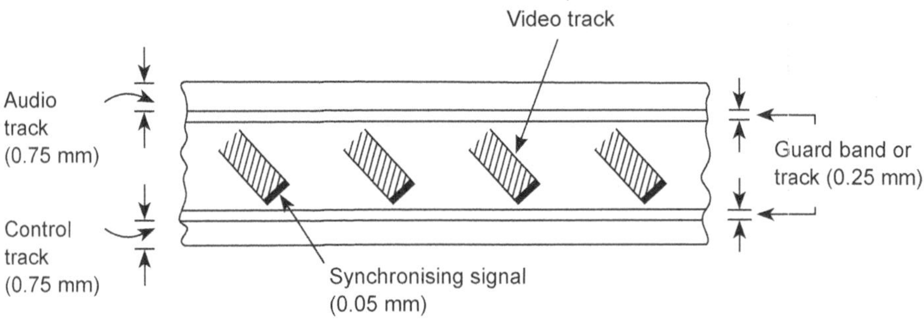

Fig. 6.11

1. *Audio track.* The sound or audio track is recorded at audio frequency upto 20 kHz. It is also called horizontal track. It is formed on one end of the tape.
2. *Video or Picture track.* This is recorded at video frequency i.e. in Magahertz. This is in middle of the tape. It is also known as *helical track,* as its shape is like helix.

3. *Control track.* It controls the above two tracks.
4. *Guard track.* The space between tracks is "blank". It contains no information. It is called guard track.
5. *Synchrous signal track.* It synchronises the two scannings occurring in the transmitter and receiver.

(d) Construction of tape (See Fig. 6.12): The thickness of the tape is about 0.05 mm. If its cross section is seen by a high power microscope, the tape is made of 4 layers.

First layer. This is stiff and non elastic.

Second layer. This is made of high quality polyster.

Third layer. This contains Iron particles.

Fourth layer. This part contains Ferric oxide particles, on this part the program is recorded.

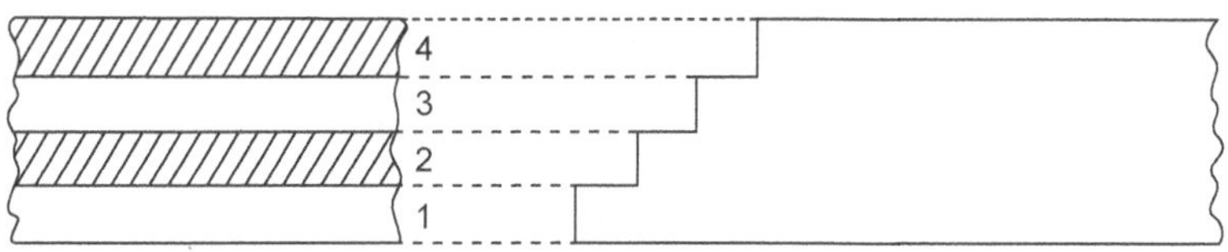

Fig. 6.12

6.1.4 Video Cassette

The handling of the video tape needs special care. It should never be touched with hands. The finger prints on the tape donot allow correct contact between play head and the tape and that leads to distortions. This is the reason that the tape is enclosed in a specially built plastic case called *cassette* (See Fig. 6.13).

There are two reels (or spools) in the cassette. They are called "supply reel" and "take up reel". While in operation, the tape unwinds from supply reel and is rewound on the take up real. This all is done by "pinch and capstan" arrangement provided in the cassette.

Looking through the "window" we can see the supply and take up reels. Different companies make cassette of different size and duration.

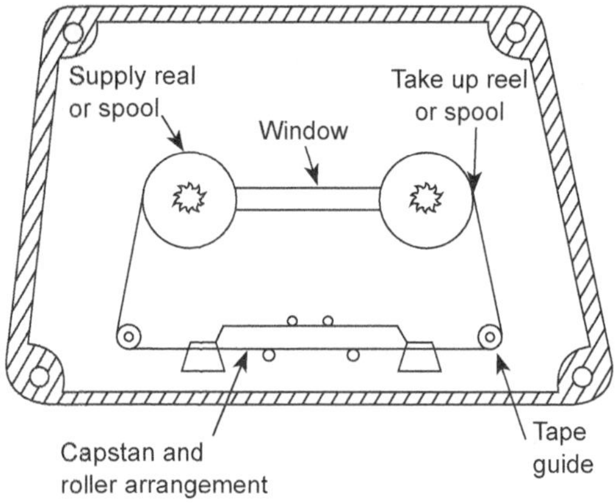

Fig. 6.13

Few companies make cassette of size 10 × 6 × 1 cm and others of 18.5 × 10.2 × 2.4 cm. The video tape is 12.5 to 12.7 mm wide. The duration of the cassettes may be ranging from half an hour to four hrs. .

Table 6.2: Total length of the tape

Duration in Hours	Total length of the tape in meters
$\frac{1}{2}$ hr	46
1	88
$\frac{v}{2f_m}$	128
2	175
3	260
4	345

Through window the tape is not visible as it is covered by a "safety cover" to protect it from dust and dirt. If the tape catches dust or dirt, the signal of that part of the tape may be "missed out". It produces "white streakes" during replay on the screen. To lift the safety cover, "Safety lock" is to be pressed.

On the backside of the cassette, there is a special construction called "Tab". If the tab is removed, recording is not possible. For example, if we have recorded some important program in the cassette and somebody by mistake puts this cassette into a VCR for recording, the VCR will not be ON and the supply will become earthed. In this way the important program is not erased.

6.1.5 Need of F.M. (Frequency Modulation) for Video Recording

The video frequency has a very high band width (5 Hz to 5 MHz). To solve this problem, frequency modulation is used in video recordings. A carrier wave is *frequency modulated* by the luminance (Black & White) signal.

Frequency modulation (F.M.) has following advantages:

(*i*) Noise is very much reduced.

(*ii*) Output becomes high.

(*iii*) A.C. Biasing is not needed as noise is automatically reduced.

(*iv*) A very low modulation index is needed.

(*v*) The F.M. improves S/N ratio.

In frequency modulation, the carrier is modulated with the signal. The carrier deviation is about 1 MHz (from 3.8—4.8 MHz). The modulating index (*i.e.,* the ratio of carrier deviation to the signal frequency) is kept 0.5, this gives a L.S.B. (Lower Side Band) of about 3 MHz, which is enough to convey all necessary information. In video tape recording *Vestigial Side Band (VSB) System* is used. (See Fig. 6.14).

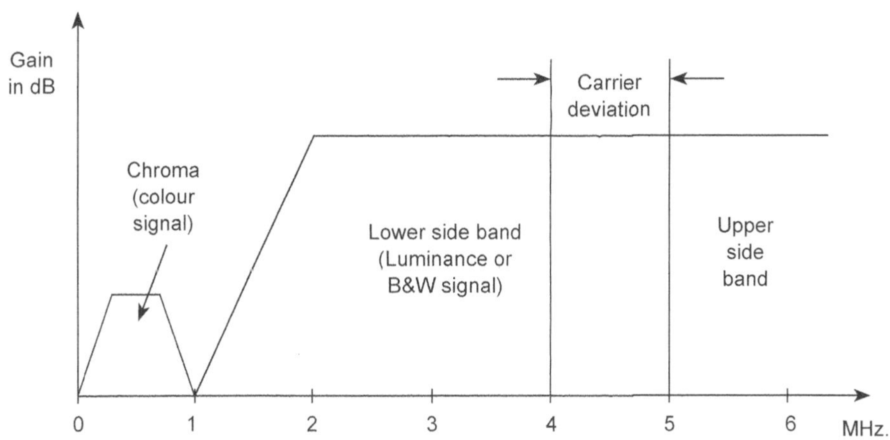

Fig. 6.14

6.1.6 Composite Video Signal (Luminance + Chrominance + Sound) (Fig. 6.15)

A video signal is composed mainly of 2 parts: Luminance (*i.e.,* Black and white) part and the chrominance (*i.e.,* colour) part. Luminance signal is also called as "Y-signal" and chrominance as "C-signal". In VCR both the signals are recorded separately and during playback "recombined" to produce a colour picture. Also the signal processing techniques for both signals are different.

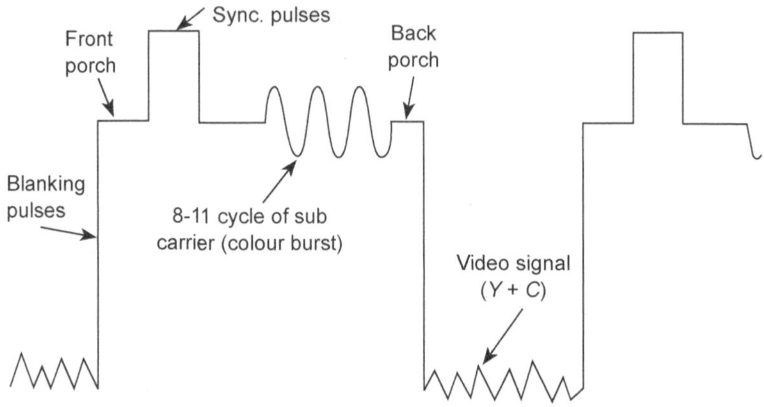

Fig. 6.15

Besides luminance (Y-signal) and chrominance (C-signal), the video signal contains (*i*) Burst signal as mentioned above, (*ii*) Blanking pulses to render the retrace period of the scanning lines blank, and (*iii*) sync pulses to synchronize the scanning taking place in transmitter and the receiver. In this way our signal is a complex (*i.e.*, a composite) video signal.

6.1.7 Recording and Replay of Luminance (B & W) Signal

(a) **Recording of Luminance (B & W) Signal:** The Fig. 6.16(a) shows a block diagram showing various stages through which a luminance signal is to pass before it is suitable for recording on the tape. The video signal to be recorded may come from a TV receiver or a TV camera or from some other source. If the original video signal is a colour signal, it will

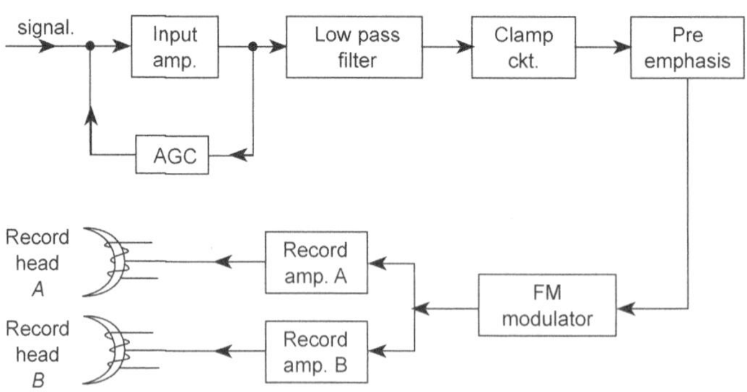

Fig. 6.16 (a)

also contain various controlling pulses like sync. pulses, blanking pulses, equalizing pulses etc. along with chrominance signal. In other words it will be a composite signal. The Luminance signal will have to be separated from the video signal in order to record it.

(b) Playback (Replay) of Luminance (B&W) Signal: The recorded tape is played back at the same speed at which it was recorded. During playback the "recording head" works as "play head" and moves on the "tracks". The signal picked up by the head is very weak and needs immediate amplification. The video signals are therefore given to pre-amplifier, through rotary transformers by operating Record/play switch. (Fig. 6.16(b))

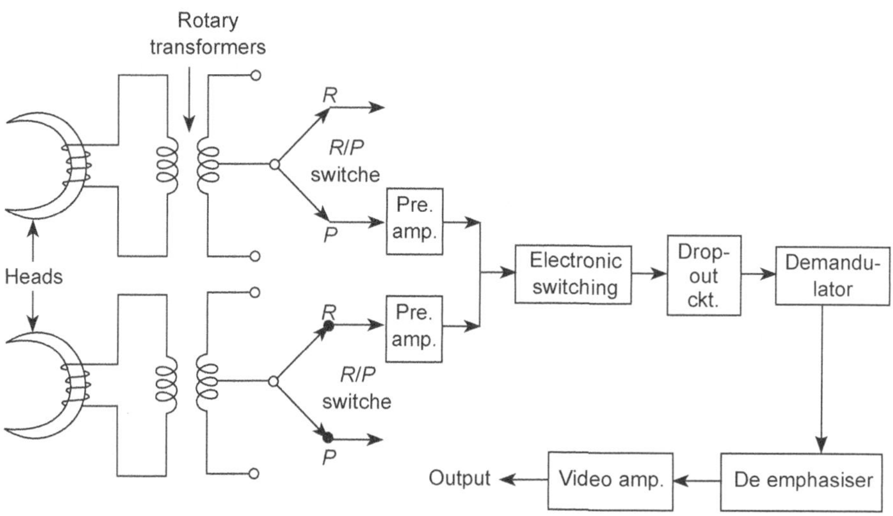

Fig. 6.16(b)

The magnetic coating on the tape may be damaged by picking dust particles. This may lead to unfaithful play back. This is known as drop outs. A drop out (compensation) circuit detects the damage and fills up these "holes" in the produced picture by picking video signal from the corresponding healthy portion of the previous line.

After drop out compensation, the signal is de-modulated and de-emphasised which are just the reverse processes of modulation and emphasisation done during recording. The signal obtained is amplified through video amplifiers and displayed.

6.1.8 Generation, Recording and Replay of Colour Signal

(a) **Generation of Colour (Chrominance or C) Signal:** For understanding the colour processing concept of generation of colour signal will be helpful here, we are briefing the same.

The colour TV camera splits the picture into three video signals corresponding to red, green and blue colours. The three video signals are then transmitted after modulation. The colour TV receiver separates the 3 signals in its picture tube which has 3 separate guns to generate 3 separate electron beams. When these beams strike the phosphor coated screen of the TV, the three colours (R, B, G) produced on the screen are integrated in our eyes and we can see the original picture.

The three video voltages produced by camera are combined or encoded to produce two main signals known as Luminance or Y- signal and chrominance or C-signal. The Y signal is obtained by mixing 30 percent red, 59 percent green and 11 percent blue video voltages obtained from colour camera. In other words $Y = 0.30\ R + 0.59\ G + 0.11\ B$. The Y signal produces a B & W picture in monochrome TV receivers. In colour receiver, the Y signal only helps to decode R, B and G signals for the colour picture tube. On the other hand the chrominance or C signal is obtained from colour difference signals *i.e.,* R-Y, and B-Y.

These two colour difference signals are transmitted after modulating two sub carriers which are in quadrature and of same frequency i.e. 4.43 MHz. If the two modulated carriers are added vectorially, we obtain chrominance or C signal as shown in Fig. 6.17. Now actually it is the C signal which modulates the main picture carrier. This process of modulation is known as Quadrature Amplitude Modulation (QAM).

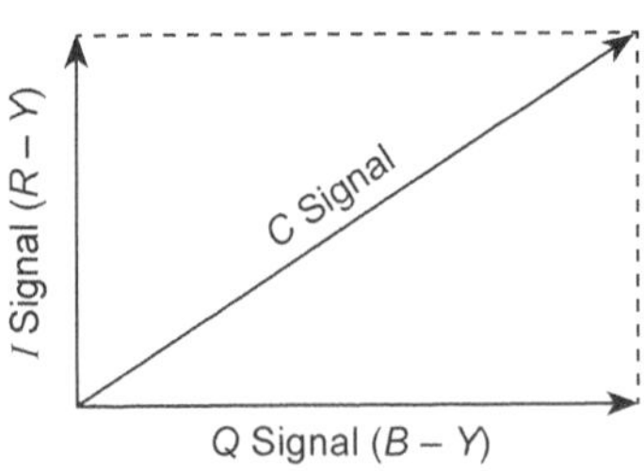

Fig. 6.17

The C-signal is therefore the resultant of two colour difference signals R-Y and B-Y. These two signals are known as I and Q signals respectively. It is also necessary to limit the bandwidth of I and Q signals before modulation, so that their sidebands do not interfere with luminance or Y signal. It is of no use to transmit colour signal beyond 1.5 MHz which is frequency of the smallest object, our eye can perceive. The I signal contains all frequencies from 0 to 1.5 MHz, while the Q signal contains frequencies from 0 to 0.5 MHz only. Practically all circuits are designed for a BW of 0.5 MHz.

The I and Q signals as mentioned above are transmitted by QAM of the sub carrier. The modulator allows only the side bands (and not the carrier) at the output. But for demodulation at the receiver, the carrier's presence is essential therefore the receiver is provided with a local oscillator which produces the sub carrier frequency. For reproduction of the colours, with its original "tint", it is necessary that *the phase* of the sub carrier produced in the receiver should be same as *the phase* of the sub carrier suppressed by the modulator. For this 8 to 11 cycles of the colour sub carrier frequency is transmitted along with the signal. This is known as colour burst signal and if the burst signal is suppressed at the receiver, we shall have a B & W picture on the screen, though colour picture is always transmitted.

(b) Recording of Colour Signal (Y + C): The recording process used is shown through block diagram in Fig. 6.18. As shown, the Y and C signals are processed separately. The Y signal is converted into a F.M. signal and the C signal is converted from its 4.43 MHz to 562.5 kHz frequencies through superhetrodyning similar

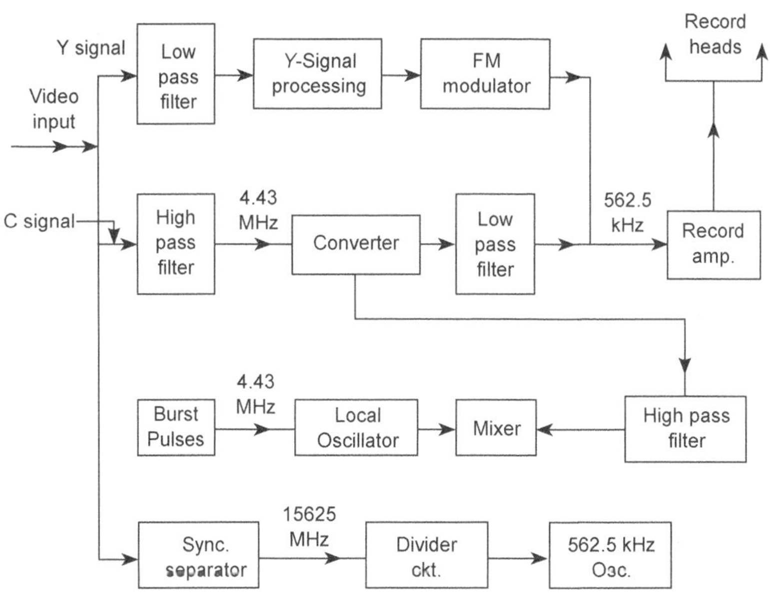

Fig. 6.18

to as done in radio and T.V. receivers. The local oscillator produces sufficient high frequencies for the purpose. It should be known that 15625 MHz is 36th sub multiple of 562.5 kHz therefore a divider circuit "divides" the oscillator frequency by 36 to get the line frequency. The frequency of 562.5 kHz combined with Y signal is suitable for recording. After suitable amplification, it is given to the recording heads.

(c) Play Back of Colour Signal (Y + C): For reproduction of the recorded signal, almost the same process is to be repeated again (as in recording) to get the original recorded signal. The scanning heads scan the video tape tracks and produce the signal which contains Y as well as C signals. This is passed through pre-amplifiers and other circuits and then separated into Y and C signals in a high pass filter circuit (Fig. 6.19). The Y signal is demodulated and C signal is reconverted into 4.43 MHz carrier frequency by an other superhetrodyning process as it was done during recording. The two signals thus obtained are added to get the original signal output and displayed on a TV monitor.

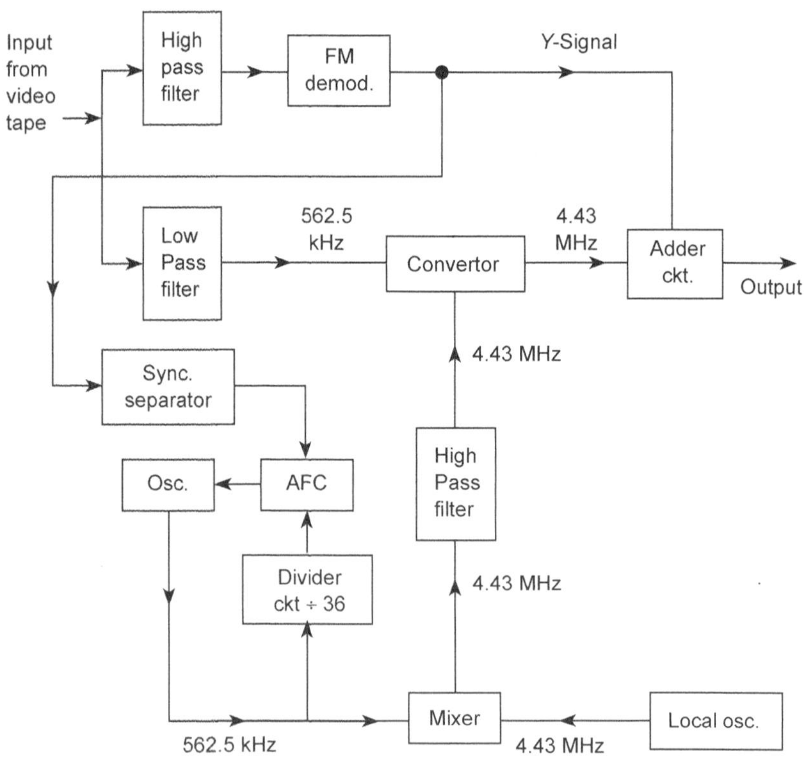

Fig. 6.19

6.2 VCR (VIDEO CASSETTE RECORDER)

This can record as well as replay a video.

6.2.1 Block Diagram of VCR

The complete block diagram of a VCR has been shown in Fig. 6.20. A VCR does not have its own monitor hence a TV receiver is generally used as a monitor. Important blocks or sections are described below:

1. *Antenna Section.* VCR and TV have a common antenna. Till the VCR is connected with the supply, the antenna remains connected with VHF and UHF antennas through a *splitter*. When VCR is connected to the supply both the antennas get connected with the output of the modulator through a selector switch. (See Fig. 6.20)

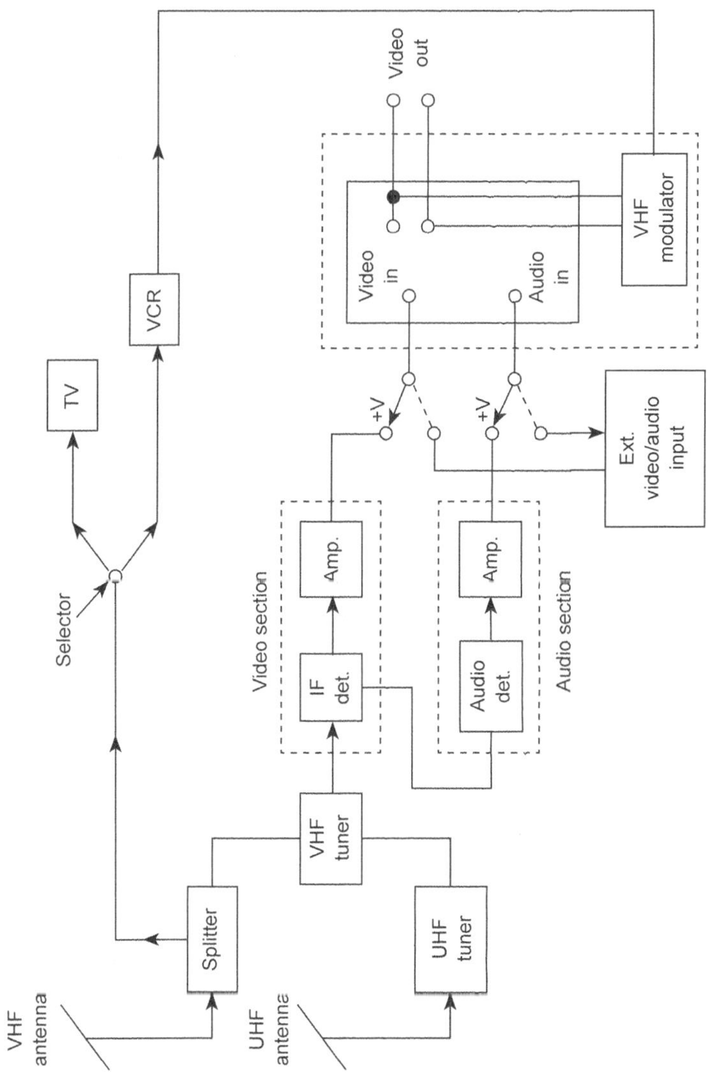

Fig. 6.20

2. *Record Section.* As discussed earlier, the VCR has a *video drum* which rotates at a constant speed by a servo-system. The tape is also moved by another mechanical system.

   The recording section records a signal on a tape. For recording a colour signal, Y and C signals are first separated. The frequency of C-signal is reduced from MHz to kHz. The *Y* signal is separately modulated. Now Y and C both signals are sent to the mixer and simultaneously recorded (Fig. 6.21)

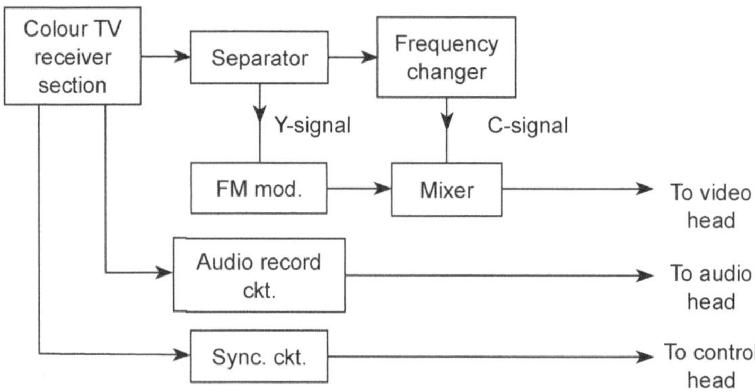

Fig. 6.21

3. *Replay Section.* For replay also, Y and C signals are separated. The C signal is reconverted into its original MHz frequencies and demodulated, similarly the Y-signal which was frequency modulated during recording is also demodulated and detected. These both signals along with the Y-signal are mixed in a mixer and given to TV for reproduction. (See Fig. 6.22)

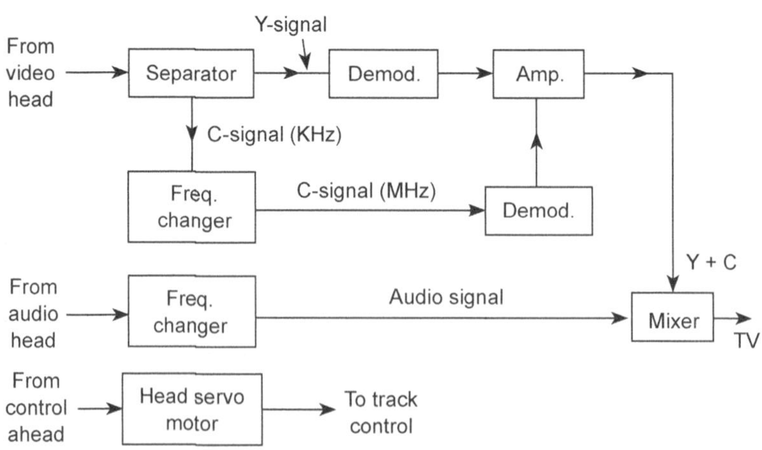

Fig. 6.22

4. *RF Tuner Section* (See Fig. 6.23). Besides playing a video cassette, a VCR can receive signal directly from the antenna and process it like a TV to get original signal. This function is performed by a RF tuner. The VCR as well as the TV both are provided with similar tuners. The RF tuner consits of a RF amplifier, mixer and a local oscillator. Like a conventional superhetrodyning process, the RF is converted into I.F. This is amplified, detected and original signals are obtained for video and audio outputs respectively.

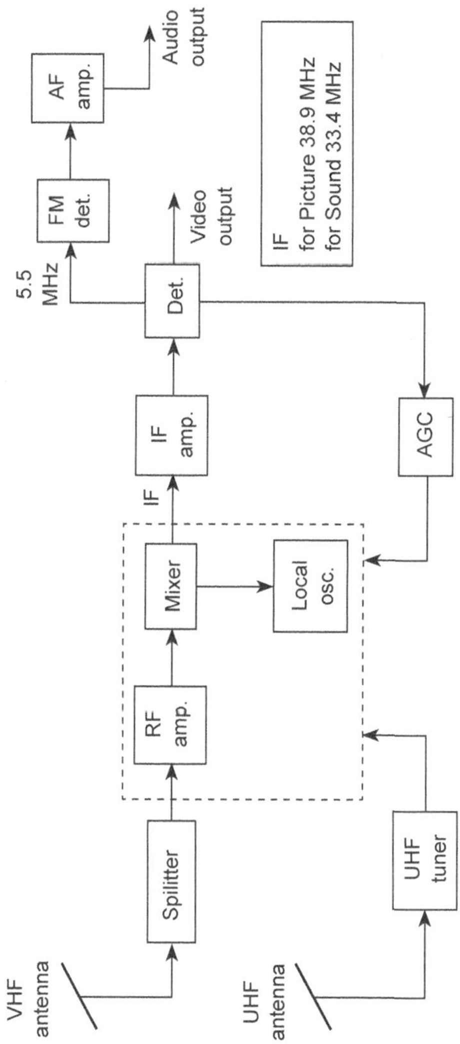

Fig. 6.23

5. *VHF modulator.* The Fig. 6.24 shows black diagram of a VHF modulator used with a domestic VCR. A crystal oscillator generates VHF frequency and a *doubler circuit* brings this frequency to the required level. The video signal obtained from VCR is amplified and sent to

AM modulator. The audio input is also amplified and it modulates a 5.5 MHz carrier frequency produced by the crystal oscillator. The FM modulated audio carrier is sent to mixer circuit and by "heterodyning", the sum of the frequencies is amplified.

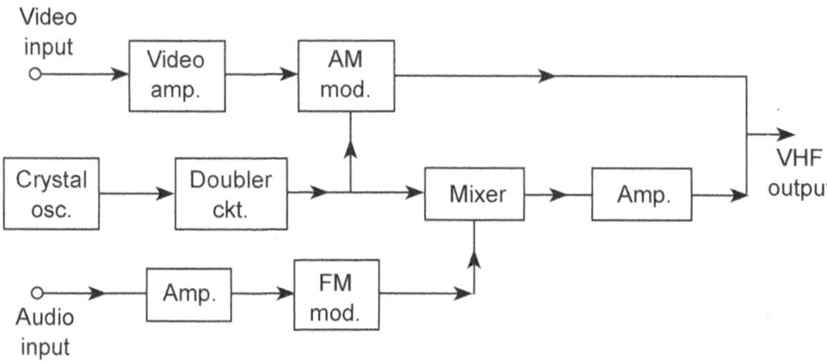

Fig. 6.24

6.2.2 Control Panel of VCR

A VCR has panel on its face as well as at the back. It has various controls. For proper control of the machine, one should have sufficient knowledge of these.

(*a*) *Control on the face panel* (Fig. 6.25). The description of only important controls provided on the face panel of a VCR is given below:

1. *Eject.* When this is pressed the cassette holder opens out and a cassette can be put or taken out whenever desired. By pressing, the holder closes itself.
2. *Rewind (FF)*. By pressing this button, the tape moves backward and rewounds. The rewind speed is about 6 times of the normal speed.
3. *Fast forward.* By pressing this control, the tape moves forward. It is required to be pressed till needed. This control is used when we donot like to see a certain part of the display *e.g.*, unwanted advertisements, songs etc. The FF speed is also about 6 times to that the normal speed.
4. *Play.* To replay the VCR.
5. *Stop.* To stop/off the VCR
6. *Record*. For recording this is pressed, along with the "play" (4) button.
7. *Pause*. To stop recording or to make the picture "standstill" for some time, this control is used.

8. *Counter*. It is a "tape counter" which counts, how much tape has moved.
9. *Cassette holder*. It contains the cassette.

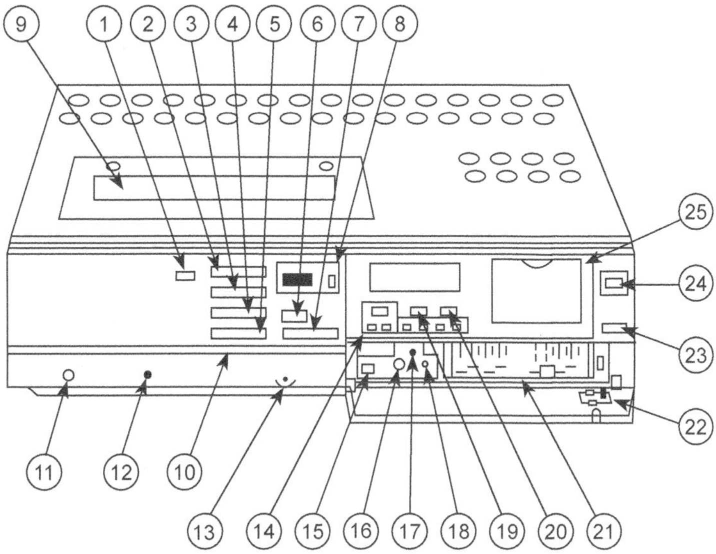

Fig. 6.25

10. *Dampness indicator*.
11. *Remote control*
12. *Track control*
13. *Sharpness*. It is used for controlling sharpness of the picture.
14. *Clock*. This is an inbuilt time clock which displays time during recording/replay. By this button, digital clock setting can be done.
15. *Input signal selector*. It selects the source of input signal, whether recording to be done through a TV or a camera.
16. *Video input connector*. This connects the recording source.
17. *Camera remote reverse*.
18. *Audio input jack*.
19. *Timer switch*.
20. *OTR*. This is "One Touch timer-Record" control. By this, a time (*e.g.* one hour) or so can be set and the VCR is left free ior automatic recording. After the preset time the VCR will become automatically off.
21. *Tuning control*. This is a channel selector.

22. *Fine Tuning.*
23. *VTR.* This can on/off the recorder.
24. *VTR/TV.* By this control it can be decided whether we are to see the recorded signal or the TV channel.
25. *Channel Selector.* For recording or to view, by this selector we can select the required channel.

(*b*) *Rear Panel of VCR* [Fig. 6.26]. At the back, VCR has the following important .controls:

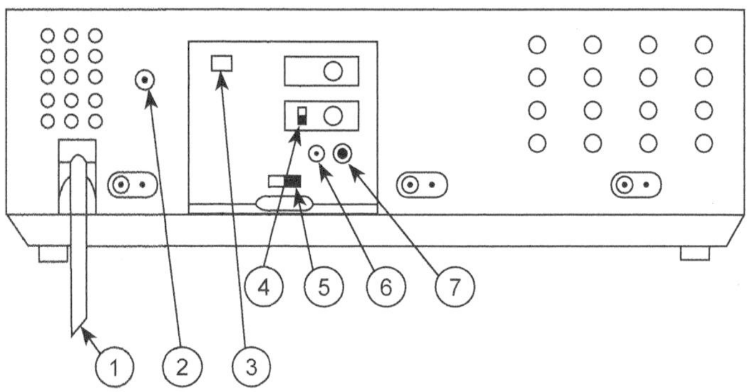

Fig. 6.26

1. *AC. Supply*
2. *Voltage selector*
3. *RF Converter*
4. *R.F. Signal Switch.* This selects the level (low or high) of the signal obtained from the antenna.
5. *Colour mode.* By this control we can select any position, *i.e.,* Auto/Test/Colour.
6. *Audio output.* For audio recording it connects the VCR with the source of recording.
7. *Video output.* For video recording this control connects the VCR with the source of video recording.

6.2.3 Mechanical System of VCR

(a) The Mechanical system moves the tape to and fro and controls its speed during recording or replay. It also moves the tape at constant speed during recording, moves it faster during rewind or FF (Fast Forward) and also stops it, when required. The (already discussed) mechanical controls on the VCR panel are:

(*i*) Ejector	(*ii*) channel	(*iii*) Rewind
(*iv*) Forward	(*v*) Play	(*vi*) Stop
(*vii*) Record	(*viii*) Pause	(*ix*) Timer.

(b) The other mechanical systems consist of

1. Video drum assembly
2. Capstan roller assembly
3. Tape drive systems
4. Heads: video head, Audio head, control head and erase head.
5. Rotating video heads.

1. *Video Drum Assembly.* The Fig. 6.27 shows the video drum assembly. In the drum, two video heads are positioned face to face and move in a circle in horizontal plane.

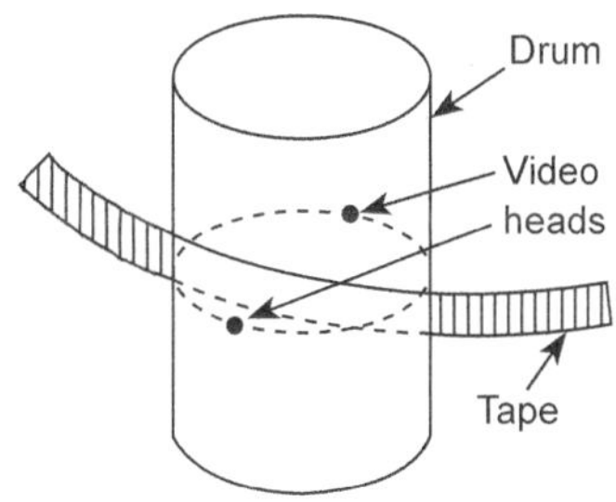

Fig. 6.27

Tape is fixed with the heads through "tape guides" which prevent any slip of the tape. The drum assembly is vertical but heads move horizontally. The tape is wound on the drum.

2. *Capstan Roller Assembly.* This has two parts: Capstan motor and the pinch roller. Capstan and roller move in mutually opposite directions. The pinch "presses" the capstan to maintain proper "tension" of the tape otherwise, it will be loose (See Fig. 6.28).

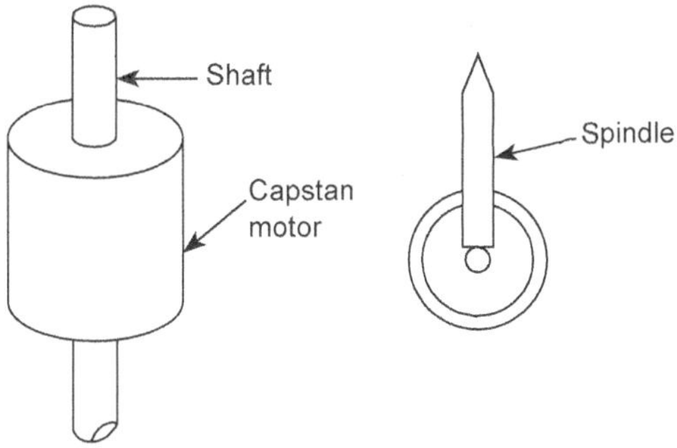

Fig. 6.28

3. *Tape Drive Systems.* Electric motors are employed to move the tape. Generally DC motors or synchronous motors are used to move the head and the capstan.

The tape drive systems used are of the following types:

(*i*) Rim drive system

(*ii*) Idle Wheel drive system

(*iii*) Belt drive system.

(*i*) *Rim drive system.* In this system a pulley is connected on the motor shaft and moves with the motor. With the pulley there is a flywheel on the middle of which there is a capstan. There is a rubber rim over the wheel, when the capstan moves, the tape is pulled from the "supply reel" to the "take off reel" [See Fig. 6.29]

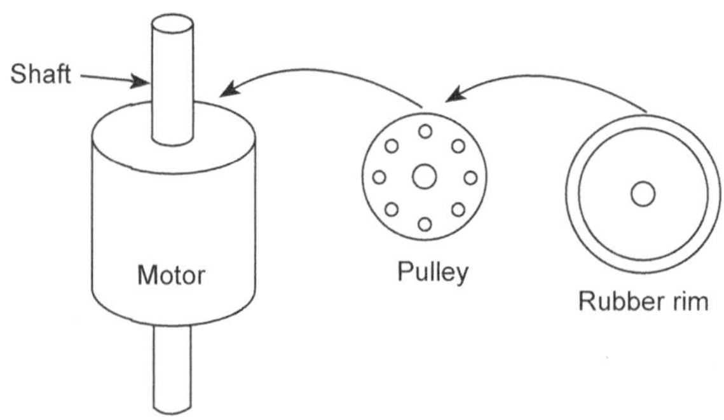

Fig. 6.29

(*ii*) *Idle wheel drive system.* This is similar to the above system but on the pulley there is an additional wheel called idle wheel. The fly wheel is over this idle wheel. When the motor moves, idle wheel, flywheel and the capston wheel move and the tape is set into motion [See Fig. 6.30]

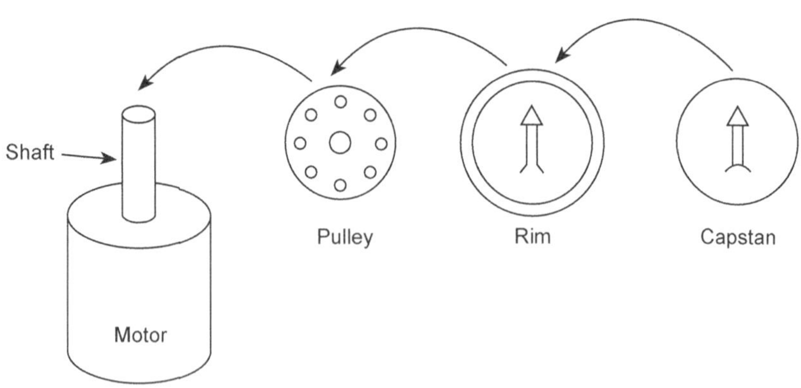

Fig. 6.30

(*iii*) *Belt drive system.* In this system, the capstan wheel is connected with the motor by a rubber belt. When the motor moves, the capstan wheel also moves and sets the tape into motion (See Fig. 6.31).

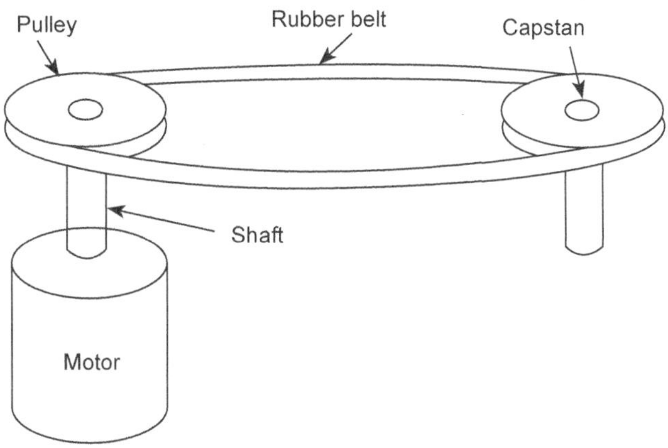

Fig. 6.31

4. **VCR Heads:** The important VCR heads are as under:

   (*i*) Video head (*ii*) Audio head (*iii*) Erase head.

   They may have any shape as shown in Fig. 6.32(a, b, c).

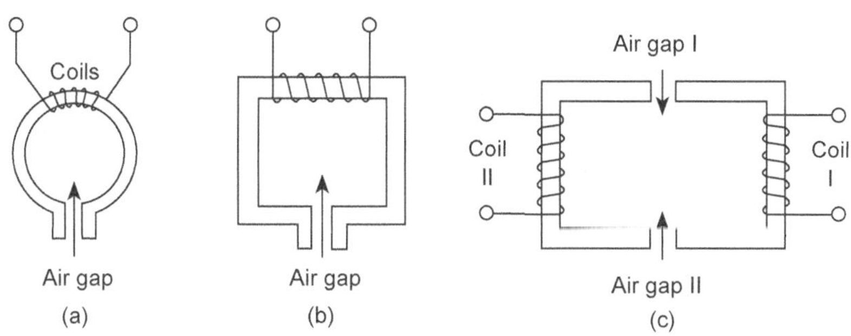

Fig. 6.32

(*i*) *Video head.* In VCR there are 2 video heads. They are use for recording and replay of the picture signals. They are basically electro-magnets and prepared by keeping coils on ferrite stamping. Both video heads are put on the video drum assembly opposite to each other.

(*ii*) *Audio heads.* In construction, they are similar to that of the video heads. The stampings, however are used of materials like Alnico (Al + Ni + Co), Mumetal (Fe + Ni + Cr), Permalloy (Fe + Ni) etc.

Special heads may have 2 coils with 2 air gaps. To get more output voltage both coils may be connected in series and for getting more current, they can be connected in parallel. These are used to record and replay audio signals. (Fig. c)

(*iii*) *Erase head.* These heads erase out the unwanted previous recorded signal from the tape. They may be:

(*a*) Video erase head. It erases out the recorded picture signal.

(*b*) Audio erase head. It erases out the recorded sound signal.

The complete tape transport system of VCR has been show in the Fig. 6.33.

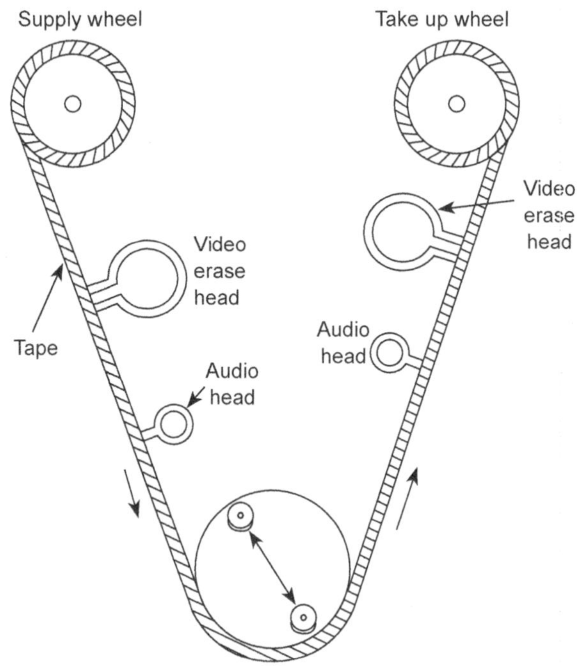

Fig. 6.33

5. **Rotating video heads:** For video recording, rotating (whereas for audio recordings, stationary) heads are used. The reason for this is simple. The video recording is done in Megahertz. For recording at such a high frequency the tape speed should be at least 10 meters per second. This speed cannot be obtained by a stationary head hence rotating becomes a necessity. The audio recording is done in kilohertz and for that a stationary head can do.

The Fig. 6.34 shows how the tape moves on the rotating heads. For this, helical former is used on which two video heads are placed 180° apart. The tape moves making helix with the heads. The heads alternately "scan" the tape and make video track on it.

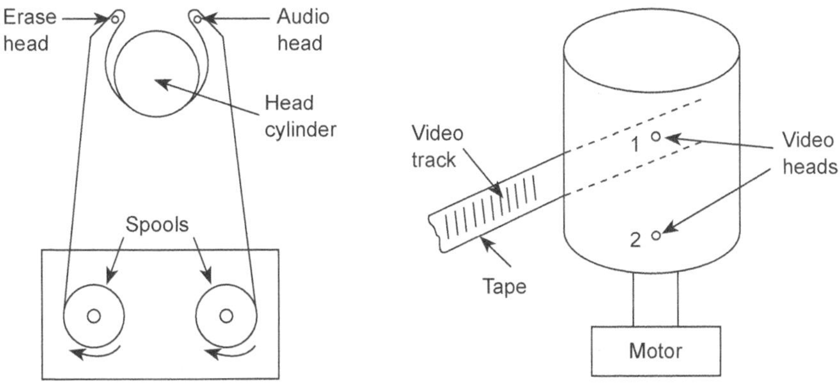

(a) (b)

Fig. 6.34

6.2.4 Servo System of VCR

We know that the process of scanning takes place in the TV transmitter as well as in the receiver. The scannings during recording as well as replay should be properly synchronised otherwise the quality of the video signals should not be good. A mechanical system can not give good results therefore an automatic control is must. This automatic electronic system is known as *Servo System*.

Actually a servo system of a VCR is an electronic cum mechanical arrangement which uses two signals called *Reference signal* and *Feed back signal*. These two signals are fed to multivibrators and the output obtained is given to a comparator circuit. The comparator compares the two signals and gives an error signal if any. This error signal after amplification is given to the motor. This is the basic principle of a VCR survo system. The error signal force acts in such a direction that leads to the reduction of the error. (See Fig. 6.35)

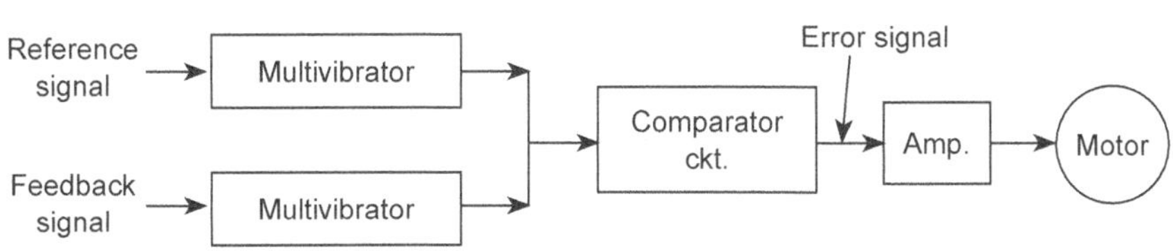

Fig. 6.35

(a) Types of Servo Systems: Two servo systems are popularly used in VCRs:

(*i*) *Drum servo system.* In this system, reference signal is in the form of vertical sync. pulses which is separated by a "separator" circuit from the composite video signal. The rotating heads have *tacho coils* which provide the feed back signal.

Both the signals are given to multi-vibrators whose output is fed to the control head (See Fig. 6.36).

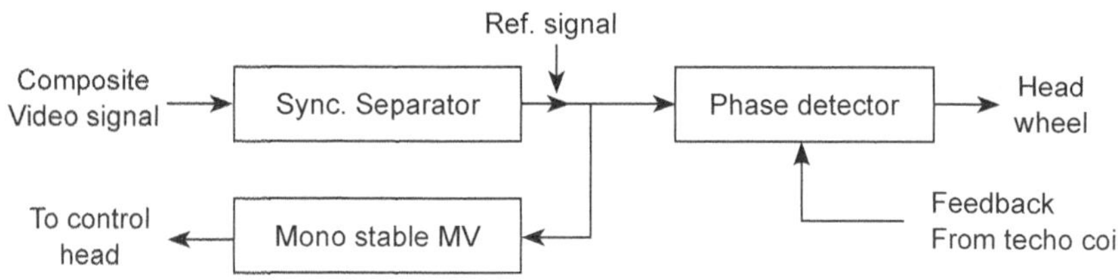

Fig. 6.36

(*ii*) *Capstan servo system.* In this system the speed of the tape motor is made constant. In this circuit, both the signals are fed to the phase detector from opposite directions. The main object of this system is to maintain a constant speed of the tape during recording as well as replay. (Fig. 6.37).

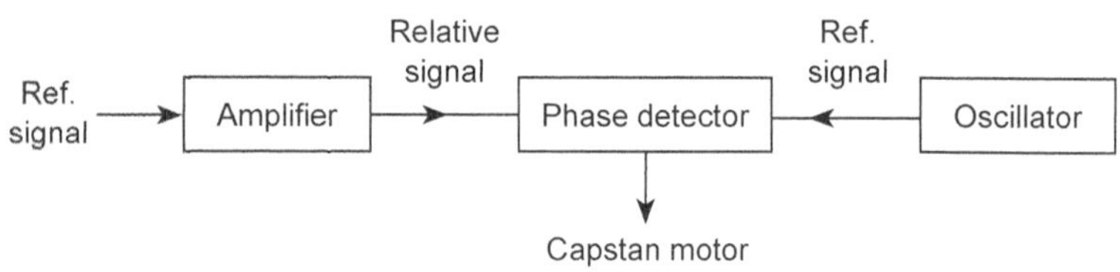

Fig. 6.37

(b) Servo Signals: In VCR, the following servo signals are used:

1. *Cyclinder frequency generated pulses.* (Fig. 6.38) These pulses are produced by a generator in a video head cylinder. The rotor of the generator has poles and the stator has a detection coil. When the rotor is moved at a speed of 25 r.p.s., a 100 Hz pulse is produced.

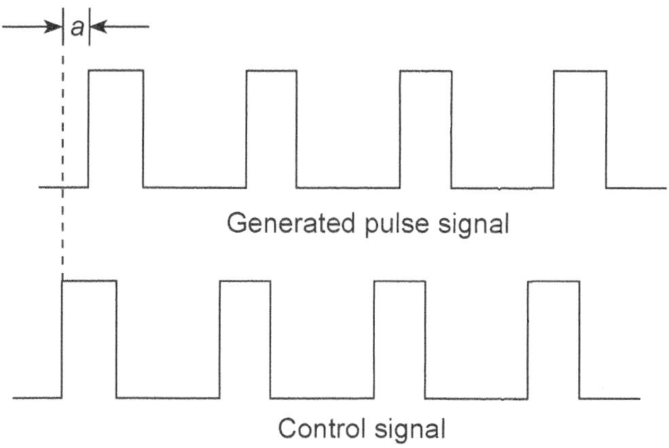

Fig. 6.38

2. *Capstan frequency generated pulses.* This pulse is produced in the capston generator. The generator has 240 poles. The frequency of pulse depends upon the speed of the capstan. Different frequencies can be generated at different speeds.
3. *Cylinder tach. pulses.* These pulses are produced in another generator in the cylinder. These pulses are 25 Hz frequencies.
4. *Reference signal.* To generate the reference signal, first square waves are generated by a schimitt trigger. The interaction of the square waves with saw tooth waves gives out a reference signal. (Fig. 6.39)

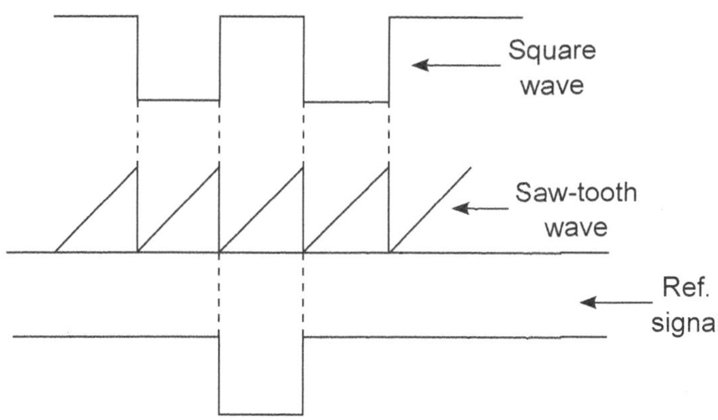

Fig. 6.39

5. *Control track pulses.* These are 25 Hz pulses and are used for controlling the track.

6.2.5 Video Systems Popular in India

The video systems popular in India are (fig. 6.40):

1. Video Home Service (VHS)
2. Betamax System (BMS)
3. Video-2000 (V-2000)

Table 6.3: Comparison of the three systems

S. No.	VHS	Betamax	V-2000
1.	Audio tape size is 1 mm	Audio tape size is 1.05 mm	Audio tape size is 0.5 mm
2.	Loaded after Playback/ Record switch is ON	Loaded immediately.	—
3.	Cassette measures 18 x 10.3 x 2.5 cm	Cassette measures 15.5 x 9.5 x 2.3 cm	—
4.	Head gap is 0.3 μm	Head gap is 0.6 μm	—
5.	Writing speed is 4.8 m/s and tape speed is 2.3 cm/s	Writing speed is 6.6 m/s. The tape speed is 1.8 cm/s	Writing speed is 5.08 m/s and the tape speed is 2.4 cm/s
6.	Head drum dia is 62 mm	Head drum dia is 74.5 mm	65 mm
7.	VHS tapes are specified by letter T followed by operating time in minutes *e.g.* T-60	Betamix tapes are specified by letter L followed by operating time in minutes *e.g.* L-60.	—
8.	Tape width is 12.6 mm	Same	Same
9.	Recording time 3 hrs	3 hrs	In this case 8 hrs
10.	Video track width 0.05 mm	0.03 mm	In this case 0.02 mm
11.	Video signal width 10.07 mm	10.2 mm	4.69 mm
12.	Shape of tape on the drum is M type	The shape of the tape is W type	The shape of the tape is V type
13.	Carrier frequency deviation is from 3.8 – 4.8 = 1 MHz	In this deviation is 3.8 – 5.2 = 1.4 MHz	deviation = 3.3 – 4.7 = 1.4 MHz
14.	It is a Japanese system	It is also Japanese system	It is an European system

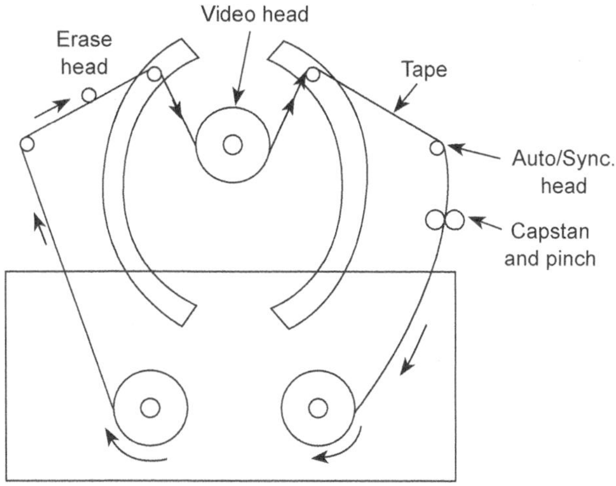

(*a*) VHS VCR

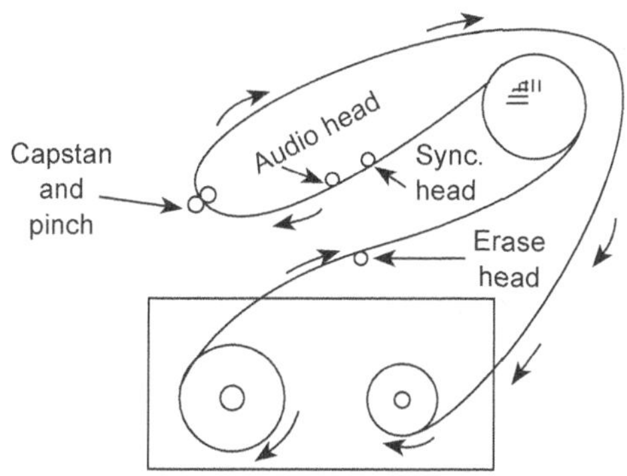

(*b*) Betamax VCR

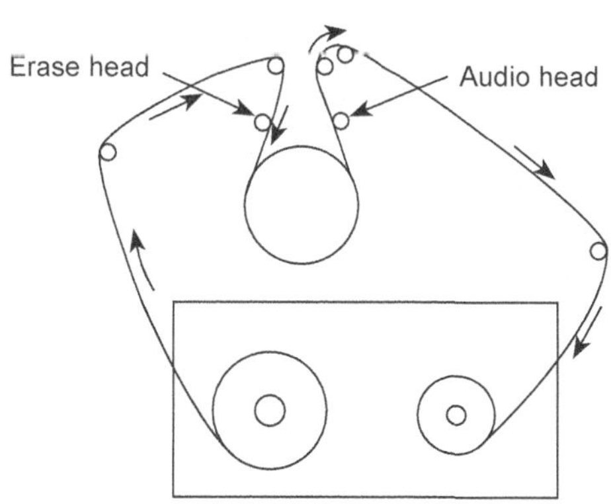

(*c*) V-2000 VCR

Fig. 6.40

Table 6.4: Trouble shooting of VCR

Faults	Reason and Remedy
1. The VCR does not become 'ON'	(*i*) Check supply, sockets, wiring etc.
2. Picture, but no sound	(*i*) Clean audio head (*ii*) Check RF converter.
3. Distortion in sound	(*i*) Clean audio head (*ii*) Check RF converter
4. During recording picture is recorded, but no sound	(*i*) May be an O.C. (open circuit) in audio circuit.
5. Sound comes but raster is clear blank.	(*i*) Check chroma signal path.
6. Snow in the picture, sound is OK.	(*i*) Clean video head (*ii*) Quality of the cassette may be poor. (iii) Check colour system.
7. Sound recording OK but picture is dull.	(*i*) Check luminance play back preset.
8. Sound is not erased.	(*i*) Clean erase head (*ii*) Check erase oscillator.
9. "Eject" does not work.	(*i*) Check eject button (*ii*) Check eject motor
10. FF and Rewind do not work.	(*i*) Reel motor does not work
11. Vow and flutter	(*i*) Check speed of capston motor with picture during play back.
12. Sound OK but picture rolls	(*i*) Check drum motor speed
13. Touch of mike gives shock	(*i*) Reverse the connections of VCR
14. Radio interference in sound	(*i*) There may be a radio station near by (*ii*) Cover the components with an aluminium foil.
15. Interference in the picture during camera monitoring	(*i*) Grounding is not proper. (*ii*) Check the cable connections.
16. Tape breaks frequently	(*i*) Check the pressure of the rollers. (*ii*) Check the break.

6.2.6 Technical Data of VCR

1. **Video recording system.** 2 rotary heads helical scanning
2. **Tape speed** Standard play (SP) 23.39 mm/s
   Long play (LP) 11.7 mm/s
3. **Tape format** VHS
4. **Video TV system:** CCIR, 625 lines

   50 fields, PAL (Phase Alternative Line) colour signal.

5. Audio Track	1 track
6. SN ratio	Video: 43 dB
	Audio: 43 dB
7. Power Source	110 - 250 V, ac 50 Hz
8. Power consumption	25 W.

Example 6.1: *Find width of air gap for a tape speed of 2.4 cm/s and frequency of 4.5 MHz.*

Solution: The air gap

$$w = \frac{v}{2f_m} = \frac{2.4 \times 10^{-2}}{2 \times \left(4.5 \times 10^6\right)}$$

$$= 0.26 \times 10^{-8} \text{ m.} \quad \text{Ans.}$$

Example 6.2: *Find frequency which is 10 octaves above 30 Hz.*

Solution:

$$f = 2^{10} \times 30$$

$$= 30720 \text{ Hz.} \quad \text{Ans.}$$

Example 6.3. *Find highest and lowest frequencies for a sub carrier signal at 629 kHz with a B.W. of ± 500 kHz.*

Solution: Highest frequency = 629 + 500 = 1129 kHz

Lowest frequency = 629 – 500

= 129 kHz. Ans.

Example 6.4: *For a VCR, find a wavelength for a tape speed of 33.35 mm/s and frequency of 3.5 MHz.*

Solution:

$$V = n\lambda$$

$$\lambda = \frac{V}{n} = \frac{33.35 \times 10^{-3}}{3.5 \times 10^6}$$

$$= 9.53 \times 10^{-9} \text{ m.} \quad \text{Ans.}$$

Note: Now VCR is almost absolete and has limited use.

6.3 VCD (VIDEO CD OR VIDEO DISC)

The VCDs (Video Compact Disks/Versatile Compact Discs) or simply, Video disks have been known for sometime for their application in the recording of audio-visual performances for subsequent playback on television screens.

But their potential for strong enormously large volumes of data representing anything from numbers, text to pictures, drawings and other forms of graphics was realized in the recent past. Also known as an *optical disk*, the technology behind the video disk for computer applications is based on the same concept as that in case of video disks developed earlier for recording audio-visual signals. Data in the form of Os and 1s is represented on a video disk by a narrow light beam (laser beam) burning tiny holes on the disk.

Due to *frame by frame* advance nature of video, they have become an attractive proposition in imparting technical training to personnel in big institutions. They are inexpensive as far as cost as per unit data item is concerned. They have enormous storage capacity. Typically, one video disk is equivalent to 8 to 10 reels of magnetic tape. Video disk storage is fast replacing microfilm storage for cataloging huge volumes of data in national level museums, art galleries, libraries and Government agencies. Unlike magnetic tapes and disks, video disk has a disadvantage that data once recorded cannot be erased and the disk cannot be reused.

The video disk technology is new and recording at home is not possible therefore its use is limited to the reproduction of the already recorded information. But the quality of reproduction is very high. It offers full *bandwidth* reproduction and also can handle color signals.

The audio/video signal is recorded on the disk in the form of "pits". It is also aluminized to retain reflective properties and is contained in a plastic cover as protection against mishandling.

The video disks are available in about 30 cm (301.6 mm) diameter and are double sided (Fig. 6.41). The track pits go from the centre towards the outside. There are about 500 tracks per mm with a total length of the track as 35 km.

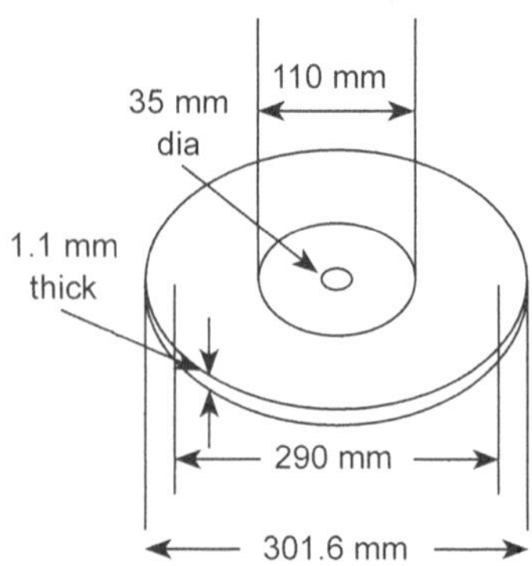

Fig. 6.41

They have many advantages over tapes like low cost, free of dust, grease, scratches and have unlimited life. We can see any programme, for which the disc does not need to be started from beginning. But a disk cannot be reprogrammed like tapes.

A video disc can also record a stereo sound.

6.3.1 Recording on VCD/Video Disc

There may be three methods of video disk recording: optical method, capacitive method and laser method.

Laser method of video disk recording which is popularly used is described below:

A laser beam modulated by sampled video signal is made to fall on the video disc which is coated with a photo resist material. The disk is rotated at a constant speed say (2000 RPM) as a result, pits are formed on the disk's surface. The pits have a constant depth but of varying length according to the variation in the video signal recorded.

The Fig. 6.42 shows track on the video disc.

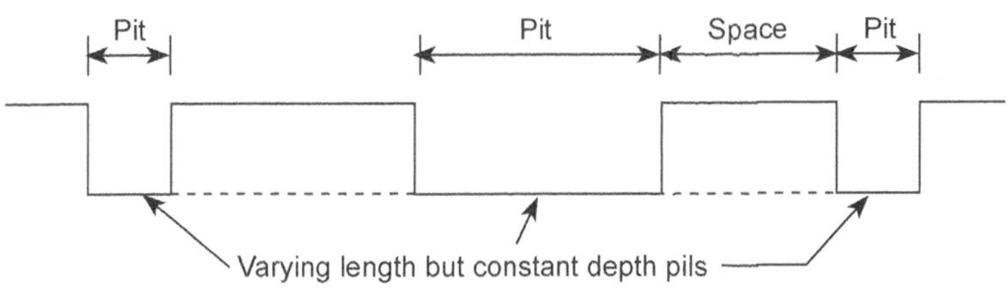

Fig. 6.42

6.3.2 Replay of VCD/Video Disc (Fig. 6.43)

A video disk player can reproduce a pre-recorded program on the video disk (sound as wells as monochrome or colour picture) through a T.V.

The production is totally optical and there is no physical contact with the video disk. The technique is more or less same as in the case of an audio CD.

A beam of laser is made incident on the track. The beam reflected from pit is weak and may be taken as zero (0) and that reflected from *space (flats)*

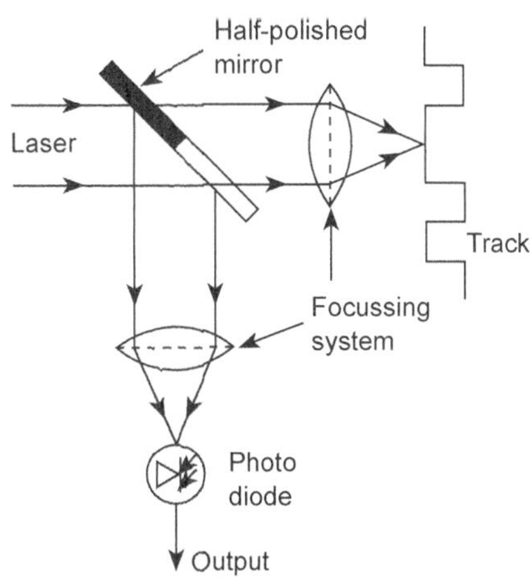

Fig. 6.43

between the pits is strong and taken as one (1); thus the output is obtained in digital form. The output is fed to a photo diode, which is processed and amplified.

6.4 DVD (DIGITAL VIDEO DISC)

A digital video (an versatile) disc is a small plastic disc which can store digital data. It has storage capacity of 15 GHz (30 percent more than a CD) and also has better resolution. It has dimensions of 12 cm diameter x 1 mm thickness. It is read (distpayed) by use of LASER, so it is covered by a protective coating. (Fig. 6.44)

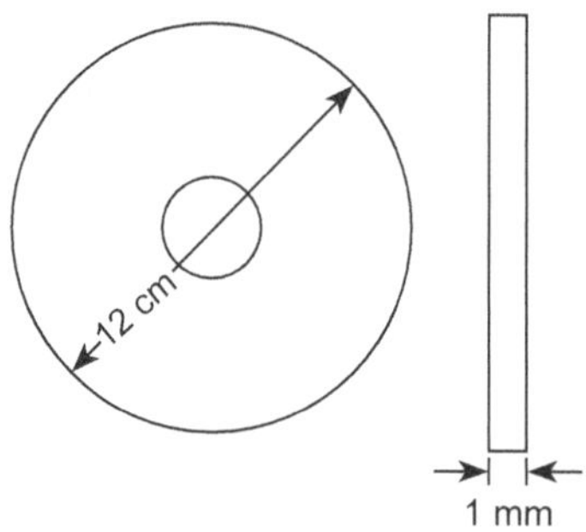

Fig. 6.44

During recording, the data is sampled at the rate of about 80,000 times a second and digitally encoded on the DVD, in the form of pits (Fig. 6.45). During reproduction, nothing touches the encoder portion and thus the DVD is not worn out.

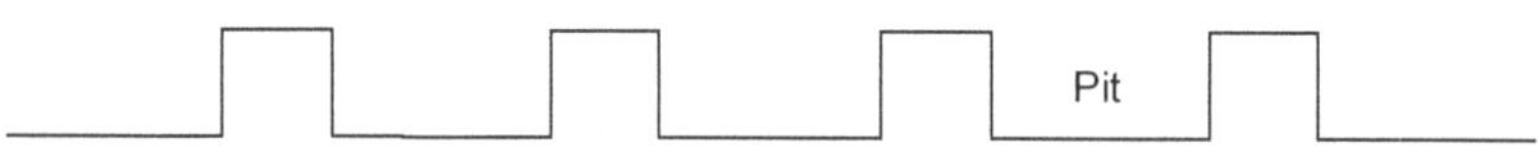

Fig. 6.45

The DVD has an access time of 1.2 MB per second. Some DVDs can store data on both sides, they have a duration of 2 hours.

6.4.1 DVD Player

These are successors of VCRs (Video cassette Recorders). These can play DVD. The working of a DVD player is almost same as of a VCD player. The Fig. 6.46 (*a*) shows the controls on Front and (*b*) on rear of the DVD player.

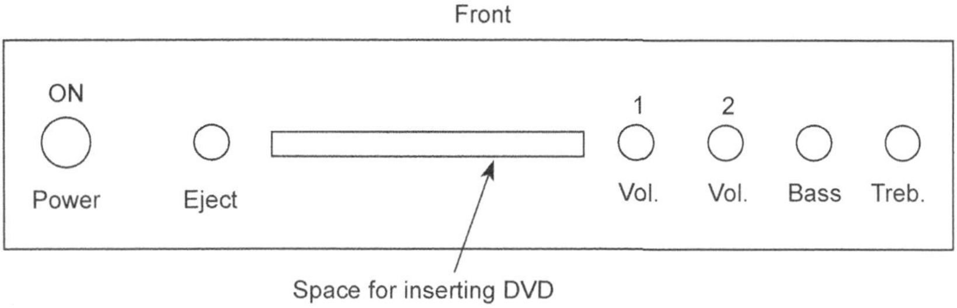

Fig. 6.46 (a) Front

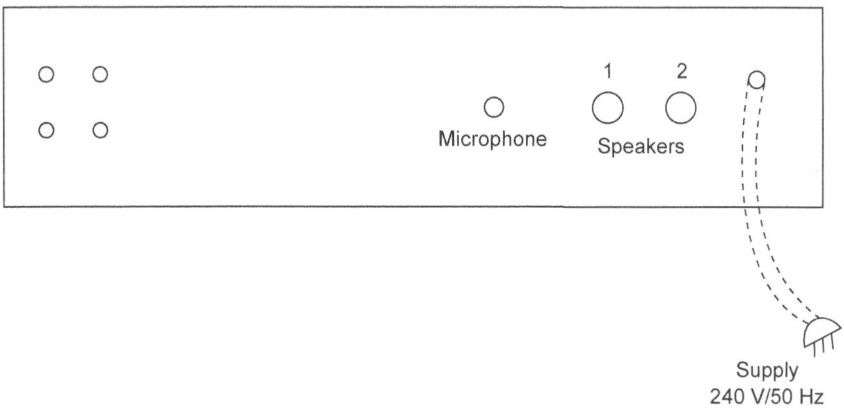

Fig. 6.46 (b) Rear

6.4.2 TECHNICAL DATA OF DVD

1.	*Recording and playback process*	Optical through LASER.
2.	*Disc diameter*	12 to 30 cm.
3.	*Disc Speed*	25 rps
4.	*Band width*	5 MHz
5.	*Record/Play time*	30 minutes each side

Table 6.5: Trouble shooting in CD players

Problems	Reason/Remedy
1. The disc tray does not close.	Place the disc correctly. Always close the disc by pressing on the unit. Do not try to press the tray with fingers.
2. The disc does not eject.	Press to cancel the CD synchro-recording. Now press the unit to eject the CD.
3. There is no picture.	Check the connections of the player and the T.V.

Problems	Reason/Remedy
4. Play does not start.	See that the disc is placed with the label side up and also that it is cleaned of dust, moisture etc.
5. The sound skips.	Wipe the disc clean. The player should be placed on a firm and vibration free stand.
6. There is severe hum or noise.	Move the system away from the source of noise or Install a "noise filter" available in the market.
7. The timer does not function.	Check the time setting and reset again.
8. The remote does not function.	Check the batteries. Point the remote toward the system's sensor.
9. There is acoustic feedback.	Reduce the volume.
10. Color irregularities on the T.V. screen.	Turn OFF the TV and make it ON after some time. If still persists, get the T.V. checked.

Table 6.6: CVD Vs DVD

CVD	DVD
1. It has a single layer/plate construction.	It has a double layer/plate construction.
2. Its quality is low.	It has high quality.
3. The CVD is written by computer.	DVD is also written by computer but differently.
4. Its storage capacity is 7000 MB.	Its storage capacity is lesser (about 4700 MB).
5. A CVD player can play ACD, CVD, MP-3 but not a DVD.	A DVD player can play ACD, CVD, DVD, MP-3 etc.

Note: The VCD and DVD can be converted into each other,

SUMMARY

1. Video recording is done on video tape.
2. Video signal is a composite signal comprising of many signals viz. luminance signal, colour signal, control signal, guard signal etc.
3. The video heads also move with the tape, while in the case of audio recording, only the tape moves and the audio heads remained stationary.

4. The colour signal comprises mainly of two parts : Luminance (*i.e.,* Black and White) part and chrominance (*i.e.,* colour) part. The Luminance signal is also called as Y signal and chrominance signal also as "C signal".
5. The VCR can record as well a replay the recorded signal.
6. The three popular VCR models are: VHS, Betamax and V-2000.
7. The three VCR heads are: Audio head, Video head and the Erase head.
8. The servo system of a VCR is an electronic cum mechanical arrangement. This gives best results in respect of recording, reproduction etc.
9. The VCD stands for Video Compact Disc. It can store video signals, which can be reproduced by VCD players.
10. The DVD stands for digital video disc. The signal can be displayed by a DVD player.

SELF TEST QUESTIONS

1. What are the basic functions performed by a VCR? Give its practical applications. How it differs from a VCP?
2. Explain, why for video recording rotating heads are used.
3. Draw a sketch of a video tape and show various tracks on it.
4. How the total length of track is calculated on a cassette?
5. What are luminance and chrominance signals? Draw and explain recording and reproduction of a luminance signal in a VCR.
6. How by QAM, the C-signal is produced from R-Y, and B-Y signals?
7. Draw block diagram and explain recording and reproduction of colour signal in a VCR.
8. Draw block diagram of a VCR. Explain the important functional blocks.
9. Explain the tape transport system of a VCR.
10. What do you mean by servo system of a VCR? Write a short note on the system.
11. Prepare a trouble shooting chart for a VCR.
12. Compare VHS, Betamax and V-2000.

13. Enlist the controls on VCR.
14. Why helican scan method of video recording is adopted?
15. Explain Servo System adopted in VCR.
16. Write short note on video tape transport mechanism.
17. What is VCD?
18. Write about DVD and DVD player.
19. Write note on: MP3, MP4, Pen drive and Blue tooth.

OBJECTIVE TYPE QUESTIONS

Fill in the blanks:

1. Recorded wavelength is dependent on ________.
2. VHS system has advantages of ________.
3. The erasing head of VCR uses ________ currents.
4. A video disc can provide an easy access ________.
5. Video recording is based on the principle of ________.
6. Video play back is based on the principle of ________.
7. Slant recording on video tape is done to ________.
8. By using rotating heads in video recording ________ is increased.
9. The frequency modulation is used in video recording to ________ bandwidth of the video signal.
10. The CED stands for ________.
11. LVD stands for ________.
12. V.C.R. stands for ________.
13. V.C.P. stands for ________.
14. The frequency of video signal is in ________.
15. VHS stands for ________.
16. OTR stands for ________.
17. ENG stands for ________.
18. Expand VCD ________.
19. Expand DVD ________.

ANSWERS

1. tape speed and signal frequency
2. good frequency response and better S/N ratio
3. R.F.
4. to any part of the program
5. electromagnetism
6. Electromagnetic induction
7. increase the duration of tape
8. relative speed
9. reduce
10. capacitive electronic disk
11. Laser vision disk
12. Video Cassette Recorder
13. Video Cassette player
14. Megahertz
15. Video Home Service
16. One touch recording
17. Electronic news gathering.
18. Video compact disc
19. Digital video disc.

❑❑❑

7

Communication Systems and Radio Receivers

The function of a communication system is to convey or send a message from one place to the another. The message may be an information or a signal. The information or a signal is obtained from a source and through a network, it is sent to the receiver.

Various communication systems are employed to transmit A.V. (audio-video) signals of telephone, radio, T.V., Radar etc.

7.1 COMMUNICATION SYSTEM

Structure of a communication system is shown below

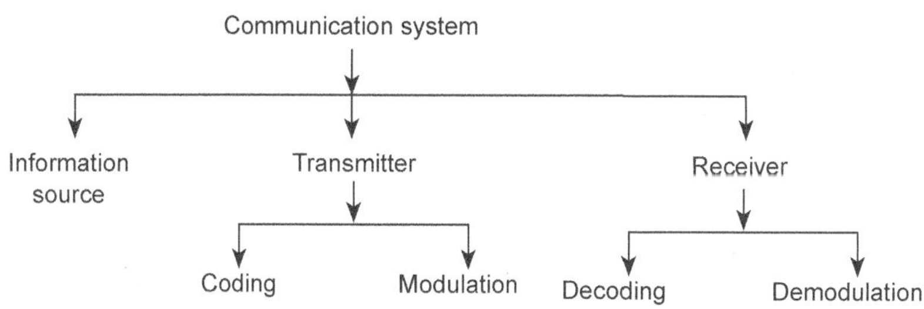

1. *The information source.* The information source is the source which generates or produces an information or signal.
2. *Transmitter.* Transmitter is the device which transmits the generated information or signal. It has the following components:
   (*a*) *Coding.* The transformation of the signal into a suitable form in which it can be transmitted.
   (*b*) *Modulation.* To superimpose the signal on a H.F. carrier, so that the signal can travel long distances.

3. *Receiver.* It is that which receives the information *e.g.* radio receiver/ T.V. receiver, telephone receiver etc. It consists of:
   (*a*) *Decoding*. It is the reverse of coding *i.e.* to re-gain the original form of the signal.
   (*b*) *Demodulation.* It is the reverse of modulation *i.e.* to separate the original signal from the carrier.

7.1.1 Types of Communication Systems

Basically communication systems are of two types:

1. *Wire communication.* Here communication is done through a wire *e.g.,* cable T.V., telephony, etc.
2. *Wireless or carrier communication.* Here communication is done without using a wire. In this system, a carrier wave is used. In other words, modulation is carried out e.g. radio, T.V., radar, radio telephony etc. Wireless Communication (radio) was invented by G. Marconi.

G. Marconi

Table 7.1: The frequency band for various communication systems

S. No.	Frequency Band	Type of communication
1.	Very low frequency (VLF) (5–30 kHz)	Long distance communication
2.	Low frequency (LF) (30–300 kHz)	Radio navigation
3.	Medium frequency (MF) (0.3–3 MHz)	Broadcasting (marine)
4.	High frequency (HF) (3–30 MHz)	FM broadcasting (television)
5.	Very high frequency (VHF) (30–300 MHz)	Radio broadcasting
6.	Ultra high frequency (UHF) (0.3–3 GHz)	Radar
7.	Super high frequency (SHF) (3–30 GHz)	Satellite communications

7.1.2 Modulation: Need of Modulation

In carrier (wireless) transmission, modulation is a necessity due to the following reasons:

(*i*) The original signal produced at mike etc. is very weak and it has a very low frequency, such a signal will die after some time moreover, it cannot travel long distances.

(*ii*) The transmitting and receiving antenna should have a height equal to a quarter wavelength of the frequency used. It will be shown by simple calculations, that at 1 MHz frequency, 75 m high antenna is required whereas at 15 kHz, an antenna with 5000 m height is required which is not practical.

The modulation is an important process in all wireless (carrier) communications. In this, the signal is superimposed on high frequency carrier waves. Some characteristic (amplitude, frequency or phase etc.) of the carrier wave is changed in accordance to the instantaneous value of the signal. A sine wave may be represented by

$$e = E_m \sin(\omega t + \varphi)$$

where e = Instantaneous value of the wave

E_m = max. amplitude

ω = angular frequency

φ = Phase angle

Accordingly, modulation is of three types:

1. *Amplitude modulation.* By changing amplitude of the carrier.
2. *Frequency modulation.* By changing frequency of the carrier.
3. *Phase modulation.* By changing phase of the carrier.

(a) Amplitude Modulation: This is the system of modulation in which the amplitude of the carrier is varied in accordance to the instantaneous amplitude of the signal. The frequency and phase remaining unchanged. (Fig. 7.1)

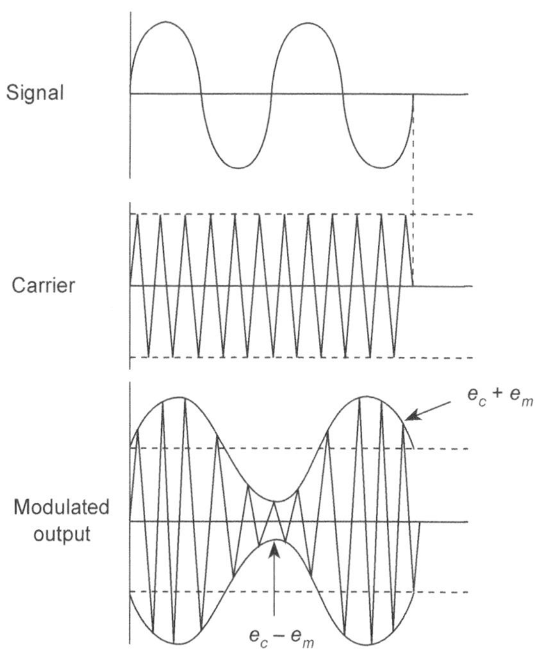

Fig. 7.1

1. AM Wave Equation

Let, the carrier is represented by:

$$e_c = E_c \sin \omega_c t$$

and signal is represented by

$$e_m = E_m \sin \omega_m t$$

Let A be the amplitude of the modulated wave. Then

$$\begin{aligned} A &= E_c + e_m \\ &= E_c + E_m \sin \omega_m t \\ &= E_c + mE_c \sin \omega_m t \; (E_m = mE_c) \\ &= E_c(1 + m \sin \omega_m t) \qquad \text{...(1)} \end{aligned}$$

Let, the voltage equation of the output modulated wave,

$$e = A. \sin \omega_c t$$

Putting the value from eq. (1)

$$\begin{aligned} e &= E_c(1 + m \sin \omega_m t).\sin \omega_c t \\ &= E_c \sin \omega_c t + \frac{mE_c}{2}\cos(\omega_c - \omega_m)t \\ &\quad - \frac{mE_c}{2}\cos(\omega_c + \omega_m)t \end{aligned}$$

This is the equation of the modulated wave.

2. AM Spectrum

In the above equation

1. First part is unmodulated carrier wave, which remains unchanged in the process. The max. amplitude is E_c.
2. Second part has the max. amplitude of $\frac{mE_c}{2}$ and its frequency is equal to the difference of carrier and the signal frequencies. This is called *Lower Side* Band (L.S.B.).
3. Third part has also max. amplitude of $\frac{mE_c}{2}$ and frequency equal to the sum of carrier and signal frequencies. This is called *Upper Side* Band (U.S.B.) (See Fig. 7.2).

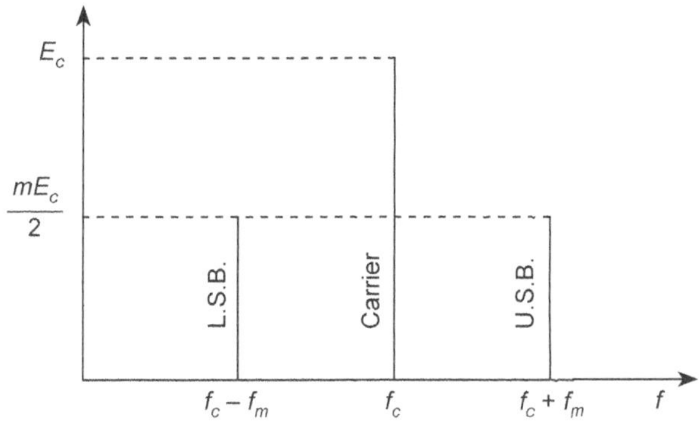

Fig. 7.2

3. Modulating Index (m)

(*a*) The modulating index may be defined as ratio of the maximum value of signal voltage to the maximum value of the carrier voltage. It is represented by

$$m = \frac{E_m}{E_c} \text{ or } E_m = m\,E_c$$

(*b*) The Fig. 7.3 shows a modulated (radio) wave.

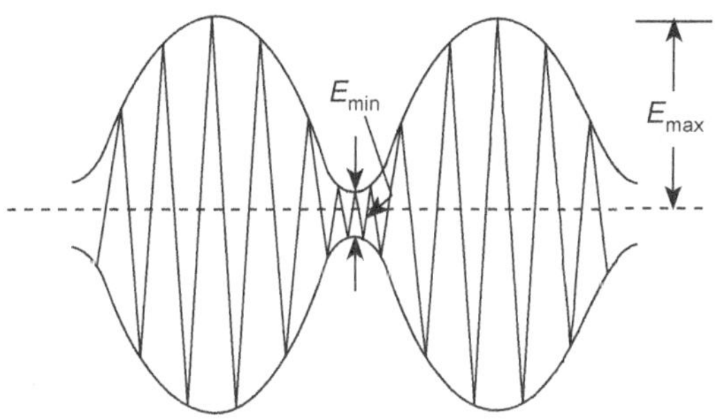

Fig. 7.3

$$m = \frac{E_{max} - E_{min}}{E_{max} + E_{min}}$$

The value of m is between 0 to 1.

(b) Frequency Modulation: In this process, frequency of the carrier is changed in accordance to the amplitude of the signal—the amplitude and phase remaining the same.

The Fig. 7.4 shows the process of frequency modulation. It can be seen that frequency of tهe modulated wave is maximum at points A & C and minimum at B.

(*a*) **FM Modulating Index**

The modulating index in FM is the ratio of *carrier deviation* to the *signal frequency*.

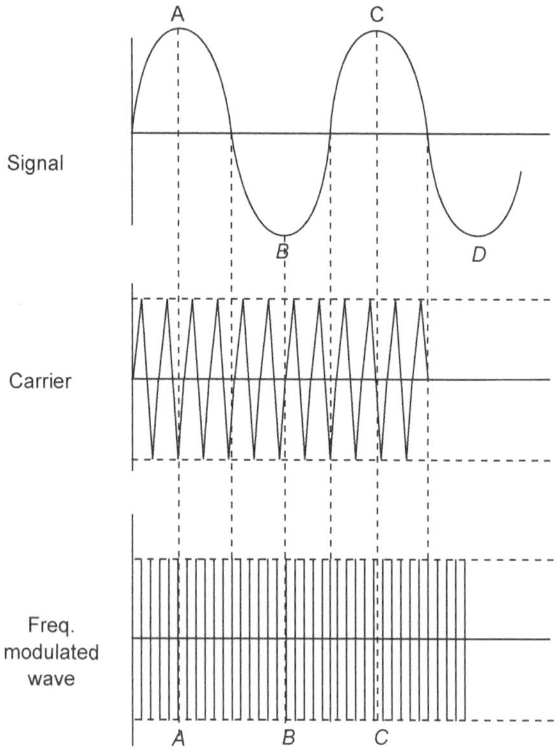

Fig. 7.4

(*b*) **AM v/s FM**

1. In AM, there are only 3 frequencies (one carrier and two side bands), but in FM there are infinite number of side bands as well as the carrier, separated from the carrier by f_m, $2f_m$, $3f_m$ and so on.
2. In AM, the modulating index (m) decides the power of the modulated wave but in FM, total transmitted power remains constant however, with higher value of m, "Bandwidth" increases.
3. In AM, the amplitude of the carrier remains constant. But in FM the amplitude of the carrier does not remain constant. Some time in FM, the carrier may disappear completely.

(*c*) **Advantages of FM**

1. The amplified FM wave is independent of the modulating index.
2. The FM reception is almost noise free as there happens to be lesser noise at frequency at which FM is used.
3. By increasing modulating index in FM, noise can be further reduced.
4. In F.M., the channel interference is less.

(*d*) **Disadvantages of F.M.**

1. In FM, a much wider channel is required.
2. The FM transmission as well as reception is complex and expensive.

(c) Phase Modulation: In this process of modulation, phase angle of the carrier voltage is changed in accordance with the instantaneous value of the modulating signal, the amplitude and frequency remaining the same.

In phase modulation, the deviation (modulating factor) is dependent only on the amplitude of the modulating signal voltage but is independent of the signal frequency, (whereas in F.M., the deviation is inversely proportional to the modulating frequency.)

Another difference is that in P.M., bandwidth is extremely large *i.e.,* the side bands do not coverage with the increase in frequency, (whereas in F.M., side bands coverage rapidly.)

This is why, the phase modulation is not used commercially for radio broadcasting and radio telephony.

7.1.3 Various Communications and the type of modulation used

1. *Radio broadcasting.* It is done on long, medium and short wave bands in the frequency range from 20 Hz to 30 MHz. The modulation used is amplitude modulation. This signal travels through ionosphere.

   The FM sound broadcasting is also done and it is done at VHF bands *i.e.,* from 20 MHz to 250 MHz. The short range of transmission, smaller wavelength, wider bandwidth and higher quality are the few important characteristics of FM transmission.

2. *T.V. broadcasting.* The T.V. broadcasting is done in VHF and UHF bands of frequencies. For picture signal, AM and for sound signal FM systems are used.
3. *Satellite broadcasting.* A satellite is a transmitter installed in space. The part of earth in its range can receive signals sent from the satellite. Signals upto 30 GHz (1 GHz = 10^9 Hz) are used in this system. For individual reception, a dish antenna of 1 m diameter maybe required.
4. *Telephony, telex, telegraph.* For transmission of these signals, fixed radio links at frequency upto 30 MHz are used. Now such links have been replaced by microwave or satellite links which give more reliability. The FM is invariably used.
5. *Communication satellites.* A satellite is installed in the space at a height of about 40,000 km. For reception, parabolic dishes of diameter 10 to 20 m are used at earth stations. The earth transmitter sends signals to the satellite which reflects it back to earth at an other carrier frequency. The T.V. and telephone signals are transmitted by this method.
6. *Mobile systems.* For radio communications between mobile units *e.g.*, marine, ships, air craft etc, Amplitude modulation (AM) is used, but above 30 MHz, FM is used.

7.2 RADIO TRANSMITTERS: CLASSIFICATION

Radio transmitters are used to process and transmit the signal.

(a) Classification: These may be classified on the following basis:

1. Type of Modulation Used: On this basis, radio transmitters are of the following types:

(*a*) *AM transmitters.* In these transmitters, carrier is amplitude modulated by the signal. These are employed for:

(*i*) Radio broadcasting—on medium and short waves

(*ii*) Radio telephony—on short waves

(*iii*) TV picture broadcasting—on very short waves.

(*b*) *FM transmitters.* In these transmitters, carrier is frequency modulated by the signal. These are used for:

(*i*) Radio broadcasting—in VHF and UHF range.

(*ii*) T.V. sound broadcasting—in VHF and UHF range.

(*iii*) Radio telephony on VHF and UHF range.

(*c*) *PM transmitters*. In these transmitters, the carrier is "phase modulated".

2. **Type of Service:** On the type of service, transmitters are of the following types:

   (*a*) *Radio broadcasting transmitters*. These transmitters are employed for transmission of sound signals for public recreation. They have low distortion and less noise. They may be AM or FM transmitters. The AM transmitters operate on medium and short waves and FM transmitters operate on very short waves.

   (*b*) *Radio telephone and telegraph transmitters*. These transmitters are employed for radio telephony i.e. to transmit telephone conversations and telegraph signals over long distances by radio means. These transmitters are also equipped with special devices such as volume compressors, privacy devices etc. Both the antenna are also specially designed. These may be AM or FM transmitters. The AM transmitters work on short waves, whereas FM transmitters work on UHF range.

   (*c*) *T.V. transmitters*. Two transmitters are used for T.V. broadcasting. One for transmitting picture and other for sound— both transmitters operate in VHF/UHF range. The picture transmitters are amplitude modulated, whereas the sound transmitters are frequency modulated. Vestigial side band transmission is used.The total BW occupied by one T.V. channel is 7 MHz.

   (*d*) *Radar transmitters*. Radar transmitters may be pulse transmitter or a continuous wave (CW) transmitters. They are pulse or frequency modulated. They operate on microwave (3 cm wavelength) frequencies of 3,000 to 10,000 MHz.

3. **Frequency Range:** On the basis of frequency range, the transmitters may be

   (*i*) *Medium wave transmitters*. They operate over a range of 550 to 1650 kHz frequency.

   (*ii*) *Short wave transmitters*. They operate at a range of 3 to 30 MHz. Ionospheric propagation is employed.

(*iii*) *VHF/UHF transmitters.* They operate in VHF (30 – 300 MHz) or UHF (0.3 – 3 GHz) range of frequencies. They are used for FM radio, T.V., radio telephony etc.

(*iv*) *Microwave transmitters.* These transmitters are used on frequencies above 1000 MHz. Their application is in radars.

(b) **Block Diagram of AM Transmitter:** The Fig 7.5 shows an AM broadcasting transmitter. An oscillator produces high frequency carrier waves which are fed into RF amplifier for amplification to the desired value. The audio signal to be transmitted is also amplified in audio amplifiers. Both amplified inputs are given to an AM modulator and the modulated output of the modulator is further amplified and fed to the transmitting antenna.

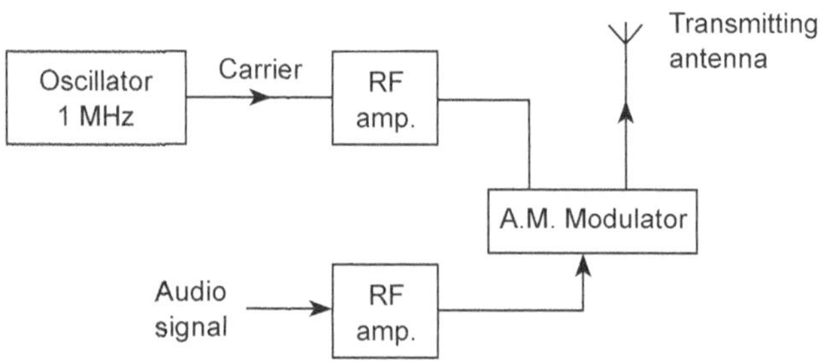

Fig. 7.5

7.3 RADIO RECEIVERS

As explained earlier, the signal to be transmitted is impressed *i.e.,* superimposed on a carrier wave, amplified and fed to a transmitting antenna. This modulated signal is radiated or propagated in the space and a part of it is intercepted by the receiving antenna. Now function of the radio receiver starts. As the signal received by receiver is very weak usually in picowatts, the receiver first amplifies the received signal. Further as the signal is likely to accompanied by lot of other unwanted signals of neighbouring frequencies, the wanted signal therefore must be selected and all others rejected, and finally the reverse of modulation that is demodulation should be carried out in the receiver to recover the original signal. Now, it is given to loudspeaker (L.S.). The Fig. 7.6 shows complete radio reception.

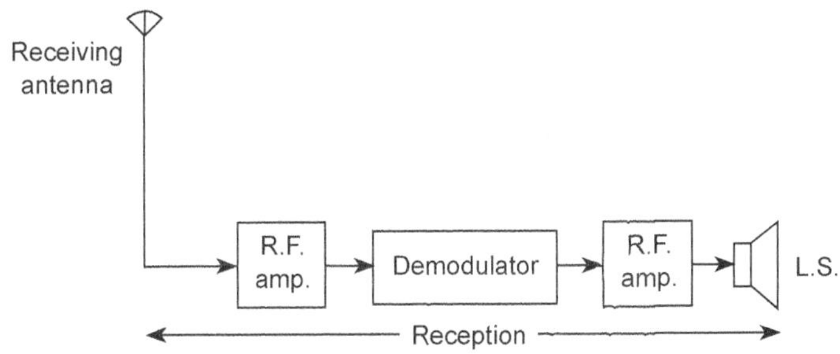

Fig. 7.6

(a) Characteristic of Radio Receivers: Here, we shall discuss only three important characteristics of radio receivers:

(*i*) *Sensitivity (Fig. 7.7).* Sensitivity of a radio receiver is the ability to amplify weak signals. It may also be defined in terms of "the voltage that must be applied to the receiver input to give a standard output power of 50 mW for all types of receivers, if the loudspeaker is replaced by a load of equal value".

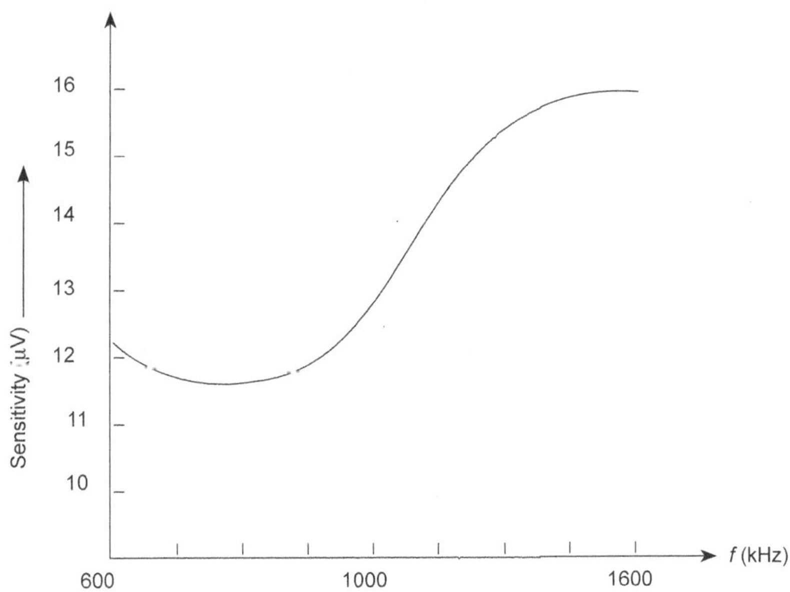

Fig. 7.7

(*ii*) *Selectivity* (See Fig. 7.8). The selectivity of a receiver is its ability to reject unwanted signals. It is expressed as a curve which shows the attenuation that a receiver offers to the neighbouring frequencies.

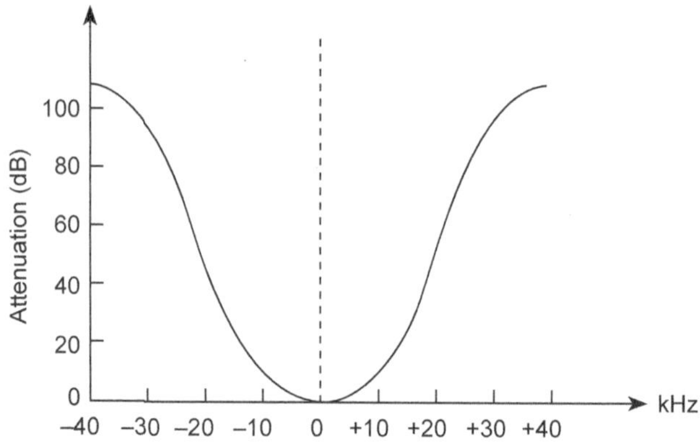

Fig. 7.8

(*iii*) *Fidelity.* By fidelity, we mean the quality of reproduction of original sound or picture. This property shows how "faithful" is a receiver in reproducing even the "minor details" of the original sound/picture.

On the same lines, "High Fidelity" (h-f) systems have been produced which possess a high degree of fidelity.

(b) AM Radio Receivers: Two types of AM radio receivers are popular and are of commercial importance, they are:

1. Tuned radio frequency receivers
2. Superhetrodyne receivers

1. **Tuned Radio Frequency (TRF) AM Receiver:** In this receiver, two or more RF amplifiers tuned together are employed to select and amplify the incoming frequency and rejecting all others. After this, the signal is detected and amplified suitably (See Fig. 7.9).

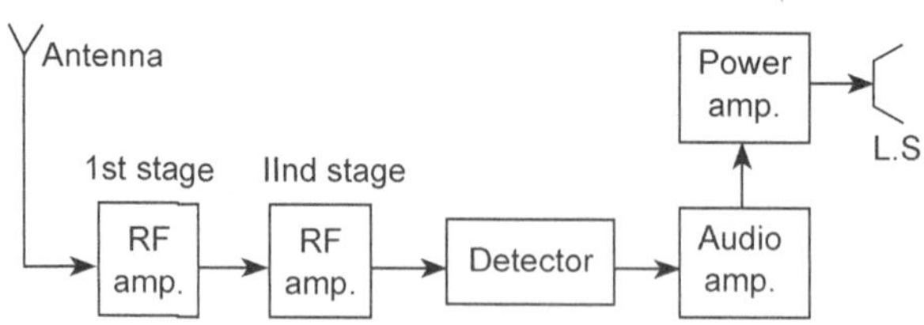

Fig. 7.9

These receivers are simple and have good performance upto 1650 kHz but not suitable for higher frequencies. They have also the problems of instability, insufficient selectivity and bandwidth variation. Due to this, they are not used, and are been replaced by the *superhetrodyne receivers*.

2. Superhetrodyne AM receiver. (Fig. 7.10): This is the most popular radio receiver. The various signals floating in the space induce their emf individually on the antenna of the radio receiver.

The particular frequency station is tuned by a tuning (or LC) circuit. This frequency signal is received by the receiver and all others are earthed. This frequency wave is amplified in a RF amplifier and output is given to a mixer. A local oscillator in the receiver generates a particular frequency, which is also fed in the mixer such that output of the mixer is always 455 KHz which is called intermediate frequency (IF). The I.F. output is given to detector (demodulator) which detects the original AF signal and the carrier is earthed. This AF signal is amplified and fed to loudspeaker (L.S.) which converts it into sound signal originally transmitted.

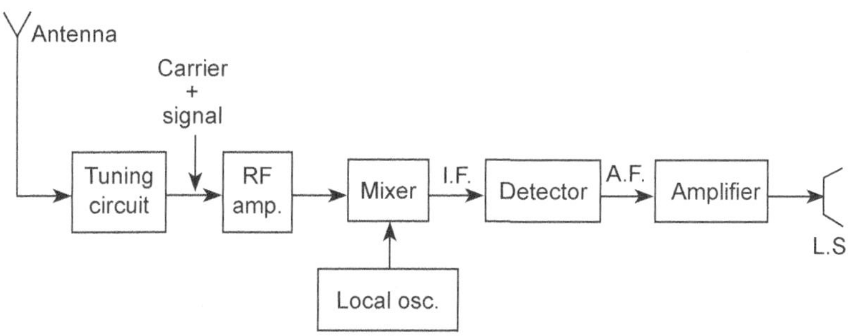

Fig. 7.10

(c) FM Radio Receiver: The Fig. 7.11 shows superhetrodyne FM radio receiver which is similar to AM receiver with the difference that a FM receiver has much more operating frequency moreover, it has a limiter and de-emphasiser network, different methods of obtaining modulation and gain control. A typical FM receiver operating at 87 MHz to 107 MHz may have an IF stage of 10.6 MHz and band width of 198 kHz.

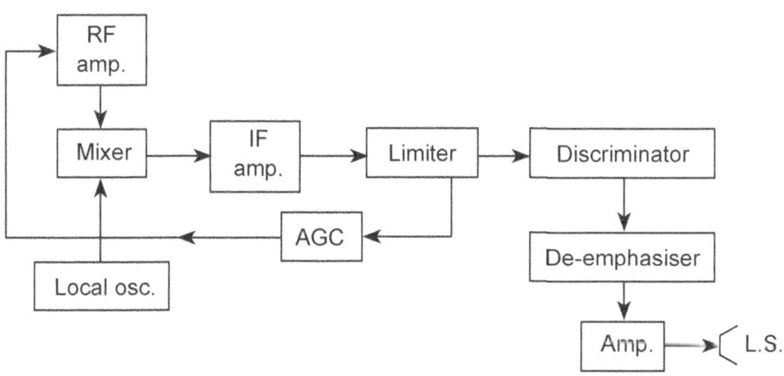

Fig. 7.11

A FM receiver always has a RF amplifier. It reduces noise level as it has to handle large BW. The input impedance of the receiver should be properly matched with impedance of the antenna.

A limiter is a type of "clipper circuit". It checks that the input to the demodulator does not contain any "spurious" signals which will be a source of distortion or noice. If need arises, two limiter circuits in cascade arrangement may be used.

The FM demodulator is bit different to the AM demodulator. A FM demodulator is a frequency-to-amplitude converter which converts the frequency deviation of the carrier into amplitude variations identical to the original signal.

7.4 DIODE DETECTOR

The diode is the most common device used for demodulation or detection. The Fig. 7.12 shows a simple diode detector circuit.

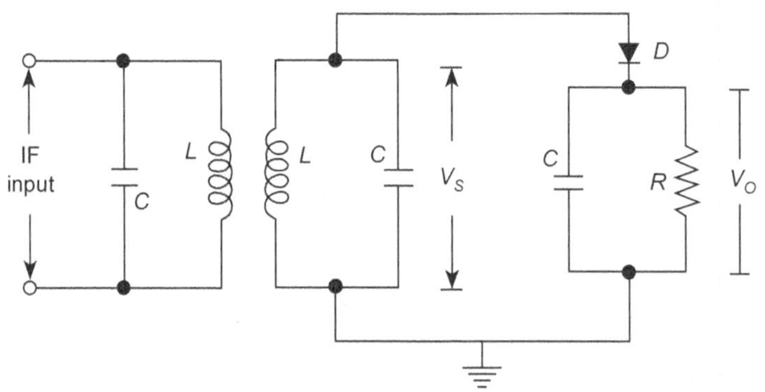

Fig. 7.12

The parallel combination of a capacitor *C* and resistance R is the load resistance across which the rectified signal voltage(V_s) is developed. The capacitor *C* charges upto a voltage equal to the peak signal voltage. The V_0 reproduces the modulating voltage accurately. The time constant RC should be suitably designed to keep RF ripples in the output as small as possible.

The circuit suffers with the disadvantage that the output contains d.c. components and also RF ripples which are unwanted.

7.5 AGC AND AFC

(a) **Automatic Gain (or Voltage) Control (AGC/AVC):** AGC is an electronic device by which gain of a radio receiver changes automatically with the changing strength of the signal, so that the output remains constant. A dc bias voltage is applied to a selected number of RF, IF and mixer stages. The overall result on the receiver output has been shown in the curves. (See Fig. 7.15)

The AGC irons out input signal amplitude variations and the gain control is not required to be adjusted everytime, whenever the receiver is tuned from one station to another.

The Fig. 7.13 shows four AGC curves. The second is ideal AGC curve, here no AGC will be applied till the strength of the signal remains within limits and after point M, a constant output will be obtained independent of the signal strength. The fourth is a "delayed AGC curve". This shows that AGC is not applied till the signal strength is within limits and after, the AGC is applied "strongly".

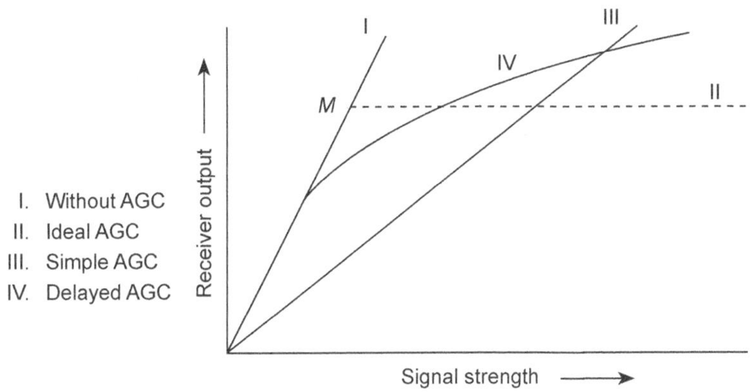

Fig. 7.13

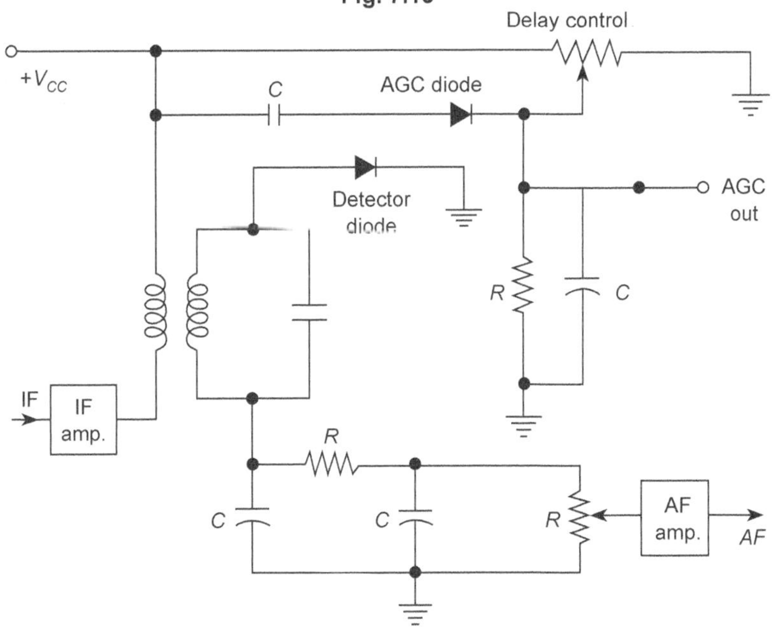

Fig. 7.14

Fig. 7.14 shows a delayed AGC circuit. There are two separate diodes—one for detector and other for AGC. A positive bias is applied to the AGC diode to prevent its conduction till the signal strength is within limits. A "delay control" as shown is provided to allow manual adjustments of the AGC, diode bias and hence on the signal level at which AGC is applied. If weak stations are to be tuned, delay control setting may be quite high to keep AGC out.

(b) Automatic Frequency Control (AFC) (FIG. 7.15): The heart of an AFC circuit is a frequency sensitive device such as *phase discriminator*, which produces a dc voltage whose amplitude and polarity are proportional to the amount of the local oscillator's frequency error. This dc control voltage is then used to vary this bias on a variable reactance device whose output capacitance is thus changed. This variable capacitance appears across the first local oscillator and the frequency of *variable frequency oscillator* (VFO) is automatically free from any variations with temperature, voltage etc.

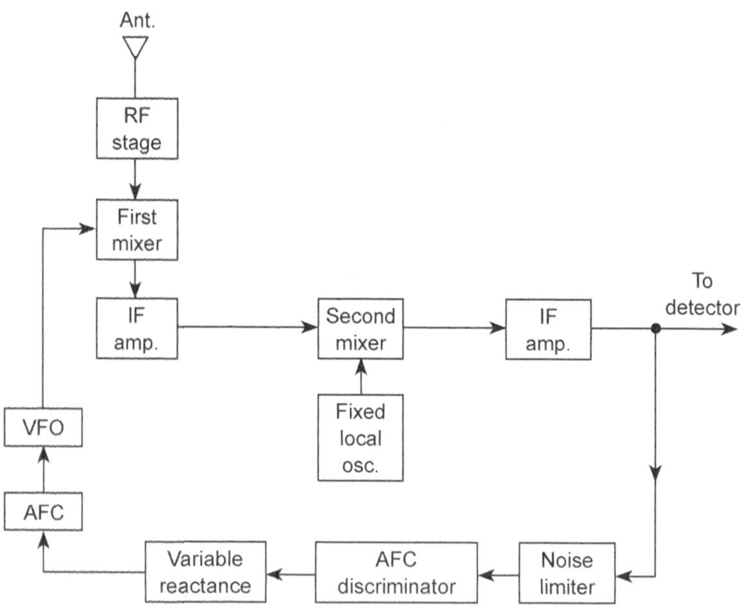

Fig. 7.15

Few receivers are equipped with a device called "Noise limiter". Though it is not possible to reduce "random noise" the receivers, however it is possible in a FM system. What does a noise limiter in an AM system is that, it reduces noise pulses created by electrical machinery, sparks etc. nearby. This is done by putting the receiver "out' for that duration. In common noise limiter circuit, a diode is used along with a *differentiating circuit* which provides a negative bias during that duration to the diode detector which is thus cuts off, and remains cut off in that period which is in milli-seconds and after becomes ON automatically.

It is to be noted that all receivers do not require an AFC.

SUMMARY

1. A communication system has the following components:
    1. Source of information
    2. Transmitter
    3. Receiver.
2. For transmission of sound, frequency modulation and for transmission of picture, amplitude modulation is used.
3. In amplitude modulation, amplitude of the carrier is changed according to the amplitude of the message.

   In frequency modulation, frequency of the carrier wave is changed according to the amplitude of the message.
4. In the radio receiver, the original sound is detected, amplified and given to loud speaker.
5. Two types of radio receivers used are:
    (*i*) Tuned radio frequency receiver
    (*ii*) Superheterodyne receiver

    Superheterodyne receivers are popularly used.

SELF TEST QUESTIONS

1. What do you mean by communication? What are the important components of a communication system?
2. What is modulation. Why it is necessary? Explain in brief the various modulation systems.
3. Describe amplitude modulation. In which audio video system it is used and for what purpose?
4. Explain frequency modulation versus amplitude modulation.
5. Name various classes of transmitters. Describe an AM transmitter.
6. Write short notes on:

(*i*) TRF radio receiver

(*ii*) Superhetrodyne radio receiver.

7. What is an AGC? Explain its function with the help of its characteristic curves.
8. What is an AFC? Explain its function.

OBJECTIVE TYPE QUESTIONS

Fill in the blanks:

1. Modulation is the process of changing ________ in accordance with the ________.
2. Wire communication finds use in ________.
3. Wireless communication finds use in ________.
4. Carriers are ________ waves.
5. Radar uses a frequency band of ________.
6. In T.V., for sound transmission, ________ is used.
7. The commercial receivers are of ________ type.
8. AGC stands for ________.
9. For T.V., ________ aantenna is used.
10. Dishes are ________ antennas.
11. The frequency band for a television is ________.
12. For F.M., a wide ________. is required.

ANSWERS

1. carrier, signal	2. telephony	3. Radio, TV
4. H.F.	5. 0.3–3 GHz	6. F.M.
7. Superhetrodyne	8. Automatic gain control	
9. Yagi	10. Parabolic	11. 0—30 MHz
12. B.W		

❑❑❑

8

Antennas

In earlier pages, we have discussed transmission and reception of radio waves. The *transmitters* transmit or propagate and receivers receive, what has been propagated. Thus in order to couple the transmitters and receivers with the space, some sort of *interface* is essential. A structure must be provided which is capable to radiate electro magnetic waves from the transmitter and to receive the same in the receiver. This structure is known as *Antenna* or *Aerial*.

8.1 ANTENNA OR AERIAL

An antenna or aerial may be defined as a "metallic structure often a wire or collection of wires/conductors which *converts h.f. currents into electromagnetic* waves and vice-versa.

Transmitting antenna and receiving antenna behave identically *i.e.*, their behaviour is reciprocal.

8.2 ANTENNA AS A TRANSMISSION LINE

A transmitting antenna or aerial can be considered as an open circuited transmission line. The space around the line may be considered as *load* with complete *mismatch* conditions. This, "transmission" line (antenna) radiates signal in the form of e.m. waves.

Consider an open circuited transmission line shown in Fig. 8.1 (*a*). It is seen that forward and reverse travelling waves combine to form a *standing wave* pattern on the line.

It is to be mentioned that all the transmitted energy is not reflected back by an open circuited line but a small portion of the electromagnetic energy escapes

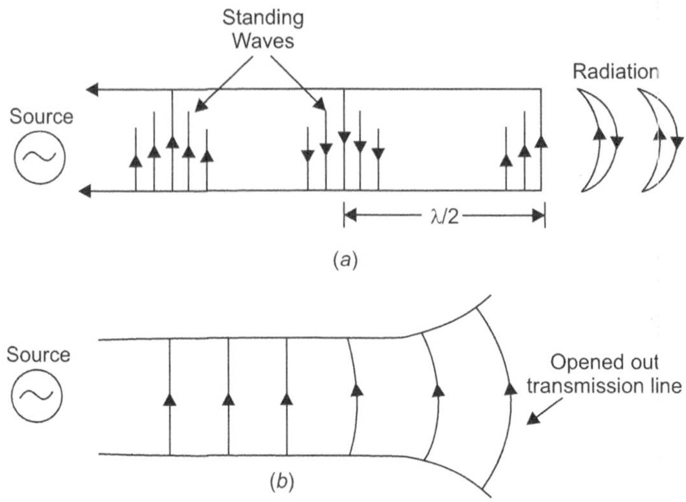

Fig. 8.1

from the line and thus radiated. The radiated energy is very small, most of the energy is returned back to the source due to the follow reasons:

(*i*) A complete "mismatch" exists between source (line) and the load (space) and thus a very little energy is dissipated to the load.

(*ii*) Two wires are close together with opposite polarity, the radiation from one wire cancels the radiation from the other.

If the lines are spread out as shown in Fig. 8.1 (*b*), due to the "enlargement" of the open circuit, the mutual cancellation of the radiation is reduced and also a better "*match*" is established of the line with the space, more energy is dissipated to the space *i.e.*, radiation increases.

8.3 EVALUATION OF A DIPOLE : THE BASIC ANTENNA

The radiation efficiency of the system is further improved if the wires are bent so as to be in the same line, see Fig. 8.2. The electrical and magnetic fields are now fully *coupled* with space, instead of being confined between the two wires. This is a *dipole antenna* which is the basic form of antenna in use today. When the total length of the two wires is *a half wave length,* the antenna is called a *half wave dipole*. By keeping half wave length ($\lambda/2$), even greater radiation occurs. The reason for this is that half wave dipole may be regarded as having the same impedance properties as a similar length of transmission line. Accordingly, we have the antenna behaving as a piece of "quarter wave" transmission line bent out and open circuited at the far end. This results as a high impedance at the far end of the antenna reflected as a low impedance at the end connected to the main transmission line. This means that a large current will flow at the input of the half wave dipole and efficient radiation will take place.

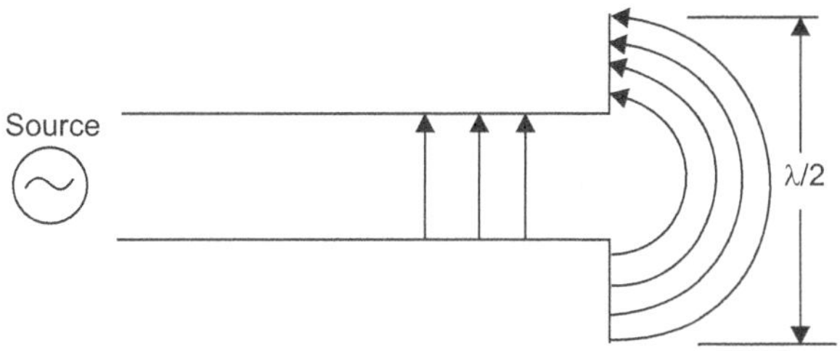

Fig. 8.2

The Fig. 8.3 shows ideal voltage and current distribution along a half wave dipole, which is the simplest wire antenna.

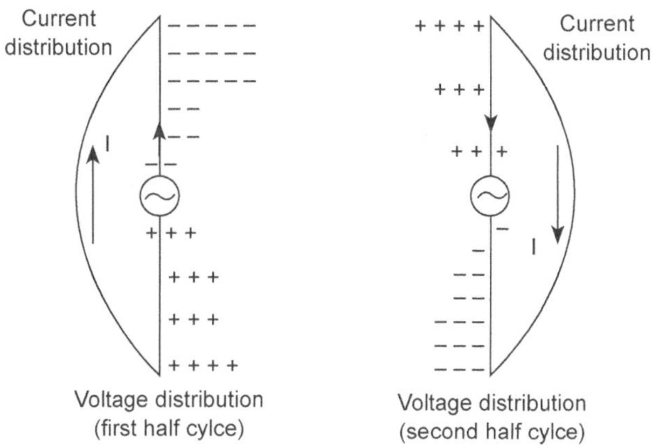

Fig. 8.3

A *current maximum* and *voltage minimum* appear and the antenna feed point, an identical situation exists λ/4 away from the open circuit on a transmission line. Also current and voltage polarity changes every half cycle as shown. (Fig. 8.3)

8.4 TRANSMISSION BY ANTENNA (SKYWAVE COMMUNICATION)

The transmitting antenna radiates signal energy in the form of em waves which travel through the layers of the "*ionosphere*". The radiation changes with season and also with day and night. These waves propagated by antenna are called *Sky waves*.

The ionosphere consists of four layers: (Fig. 8.4)

(*i*) ***First layer:*** exists at a height of 70 km from grounds and is 10 km thick. It disappears at night and reflects low frequency signals.

(*ii*) ***Second layer :*** exists at a height of about 100 km and is about 20 km thick. It also disappears at night.

(*iii*) ***Third layer :*** exists at a height of about 185 km. Its thickness is about 50 km, but at night it mixes with fourth layer and thus disappears. It absorbs high frequencies.

(*iv*) ***Fourth layer :*** It exists at about 250 km with 150 km thickness which becomes 200 km at night.

Sky waves are radiated by transmitting antenna at various angles, (see Fig. 8.4).

The distance between transmitting antenna and point where sky wave is first received on earth after reflecting from ionosphere layers is known as *skip distance*. If the signal is strong it can be reflected back by the earth which is again reflected by ionosphere to the earth thus, through series of reflections, signal can be transmitted *around the world*. The angle above which sky wave no longer returns to earth and escapes into space is called *critical angle*. At high frequencies, ionosphere becomes less effective, at frequencies above 30 MHz, the signal does not return to the earth.

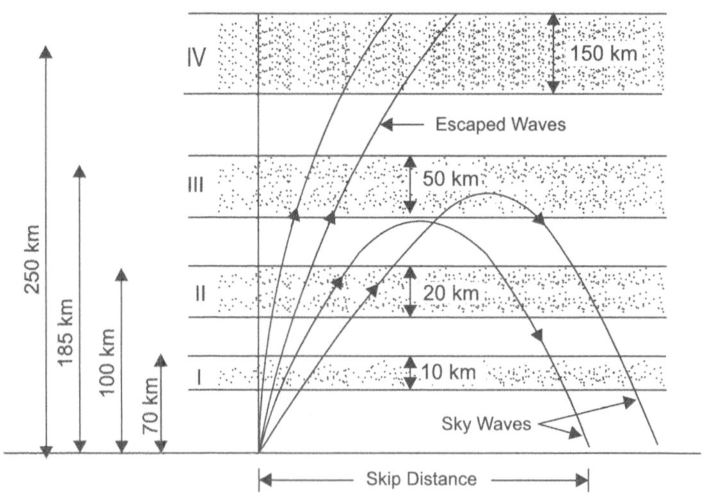

Fig. 8.4

The maximum frequency that can be used for "sky wave communication" is known as *maximum usable frequency* (MUF). There is a different value of MUF for each pair of points on the earth, its normal value being between 10-30 MHz.

8.5 TERMS RELATED TO ANTENNAS

The important terms related to antennas are defined below:

8.5.1 Gain

The gain of an antenna is not same in all directions, no doubt its radiation and the gain in one particular direction will be maximum. Accordingly an antenna have the following gains:

(*a*) ***Directive gain or directivity :*** This is defined in a particular direction as the "ratio of the power density radiated in that direction by the antenna to the power density that would be radiated by an *isotropic* antenna."

Here we should note that an **isotropic antenna** may be assumed as an antenna which is "omini-directional", *i.e.*, which radiates (equally) in all directions. It is just an imaginary antenna used as a *reference*, through practically no such antenna exists.

Directive gain or directivity is a ratio of powers and its value (measured at some standard distance from antenna) is more than unity in all cases. The value of this gain for **resonant antenna** is from 1.64 to 7.2 in free space and for **non-resonant antenna** it is from 3.2 to 17.5.

The non resonant antenna has more directive gain than the resonant antenna of equal length. The directive gain is expressed in *decibel.*

(*b*) ***Power gain :*** Power gain is "the ratio of the power that must be radiated by an isotropic antenna to develop a certain field strength at a certain given distance to the power fed to the antenna to develop the same field strength at the same distance in the direction of maximum radiations".

Thus this definition is similar to that of the "Directive gain", the only difference is that in case of "directive gain or directivity" the *power radiated* by the antenna is considered, while in the case of *power gain* "the power fed" to the antenna is taken. Thus the two terms are identical except that power gain also takes into account the antenna losses.

Directivity is of theoretical, whereas power gain is of practical importance.

8.5.2 Antenna Resistance

An Antenna has Two Resistances

(*a*) ***Radiation Resistance:*** Radiation resistance may be defined as the "ratio of power radiated by the antenna to the square of the currents at its feed point". It is an ac resistance like an equivalent resistance of a parallel tuned circuit. In simple words, we may say that this is the resistance, which if replaces the antenna, would dissipate same power as the antenna itself radiates.

(*b*) ***Loss Resistance :*** In addition to the energy radiated by an antenna, some energy is also dissipated as a result of **antenna and ground resistance**, corona and eddy currents induced in the antenna. This all *lumped resistance* is called as *loss* resistance.

8.5.3 Bandwidth

The bandwidth of an antenna is the frequency range over which the antenna will give satisfactory performance. In case of antennas, there are two bandwidths. One referring to the *radiation pattern* and the other to the *input impedance*. Everything being same, the radiation pattern bandwidth is equal to the difference between the frequencies, at which the received power falls *to half of the maximum*.

8.5.4 Beamwidth

Beamwidth of an antenna is the angular separation between its two 3 dB *down points* on the field strength radiation patterns. This is given in degrees. The term is used for narrow beam antennas.

8.5.5 Law of Reciprocity for Antenna

According to this law, the same antenna can be used as *transmitting* as well as *receiving* antenna.

8.6 DESCRIPTION OF VARIOUS ANTENNAS

The important antennas are described below:

8.6.1 Basic Antenna—A Simple Dipole [Figure 8.5 (a)]

A basic antenna is a metal rod having a length $\lambda/2$ (half bandwidth) at the frequency of operation. This is also called *half wave dipole*. Its two ends are at equal potential w.r.t. to its mid-point.

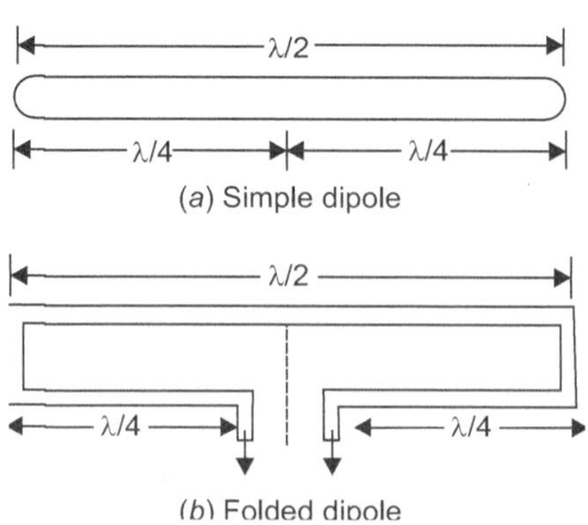

Fig. 8.5

A dipole may be a "folded dipole" which consists of two $\lambda/2$ dipoles: one a continuous rod and other splitted at centre and both connected in parallel as shown in Fig. 8.5 (*b*). The transmission line is connected to the splitted rod.

8.6.2 Resonant Antenna

A resonant antenna corresponds to a resonant transmission line and the **dipole antennas** are the example of a resonant antenna. Such an antenna can be taken as an opened out transmission line, open circuited at the far end of the resonant length *i.e.*, a multiple of a quarter wavelength so that the length of the antenna is a multiple of half wave length. (Fig. 8.6a)

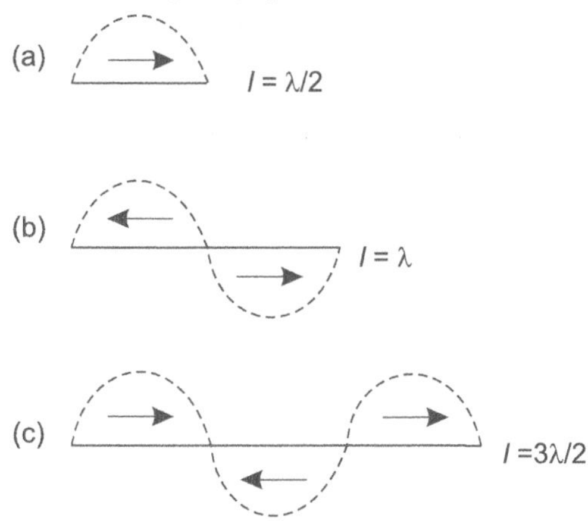

Fig. 8.6

When length of the antenna is a full wave length, the polarity of current in one half of the antenna is opposite to that on the other half as shown Fig. 8.6 (*b*). It is obvious that radiations at right angles from this antenna will be zero, because the field due to **one half** fully cancels the field due to **other half** of the antenna. The direction of maximum radiations now for a full wave dipole is not at right angles but at about 55° from the antenna. As the length of the dipole increased to $3\lambda/2$, current distribution is changed as shown. (Fig. 8.6 c)

The Fig. 8.7(a), (b), (c) shows polar diagrams (radiation patterns) of the resonant antenna in the above three cases.

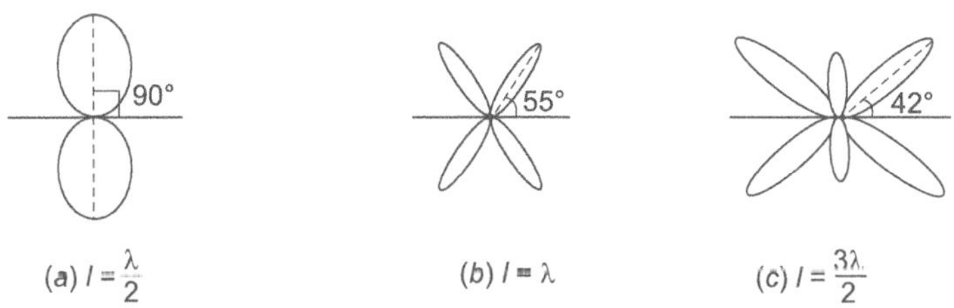

Fig. 8.7

8.6.3 Non Resonant Antenna

This antenna is like a non resonant transmission line, on which there are no *standing waves*. In a correctly matched line all the transmitted power is dissipated in the terminating resistance (Fig. 8.8). When antenna is terminated, about 2/3 of the forward power is radiated, the remainder is dissipated in the antenna and none is reflected back.

The non resonant antenna is unidirectional, since there are only forward travelling waves on it.

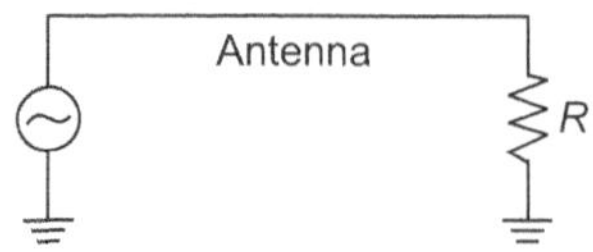

Fig. 8.8

These are of two types:

(*a*) **A non resonant (or a long wire) or V-antenna:** Basically is a long wire of several wave lengths suspended at a height from the ground. No reflection of the transmitted wave takes place, about half of the energy is radiated into space. This antenna has low efficiency and requires more space to install. It is cheap and can be used for *point to point communication*. It has the advantage of increased signal pickup and share directivity along the wire itself.

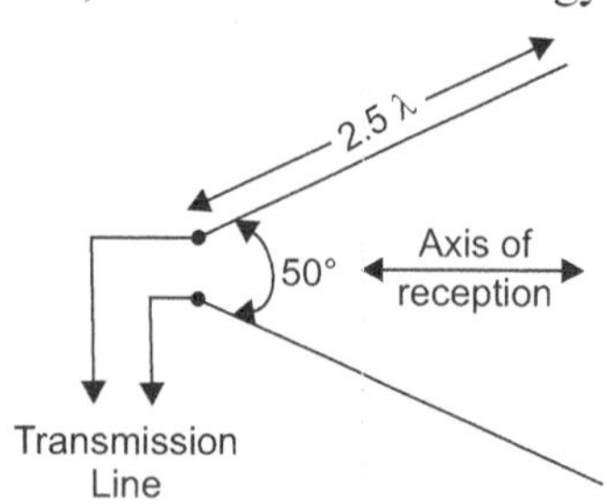

Fig. 8.9

The Fig. 8.9 shows long wire *V* antenna; in which two long wires are bent to make a *V* shape. For maximum gain the included angle should be minimum. The gain of this antenna is about 7 dB.

(*b*) **Long wire Rhombic Antenna:** It is a network of interconnected long antenna in the shape of a rhombus. The length and angles are to be designed properly to get the best results. It is a highly directional antenna and used for *point to point* communications with higher gain reception. Actually, it consists of "Two horizontal *V* antennas". The gain of this antenna is about 10 dB. (See Fig. 8.10).

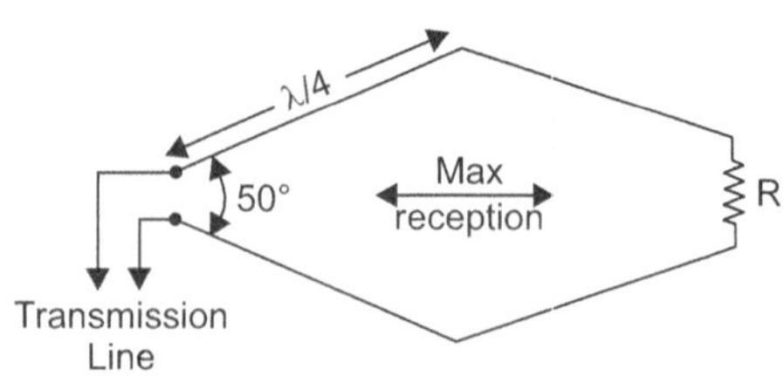

Fig. 8.10

8.6.4 Broadside Antenna

It is a network of number of dipoles of equal size spaced by λ/2 from each other. All the dipoles are fed from the same source. (See Fig. 8.11).

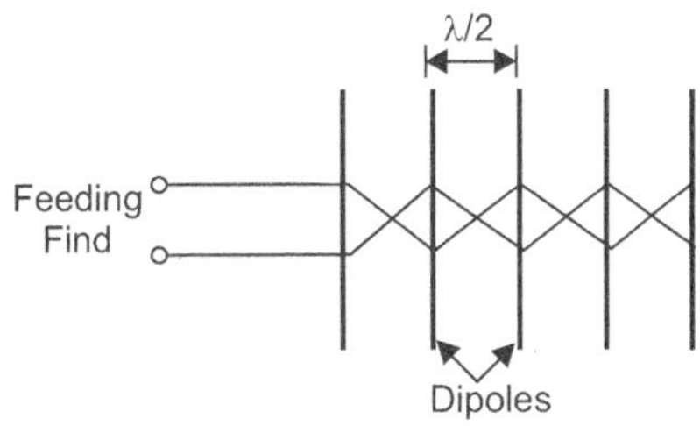

Fig. 8.11

8.7 T.V. RECEIVER ANTENNAS (PARASITIC ARRAYS)

This is the most popular antenna used for T.V. reception. When current flows in the receiving antenna, it radiates a part of the intercepted signal energy that may be wasted out. But, if a conductor approximately of λ/2 length is placed parallel to the main receiver element (half wave dipole) but not connected to it, this conductor will intercept some of the energy radiated out by dipole, therefore the lost signal is regained thus increasing the efficiency of the antenna. The conductor used for this purpose is called **parasitic element** as it is free and *not connected* to the dipole. A parasitic element placed behind the antenna is called **reflector** and if placed in front of the antenna, it is called a *director*. The dipole itself is the main antenna and called the **driven** element. This can be a straight or folded dipole.

A dipole with one or more parasitic elements is called as a *parasitic array* (*i.e.*, net work). This is the most popular TV receiver antenna as it is simple and cheap in construction, easily oriented and increases directivity and gain. The main **disadvantage** is that it is **unidirectional** *i.e.*, it can intercept signal from one direction only.

The parasitic antennas may be of the following types:

8.7.1 Antenna with a Dipole and Reflector

In this, only one reflector is placed behind the dipole at a distance of 0.2 λ. The length of the reflector is 5% more than the dipole. The spacing is designed carefully to get the required results. The gain of this antenna is about 5 dB. (See Fig. 8.12)

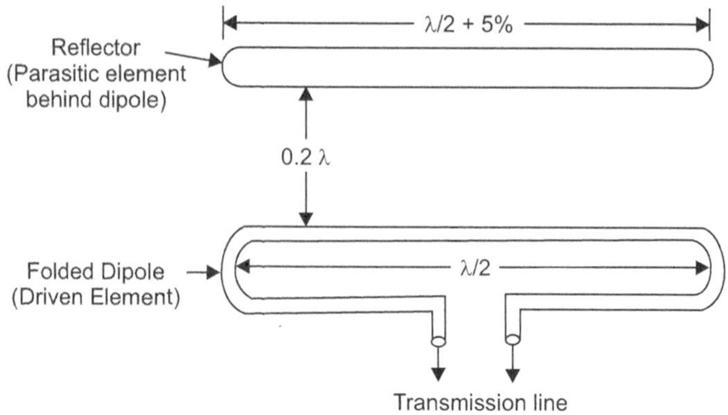

Fig. 8.12

8.7.2 Antenna with a Dipole and Director (Fig. 8.13)

The antenna has one dipole and one director placed infront of the dipole. The spacing of the director from dipole is about 0.15 λ. The length of the director is about 5% less than that of the dipole. As the dipole is λ/2 in length, the length of director is λ/2 – 5%. The antenna with one parasitic element uses one "reflector" (instead of director) because director needs close spacing for the same gain which narrows the frequency response of the antenna.

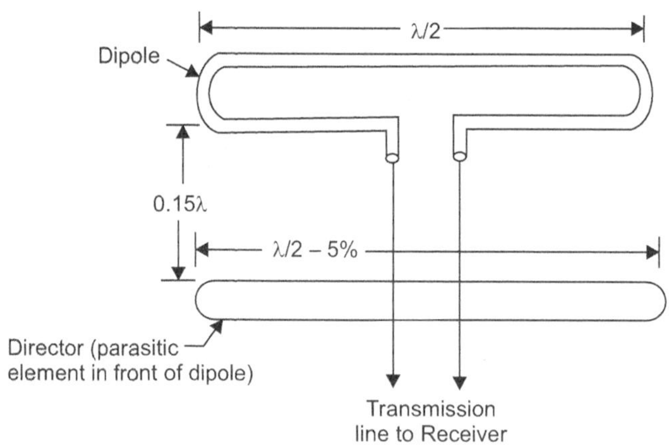

Fig. 8.13

8.7.3 Antenna with one Reflector and one Director

Such antenna uses one reflector at the back and one director at the front. They operate at narrow frequencies. Such an antenna has a gain of about 7 dB. (See Fig. 8.14).

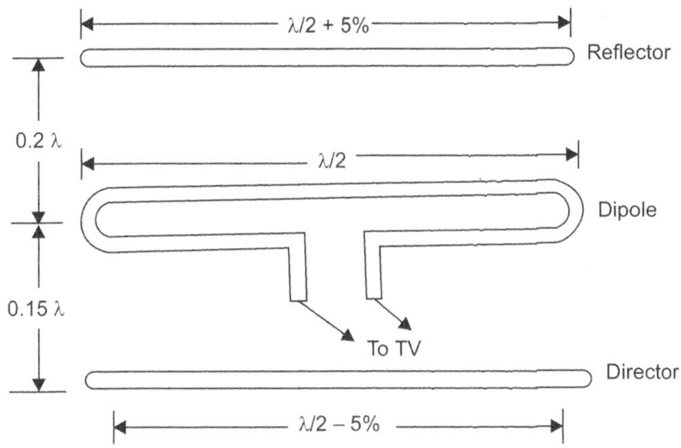

Fig. 8.14

8.7.4 Antenna with one Reflector and More than one Directors *i.e.*, Yagi-uda Antenna

This is a compact, high gain, narrow bandwidth network antenna used in low signal areas to cover several adjacent TV channels. The gain of yagi-uda antenna with 3 parasitic elements is about 10 dB. A high impedence folded dipole is used so that the antenna impedence can be about 300 Ω. We can increase the number of directors but generally one reflector is sufficient as additional reflectors do not improve performance of the antenna. Each director reduces its length by 5% of the previous, so that antenna is *tapered* in the direction of transmission. The elements get currents induced from the main element (folded dipole) or directly from the transmitted waves. (See Fig. 8.15)

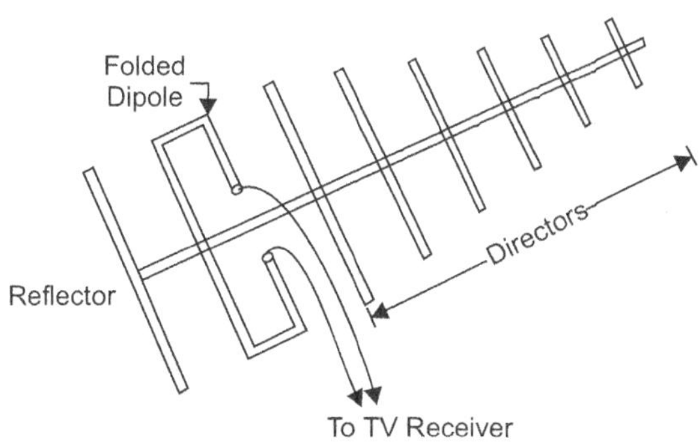

Fig. 8.15

8.7.5 Antenna with Dipole and Corner Reflector

As shown in Fig. 8.16, this has a reflector constructed as a corner conducting sheet behind the dipole. The reflector can be a solid metal sheet or a wire. The dipole insulated from the reflector is mounted along the line bisecting 90° angle. This antenna is useful for VHF bands.

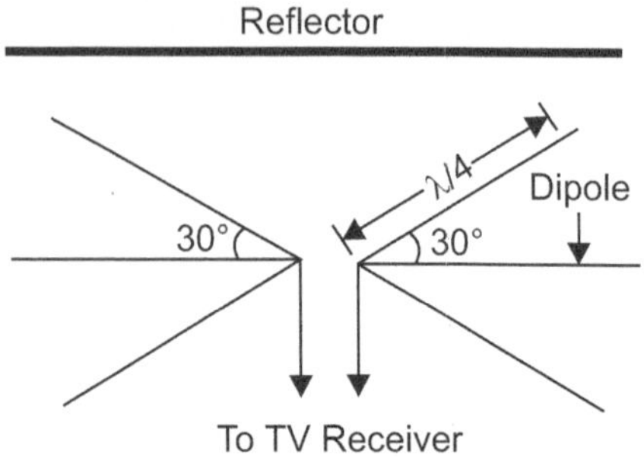

Fig. 8.16

8.8 TV TRANSMITTING ANTENNAS

The television signals are transmitted by space wave propagation, so the height of the antenna should be as high as possible in order to increase the "line of sight distance". Horizontal polarization is standard for television broadcasting, as signal to noise ratio is favourable for horizontally polarized waves, when antennas are placed quite high above the surface of the earth.

The following antennas are used for transmission of TV signals:

8.8.1 Turnstile array

In order to obtain equal TV signal radiation in all directions, an arrangement known as **Turnstile array** is used. In this antenna, two crossed dipoles are used in "turnstile" arrangement. These are fed in **quadrature** from the same source by means of an extra $\lambda/4$ line. Each dipole has a pattern of figure "Eight" in

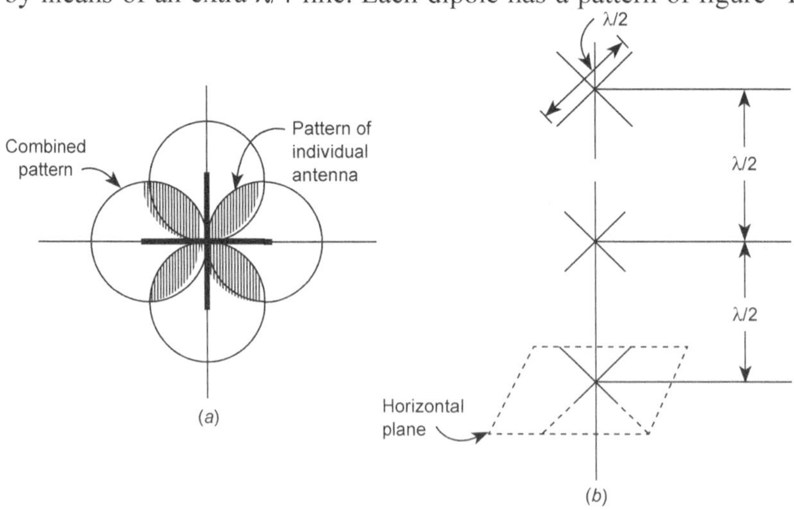

Fig. 8.17

the horizontal plane but crossed with each other and the resultant pattern is as shown in Fig. 8.17 (a). The Fig. (b) shows stacked turnstile array for vertical directivity.

8.8.2 Dipole Panel Antenna System

Each panel consists of an array of full wave dipoles mounted as shown in Fig. 8.18 (*a*). For obtaining unidirectional pattern, the four panels are mounted on the four sides of the antenna tower as shown in Fig. (*b*).

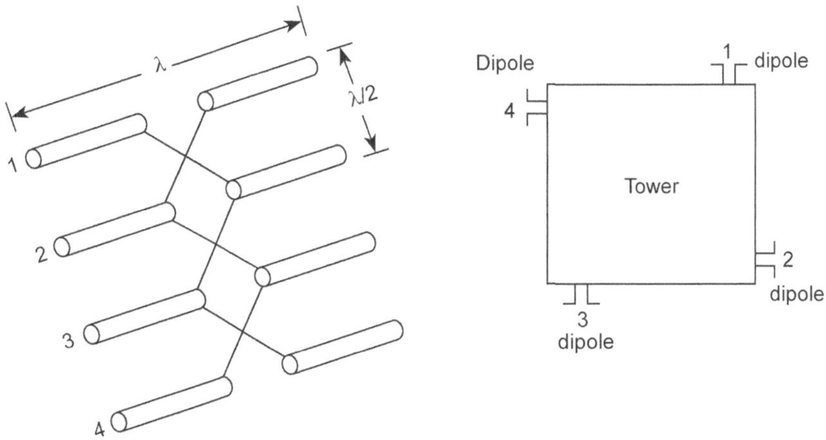

Fig. 8.18

8.8.3 Combining Network

The picture signal (amplitude modulated) and sound signal (frequency modulated) from corresponding TV transmitters are fed to the same antenna through a balancer unit called diplexer. The antenna combining system is a "bridge" configuration", in which two arms are formed by two radiators of turnstile antenna and the other two arms are formed by two capacitive reactances. Under balance conditions, picture and sound signals radiated by the same antenna do not interfere with the functioning of the transmitter other than their own. The Fig. 8.19 shows equivalent bridge circuit of a diplexer for feeding picture and sound transmitters to a common turnstile array.

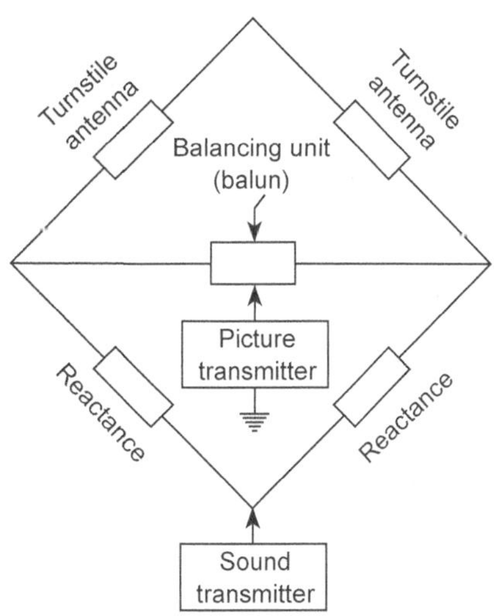

Fig. 8.19

8.9 UHF AND MICRO WAVE ANTENNAS

The transmitting and receiving antenna used at UHF *i.e.,* ultra high frequency (0.3 to 3.0 GHz) and micro waves (1–100 GHz) must be directive, because they are used in radars and such like purposes, where their function is *direction finding*, hence the need of a *directive* antenna is must. The other fields of their use is *micro wave communications* where a directional antenna helps a lot due to its "high gain".

The *yagi uda* antenna mostly used as a TV receiving antenna discussed previously is a VHF (30– 300 MHz) antenna. The VHF region is an "overlap" region, therefore VHF antennas can also be used in UHF region.

Here we will discuss some antennas used in UHF/micro wave regions:

8.9.1 Antenna With Parabolic Reflectors (Dishes)

A "parabola" may be defined as the "locus of a point which moves such that its distance from a straight line (*called directrix*) and from another point (*called Focus*) remains constant".

The Fig. 8.20 shows a parabola whose focus is at *F* and axis is *AB*. If a source of radiation is placed at *F*, all waves coming from the source will be reflected by parabola as a result radiation will be very strong and concentrated along axis *AB*. The practical reflector is a three dimensional surface called *paraboloid or parabolic reflector or a micro wave dish.* According to the "principle of reciprocity" the properties of an antenna are independent whether it is used for transmission or reception purposes. The gain of such an antenna is also very high as the "dish" acts as a strong reflecting mirror and collect radiations from a large area and concentrates at the focal point.

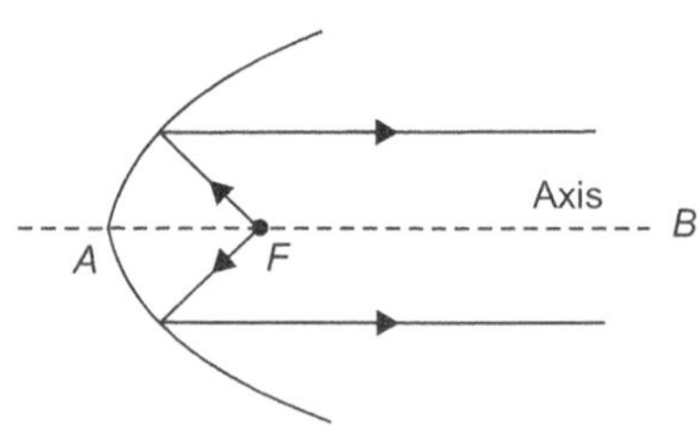

Fig. 8.20

The *antenna feed* is placed at the focus of the parabolic reflector F, the arrangement gives best results in transmission as well as reception.

The radiations from the "feed" which are not reflected by the parabola, spread out in all directions and spoil the directivity of the antenna. For this, a spherical reflector can be put (See Fig. 8.21) which will redirect all such radiations back to parabola. At the place of a reflector, a yagi"dipole" can also be used. The another way to deal with the problem is to use a "horn antenna' pointing at the main reflector.

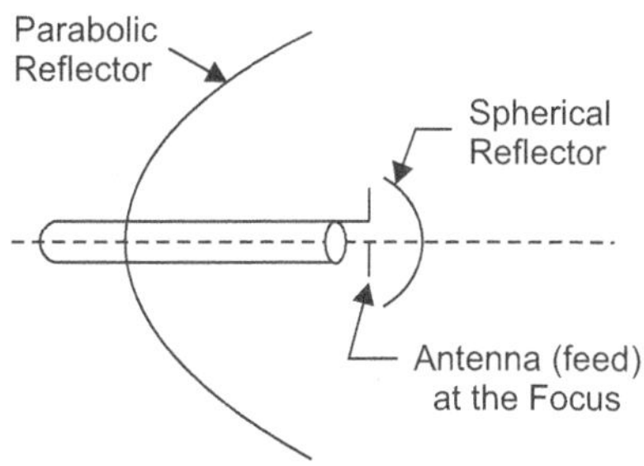

Fig. 8.21

8.9.2 Cut Parabolic Reflectors

A part (cut piece) of a parabola can also be used in place of full parabola. The only disadvantage is that the beam is not so directional as in the case of full parabolic reflector. In many applications like in "ship to ship" radar, directivity is immaterial. Figure 8.22 (*a*), (*b*), (*c*) show various shapes of cut parabolic reflectors.

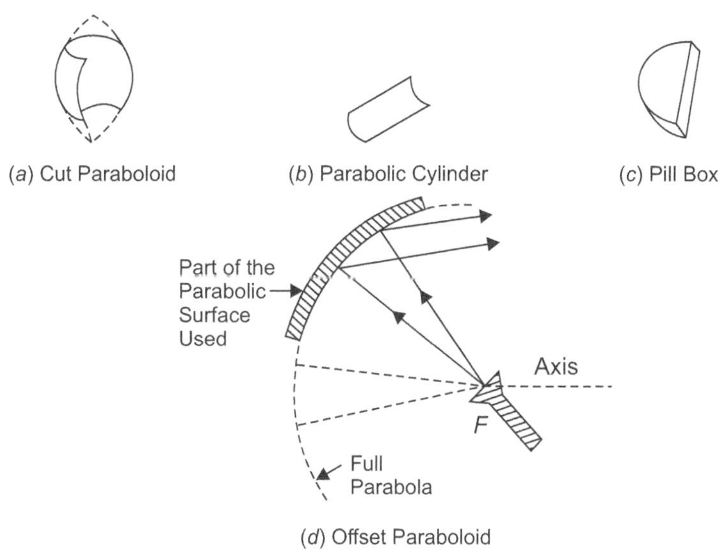

Fig. 8.22

The another form of cut parabolic is an "offset paraboloid" Fig. (*d*), in which focus (F) is located outside the aperture. If an antenna feed placed at the focus, the reflected rays will pass above it, without any interference.

8.9.3 Horn Antennas (Fig. 8.23)

In the previous pages, we have explained that electromagnetic waves will travel from one point to another if suitably radiated. It has also been explained how it is possible to "guide" radio waves from one point to another by the use of "transmission lines". Now we introduce a new term *i.e.*, "*wave guide.*

Any system of conductors (like transmission lines) for carrying electro magnetic waves can be called a wave guide, though in practice this name is used for specially constructed hollow metallic pipes-rectangular or circular. As transmission lines are used at low frequencies, in the same way, wave guides are used at microwave frequencies but wave guides are preferred as they incur less loss.

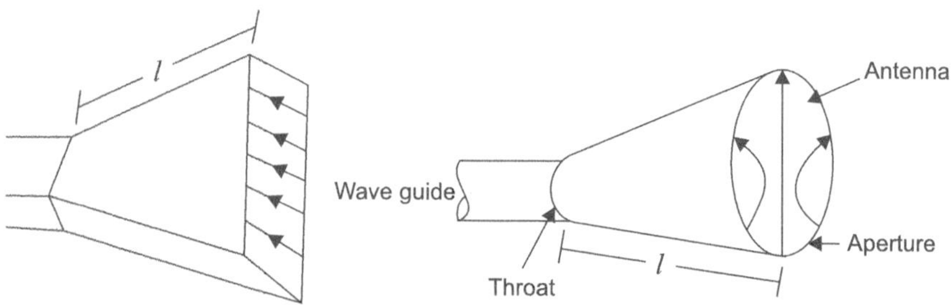

Fig. 8.23

As mouth of the transmission lines is opened out to get a dipole structure, similarly mouth of a wave guide is opened out to get a horn antenna.

Thus a horn antenna is nothing but a "wave guide terminated by a horn".

A horn antenna provides a better impedance match with the space around. Hence all the energy travelling in the wave guide is radiated as a result, directivity is improved, diffraction reduces, and a very little energy is returned back.

8.9.4 Lens Antennas

In a parabolic reflector, we have seen how principle of optics were utilised to receive or transmit a signal. Lens antenna also utilises the same principles and is very useful at 3 GHz or at even higher frequencies.

Figure 8.24 (*a*) shows the basic principle of a lens antenna. It works in the same manner as a glass lens used in optical instruments.

A lens antenna also proves the principle of reciprocity of antennas. If a source is kept at the *focus* of a convex lens, the rays are diverged that gives us a *transmitting lens antennas*. On the other hand, if it is used as a receiving antenna, it will converge all the signal waves at its focus. (See Fig. 8.21 (*a*), (*b*).

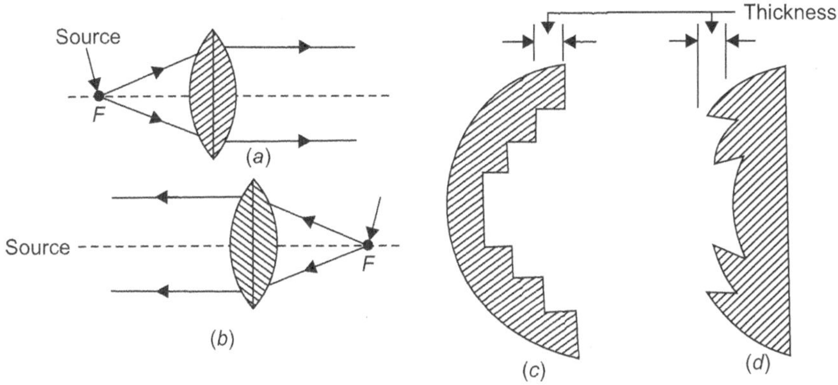

Fig. 8.24

Lens antenna suffers from a serious problem that it requires excessive thickness at frequencies below 10 GHz. To remove this problem, the lens antenna in practice are *stepped* that cures the problem of greater thickness at the centre of the antenna, when used at lower micro wave frequencies.

Fig. 8.21(*c*), (*d*) show two stepped lens antennas.

8.10 SPECIAL ANTENNAS

These are wideband antennas *i.e.*, which can be operated over a wide range of frequencies. These antenna have special applications in TV broadcasting, radar and satellite communications. Some are described below:

8.10.1 Helical Antenna

It is a broadband VHF and UHF antenna.

It consists of a loosely wound helix backed up by a "ground plane" which is a *screen* made of "chicken wire". There are two modes of radiations *i.e.*, *towards* axis and *normal* to the axis.

The helical antenna is used for transmission and reception of VHF signals through "ionosphere". It is widely used for satellite communications. (See Fig. 8.25)

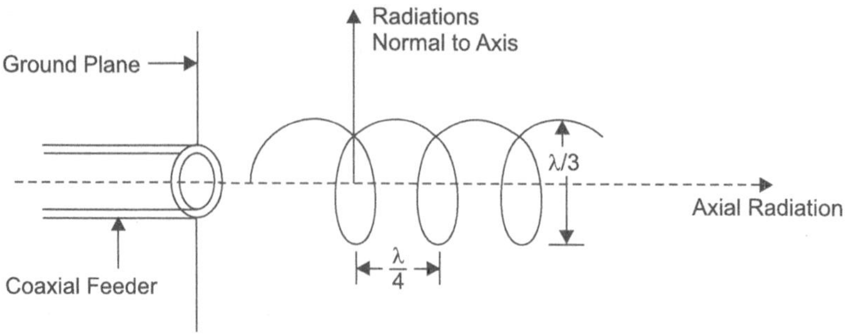

Fig. 8.25

8.10.2 Discone Antenna

A discone antenna is a combination of a *disc* and a *cone* as shown (See Fig. 8.26). The disc acts as a "reflector" and the antenna has an enormous bandwidth. However, it is a low gain but "**all directional**' antenna. It is used as VHF and UHF receiving antenna at the airports, where it maintains communication with the aircrafts coming from all directions.

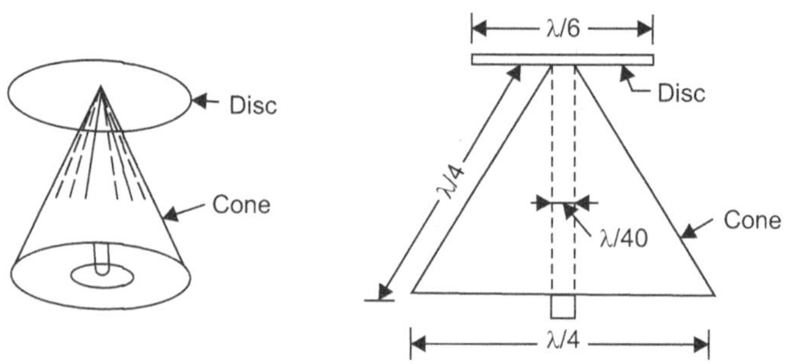

Fig. 8.26

8.10.3 Direction Finder (DF) Loop Antenna

The DF loop antenna is used for "direction finding" purposes and may be mounted on a portable radar receiver.

Basically, it is a single turn coil carrying RF currents. The loop of the wire is surrounded by magnetic field everywhere perpendicular to the loop. It acts as a dipole and its working is independent of its shape. The Fig. 8.27 shows a circular and a rectangular loop antenna.

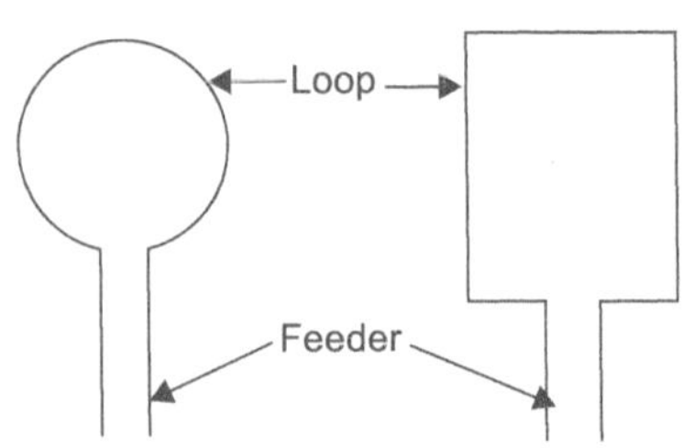

Fig. 8.27

No radiation is received normal to the plane of the loop, this makes it useful for direction finding purposes. Due to its being small and simple in size it is very much suitable to be mounted on portable radar receivers, and the output of the receiver may be connected to a meter and thus distance of the *target* can also be calculated.

Having tuned to the desired transmission, the loop is rotated till the received signal is *maximum*, the plane of the loop is now perpendicular to the direction of radiations. Loops are some times provided with several turns and also with a fertile core.

SUMMARY

1. An antenna or aerial is defined as a metallic structure or wire, or collection of wires/conductors, which converts h.f. currents into electromagnetic waves or vice versa.
2. There should be a complete impedance matching between antenna and the surrounding atmosphere so that maximum energy may be delivered from antenna into the atmosphere.
3. Important antennas are: Resonant antenna, antenna with dipoles and reflectors, dish antenna, cone antenna etc.
4. Dish antenna is used for cable TV/direct to home (DTH) service etc..

SELF TEST QUESTIONS

1. What do you mean by an antenna? Do you think it as an Audio-Video system?
2. Explain an antenna as a transmission line.
3. Write short note on "Sky wave" communication.
4. Define for an antenna system: Gain, Resistance, Band width, Beam width and Law of reciprocation.
5. What are the various antennas? Give their names.
6. Write short note on:
   (*i*) Dipole anetenna.
   (*ii*) Parasitic arrays and its various combinations.
7. Explain the Yagi antenna.
8. Explain a parabolic (dish) antenna
9. Write short notes on:

(*i*) Cut parabolic dishes

(*ii*) Horn antenna

(*iii*) Lens antenna

(*iv*) Discone antenna

(*v*) Loop antenna.

OBJECTIVE TYPE QUESTIONS

I. Fill in the blanks

1. The another name for antenna is____________
2. An antenna converts HF currents into____________.
3. An antenna can be assumed as an____________
4. An antenna is an____________ system.
5. The antenna, popularly used for TV reception is ____________
6. A dish is____________.

ANSWERS

1. aerial
2. electromagnetic waves
3. open circuited transmission line
4. audio video system
5. yagi antenna
6. parabolic reflector

❑❑❑

9

Basics of Television and Monochrome TV

Television is a very popular audio video device. It is now a basic need of every house. It is a source of entertainment and education.

9.1 TELEVISION

"Tele-Vision" means to "see at a distance". The visual information in the picture is converted into electrical signal for transmission. At receiver it is reconverted into its original form. In monochrome TV, the picture is reproduced in *black and white* (whereas in colour TV, it is reproduced in its original colours).

The TV was invented by J.L. Baird.

J.L. BAIRD

9.1.1 TV Applications

Few Application of TV are Discussed Below

(*a*) *Cable TV (CATV)*: When signal is to be sent to shorter distances, a cable is connected from transmitter to the receiver. Generally modulation is not required. The signals are distributed by co-axial cables to the consumers.

(*b*) *Close circuit TV (CCTV)*: In this system, video signal output of camera is connected by cable directly to the monitors placed at distance. Thus picture is reproduced on each monitor. (A monitor is a video display device, it is a TV receiver without RF and IF circuits). The CCTV equipment is available for monochrome (B & W), as well as for colour TV. A network of co-axial cable is used for connection.

(*c*) *Picture(Video) phone:* This is a *"telephone plus TV"*. By this phone we can see as well as hear each other.

(*d*) *Fax* (*Facsimile*): This is electronic transmission of visual information over telephone lines. It is also called *"slow scan TV"* as scanning in this case is slow. Only still pictures can be transmitted on fax.

(*e*) *Satellite TV*: By installing a satellite in space we can transmit signals on TV receivers over large areas. Satellites are used as *relay stations* to provide world wide TV broadcasting.

(*f*) *TV Games*: The earlier TV games were using TTL (Transistor-Transistor Logic) circuits. Now with the development of *microprocessors*, more sophistication has been obtained. Basically all TV games are *logic circuits*.

A TV game has two important parts. The *first* part is the "game unit" (which further includes control, logic circuit and RF oscillator circuit) and *second* part is the TV receiver. The system has been shown through a block diagram in Fig. 9.1 Now microprocessor controlled games are also available.

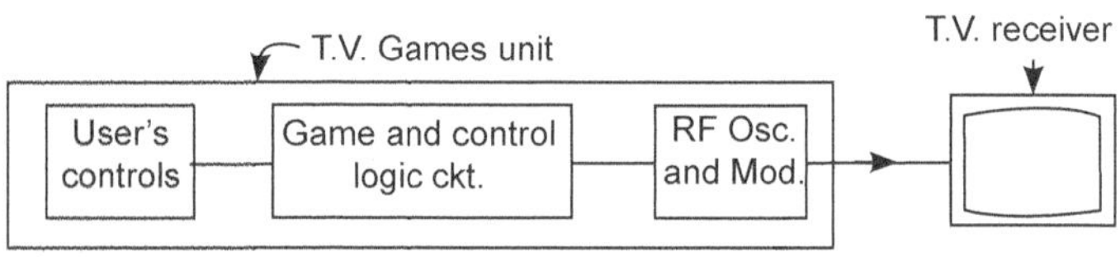

Fig. 9.1

9.1.2 Broadcasting, Transmission and Reception of a TV

(*i*) "*Broadcasting*" means to *"cast broad' i.e.*, to send a massage in all directions. The TV signal has audio as well video information. The transmitting antenna radiates TV signals in the form of electro magnetic waves that can be picked up by a TV receiver placed in the range. The amplitude modulation is used for the picture signal and frequency modulation for the sound signal.

(*ii*) *Transmission* (*Fig.* 9.2*a*): In a TV studio, audio signal produced at mike is converted into corresponding electrical pulses and fed to amplifier. For picture transmission, the picture signal (light energy) is converted through camera into electrical pulses. This video signal (electrical pulses) are given to a video amplifier. The both *i.e., sound as well as the picture* are sent to the transmitting antenna for transmission.

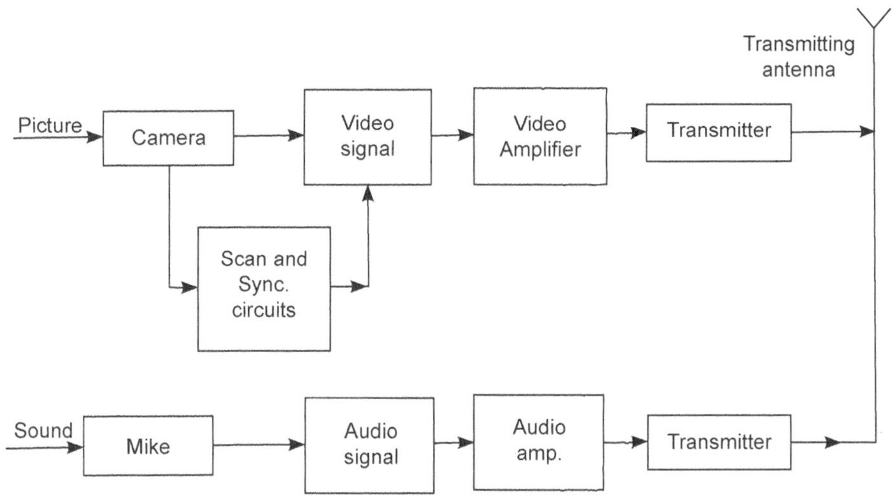

Fig. 9.2 (*a*)

(*iii*) *Reception* (*Fig. b*): Separate carrier waves are used for sound and picture signals but they are radiated by one transmitting antenna. At TV receiver also, the same antenna is used to pick up both signals. The signals in the receiver are amplified and separated. The picture signal is given to picture tube, which by its transducing action converts it into original picture produced in the camera. The sound signal is fed to the loudspeaker which gives the original sound produced before mike.

During transmission, to convert picture into video signal (electrical pulses), the camera *scans* the picture into horizontal and vertical lines (the picture is divided into 625 lines).

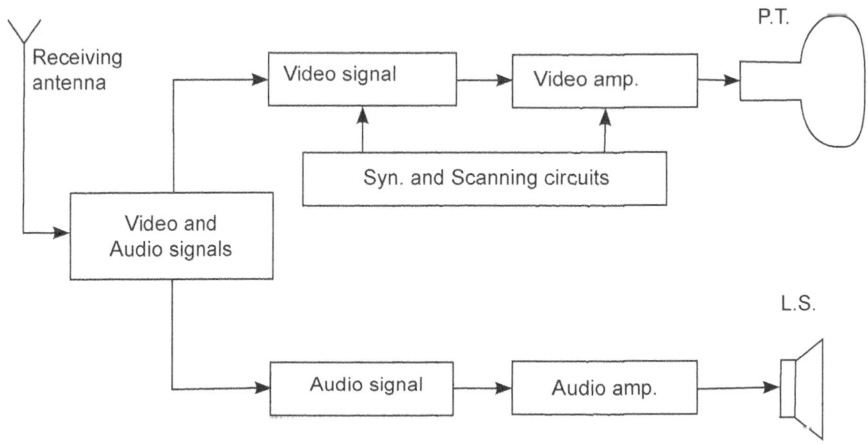

Fig. 9.2 (*b*)

This process is known as *scanning*. Similarly during reception, the picture tube re-assembles these lines to obtain the original picture. It is important that both the scanning processes taking place in camera as well as in the picture tube should be property *synchronized*. This is done by "scanning and synchronizing circuits".

9.1.3 TV Cameras

Camera is the first and basic equipment in a TV. The input to a camera is the light from the picture or scene to be televized and output obtained from camera is the electrical pulses corresponding to the information contained in the picture.

The TV camera is just *analogous* to the human eye. The basic principle of all TV cameras is based on the fact that each picture may be assumed to be composed of small elements with different light intensity. The camera picks up each element and by *transducing action* converts it into "electrical signal" proportional to its brightness. There is a photosensitive layer called *target or image plate* in each camera which performs this job. At the same time simultaneous pick up of this information is also necessary. For this purpose, there is an *electron gun* (which produces an electron beam) which *scans* the image plate at a fast speed. Thus in the camera, *opto-electric conversion* as well as *pick up* of the signal takes place simultaneously and at a fast speed.

(a) **Working Principle of Camera:** *The Photo emission and photo conductivity effects* are utilised for opto-electrical conversion. Certain metals emit electrons when light falls on them. The property is called photo emission. See Fig. 9.3 (*a*). The emitted electrons are called *photo electrons* and the surface is called a *photo cathode*. Generally, *Cesium* or *Bismuth* are used for making photo cathodes of cameras.

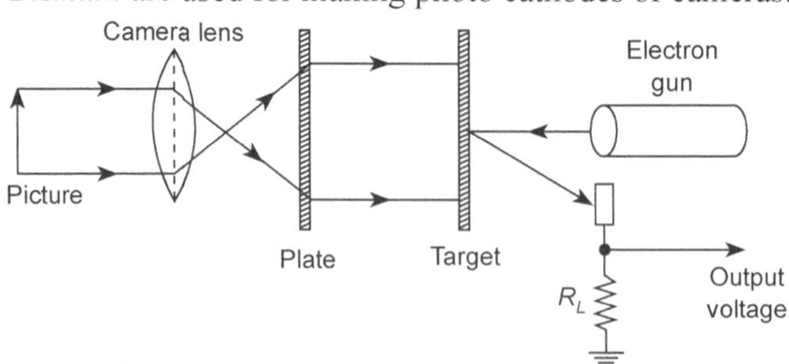

Fig. 9.3 (a)

The *Photo conductivity* is the property, according to which a metal loses its resistance when light is incident upon. Generally *selenium* and *lead* make good photo conductive surfaces and hence used in cameras. (Fig. 9.3 b)

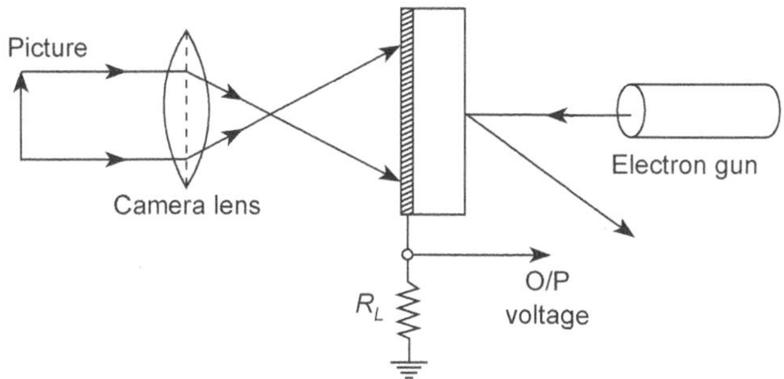

Fig. 9.3 (b)

Resolution of a Camera

By resolution of a camera we mean, how efficiently a camera can *scan* the picture element. If resolution of camera is high, we can see finer details of the picture. Resolution is given in percentage.

Persistence of Vision

Our eye can retain a picture only for 50 ms after the picture is removed. This property of eye is known as persistence of vision. Video systems (TV etc.) work on this principle.

(b) Various T.V. Cameras: The various TV cameras are described below:

1. Image Orithicon (I.O.) Camera, (See Fig. 9.4): This camera tube can be divided into 3 sections—image section, scanning section and the multiplier section.

Working

Light from the scene is focussed on the photo cathode. This gives an *electron image* corresponding to the picture. Needless to tell, that the *cathode* is made of photo emissive material like cesium oxide. This electron image is accelerated to the *target* to produce secondary emission. The secondary emission produces on the target a pattern of positive charges corresponding to the picture, the *white* colour being most positive. The scanning beam from the electron gun neutralises these positive charges and the excess electrons of the beam return back to the gun.

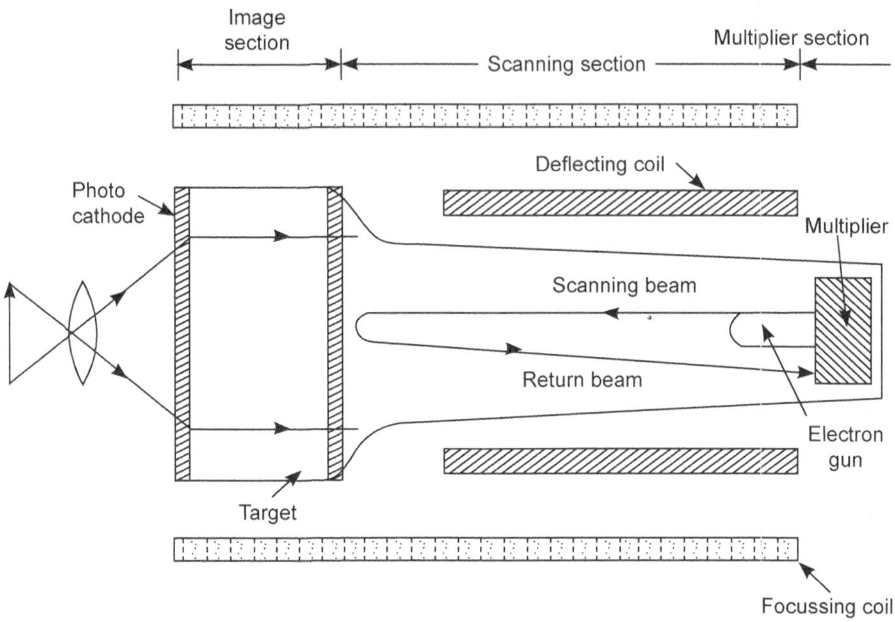

Fig. 9.4

As the beam *scans* the target, the electrons turning back provide a signal current in accordance with the picture. The signal current is maximum for the *black* colour of the picture. Now the signal current passes through the *multiplier* section, where it is amplified. The amplified current passes through the load and produces the camera signal output voltage. If signal current is assumed as 5 μA and load as 20 K, the output voltage is 0.1 V.

2. Vidicon Camera: Vidicon is a very small camera tube of about 8″ long and 1.5″ dia. It is simple in construction as it has only *target plate* and the *gun*.

(*a*) ***The target*** (See Fig. 9.5). It has two layers, one layer is of transparent glass coated with some conducting material usually *Tin oxide*. The second layer is coated with thin photo conductive material like Antimony sulphide (Sb_2S_3). The Tin oxide acts as positive terminal for the target.

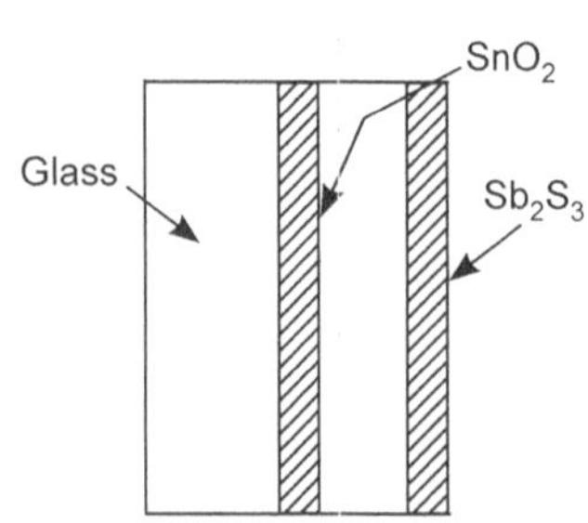

Fig. 9.5

(*b*) ***Electron Gun:*** The electron gun has a heated cathode, a control grid and a focussing grid. The deflection of the beam produced by the gun is used for scanning, and the direction of the beam is controlled by *deflecting* coils. [See Fig. 9.6 (*a*)]

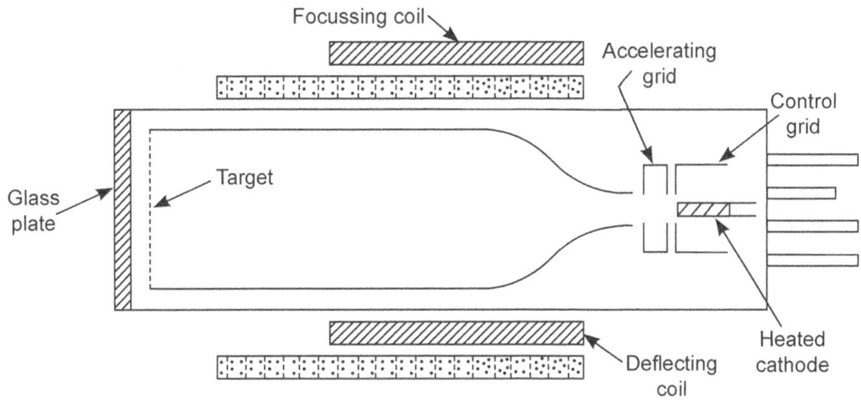

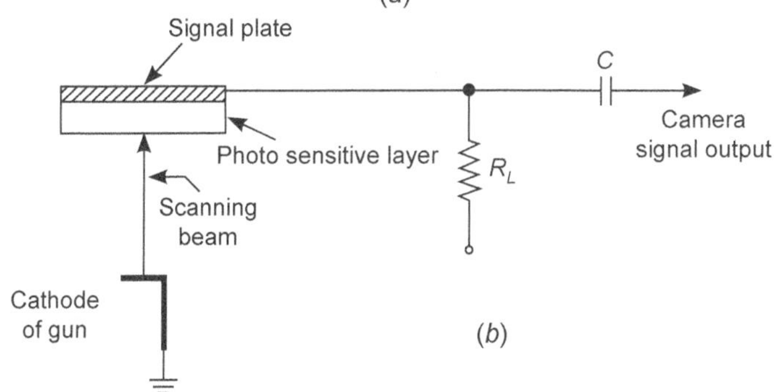

Fig. 9.6

(*c*) ***Working:*** With the optical image focussed on the target (or signal plate), it produces a *charge image* which is scanned by the electron gun. Each point in the charge image has a different positive potential on the side of the target facing the gun. The beam from the gun deposits its electrons on the photolayer surface of the target reducing the positive potential to zero. Excess electrons are returned back but they are not used in vidicon, though they were used in the orithicon. This change in potential of the target plate causes signal current to flow in the circuit. This has been shown in Fig. 9.6(*b*) producing voltage across R_L. For black in the picture where photo layer is less positive than the *white*, the deposited electrons cause a small change in signal current, hence the signal current results from the change in the potential difference between the two surfaces of the photo layer.

(d) *Characteristics of Vidicon Camera Tube*

(*i*) *Dark current sensitivity*: When the camera lens is closed, very small signal current flows in the target circuit of the tube as the target is in total darkness. This small current is known as *dark current*. The value of dark current in vidicon is about 20 nano amp. The dark current affects the *sensitivity* of the tube. For higher value of dark current, sensitivity is more. A typical value of sensitivity of vidicon is 120 nA per lumen illumination on the target.

Dark current: This is the output signal current even when there is no illumination. It is due to electron hole pairs generated by thermal energy.

(*ii*) *Resolution*: As defined already resolution offers the smallest details of the picture that can be *resolved* by the camera. The resolution of vidicon is about 55% which can be said as satisfactory.

(*iii*) *Signal/Noise Ratio*: The S/N ratio of vidicon is about 47 dB. It is a quite high value.

(*e*) ***Applications***: Vidicon is compact in size and simple in operation. It is just a target and gun assembly. It is used in outdoor and indoor shootings with black and white TV.

3. Plumbicon Camera: Plumbicon is similar to vidicon. The electron gun is also similar to that of vidicon but with a different target plate which is basically a PIN (*P*, *I* and *N* type) semi conductor diode. In this diode, on one side of the plate, PbO (*P* type semi conductor material) is deposited and on the other side a layer of SnO_2 (*N* type material) is deposited—both separated again by a PbO layer which is an intrinsic (I type) material. Thus a *PIN* diode is formed (Fig. 9.7(a)). The SnO_2 coated side works as a signal plate.

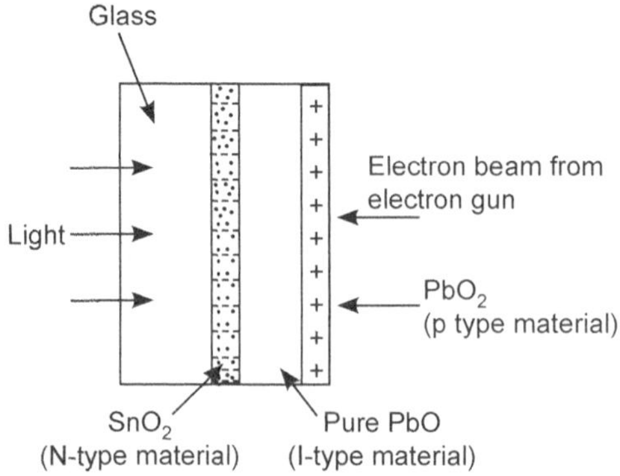

Fig. 9.8 (*a*)

(*a*) Construction of Plumbicon [See Fig. (*b*)]

The construction of plumbicon is just similar to the vidicon. It is also a target and gun assembly. It also has a cathode, that emits electron beam which is controlled by deflecting and focussing coils.

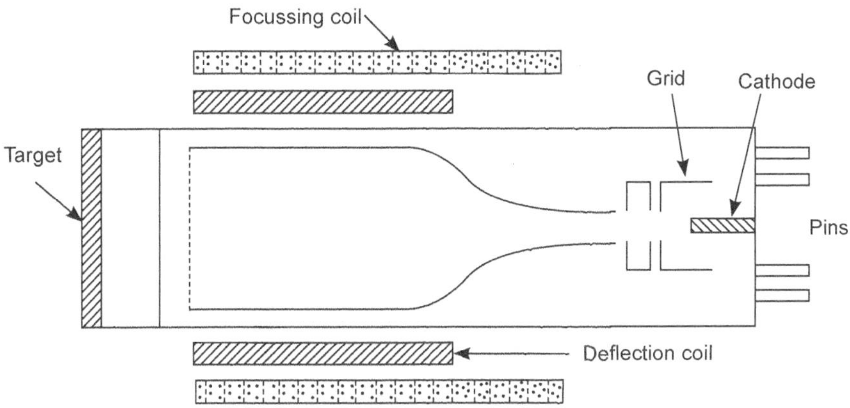

Fig. 9.7 (*b*)

(b) Working

The SnO_2 side of the target is connected to a supply of 40 V through a load R_L (Fig. 9.8), across which output signal voltage is developed. When the electron beam *scans* the target, the signal current varies in accordance with the light of each part of the picture.

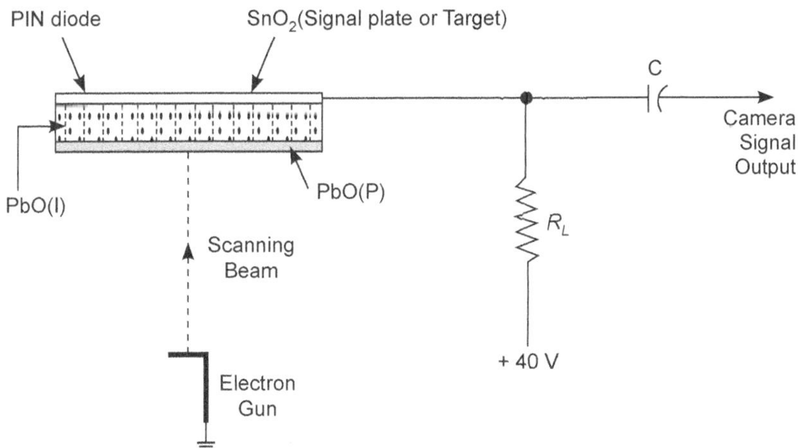

Fig. 9.8

When there is no light, the *PIN* diode is reverse biased, and a negligible current flows. When light falls on the PIN diode (target), it becomes forward biased in accordance to the intensity of light. The bias is the result of photo excitation of *P* layer and intrinsic layer of the target, thus the target behaves as a capacitor in series with the *PIN* diode, and output is obtained.

(*c*) Characteristics of plumbicon

(*i*) *Dark current and sensitivity*: The dark current of plumbicon is nearly 1 nano amp. and sensitivity is better than vidicon. The value being 400 μA per lumen illumination of the target.

(*ii*) *Resolution*: The resolution of plumbicon is poor than vidicon. The typical value is 45%.

(*iii*) *S/N Ratio*: The SN ratio of plumbicon is higher than vidicon. The typical value is 52 dB.

(*d*) Application

It is also simple in construction and working, therefore, used in studio as well as outdoor shootings specially with colour TV.

9.1.4 Human Eye

Human eye is just like a colour camera. An eye is an important video system of human human beings. Simplified diagram of the human eye is given in Fig. 9.9. The main parts of the human eye are; cornea (*C*), iris (*I*), pupil (*P*), ciliary muscles (*M*), convex lens (*L*), retina (*R*) and optic nerve (*N*).

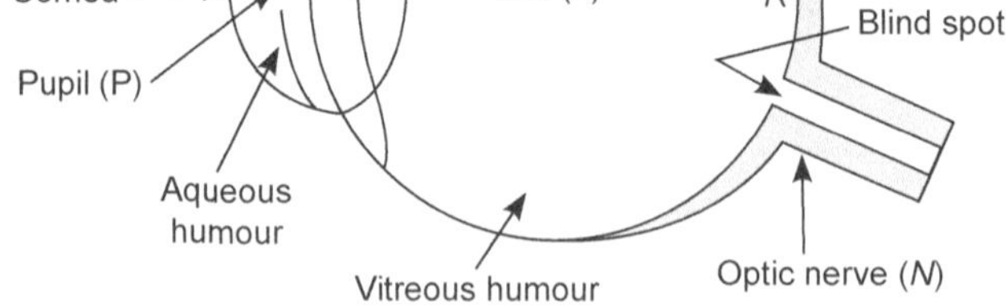

Fig. 9.9

The front part of the eye is called cornea. It is made of a transparent substance and it is *bulging outwards*. The light coming from objects enters the eye through cornea. Just behind the cornea is the iris (or *coloured diaphragm*).

There is a hole in the middle of the iris which is called pupil of the eye. The pupil appears *black* because no light is reflected from it.

The eye-lens is a *convex lens* made of a transparent and flexible material like jelly made of *proteins*. The eye-lens is held in position by ciliary muscles. *The focal length* of eye-lens (and hence its converging power) can be changed by the action of *ciliary muscles*.

The screen on which the image is formed in the eye is called retina. The retina is behind the eye-lens and at the back part of the eye. The retina of an eye is just like the *film* in a camera. It should be noted that there is an *eye-lid* in front of the eye which is just like the shutter in a camera. When eye-lid is open, light can enter the eye but when eye-lid is closed, no light enters the eye. The space between cornea and eye-lens is filled with a viscous liquid called *'aqueous humour'*. And the space between eye-lens and retina is filled with another liquid called 'vitreous humour'.

Working of the eye: The light rays coming from the object enter the pupil of the eye and fall on the eye lens. The eye-lens is a convex lens, so it converges the light rays and *produces a real and inverted image of the object on the retina*. The image formed on the retina is conveyed to the brain by the optic nerve and gives rise to the sensation of vision. Actually, the retina has a large number of *light-sensitive* cells. When the image falls on the retina, these light-sensitive cells get activated and generate *electrical signal*. The retina sends these electrical signals to the brain through the optic nerve and gives rise to the sensation of vision. Although the image formed on the retina is inverted, our mind interprets the image as that of *an erect object. A small region of the retina where the optic nerve enters is insensitive to light and it is called blind spot.*

Table 9.1: Comparison Between Human Eye and the Camera

The eye	The camera
Points of similarity	
1. In an eye, the image is formed by a convex lens made of a transparent and flexible substance.	In a camera, the image is formed by a convex lens made of glass.
2. In the eye, a real and inverted image is formed on the retina.	In a camera, a real and inverted image is formed on the photographic film.
3. Pupil in the iris of the eye controls the amount of light entering the eye.	The diaphragm controls the amount of light in a camera.
4. The time of exposure in the eye is controlled by the eye-lids.	The time of exposure in a camera is controlled by a shutter.

The eye	The camera
Points of difference	
1. The focal length of the eye lens can be changed by the action of ciliary muscles.	The focal length of a camera lens is fixed. It cannot be changed.
2. The focussing in the eye is done by changing the focal length of the eye-lens.	The focussing in a camera is done by changing the distance between the lens and the film.
3. The retina of the eye retains the image only upto 1/20th a second after the object is removed. Thus, the image formed on the retina of an eye is not permanent.	The photographic film of a camera retains the image of the object permanently.
4. Retina can be used again and again for forming the image.	A photographic film can be used only once for forming the image.

9.1.5 Picture Tube

A picture tube that provides a screen for a TV receiver is basically a cathode ray tube (CRT). It consists of an evacuated glass envelope. At its neck, there is an *electron gun* which supplies the electron beam. The inner surface of its face plate has a phosphor coating which produces light when the electron beam strikes. A monochrome (B & W) picture tube has one electron gun and a *continuous* phosphor coating that produces a picture in black and white. For colour picture tubes, the screen is formed of three different phosphors (red, green and blue). The neck of the colour picture tube may have one gun emitting three beams for the three phosphors. These 3 phosphors by combination can produce any colour; or the PT (picture tube) may have three guns each emitting different colour.

Table 9.2: Screen phosphors for Picture Tubes

S.No.	Phosphor colour	Applications
1.	White	Monochrome picture tubes
2.	Red, Green, Blue	Tricolour picture tubes
3.	White, Yellow	Two layer screen
4.	Green, ultra violet	Flying spot scanner

(a) **Monochrome (B & W) Picture Tube (Fig. 9.10):** The monochrome picture tube is used in B & W-TV. These tubes employ *electrostatic focussing* and *electromagnetic deflection*. The composite video signal obtained is fed to cathode of the picture tube. The picture is constructed *bit by bit* but the same is perceived by the tube as a *complete* and *continuous* due to *persistence of vision*.

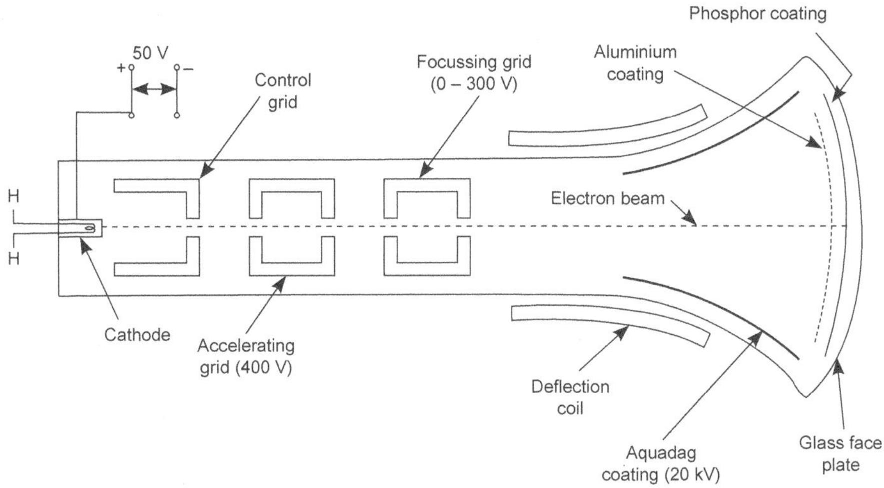

Fig. 9.10

The important parts of a PT are described below:

(*i*) *Electron gun*: The electron gun that produces high velocity electron beam consists of an indirectly heated cathode (of tungsten) and 3 grids. These grids are control grid, accelerating grid and focussing grid constitute the *electrostatic focussing* system of the tube.

(*ii*) *Focussing grids*: The grids control the movement of the beam through a smaller area and hence they 'focus' the beam at the screen. The *control grid* is maintained at a negative potential w.r.t. cathode. The *accelerating and the focussing* grids are maintained at positive potentials between (+200 to 600 V) w.r.t. cathode.

(*iii*) *Aquadag coating*: Starting from the half way into the neck to about 3 cm of the screen there is a special material coating called *aquadag*. This is a conducting coating generally of graphite. It is connected with a very high potential. Generally a 48 cm monochrome TV has an aquadag coating at 18 kV. Due to such a high positive potential, it increases the velocity of the electron beam to a very high value.

(*iv*) *Electromagnetic deflecting coils*: (See Fig. 9.11) Two pairs of coils are mounted outside and close to the neck of tube. These coils provide desired deflection (horizontal as well as vertical) to the beam.

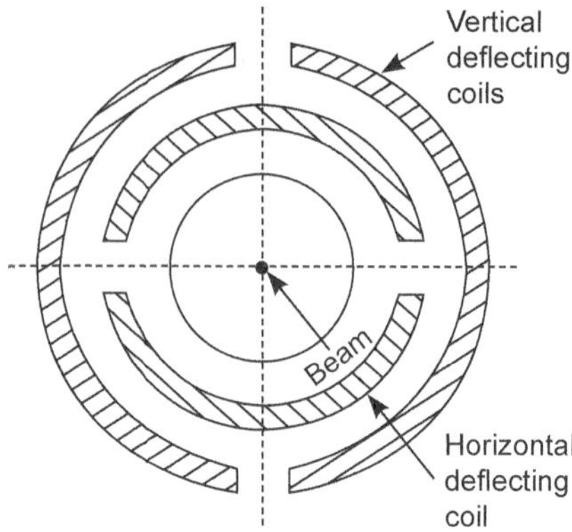

Fig. 9.11

These coils maintain a deflection angle (α) between 55° to 110°. The deflection angle decreases with length of the tube. (See Fig. 9.12)

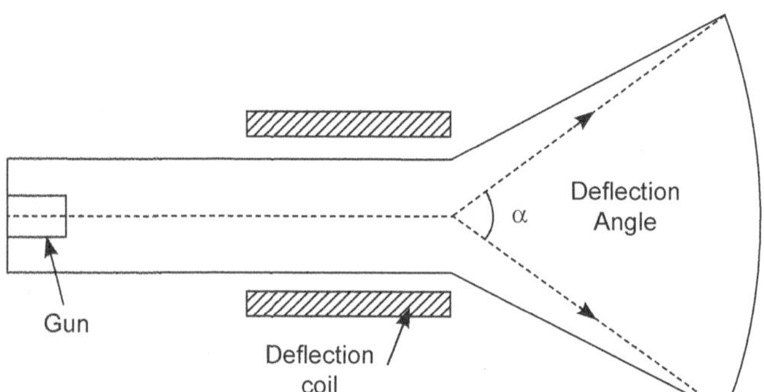

Fig. 9.12

(*v*) *Face Plate/Screen*: Generally picture tubes have rectangular face plate with a length to breadth ratio 4 : 3 (called aspect ratio). A 20″ screen means distance between two diagonal points of the screen is 20″ (See Fig. 9.13). The thickness of about 1 to 1.5 cm of the screen is sufficient to provide it the mechanical strength to withstand the air pressure on the evacuated glass envelope.

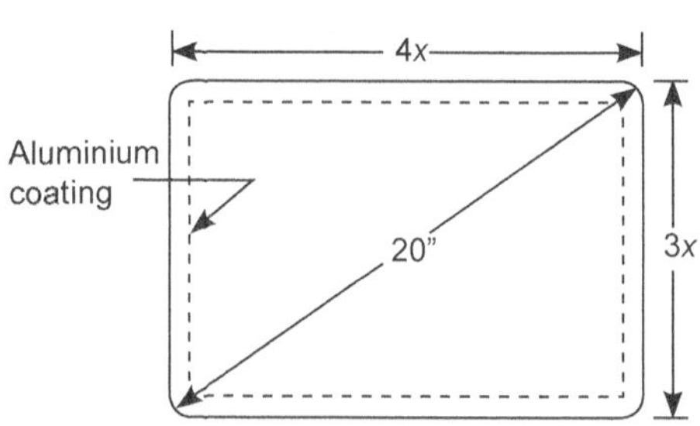

Fig. 9.13

(*vi*) *Aluminium coating*: An aluminium coating is provided at the back surface of face plate of the tube. The coating is very thin and high velocity beam can easily penetrate it to reach the phosphor screen.

After striking the screen, about 50% of the light emitted is returned back into the tube. Another 20% is lost in internal reflections within the tube thus, only 30% light is utilised. The aluminium coating reflects back to the screen much of the light lost in the tube, thus it improves *brightness* on the screen.

9.1.6 Scanning

In order to convert a picture into a *video signal* through a camera, the picture is divided into number of horizontal lines. The *picture tube* in turn re-assembles these lines into the original picture. These horizontal lines are produced by making the electron beam to *scan* the picture *line by line*. There may be 525 (or 625) lines per picture frame. The electron beam scans these lines horizontally as well as vertically. The process is known as *Scanning*.

The scanning in camera and scanning in the picture tube, the both should be *synchronised*. This is essential to re-assemble the picture on the correct lines. This is known *Synchronisation*.

For both the purposes *scanning* and *synchronising circuits* are provided in TV transmitter as well as in the receiver.

The scanning is of two types:

(a) Horizontal Scanning: For scanning, saw tooth currents (Fig. 16.14(a)) are used. This current flows through horizontal deflection coils of the picture tube for horizontal scanning.

The linear rise of the saw tooth wave deflects the beam across the screen with a continuous uniform motion for the *trace* from left to right. At peak the wave reverses and decreases rapidly to its initial value. This fast reversal produces *retrace* or *flyback*. (See Fig. 9.14(b) and (c))

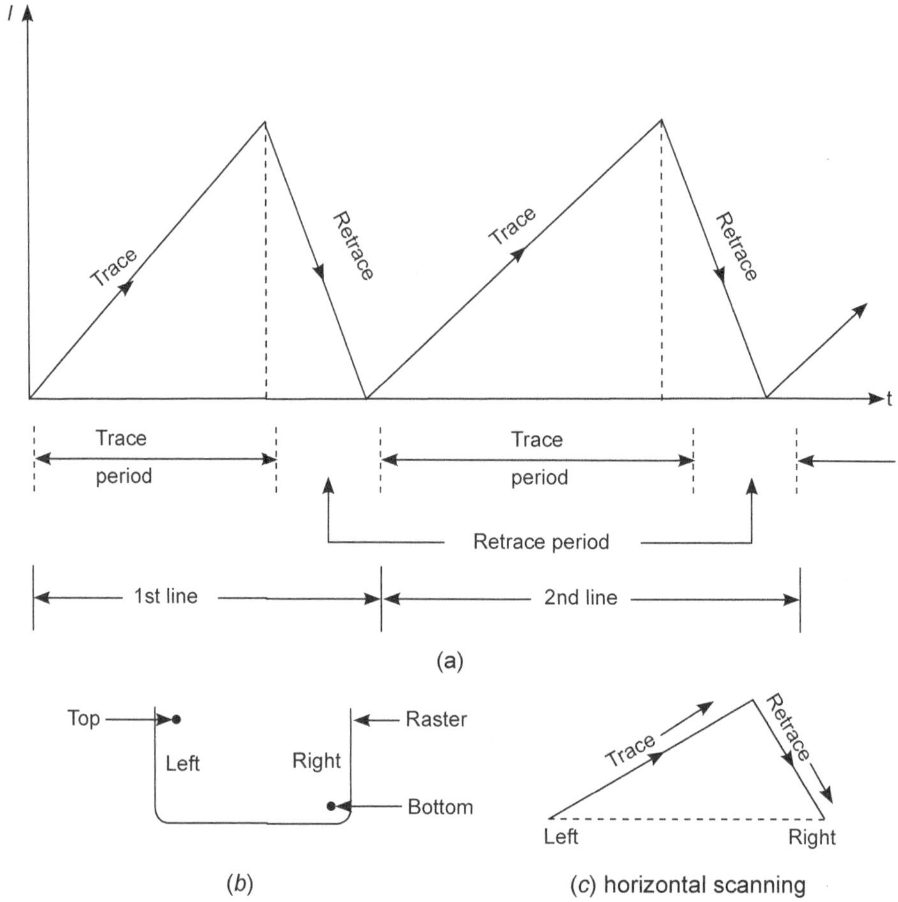

Fig. 9.14

The *start* of the horizontal trace is at the left edge of the raster and the *finish* is at its right edge. The flyback produces retrace back to the left edge. (See Fig. 9.15)

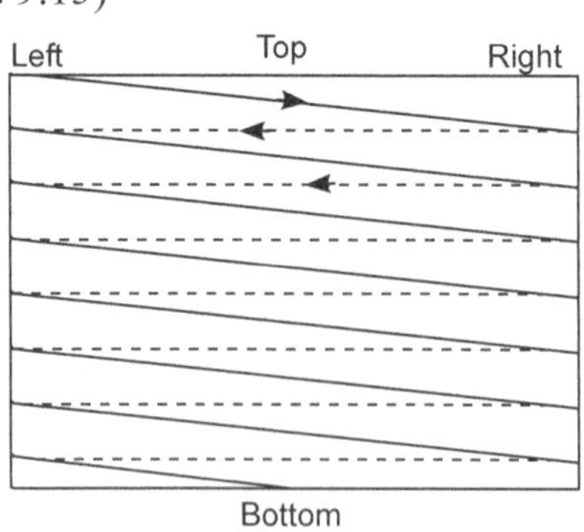

Fig. 9.15

(b) Vertical Scanning: The saw tooth current when made to flow through vertical deflecting coils of the picture tube, the current moves the electron beam from top to bottom of the raster. While the electron beam is being deflected horizontally, the vertical deflection coils move the beam downward with uniform speed. Thus the beam produces complete horizontal lines one under the other. The trace part of the wave for vertical scanning deflects the beam to bottom of the raster, then the rapid vertical retrace returns the beam to top. [See Fig. 9.16 (*a*), (*b*) and (*c*)].

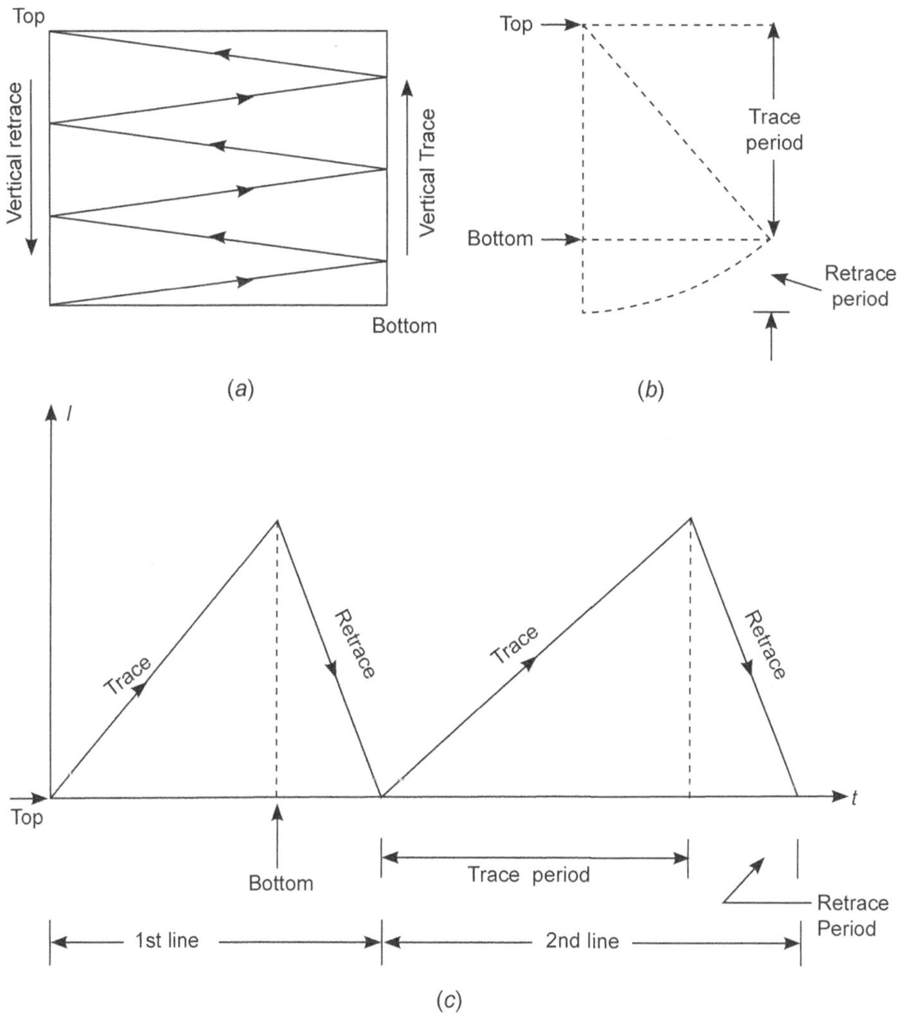

Fig. 9.16

(c) More About Scanning:

(i) Scanning Frequencies: Both trace and retrace are included in one cycle of the saw tooth wave. Since number of complete horizontal lines scanned in 1 second are 15,750 for horizontal deflection, the frequency of the saw tooth waves is 15750 cycles per second. For vertical scanning, the frequency of saw tooth wave is 60 cycles per second. As vertical scanning frequency is much lower than of the horizontal scanning frequency, many horizontal lines are scanned during one cycle of vertical scanning.

(ii) Number of Scanning Lines: How many scanning lines should be there in a picture to produce it effectively depends upon many factors. Greater the number of lines into which the picture is divided, better will be the resolution. It also very much depends upon the *resolving capacity* of human eye. There are other factors also which decide the total number of lines. For an ideal case the picture should be divided into 800 lines but above 500, the improvement is not significant. Morever with more number of scanning lines, the bandwidth also increases and this adds to the cost of the system. As a compromise between cost of the system and the quality of the picture, 625 lines have been fixed in India whereas in America this figure is 525 lines.

(iii) Scanning Period: The horizontal as well as vertical scanning currents are shown in the Fig. 9.17 (*a*) and (*b*) respectively. The nominal time for horizontal line = $10^6/15625 = 64\ \mu S$ out of which trace period is 52 μS and retrace period is 12 μS which is also called as *blanking period.*

Similarly nominal duration for vertical trace is 20 ms (1/50 = 20 ms). Out of this, 18.72 ms is spent to bring the electron beam from top to bottom and remaining 1.28 ms is spent by the beam to return back.

As horizontal and vertical sweep oscillators operate continuously, 1280 μS/64 μS = 20 lines are traced during each vertical trace interval. In this way, a total = 20 × 2 = 40 lines are lost per frame; as *blank lines* during the retrace of the two fields. In this way, the lines actually scanned per frame left = 625 – 40 = 585 lines.

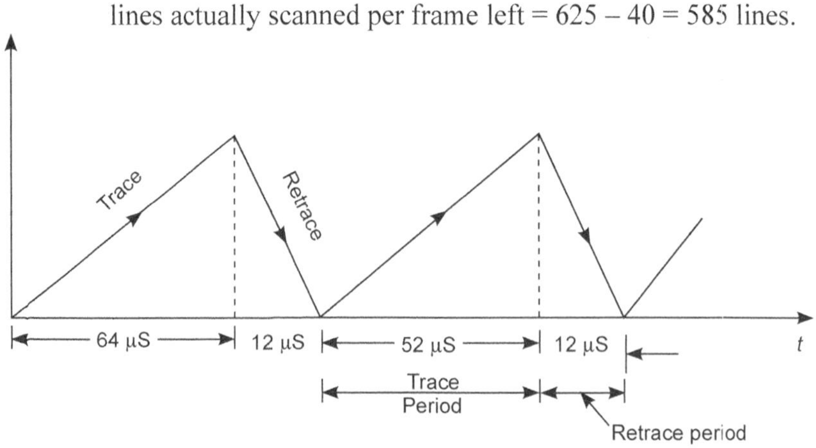

(*a*) Horizontal scanning currents

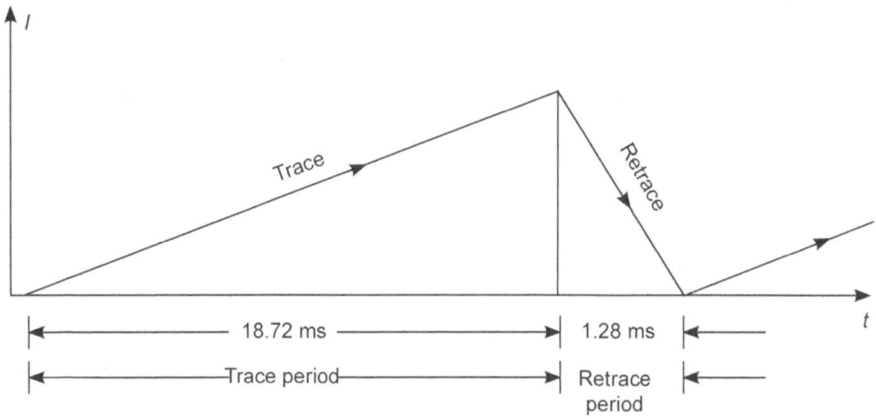

(*b*) Vertical scanning currents

Fig. 9.17

(iv) Flicker in Scanning: Interlace Scanning: Though scanning at the rate of 25 frames (pictures) per second produces illusion of continuity in TV system but when the screen is made alternately bright and dark in between two successive frames it causes a flicker of light which is very annoying to the viewer. To reduce the flicker, an improved method of scanning known as *interlaced* scanning is used.

(d) Interlaced Scanning: In this scanning alternate lines are scanned, *e.g.*, all the *odd* lines from top to bottom of the frame are scanned first, leaving out the *even* lines. After this, a rapid vertical retrace moves the electron beam back to top of the frame and all the even lines on the (frame which were omitted in the previous scanning) are scanned from top to bottom.

This reduces *flicker* since the area of the screen is covered at a double rate. This is similar to reading alternate lines of the page of a book from top to bottom once and then going back to the top to read the remaining lines down the bottom.

This is illustrated in the Fig. 9.18: In this Fig.

(*i*) Vertical retrace period has been assumed as zero.

(*ii*) Retrace lines not shown.

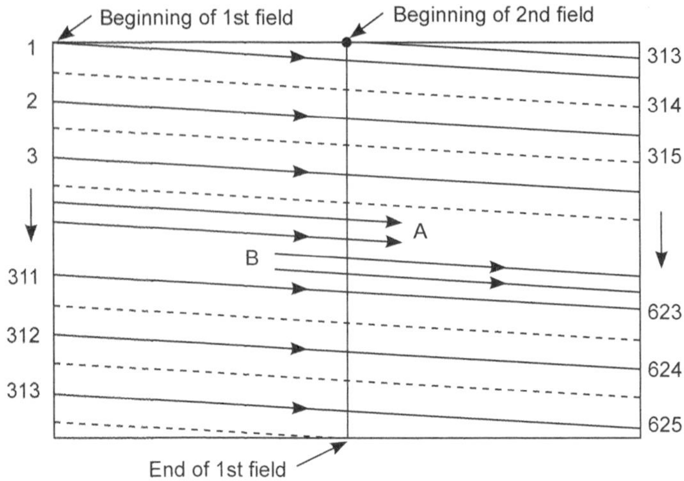

Fig. 9.18

In a 625 lines monochrome TV system, the lines of each picture (or frame) are divided into two sets of 625/2 = 312.5 lines, each set is called a *Field.* The each field is scanned alternately to cover the entire picture area. For this purpose, a horizontal sweep oscillator is used at a frequency of 15,625 cycles per second (312.5 × 50 = 15625) to scan the same number of lines per frame (15625/25 = 625) but a vertical sweep oscillator is made to work at a frequency of 50 Hz. Since the electron beam is now deflected from top to bottom in half the time and the horizontal oscillator is working at 15625 Hz, only half of the total lines *i.e.*, 312.5 are scanned during each vertical sweep. The first field ends in half line (point A), the second field starts from middle of the line (point B) on the top of the screen and the electron scanning beam is able to scan remaining 312.5 lines during its downward journey. In all, the electron beam scans total 625 lines at a rate of 15625 lines per second. Thus we are able to reduce the flicker effect without increasing the speed of scanning and thus no need to increase *band width* or cost of the system.

Actual Process of Interlaced Scanning

The process of *odd lines scanning i.e.,* interlaced scanning has been illustrated in Fig. 9.19. To understand, start from point *A*. For line 1, the electron beam sweeps across the frame with uniform velocity to cover all the picture elements contained in the line. At the end of the line, the beam retraces rapidly to the left side of the frame with more velocity to scan the next line. After line no. 1, the beam is at the left side ready to scan line 3, omitting line no. 2. This is the reason, that the process is also known as "*odd lines interlacing*". In this way, the beam scans all the *odd* lines reaching at point *B* at the bottom of the frame (Fig. a). At *B*, vertical retrace begins and the beam comes back at top of the frame to scan the *even* lines. Now the beam moves from *B* to *C* traversing the whole number of horizontal lines. (Fig. b)

The horizontal scanning during the second field begins with the beam at *C* [Fig. 16.22 (*c*)]. This point is at the middle of horizontal line as the first field contains 312.5 lines. After scanning, a half line from *C* the beam scans line no. 2 (even line) in the second field. The beam now scans all the even lines (2, 4, 6 ....) which were earlier skipped. The vertical motion is exactly same as in the previous field. Now the vertical retrace in second field starts at point *D*. The beam is now returned to the top. The points *C*, *D* and *B* are only half a line apart.

Back at point *A*, the beam has just completed *two fields* or *one frame* and is ready to start the third field [Fig. 16.22(d)]. This process is continued at a rate of 50 times a second, which creates an illusion of *image continuity* as well as solves the problem of flicker. The Fig. 9.22(e) explains the total process schematically.

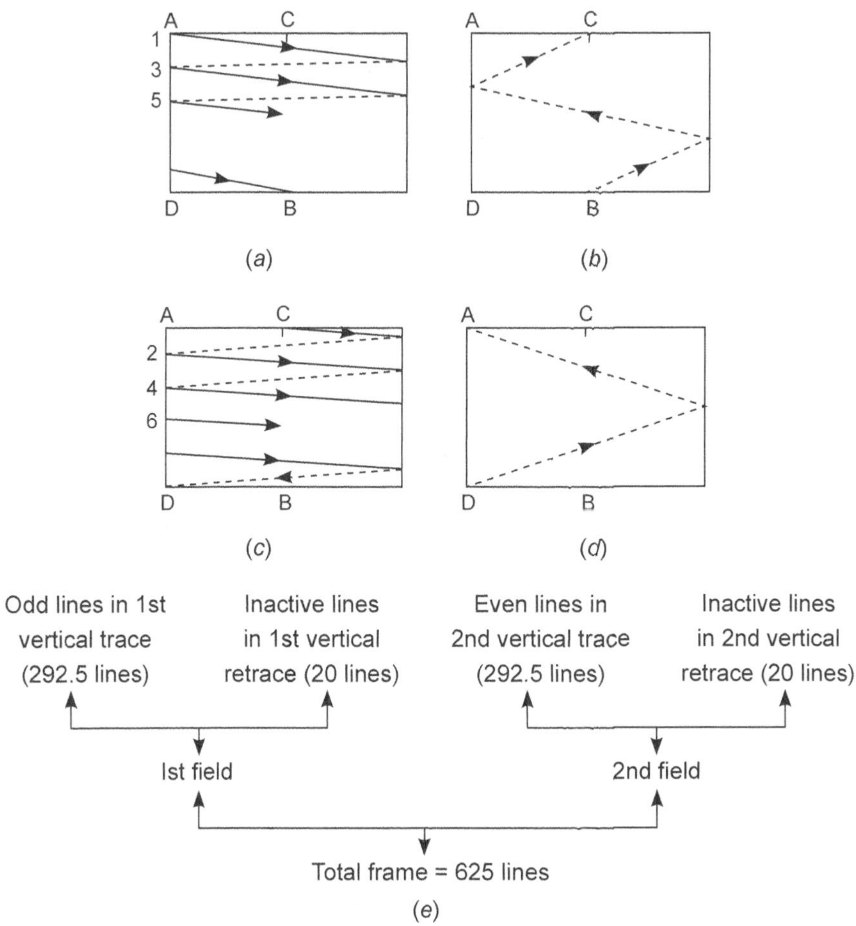

Fig. 9.19

9.1.7 Synchronization & Blanking

(a) Synchronization: At the transmission end, camera scans the picture and the same process is carried out at the receiving end by picture tube. At the receiver the picture tube should reassemble the picture elements on each horizontal line with the same left-right position as the image at the camera. As the beam scans the successive lines in the picture tube vertically, the screen should show the same picture elements in the corresponding lines as at the camera. Therefore, a *horizontal synchronizing pulse* is transmitted for each horizontal line to keep the horizontal scanning synchronized and a *vertical synchronizing pulse* is transmitted for each field to synchronize the vertical scanning. Accordingly the horizontal synchronizing pulses should have a frequency of 15750 cy/s and vertical synchronizing pulses a frequency of 60 cy/s. The term "sync." is generally used as abbreviation for "synchronizing". The sync. pulses are transmitted as a part of the picture signal but are sent only during the *blanking period*, when the signal is to be transmitted.

The Fig. 9.20 shows the waveform of the sync. pulses for the purpose of synchronization. All pulses have the same amplitude but differ in pulse width or waveform. The sync. pulses include—3 horizontal pulses, 6 equalizing pulses, 6 additional equalising pulses and then 3 horizontal pulses. There are many additional horizontal pulses till *equalising* pulses occur again for beginning of the next field.

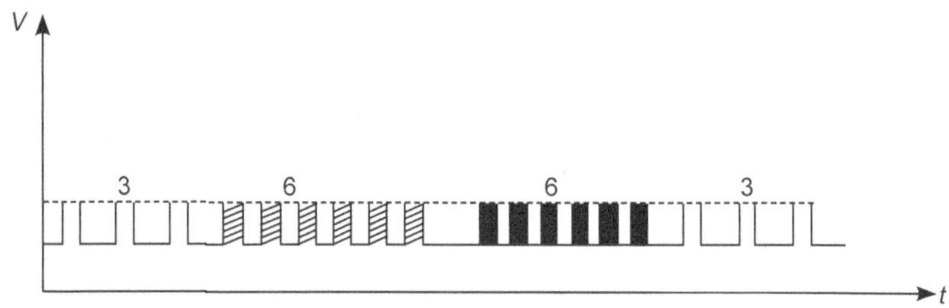

Fig. 9.20

The sync. pulses do not produce scanning. For scanning, saw tooth pulses must be produced as explained earlier. However, the sync. pulses hold the picture on the screen in still position. If horizontal sync. is not provided, the picture *drifts* to the left or right on the screen and the picture is torn into diagonal segments. If vertical sync. is not provided, the picture appears *rolling up* and down on the screen.

(b) Blanking: The purpose of blanking pulses is to make the retrace as black *i.e.*, invisible.

In TV, “Blanking” means “going to blank or black”, for a part of the video signal, the blanking voltage is at the black level. Video voltage at the black level cuts off the beam currents in the picture tube to black out the light from screen. The purpose of providing the “blanking pulses” is to make invisible the retraces of the scanning process. The horizontal blanking pulse at a frequency of 15750 Hz, blanks out the retrace from right to left for each line. The vertical blanking pulses at 60 Hz blank out the retrace from bottom to top for each field.

The time period of blanking pulses is 16% of each horizontal line *i.e.*, = 16 per cent of 63.5 μS = 10.2 μS. In other words, retrace from right to left must be completed in 10.2 μS.

The time period of vertical blank pulses is 8 per cent of each vertical field. It comes equal to 8 per cent of 1/16 S = 0.0013 S. In other words, the vertical retrace must be completed within 0.0013 S.

A blanking pulse comes first to put the video signal at black level, then a sync. pulse comes to start the retrace. This sequence applies to blanking, horizontal and vertical retraces.

Table 9.3 : Frequencies of Scanning, Synchronising And Blanking Pulses

Particulars	Frequency (Hz)
1. Horizontal scanning pulses	15,750
2. Horizontal sync. pulses	15,750
3. Horizontal blanking pulses	15,750
4. Vertical scanning pulses	60
5. Vertical sync. pulses	60
6. Vertical blanking pulses	60

9.1.8 Composite Video Signal

The video signal is not simple but it is a complex signal. It is a composite of the following signals:

(*i*) Camera signal—corresponding to the picture information.

(*ii*) Blanking signal—during retrace period to make it invisible.

(*iii*) Synchronizing signal—to synchronise the two scannings occurring in transmitter and receiver.

How this signal is formed is shown in Fig. 9.21

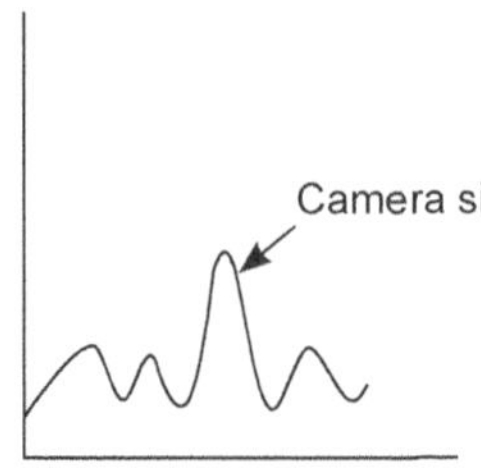

(a) Camera signal

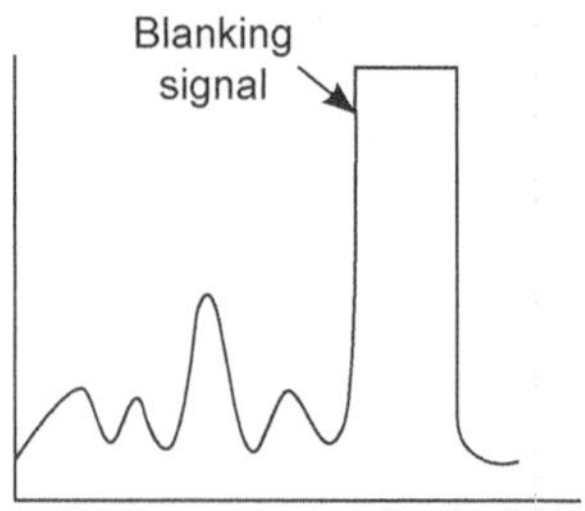

(*b*) Camera + blanking signal

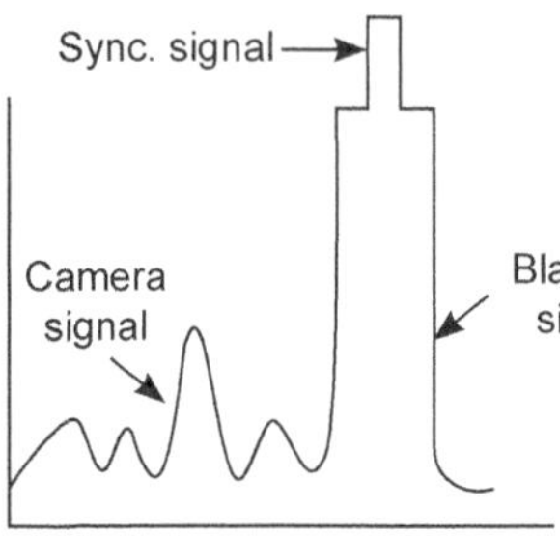

(*c*) Camera + blanking + sync. signal = Composite video signal

Fig. 9.21

(a) Video Signal Dimensions: The Fig. 9.22 shows detailed dimensions of the composite video signal. The level of the video signal corresponding to the maximum whiteness to be handled is known as *peak white level*. This is about 10 per cent of the maximum value of the signal, while the black level is about 70 per cent. The sync. pulses are added at about 75 per cent level known as *black level*. In actual practice, these two levels are very close and are almost merged with each other.

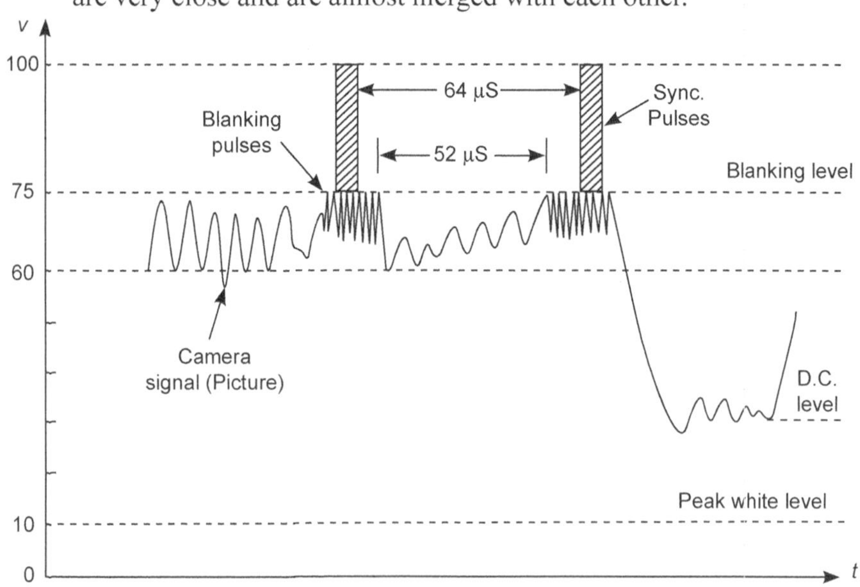

Fig. 9.22

In addition to continuous variation in amplitude for the picture elements, the video signal has a d.c. component corresponding to the average brightness of the scene. In its absence, grey picture on a black background will be same as a white picture on a grey background.

The video signal also contains *blanking pulses* (or blanking signal) which raises the signal amplitude slightly above the black level (75 per cent) so that the retrace is not visible. This signal contains horizontal as well as vertical blanking pulses. The frequency of horizontal blanking pulses is same as that of the horizontal scanning and frequency of the vertical blanking pulses is same as that of vertical scanning that is, 15625 Hz and 50 Hz respectively.

The *picture to sync. signal* ratio (P/S) is kept as 13:5 *i.e.*, 65 per cent of the maximum carrier amplitude is occupied by video signal and about 25 per cent by sync. pulses. This gives best results in reducing the noise level.

9.1.9 TV Signal modulation techniques

In most of the TV transmission systems, the picture signal is *amplitude modulated* and the sound signal is frequency modulated.

Need for modulation has already been described. Just to remind that to transmit a signal directly (without modulation) a very very long antenna is required which is impractical. Further both picture and sound signals from different stations are concentrated within the same range of frequencies, therefore, both will be mixed up (if they are unmodulated) and it will be difficult to separate them in the receiver. Thus in order to make separation of "intelligence" (signal) from the different stations, it is necessary to keep them at different portions of the spectrum depending upon the carrier frequency assigned to each station.

(a) **Amplitude Modulation of Video Signal (Fig. 9.23):** Modulation of video signals is possible only by amplitude modulation because of their large bandwidth. The modulation can be of low level or high level. The high level amplitude modulation is done at final (power) stage. The intermediate frequency modulation is a low level modulation in which modulation of video signal is carried out at 38.9 MHz. The modulated IF is then changed to channel frequency by heterodyning.

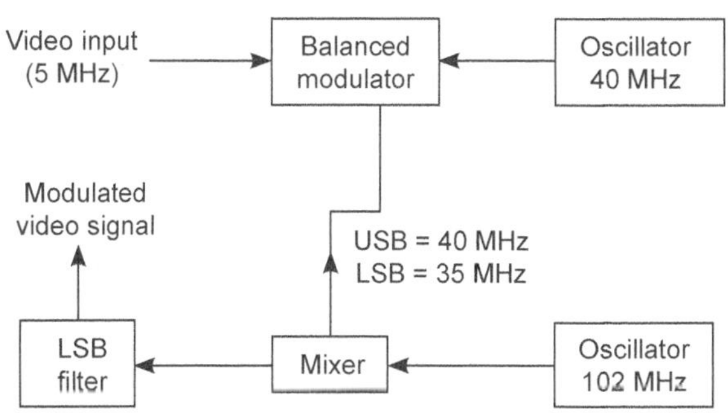

Fig. 9.23

To reduce the BW, the upper side band is transmitted fully and lower side band is transmitted partially. This is called *vestigial side band (VSB) transmission*. The LSB filter is used to allow only the desired frequency range.

(b) Frequency Modulation of Sound Signal: The output of all microphones terminate on the sound panel in the control room. Each microphone output is amplified before being fed to mixers. *Pre-emphasis* and *de-emphasis* are done to improve the quality of the sound. The modulated audio signal is transferred to the assigned channel sound carrier frequency by the use of multipliers.

The "Pre emphasis" is the process of emphasing or increasing the amplitude of high frequency components of modulating signals before modulation. This is done to increase the signal-noise ratio of the high frequency components. This makes the reception noise free.

By changing the amplitude of high frequency components some distortion is introduced in the signal. This is avoided by doing de-emphasis (i.e., decreasing the amplitude) during demodulation.

It can be seen that the noise has a greater effect on higher modulating frequencies than on the low modulating frequencies. Therefore, if higher modulating frequencies are amplified at the transmitter and accordingly cut at the receiver, an improvement in the noise immunity can be obtained. The process carried out in the transmitter is known as "*Pre-emphasis*" and the process carried out in the receiver is called *de-emphasis*.

If two modulating signals have same initial amplitude and if the high frequency is amplified thrice of its amplitude and the low frequency remains uneffected, the receiver has to de-emphasis the first signal to one third so that both the signals have the same amplitude at the output of the receiver. When a signal is de-emphasised, any noise sideband voltages along with the signal are also de-emphasised and thus the effect of the noise is reduced.

The Fig. 9.24 (*a*) shows a circuit for pre-emphasis, and (*b*) a circuit for de-emphasis.

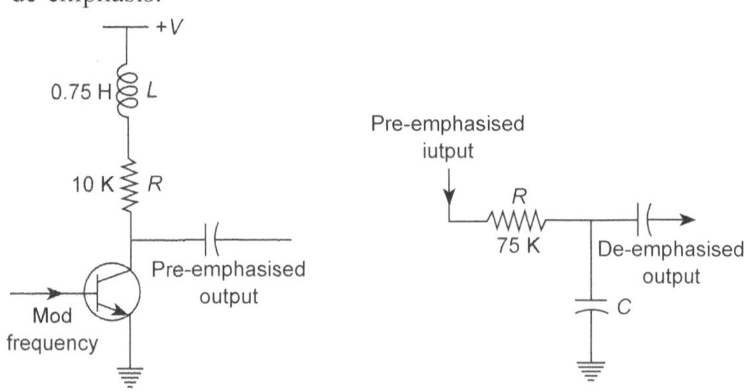

(*a*) Pre-emphasis (*b*) De-emphasis

Fig. 9.24

(c) Preference of Modulation Techniques:

1. *Preference of FM over AM for Sound Transmission*: Because of crowding in *medium* and *short wave* bands in radio transmission, only 5 kHz frequency is used for signal as maximum. This is done to limit the BW to 10 kHz as a maximum. This enables to accommodate large number of radio stations.

   Further FM provides almost noise free and high fidelity output. The maximum apace it needs is 200 kHz on each side of the picture carrier. As 7 MHz is allotted to the BW, it has no problem to accommodate the sound signal.

   Hence FM is preferred for sound transmission in TV channels due to

   (*i*) It reduces noise.

   (*ii*) It increases, efficiency of transmission.

   (*iii*) There is no inter-channel or inter-signal interference (I.S.I.).

2. *Preference of AM over FM for picture transmission*: The distortion which is produced due to interference between signals is more objectionable in FM then it is in AM, because frequency of the FM signal changes continuously. If FM is used for picture, it will produce the *shimmering effect* and the picture will not be steady. Further, the BW requirement in AM is lesser as compared in FM and also in FM, complexity of the circuitry increases. This is the reason, that for picture, AM is preferred.

9.1.10 TV Signal Transmission Techniques

Following three techniques are popularly used for TV signal transmission

1. **Double Side Band (DSB) Transmission:** In a 625 line TV transmission in which frequency components ranging from zero to 5 MHz are present, a *double side band* AM transmission will occupy a total BW of about 10 MHz. In addition, a *slope* of 0.5 MHz is to be added on both sides. Thus the total BW becomes 11 MHz. Further, each TV channel has an FM *sound signal* which is situated just out side the upper limit of the picture signal. A 0.25 MHz *guard* band is also added to this signal, which makes the BW equal to 11.25 MHz. (See Fig. 9.25)

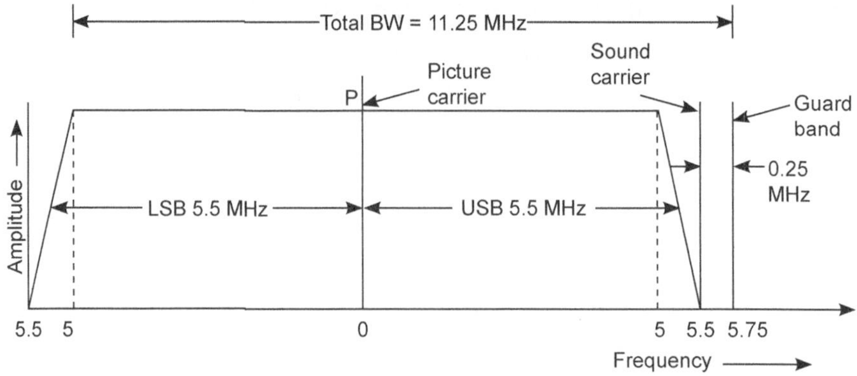

Fig. 9.25

Thus in a *double side band* (DSB) transmission, the BW is very large and would limit the number of channels for a particular station. In order to reduce this BW, single side band transmission is preferred.

2. **Vestigial Side Band (VSB) Transmission (Fig. 9.26):** The VSB techniques is most popularily used in TV transmission.

   Practically it is not possible to filter out LSB completely as it has been seen that the lower frequencies contain the most important information of the picture and filtering out the LSB (Lower Side Band) as a whole gives rise to distortions. Therefore, as a compromise full USB (Upper Side Band) and some part (Vestigial part) of the LSB (frequencies up 0.75 MHz) are transmitted. This is known as VSB transmission which is .practically done in 625 lines TV system. Now the composition of the SSB bandwidth comes to be as 7 MHz as shown.

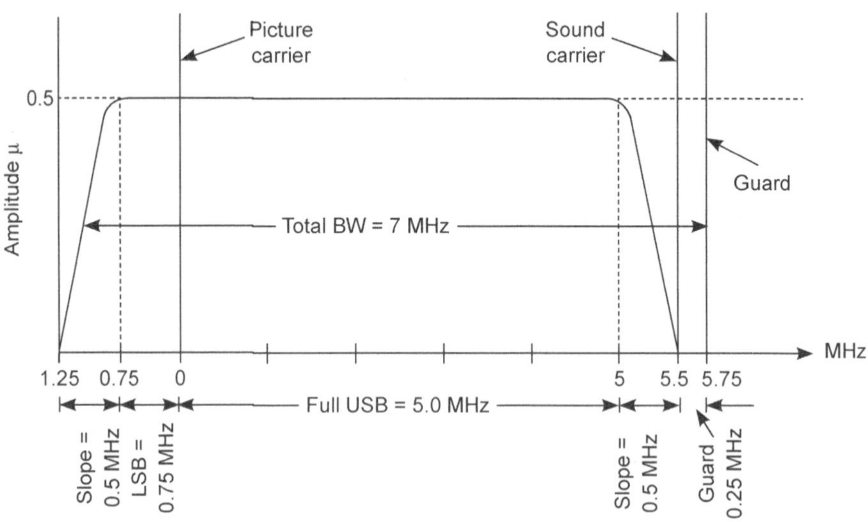

Fig. 8.26

9.2 TV RECEIVERS

(a) The TV receivers may be of the following 3 types:

(*i*) *All tube receivers*: The type mainly applies to monochrome and few old colour TV receivers. All the functions are carried out by electron tubes. A colour receiver may employ more number of tubes as compared to monochrome receive. They are obsolate now.

(*ii*) *Solid state receivers*: In this type, all the stages use solid state devices, except the picture tube (P.T.). The devices include semi conductor diodes, transistors and IC etc.

(*iii*) *Hybrid receivers*: This is a combination of the above two types. The deflection circuits use power tubes, while the signal circuits use solid state devices.

In fact, a TV receiver is a combination of *AM receiver* (for the picture signal) and *FM receiver* for sound signal. In addition, it also contains other circuits for scanning, synchronising etc.

(b) The TV receivers may be classified also as:

(*i*) Monochrome TV Receiver

(*ii*) Colour TV Receiver

Here we will discuss monochrome TV receiver.

9.2.1 Monochrome TV Receiver

The important sections/stages of a monochrome TV receivers are discussed as under: (See Fig. 9.27)

(*i*) *Antenna*: The picture and sound signals are intercepted by antenna of the receiver. A wire connects the antenna to the input of the receiver. Twin lead is generally used. There are various antennas used but the most popular is "yagi-uda" which gives good output in fringe areas. If the signal is weak a *booster* can be used and, if signal is strong or if the transmitting station is near, either no antenna is required or a telescopic indoor antenna may be sufficient. The twin lead from antenna to the receiver has an approximate impedance of 300 ohm. A folded dipole has also the same impedance at its resonant frequency, hence both provide a good *impedance matching*.

(*ii*) *VHF Tuner*: The antenna input provides r.f. picture and sound signals for the r.f. amplifier stage. The amplified output of amplifier is given to the mixer. Also output of a local oscillator is given to the mixer to heterodyne with signals as in the case of radio. When the oscillator frequency is set for the particular channel to be tuned, the signals are converted into intermediate frequency.

There are two I.F. outputs from the mixer—one for the picture and other for the sound. The standard value for I.F. in a 625 line TV system is 38.9 MHz for picture and 33.4 MHz for sound.

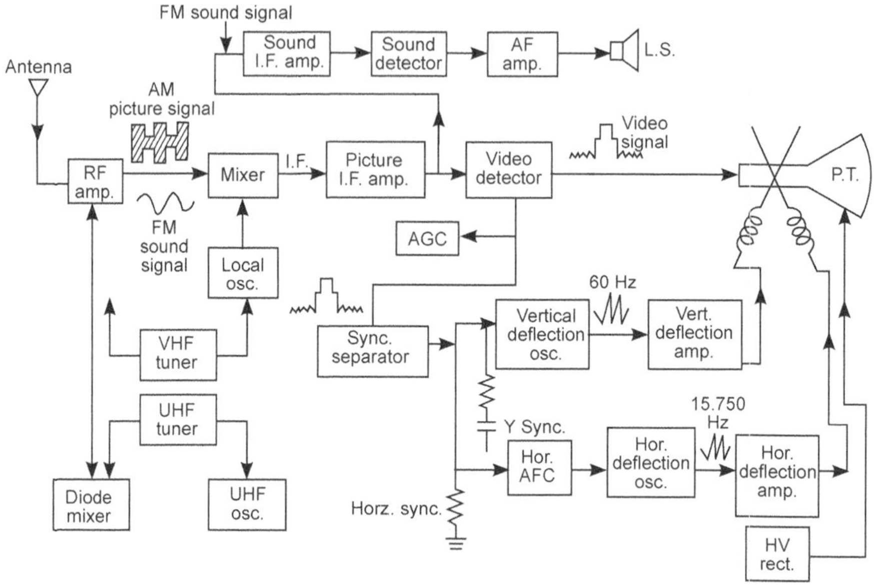

Fig. 9.27

The tuner selects the channel to be received by converting its picture and sound R.F. carrier frequencies into intermediate frequencies which can be amplified in I.F. amplifiers. The station selector is a gang switch that changes value of capacitor of the tuned circuits of the R.F. amplifier, mixer and local oscillator simultaneously.

(*iii*) *UHF Tuner*: When channel selector is set to UHF tuner position, the antenna input is obtained from a separate UHF antenna. This heterodynes the UHF input to the intermediate frequencies. These frequencies are amplified by required number of amplifier stages.

(*iv*) *I.F. Amplifiers*: These amplify output of the mixer in number of stages to get sufficient voltage for the video detector. The gain of the I.F. amplifier is controlled by an AGC.

(*v*) *Video detector*: The modulated I.F. picture signal is rectified and filtered in the detector to detect (recover) the amplitude modulated picture signal which is required for driving the picture tube.

(*vi*) *Sync. separator*: A sync. separator is basically a *clipper* circuit that separates the sync. pulses from the camera signal contained in the composite video signal. Since there are sync. pulses for horizontal as well as vertical scannings, output of the sync. separator is divided into two parts.

(*vii*) *Deflection circuits*: These include the vertical and horizontal oscillators for vertical and horizontal scannings. The deflection circuits produce the required scanning currents. The deflection oscillators basically are *A-stable multivibrators*, which do not need any external triggering pulse for operation.

(*viii*) *L.T. (Low Tension) supply*: This is needed for tubes and transistors. The dc output upto 280V is needed to vaccum tube amplifers. A dc supply upto 90 V is needed for rectifier and other circuits. The heater of the vaccum tubes may need dc supply of 6.3 V.

(*ix*) *EHT* (*Extra High Tension*) *supply*: The high voltage supply needed to rectifiers is given from horizontal amplifiers. The H.V supply is also needed to picture tube for its suitable operation. Approximate value of EHT supply for anode of the picture tube is 15—18 kV. This is needed for sufficient brightness. A voltage of 6 to 9 kV is developed across primary winding of horizontal output transformer. This is stepped up by an auto transformer to 15–18 kV (See Fig. 9.28).

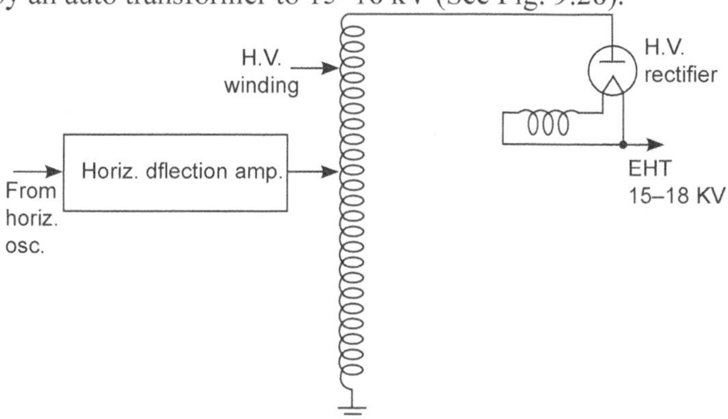

Fig. 9.28

(*x*) *AGC*: It stands for "Automatic Gain Control": It helps in getting constant amplitude of the video signal for different carriers. For the picture signal, AGC provides an automatic control of contrast in the reproduced picture. An AGC circuit has been shown in Fig. 9.29.

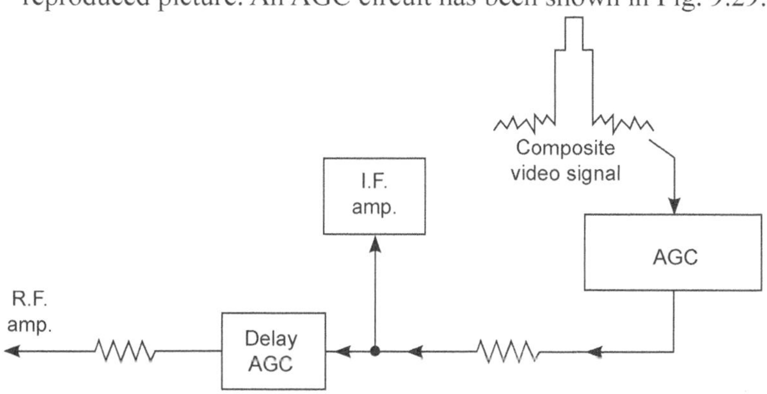

Fig. 16.29

(*xi*) *Video amplifier*: Before giving composite video signal to picture tube, it is amplified in video amplifier sufficiently. It may have number of stages as per requirement. The amount of video signal required for the picture tube is about 100 V for strong contrast. More strong video signal means more contrast. (Fig. 9.30)

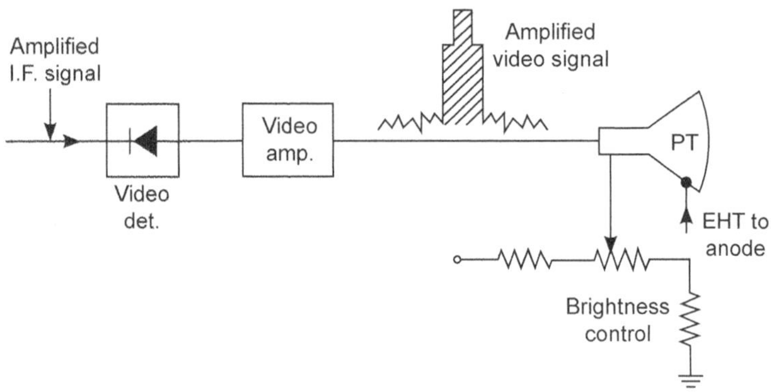

Fig. 9.30

9.3 THE TV STANDARDS

9.3.1 American Standards (Fig. 9.31)

The band of frequencies assigned to a station for transmission of the signal is called a *channel*. Each TV station has a 6 MHz channel within specific bands for commercial broadcasting:

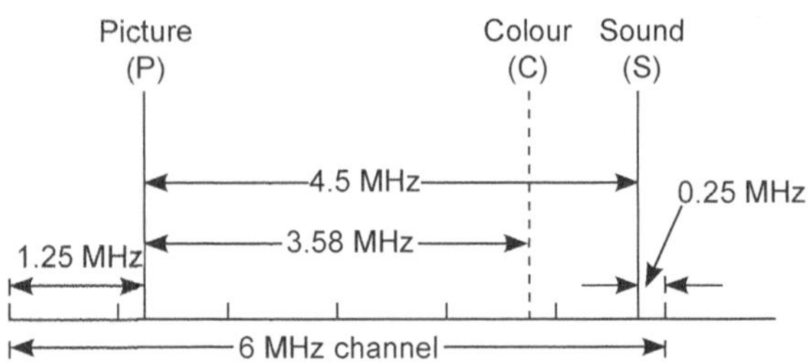

Fig. 9.31

(*i*) *Video modulation*: The 6 MHz bandwidth is needed for picture carrier signal. The carrier is amplitude modulated by the video signal.

(*ii*) *Chrominance modulation*: For colour broadcasting 3.58 MHz chrominance signal has the colour information. The colour signal (C signal) is combined with luminance signal (Y signal) to form the video signal, that modulates picture carrier wave for the transmission.

(*iii*) *Sound signal*: In 6 MHz channel, sound carrier signal for the picture is also included. The sound carrier is a "frequency modulated" signal by audio frequencies between 50 Hz to 15 kHz.

(*iv*) *Carrier frequencies*: The shows how different carrier signals fit into the standard 6 MHz channel. The picture carrier frequency is always 1.25 MHz above lower end of the channel. At opposite end, the sound carrier frequency is 0.25 MHz below the higher end.

9.3.2 Indian standards

The total channel width used in India is 7 MHz. The spacing between picture and sound is 5.5 MHz and between picture and colour signal is 4.43 MHz. The Fig. 9.32 shows the complete spectrum.

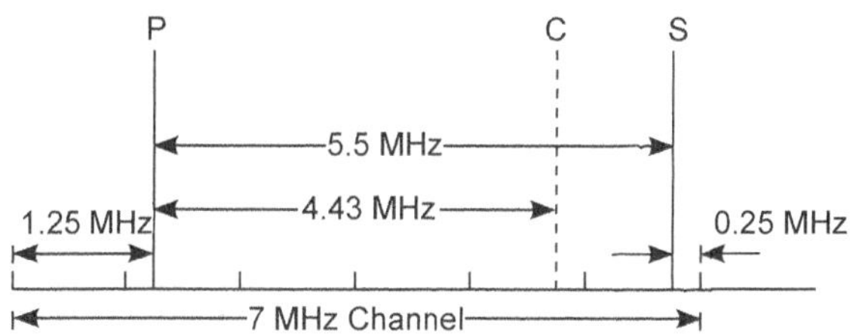

Fig. 9.32

Table 9.4: Popular TV standards in the World

S. No.	Particulars	India, Europe and Asian countries	America, Canada, Mexico and Japan	England	USSR	France
1.	Lines per frame	625	525	625	625	625
2.	Frames per second	25	30	25	25	25
.5.	Field frequency (Hz)	50 Hz	60	50	50	50
4.	Line frequency (Hz)	15,625	15,750	15,625	15,625	15,625
5.	Channel BW (MHz)	7	6	8	8	8
6.	Picture signal modulation	AM	AM	AM	AM	AM
7.	Sound signal modulation	FM	FM	FM	FM	FM

SUMMARY

1. The "Television" means to see at a distance.
2. The few TV applications are: cable TV, close circuit TV, Picture phone, TV games etc.
3. The TV camera converts picture signal into electrical signal. The popular black and white TV camera tubes are: Orthicon, Vidicon and Plumbicon.
4. The B and W cameras are used in different configurations.
5. The picture tube converts the electrical signal into picture signal. It may be a monochrome picture tube or a colour picture lube.
6. The picture is "scanned" by a camera before converting it into electrical signal. Usually "Interlaced Scanning" is used; in which the picture is divided into 625 lines and is then scanned line by line.
7. The Vestigial side band (VSB) transmission is generally used for TV signals. In this technique full upper side band and some part (called Vestigial) of lower side band are transmitted.
8. Few blocks in monochrome and colour TV are common.
9. In a TV transmitter microphone converts sound into electric waves, these waves are processed and transmitted into space through an antenna. In camera, the picture is also converted into electrical waves, which are processed and transmitted into space through the same antenna.

   At the receiver, sound and picture both are detected separately. Sound after processing, is given to loudspeaker, which converts it into the original sound and the picture signal after processing is given to the picture tube, which converts it into original picture.
10. *The important TV processes are*: scanning, synchronization and blanking.
11. The picture signal along with blanking and synchronizing signals is called "composite video signal".
12. In TV, the sound signal is *frequency modulated* and the picture signal is *amplitude* modulated.
13. The TV receiver may be monochrome (Black and white) or a colour TV.

SELF TEST QUESTIONS

1. What are the various commercial applications of a TV?
2. On what principle TV cameras work? Explain a vidicon camera.
3. Explain a picture tube used in B & W television.
4. Explain the principle, purpose and process of scanning.
5. Explain interlaced scanning with a special reference to Indian standards.
6. Explain synchronising and blanking pulses.
7. Explain composite video signal and how is it produced?
8. Explain "Bandwidth" with special reference to Indian standards.
9. Explain DSB transmission used in TV, Also give its significance
10. Write short note on SSB and VSB transmission.
11. Draw block diagram of a monochrome TV transmitter and explain.
12. Draw block diagram of a mono-chrome TV receiver. Explain its important blocks.

OBJECTIVE TYPE QUESTIONS

I. Fill in the blanks:

1. Target plate of vidicon is made of _______ .
2. Target plate of plumbicon is made of _______ .
3. The dark current in cameras is in _______ .
4. The basic principle of camera is _______ .
5. Scanning is required only in _______ .
6. Photo sensitive material converts brightness into _______ signal.
7. Usually _______ scanning is used.
8. Frequency of horizontal scanning is _______ .
9. *C* signal means _______ .
10. *Y* signal means _______ .
11. Practical transmission system used in TV is _______ .

ANSWERS

1. Antimony compound
2. Lead oxide (PbO)
3. Nano amperes
4. photo effect
5. Video systems
6. video
7. Interlaced
8. 15750 Hz,
9. chrominance or colour signal
10. monochrome, luminance or white signal
11. Vastigal Side Band System.

❑❑❑

10

Colour Televisions

The colour television produces transmitted information in the original coloured form. All stages described in the monochrome TV are used in a colour TV. In addition, few stages, which are required for treatment of colours are also included.

10.1 COLOUR TELEVISION

A colour picture is nothing but "a monochrome picture on a white raster with colours added". The required information is in the chrominance (or *C* signal) broadcasted along with monochrome signal. If we turn down the colour control on the receiver to eliminate the colour signal, what we get will be a *black & white* picture. With *C* signal; the picture is reproduced in natural colours. All colours can be produced as a combination of red, green and blue colours, which are known as **Primary Colours**.

10.2 COLOUR SCHEME

10.2.1 Primary, Secondary and Complementary Colours

The colours which cannot be produced by mixing other colours are called **primary** colours *e.g.,* **Red, Green and Blue** are called primary colours. Sometimes Yellow is also taken as primary colour. There are two types of primary colors.

(*i*) *Additive Primary Colours*: The red, green and blue are called **Additive Primaries** and are used when coloured light sources are blended to produce the required colour.

(*ii*) *Subtractive Primary Colours*: The red, blue and yellow are called **Subtractive primaries** and are used when a picture on print is viewed by reflected light from a white source.

The colours, which can be produced or obtained by mixing primary colours are called **secondary colours.**

The two colours, which on mixing give white colour are called **complementary colours** to each other *e.g., yellow* is a complementary to blue, *magenta* is complementary to green and *cyan* is complementary to red.

10.2.2 Additive and Subtractive Mixing of Colours

The mixing of different colours to get new colour is known as **additive mixing** of colours.

When we mix pigments, this is known as **subtractive mixing** of colours.

e.g. the additive mixing of red, green and blue gives white colour, while mixing them subtractively gives black colour. The Fig. 10.1 shows additive mixing of colours:

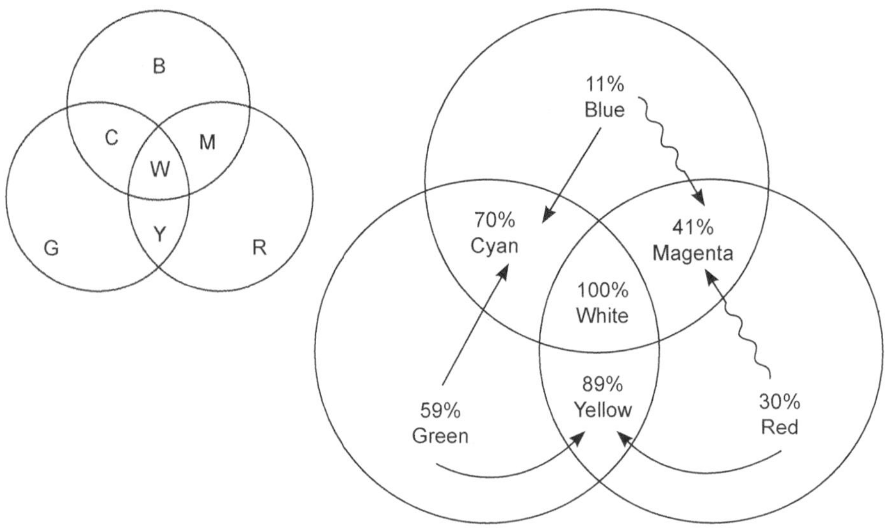

Fig. 10.1

e.g., Red + Green = Yellow $R + G = Y$

Red + Blue = Magenta $R + B = M$

The Fig. 10.2 shows subtractive mixing of colours *e.g.,* White – Green – Blue = Red ($W - G - B = R$) and White – Green = Magenta ($W - E = M$) etc.

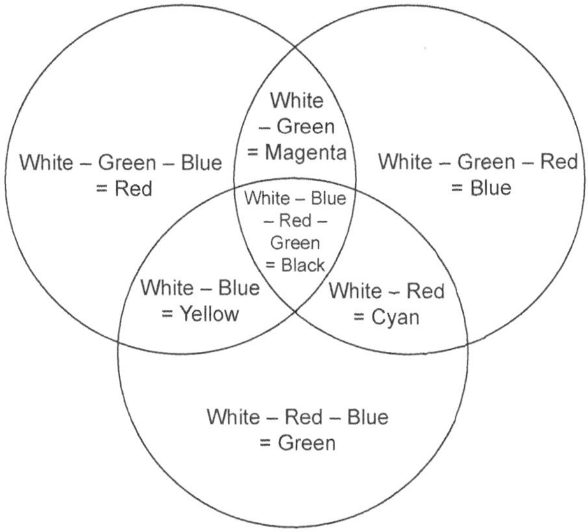

Fig. 10.2

As told earlier, the two colours which on mixing produce white colour are called **complementary colours**. This can be shown that

$$\text{Red} + \text{Cyan} = \text{White}$$

$$\text{Green} + \text{Magenta} = \text{White etc.}$$

So, (Red, Cyan), (Green, Magenta) are pairs of complementary colours.

In colour TV system red, green and blue are used as Primary colours, By mixing these colours in required proportion any colour can be obtained.

10.2.3 Tristimulus Values

The primary colour components values in a secondary colour are called **Tristimulus values.** The natural white has 30% red, 59% green and 11% blue. In other words, one lumen of white light has 0.30 lumen of red + 0.59 lumen of green + 0.11 lumen of blue.

10.2.4 Trichromatic Values (Or Coefficients) and Colour Triangle

The Tristimulus values (0.30, 0.59 and 0.11) are not very convenient. So we use Trichromatic (T) units or (T) Coefficients for obtaining secondary colours. The Trichromatic units of white colour consist of 1/3 of green, 1/3 of red and 1/3 of blue. These coefficients are represented as x, y and z. Note that $x + y + z = 1$, so the T coefficients of x, y and z can be represented by a triangle called *colour triangle*, as shown in Fig. 10.3.

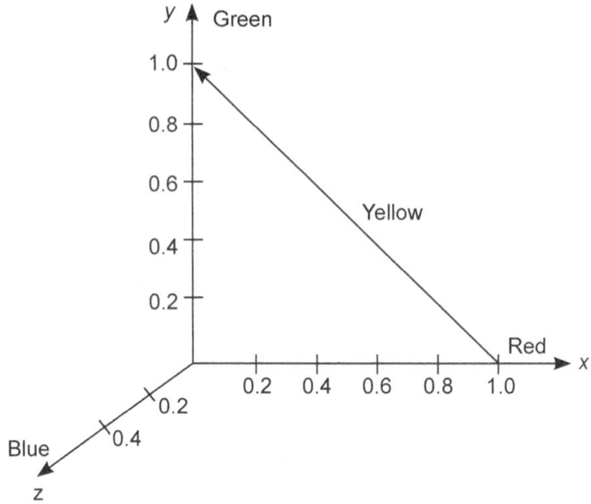

Fig. 10.3

The red is shown on x axis, green is shown on y axis, the third axis z is perpendicular to xy plane and shows blue. Since $x + y + z = 1$, the third axis need not to be shown.

The coordinates of the colours are as given below:

Magenta:	$x = 0.5$,	$y = 0$,	$z = 0.5$
White:	$x = 0.333$,	$y = 0.333$	$z = 0.333$
Yellow:	$x = 0.5$,	$y = 0.5$,	$z = 0$
Cyan:	$x = 0$,	$y = 0.5$	$z = 0.5$
Note that: x = red,		y = green,	z = blue and so on.

10.2.5 Colour Circle

The colour triangle can be converted into colour circle where red, green and blue are radial vectors spaced at 120° from each other. The distance from centre to the circumference represents **saturation (colour concentration)**. The vector *reference axis* is R (Red). The colours are represented by the phase angle w.r.t. this axis (Fig. 10.4).

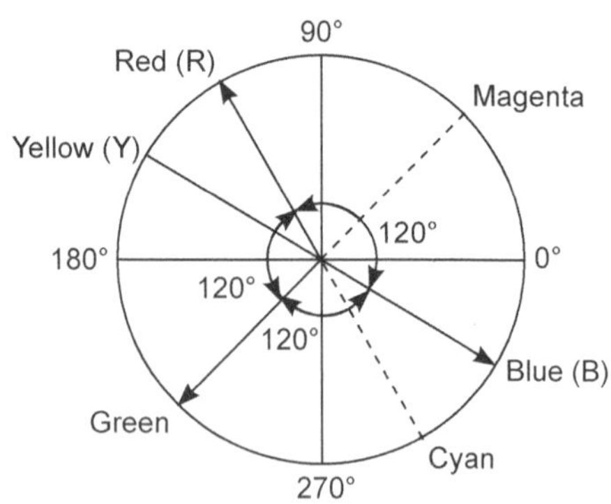

Fig. 10.4

If one end of a vector represents one primary colour, the other end represents complementary colour. Yellow is complementary to blue, magenta is complementary to green and cyan is complementary to red.

10.3 TYPES OF COLOUR VIDEO SIGNALS

The important types of colour video signals are three: Red, Green and Blue (*R, G,* and *B*), as a TV system starts with *R, G, B* at the camera and finishes at *R, G, B* in picture tube. However, colour mixtures are used for *coding* and *decoding,* because mixing of two colours can have all the *colour* (Chrominance) information of the three colours, allowing the third signal to be *Y* (luminance) signal.

The important colour video signals and their combinations is given below.

1. ***I* Signal** = 0.60 *R* – 0.28 *G* – 0.32 *B*.

   It is a combination of 60 per cent red, 28 per cent green and 32 per cent blue video signals. The minus sign indicates the addition of video voltages of negative polarity *e.g.*, – 0.32 *B* means 32 per cent of the total blue video signal, but with inverted polarity, which reproduces blue.

   Similarly

2. ***Q* signal** = 0.21 *R* – 0.52 *G* + 0.31 *B*.
3. ***B* – *Y* signal** = – 0.30 *R* – 0.59 *G* + 0.89 *B*.
4. ***R* – *Y* signal** = 0.70 *R* – 0.59 *G* – 0.11 *B*.
5. ***G* – *Y*** signal = – 0.30 R – 41 G – 0.11 *B*.

This has been explained earlier.

10.4 IMPORTANT TERMS

Any colour has 3 characteristics: First is *hue or tint,* which we generally call the "colour", the second is *saturation* and the third is *luminance*. The saturation describes the *concentration* of the colour, while the luminance indicates brightness or the "shade". The important terms are described below:

10.4.1 White

Actually white light can be considered as a mixture of red, green and blue (primary colours) in right proportions. A white colour for colour TV is a mixture of 30 per cent *red*, 59 per cent *green* and 10 per cent *blue*. The percentage of the *luminance* signal is based on sensitivity of the eyes for different colours.

10.4.2 Hue (or Tint)

The colour itself is known as its hue *e.g.*, green leaves has a green *hue* and a red flower has a red hue.

10.4.3 Saturation

Saturated colours are deep and strong. Weak colours have no saturation. By saturation we mean how a colour can be diluted by white. When a saturated (strong) red colour is mixed with white, we get *pink* colour, which can be called as "desaturated" red.

10.4.4 Brightness

Each colour produces a certain amount of brightness. It is determined by the amount of light contained by it.

"Brightness" is different from "Saturation", one can be changed keeping the other constant. The colour TV has two different controls: Saturation and Brightness. When saturation control is operated, amount of white light contained in the colour is changed, when the Brightness control is operated white as well as coloured lights are changed.

So brightness is the measure of white and colour lights, whereas saturation is the measure of white light only.

10.4.5 Luminance

By luminance, we mean amount of light intensity or brightness sensed by our eyes. In B & W picture, the lighter parts are more luminous than its darker parts. Some colours appear brighter than others. The luminance indicates how a particular colour will look in black and white reproduction. For example, a monochrome picture will show yellow colour as white, light blue colour as grey and dark red colour as black. Luminance signal is written as *Y*-signal. In brief, luminance means brightness.

10.4.6 Chrominance

The term is used to indicate *hue* as well as *saturation* of a colour. The chrominance signal includes all the colour information except the *brightness*. Chrominance and luminance together give complete information about the pricture. Chrominance is abbreviated as *chroma* and is also written as "*C*-signal" . As colour TV is concerned, chrominance means 3.58 MHz modulated sub carrier before modulation, and after demodulation it is the colour information in red, green and blue signals.

10.5 SENSITIVITY OF HUMAN EYE

All types of radiations do not produce sensation of light on human eye. The radiations between wave lengths of 4000 Å and 7000 Å (Angstrom, $1 \text{ Å} = 10^{-10}$ m) only produce sensation. The human eye is the most sensitive to radiations of 5500 Å, though it differs from man to man and from age to age. The curve between sensitivity of human eye and the wave length is shown in Fig. 10.5.

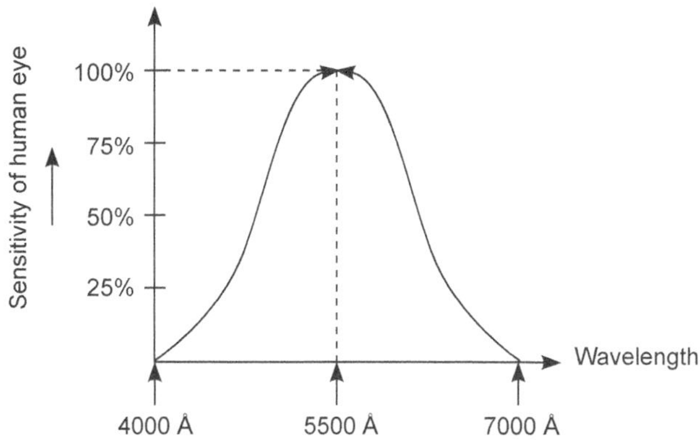

Fig. 10.5

The colour corresponding to 5500 Å is **yellowish green** which is the most suitable colour tor all purposes.

The retina of the human eye consists of large number of minute cells called rods and cones, each of which is connected to the brain by optic nerve. These cells are sensitive to the quantity and colour of light falling on them and transmit the colour information to the brain.

The 'cones' and 'rods' are light sensing cells. 'Rods' help us to know about 'light' and 'darkness' while the 'cones' give an idea of colours. The 'cones' have special attraction for red and yellow wave lengths. This is why we like these colours.

10.6 COMPATIBILITY

Black and white television was established well before colour television. After the invention of colour TV, it was necessary that existing system (black and white) should be modified. In other words colour system must be compatible with the existing black and white TV system, which needs following requirements:

1. The colour TV signal must produce normal black and white picture on a monochrome receiver without much modification in the receiver.
2. The colour TV must be able to produce black and white picture from normal monochrome transmission.

   So to make the system compatible, following changes were needed:

   (*i*) The colour signal should occupy same bandwidth as monochrome signal.

   (*ii*) The location and spacing of picture and sound carrier frequencies should be same.

   (*iii*) The colour signal should have the same (brightness) information as would a monochrome signal, transmitting the same scene.

   (*iv*) The composite colour signal should contain colour information together with synchronising signal.

(*v*) This extra colour information should not affect the picture reproduced on the screen of a monochrome receiver.

(*vi*) The same deflection frequencies and sync. signals should be used as were used for monochrome transmission and reception.

10.7 POPULAR TV SYSTEMS

The popular TV systems are:

10.7.1 NTSC System

It stands for "National Television System Committee system". In this system, the two colour difference signals created by subtraction of red and blue from the total signal are transmitted in quadrature (with one quarter cycle behind). The signals are then added together to get the chrominance signal.

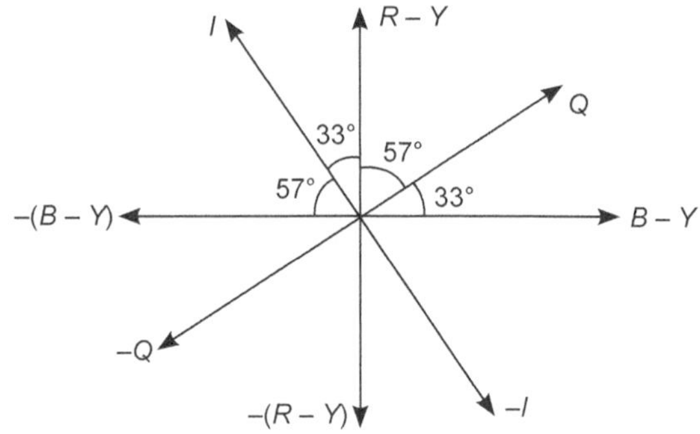

Fig. 10.6 (a)

The **Fig. 10.6** shows phase relationship in NTSC transmitter system and the **Fig. 10.7** shows NTSC transmitter.

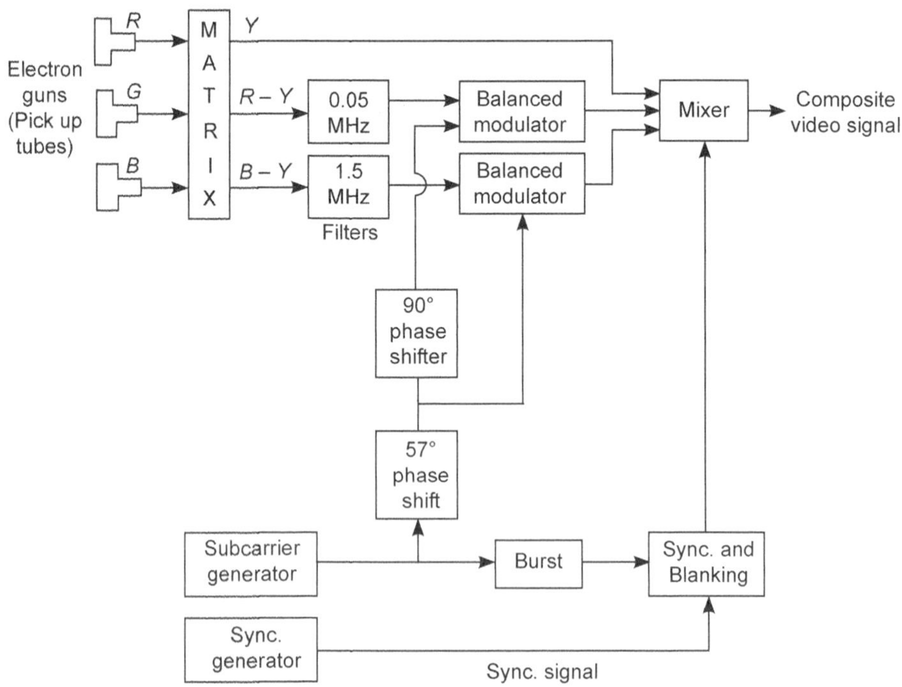

Fig. 10.6 (b)

The three primary signals generated by three electron guns or pick up tubes (*R*, *G*, *B*) are amplified and applied to a **Matrix** where these are combined algebraically to produce one **luminance** signal (*Y*) and two **chrominance** signals (*R–Y*, *B–Y*).

The chrominance spectras are clipped by **Filters**. The chrominance signals are modulated on the subcarrier. It accepts all the signals that make up a composite video signal. The sub carrier generator and the sync. generator must be linked in a suitable manner; so that the sub carrier may be an odd harmonic of half the frame (line) scanning frequency.

A colour subcarrier burst signal is inserted in the composite video signal during the line blanking interval. The frequency and phase of the burst should be equal to the chrominance subcarrier at the transmitter.

The **Fig. 10.7** shows NTSC receiver. After amplification and detection of R.F. and and I.F. signals, the composite signal is applied to the filters, demodulators, amplifiers and through matrix, it is given to the picture tube.

(*i*) **Amplifier:** tuned to sub carrier and extracting the colour burst. The phase of the burst should be same, as of the output of the local oscillator.

(*ii*) **A filter.** Of 2 to 4 MHz: From filter, the signal goes to the synchronous detector. The filter blocks all frequencies lying outside the spectrum of the chrominance signal.

(*iii*) Three parallel connected cathodes of the picture tube.

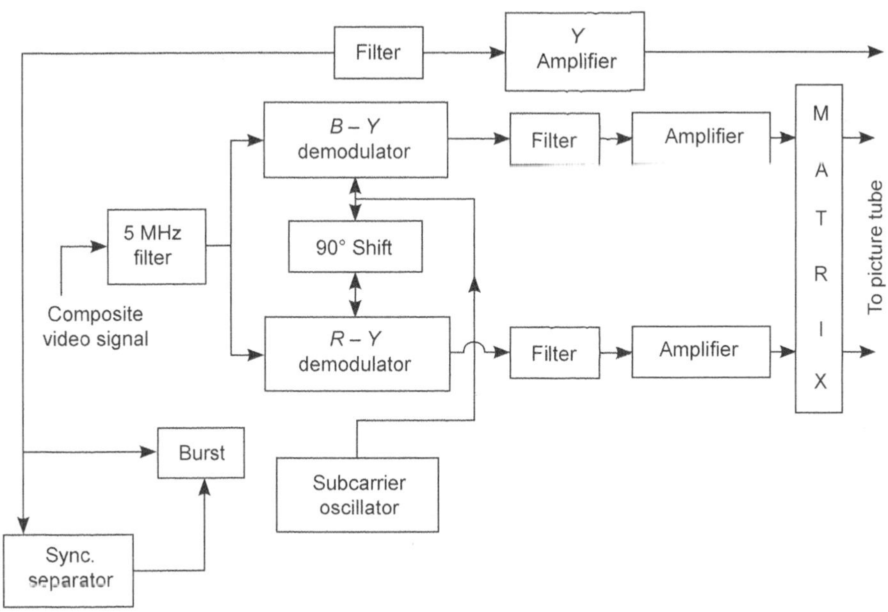

Fig. 10.7

Limitations of NTSC System:

(*i*) The difference of phase between the sub carrier and local oscillator produces incorrect hues.

(*ii*) The cross talk between demodulator outputs at the receiver causes colour distortions.

10.7.2 The SECAM (Sequential) System

It stands for "sequential colour a memory' system The system has a different mode of transmission. The colour difference signals are not arranged a quarter of cycle apart but are kept separate by transmitting them on alternate lines of the picture. Delay lines inside the receiver hold up one set of signals so that they can be recombined to form a picture from alternate lines of the signal.

The system avoids the problem of phase error. In SECAM, the two chrominance signals (*R*-*Y* and *B*-*Y*) are transmitted on alternate lines in sequence. Though half the colour information is lost but the human eye cannot distinguish this. The principle of SECAM has been shown in the **Fig. 10.8**.

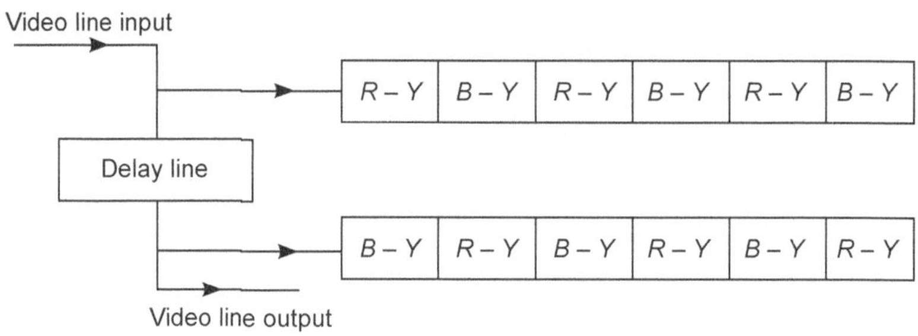

Fig. 10.8

The delay time is exactly one line scan interval, which is about 64 μs. For first time, the signal is taken from input (Video line) to the delay line and second time from output of the delay line.

The **Fig. 10.10** shows block diagram of SECAM **Transmitter**. The colour difference signals (*R*-*Y* and *B*-*Y*) are band limited to 1.5 MHz by filters. The two signals are applied to subcarrier FM modulator alternately by using an **electronic switch**. The switch operates at line frequencies. The colour sub carrier is, therefore, modulated alternately. The channel B.W. is 8 MHz.

To obtain the composite signal, the modulated subcarrier is mixed with luminance, sync. and blanking signals.

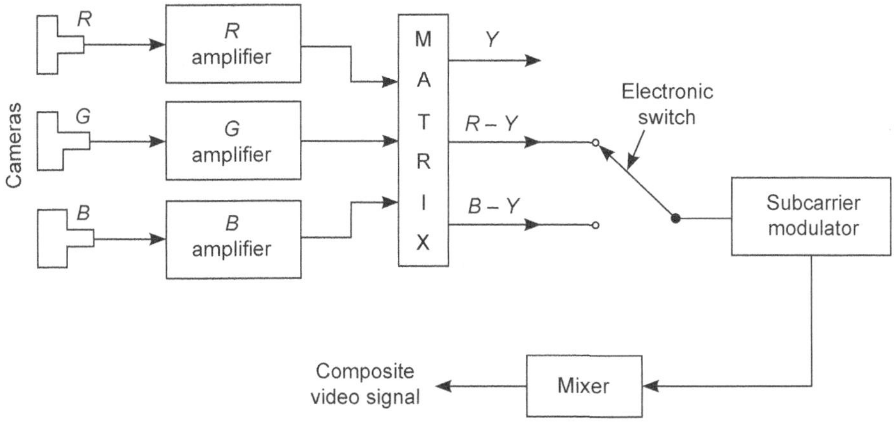

Fig. 10.9

The **Fig. 10.10** shows the SECAM receiver. In the receiver, the colour difference signals existing at input and output from the delay line are routed by an electronic switch to the two inputs of a MATRIX which drives the third colour difference signal (*G-Y*). With the chrominance signal transmitted sequentially (on alternate line scans), the delay line enables the three chrominance signals to be generated in receiver at the same time. This is the reason, that the SECAM system is also called **sequential simultaneous**.

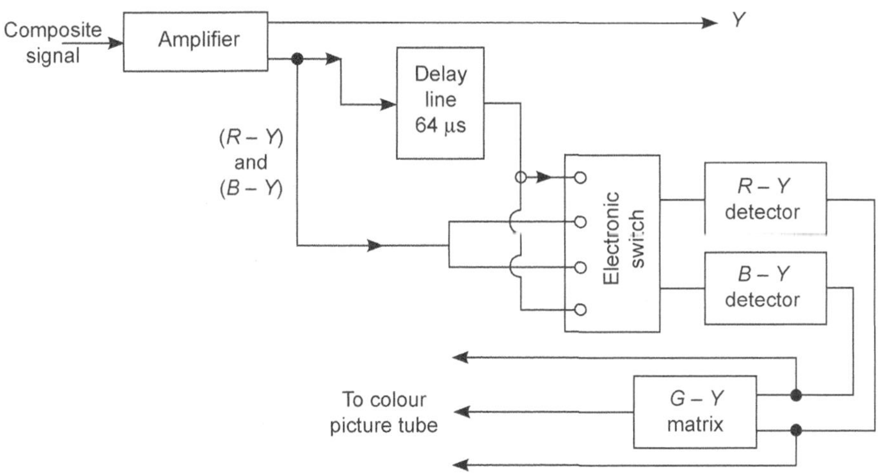

Fig. 10.10

Advantages of SECAM

(*i*) As FM is used, the SECAM receiver is free from phase distortion.

(*ii*) There is no **cross talk**, as the signals are not present simultaneously.

(*iii*) The SECAM receiver is simpler and cheaper than the NTSC receivers.

Disadvantages of SECAM

(*i*) The vertical resolution of SECAM system is inferior.

(*ii*) The colour is more saturated during fade to black. During fade, the pink colour changes to red.

10.7.3 PAL System

PAL stands for "Phase Alternate Line" system and is adopted in India. It is an improved NTSC system and stands in between NTSC and SECAM systems.

In this system, the signal is transmitted in the same way but difference lies in reception. In receiver, the information is delayed at every line *e.g.*, if a certain line of the picture signal has a strong green signal, the receiver reverses the polarity of alternate lines, it assures that the next line contains very low green signal. The input sent to the picture tube is the average of the delayed line and corrected line; thus eliminating the error.

The Fig. 10.11 shows PAL transmitter. Camera tube converts light into video signals having three primary colours. These colour signals are converted into luminance signal Y and colour difference signals $B\text{-}Y$ and $R\text{-}Y$.

In this system, $B\text{-}Y$ modulates the sub-carrier in phase and $R\text{-}Y$ modulates the subcarrier with phase = +90° on one line and –90° on the next line and so on. So, the phase of the subcarrier changes are automatically corrected.

The phase shifter stage uses an electronic switch to change the phase from + 90 degree on one line to – 90 degree on the next line.

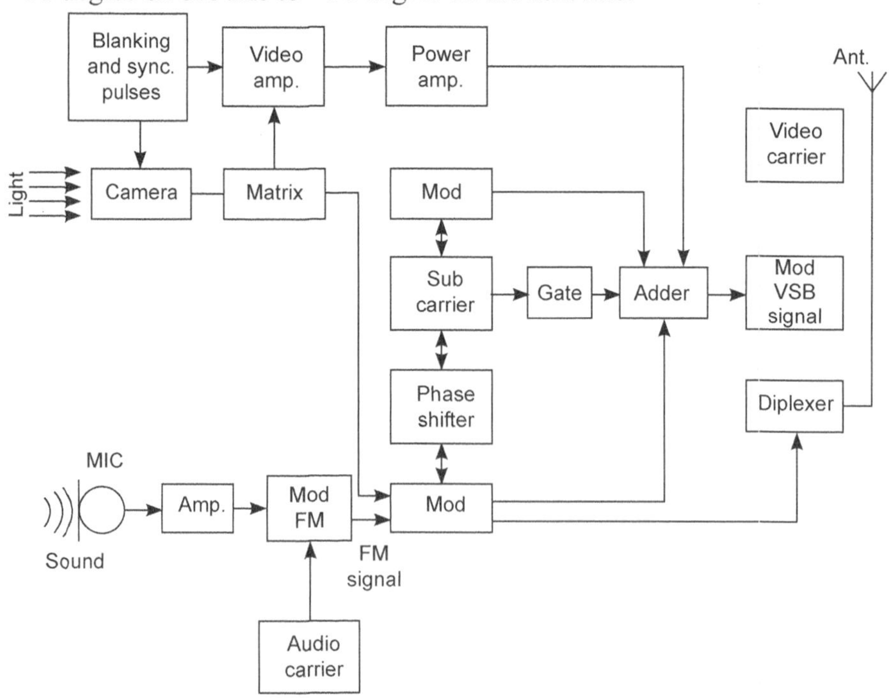

Fig. 10.11

The encoded signal is called chroma signal. The *Y* signal alongwith its control signal is added to the chroma signal by an adder. This colour video signal (CVS) modulates the main video carrier. Audio signal is converted into electrical signal by the microphone. After amplification, the audio carrier is frequency modulated, which is located within the 7 MHz channel width.

Both modulated signals (video modulated signal and audio modulated signal) are sent to the transmitting antenna with the help of a diplexer circuit. The transmitting antenna transmits the signal into space.

The Fig. 10.12 shows PAL receiver. Here, the receiving antenna picks up the signal and feeds it to the tuner stage. The tuner contains a low noise RF amplifier. The signal is converted into intermediate frequency signal by superheterodyning process. Conversion of RF and IF results in better selectivity, higher gain and better stability.

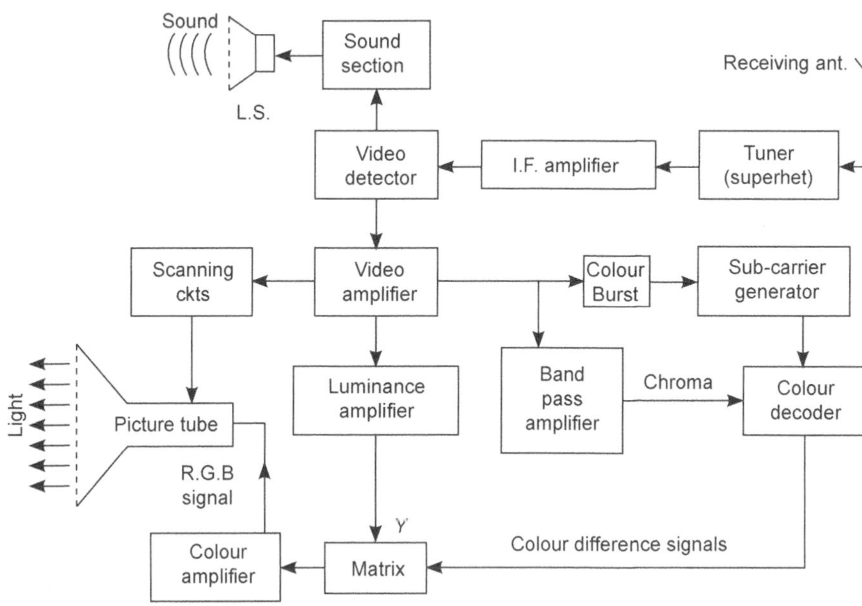

Fig. 10.12

The IF signal is amplified by three or four stage of amplification and then goes to video detector.

In the output of video detector, we get:

(i). Intercarrier frequency signal (5.5 MHz)

(ii). Chroma signal

(iii). Colour burst (CB) signal

(iv). Luminance (Y) signal

(v). Sync. pulses.

Now sound IF is amplified and is fed to the FM detector to recover audio signal. After amplification it is fed to the loudspeaker which converts it into sound signal.

Chroma signal is decoded to retrieve original colour difference signals. The decoding is done by synchronous demodulation process. Luminance (*Y*) signal is added to the colour difference signals to get *R*, *G* and *B* signals in the same proportion of amplitude as in the output of the camera tubes.

R, *G* and *B* signals are fed to a colour picture tube which produces three beams of electrons. These beams are focussed on three vertical stripes of red, green and blue phosphors, respectively, through vertical slits.

The closely spaced phosphors produce red, green and blue light intensity in the same proportion as contained in the original light and our eyes integrate these colours to give the resultant colour of the original scene. .

10.8 COLOUR TV CAMERA

Basically a colour TV camera consists of three camera tubes correspondingly for the three primary colours red, green and blue. (Fig. 10.13)

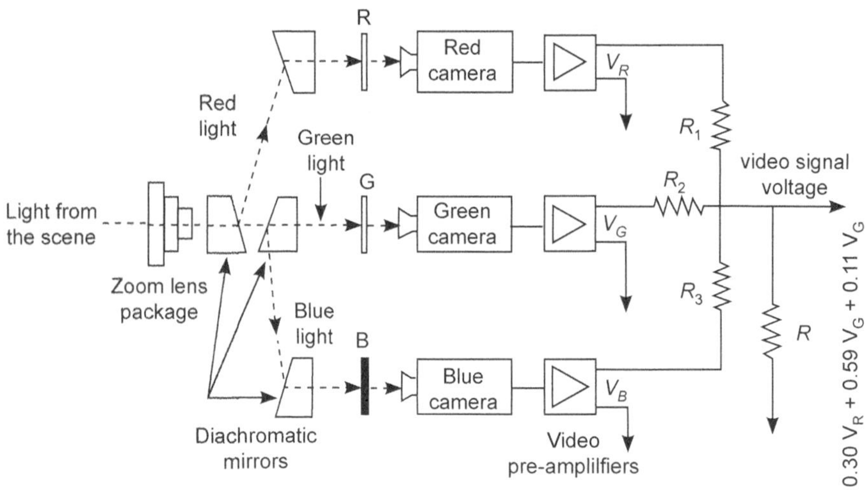

Fig. 10.13

Basically, colour TV camera consists of

- Diachromatic Mirrors
- Trimming Filters
- Camera Filters
- Video pre-amplifiers
- Resistive networks (R_1, R_2, R_3)

Light from the scene is processed by the lens system. Diachromatic mirrors are specially designed which allow a particular colour to pass through and reject all other colours. So these are also a type of colour filters. Diachromatic mirrors are designed to pass one wavelength and reject other wavelengths. So the light is filtered into three colours red, green and blue. These colours also pass through colour filters (R, G, B) which finely tune each colour and provide highly precise colours. Now these colour signals are converted into corresponding electrical signals by cameras. So in this way three colour video signals are generated. As the magnitude of video signals is less, so these are amplified by video pre-amplifiers. Now these amplified video signals are passed through resistive network. All the resistors chosen are of different values. These are chosen to produce 30% Red, 59% Green and 11% Blue colour signal.

The output of colour TV camera is $0.30\ V_R + 0.59\ V_G + 0.11\ V_B$.

A colour TV camera is basically an arrangements of three black and white camera tubes. In case of three tube configuration, light falling on the target is split into three primary colours: *R*, *G* and *B*. (Fig. 10.14). Each tube now processes a particular colour. The two tubes or single tube arrangement can also work.

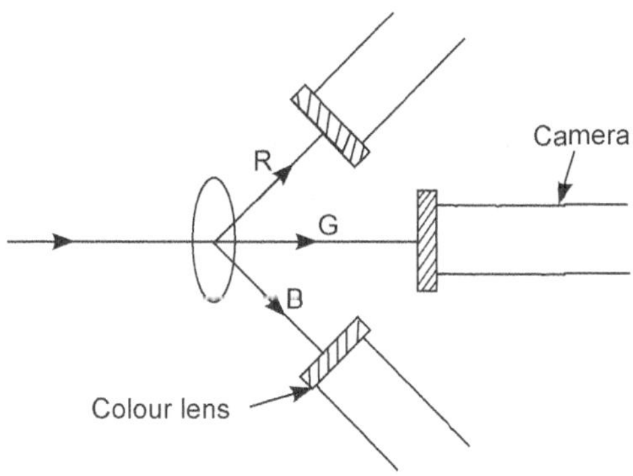

Fig. 17.14

10.8.1 Solid State Video Camera

The solid-state camera does not use a camera tube described in the earlier pages. In turn, it uses a fully solid-state device in the integrated circuit (I.C.) package as the device for converting light variations from the object into corresponding electrical signals.

The CCD (charge coupled device) photosensitive array package in the IC form is the device used. The array (Fig. 10.16) consists of a matrix of CCD elements arranged in rows and columns. Each CCD element, comprising of a photoconductor whose output is coupled to a capacitor through a FET switch, represents one picture element. One row of CCD elements is represented in Fig. 10.15.

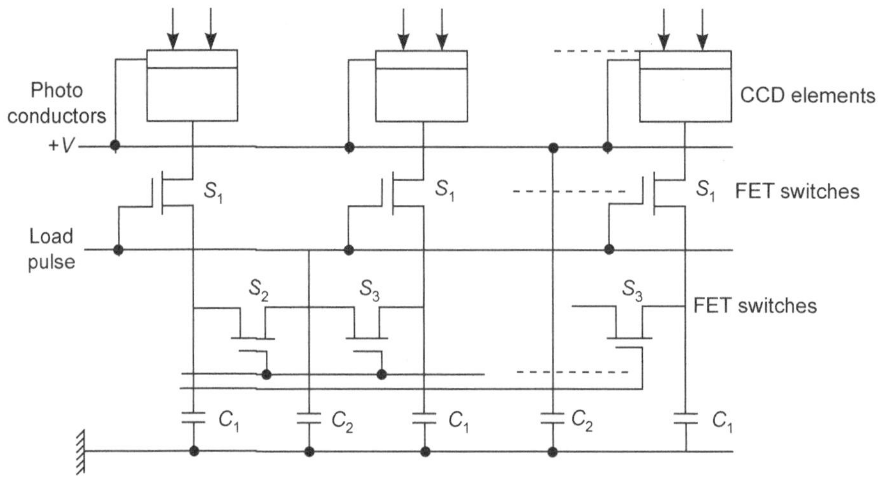

Fig. 10.15

The circuit operates as follows: If the FET switches (S_1) for all the elements of a particular row are switched on for a short-time, capacitors (C_1) will charge to a voltage depending upon the light falling on the respective photoconductor. Turning on switches (S_2) transfers these charges to capacitor (C_2). Now if switches represented by (S_3) are switched ON, the charge corresponding to a particular picture element passes on to the following adjacent CCD element. Thus by alternately switching ON (S_2) and (S_3), the charge can be shifted along the row and finally taken out. This is equivalent to a line scan in electron tube based cameras. By taking the output from each row in sequence, a field scan can also be achieved and thereby we can generate a video output signal that represents light intensity falling on the CCD-array. In a colour CCD video camera, a colour striped filter is bonded to the surface of CCD array.

Solid-state video cameras are more rugged and compact than their camera tube counterparts. They donot suffer from the lag problem of vidicon tube based cameras. They do suffer from 'blooming' problems. Blooming occurs due to saturation of individual photocells when illuminated by very bright light.

10.9 COLOUR PICTURE TUBES

In colour TV, colour picture tube is used.

The screen of the colour TV has three *i.e.*, red, green and blue phosphors. These are three primary colours and their combination can produce any desired colour.

There are 3 electron beams to excite these colours. A metal plate called *shadow mask* is also provided which has holes to allow the beam to reach the screen and the electrons that do not have the required angle are blocked. It is important to mention that each beam excites its respective phosphors on the screen. The screen phosphors produce colours, and *the electron beams have no colour of their own*. If only red gun is operated the whole screen will be red, similarly by operating green gun, the screen can be made all green and so on.

In normal operation when the three guns are operated, they excite their respective phosphors. We get a picture on screen by *superimposition* of the three primary colours. For example, *yellow* colour is obtained by combination of red and green. *White* colour is obtained by proper combination of red, green and blue in the picture is obtained when all the three beams are off.

According to configuration of guns and the phosphors, three types of colour picture tubes are popular:

1. Delta gun tube (Delta tube) [Fig. 10.16 (*a*)]
2. Guns in Line tube (GIL tube) [Fig. (*b*)|
3. Single gun tube (S.G. tube) [Fig. (*c*)]

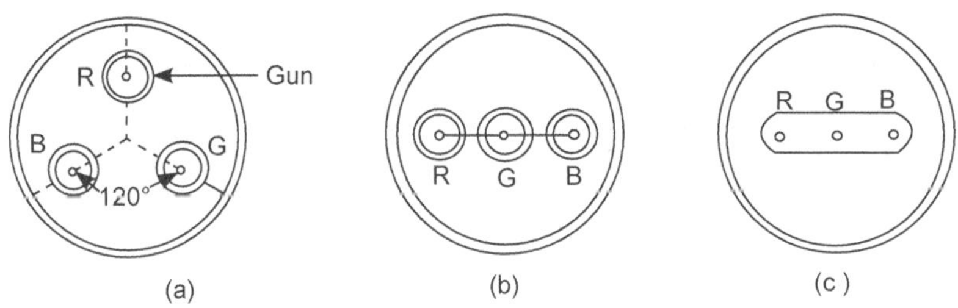

Fig. 10.16

The basic construction of all the above tubes is same, the difference lies only in their guns configuration. In delta tube, the guns are arranged in a *triangular* (delta) form, *i.e.* 120° apart, in GIL tube the 3 guns are arranged in a horizontal line and in single gun tube, as the name suggest, there is only one gun having 3 cathodes (*R*, *G*, *B*) in a line.

Here **Delta gun tube** will be discussed.

10.9.1 Delta Gun Colour Picturc Tube

This is the most widely used colour tube.

The Fig. 10.17 shows delta picture tube. There are 3 separate cathodes and 3 separate control grids. Each gun has a separate screen grid, focus grid and accelerating grid.

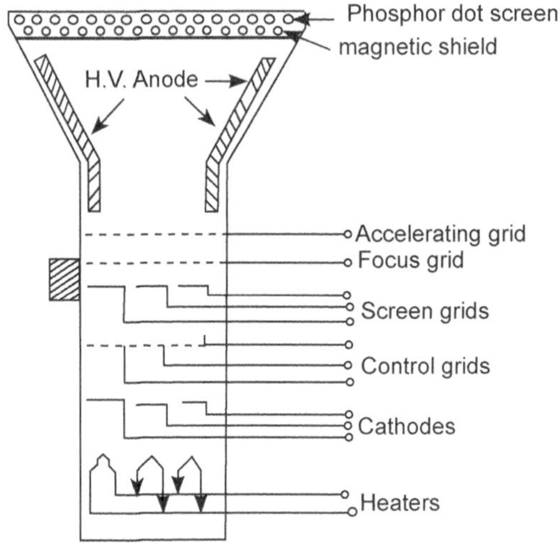

Fig. 10.17

At the neck of the tube, there are following components: (See Fig. 10.18)

(*i*) *Deflection yoke*: Its horizontal and vertical deflecting coils deflect the three beams.

(*ii*) *Convergence yoke*: This yoke possesses individual adjustment for *R*, *G* and *B* beams to make them *converge* through the openings of the *shadow mask*. For each beam there is a permanent magnet and coil.

(*iii*) *Purity magnet*: This magnet adjusts the three beams to produce pure red, green and blue colours.

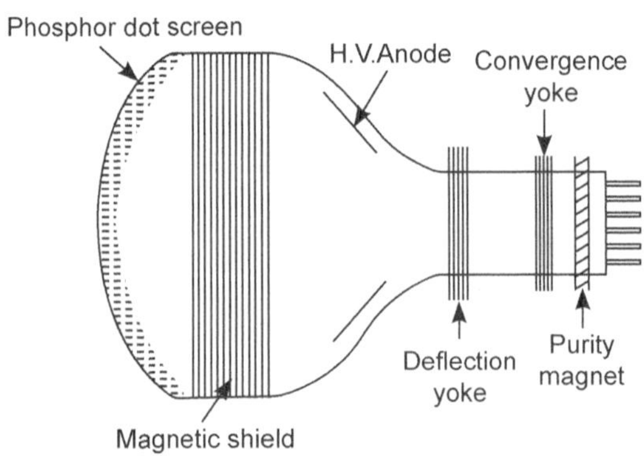

Fig. 10.18

A colour P.T. is different from monochrome P.T. in the following ways:

1. It has 3 guns, which provide three electron beams, one each of the three primary colours.
2. The screen of the picture tube is coated with three types of phosphor colours. When light from the three guns is incident on the phosphors, they separately emit red, blue and green lights.
3. The phosphors are embedded on the screen in triangular dot pattern (Triads). Each triad has a group of three phosphor dots.
4. In colour picture tube, a shadow mask is mounted, about 1/2 inch behind the screen.

10.10 COLOUR SCREENS

There are 3 colour screens used by the colour picture tubes:

(*a*) Delta gun tube uses a screen with *R*, *G* and *B* phosphors in dots, forming *triads*. With this screen, shadow or aperature mask has holes opposite the triads through which beam passes and excites the traids. [See Fig. 10.19 (*a*)]

(*b*) GIL type tubes use screen phosphors in vertical strips in *R*, *G* and *B* colours, the green being in the centre. With the stripped screen, the mask has vertical slot for the beam. [See Fig. 10.19 (*b*)]

(*c*) Single beam type tube uses a vertical stripped three colour phosphor screen and opposite to it is *grill* shaped aperture mask. [See Fig. 10.19 (*c*)]

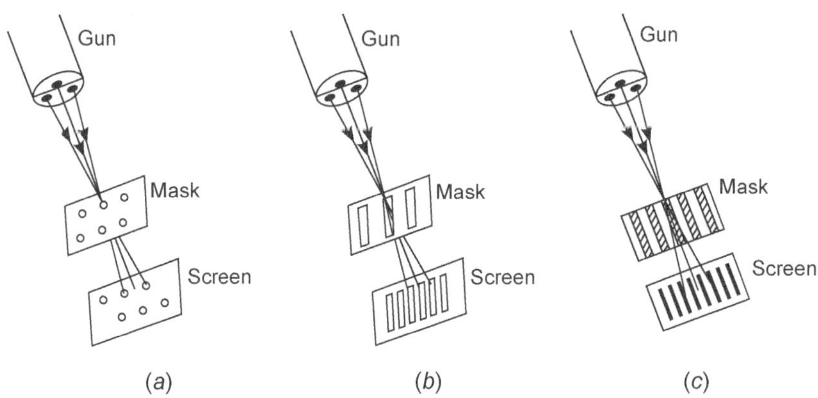

Fig. 10.19

Whatever may be shape of screen and the mask, only 20% of the beam is allowed by mask to reach the screen. This is the reason, why colour P.T. needs higher value of anode voltage, cathode current and heater power as compared to monochrome tubes.

10.11 TRANSMISSIN AND RECEPTION OF COLOUR TV SIGNALS

10.11.1 Transmission of Colour TV Signals

Most of the T.V. signals are broadcasted through air. At the TV station, a device "Transmitter" is used to produce the composite TV signal from separate audio and video signals. This signal is then transmitted in the form of electromagnetic waves.

The signals can be received clearly upto 160 km distance from the transmitting station. To send TV signals farther, other means of transmitting like co-axial cables, fibre optic cables, microwaves and satellites may be used.

1. *By Co-axial and Fibre Optic Cables*: Co-axial and fibre optic cables are used to carry TV signals for long distance or to areas, which have difficulty in receiving the signal through air. The TV network sends programs to their affiliated stations through co-axial cables. The cable TV system use co-axial or fibre optic cables to carry signals to the individual TV receiver.
2. *By Microwaves*: These are also electromagnetic waves. A row of towers at about 50 km span carry program to the affiliated stations through these waves. The equipment in a tower automatically receives, amplifies and then retransmits the microwave signals to the next tower.
3. *By Satellites*: The satellite is used between stations where cable or microwave towers cannot be built. The satellite receives coded signal from an earth station, amplifies and sends to another earth station. The two stations may be thousands of kilometers apart.
4. *By air* (*through space*): This is the most popular method used for broadcasting (transmission) colour TV signals detail:

In the transmission by air (through space), the transmitter at the broadcasting station amplifies the signal so that it has enough power to reach large distances. The transmitter increases the frequency of both the audio and video signals through "modulation". For this purpose, high frequency electromagnetic waves called "carrier waves" are generated by the transmitter. The video signal varies the amplitude of the carrier waves to carry video part of the TV signal. The process is called *"amplitude modulation"*. The video signal is then amplified to a power of 10^3 to 10^6 watts.

Another carrier waves modulate the audio part of the TV signal. The process is called *frequency modulation*. The readers know that a TV signal is a composite signal comprising of audio and video parts.

The transmitter then combines the modulated video and audio parts to form the composite TV signal. This signal is transmitted into space.

The Fig. 10.20 shows the process of transmission through air. The colour TV transmission begins with a TV camera, in which a mirror (or prism) system breaks light from the picture (scene) to be transmitted into 3 primary colours: red, blue and green. At the same time, the camera tubes change these colour signals

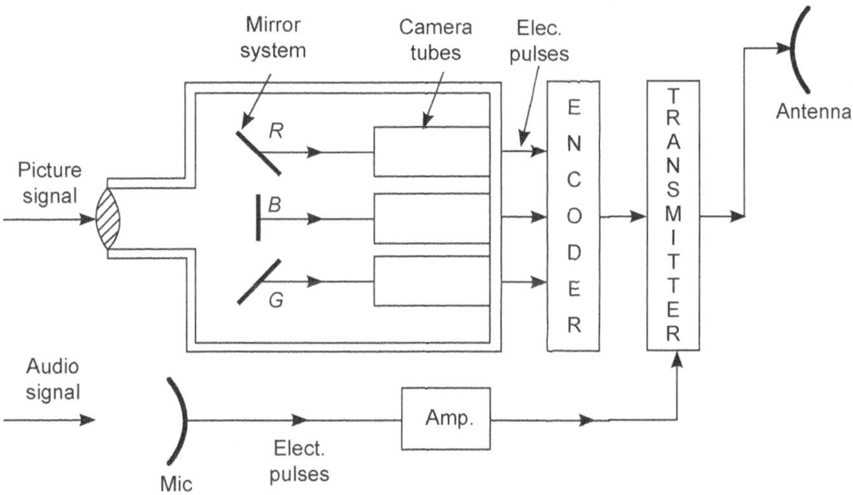

Fig. 10.20

into electrical pulses. These colour signals then go to "encoder" which produces a compatible colour signal and sends it to the transmitter. A microphone on the other side changes the audio (sound) signal into electrical pulses. The electrical pulses obtained from microphone are amplified and are also fed to the transmitter. The transmitter combines the audio and video signals for broadcasting from the transmitting antenna.

10.11.2 Reception of Colour TV Signals (Fig. 10.21)

The TV signals can be received by a TV receiver through a receiving antenna.

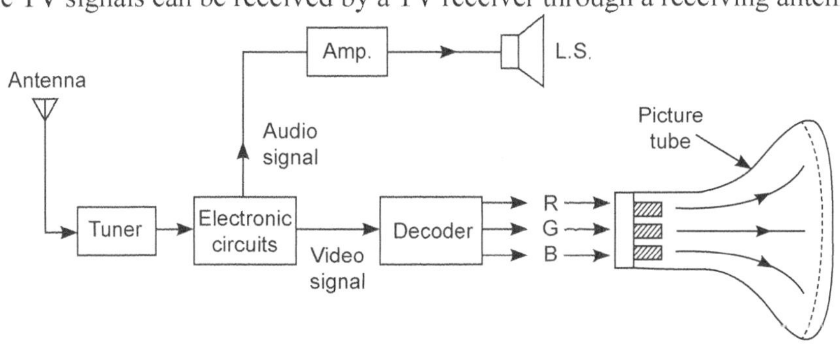

Fig. 10.21

The TV tuner selects only the signal, which is desirable. Few TV receivers have two tuners—one for selecting VHF channels and other for UHF channels. The electronic circuits in the receiver *separates* the received composite TV signal into audio and video signals. The audio signal is amplified and is changed into sound by a loud speaker. A *decoder* changes the video signal again into primary colour signals. The three electron guns of the picture tube—one for each primary colour scan the screen. The screen is covered with red, blue and green dots. When struck by a beam from the guns, the coloured dots glow and blend together in the viewer's mind to produce the original colour image.

10.12 COLOUR TV RECEIVERS

In colour TV receiver all stages described in monochrome TV receivers are used. In addition, some stages which are required for treatment of colours are also included in a colour receiver.

A colour picture is actually a "monochrome picture on a white raster with colours added". The required colour information is in the chrominance (C signal) broadcast along with monochrome signal. If we turn down the colour control at the receiver to eliminate the colour signal what we get will be a *black and white* picture. With C signal, the picture is reproduced in natural colours. As explained earlier all colours can be produced as a combination of red, green and blue colours, which are known as basic or *primary* colours.

The simplified block diagram of a colour receiver is shown in Fig. 10.22. The (Picture + Sound : P + S) signal from the selected channel is processed in the same way by the tuner, I.F. and video detector stages as in case of a monochrome receiver. The sound signal (S) is separately detected, amplified and fed to the loudspeaker. The AGC, sync., separator etc., have the same form and function as in monochrome receiver. In addition, the picture tube has *purity magnet*, *convergence* and *deflection* yokes etc. (described earlier) for colour treatment.

At the output of video detector, the composite video signals appear in their premodulated form. The Y-signal is processed in the same way as in black and white TV receiver, except that the video amplifier has a *delay line* which introduces a delay of about 490 ns (1 second = 10^9 ns) which is necessary to ensure coincidence of luminance and chroma signals.

The circuits common to monochrome and colour TV receivers have already been discussed. Here we describe important circuits of colour TV receivers.

10.12.1 Description of Important Circuits of Colour TV Receivers (Fig. 10.22)

(*i*) *Chrominance Band Pass Amplifier*: This separates the chrominance signal from the composite video signal, amplifies it and passes to the demodulator. The *colour burst* is prevented from appearing at its input by horizontal blanking pulses.

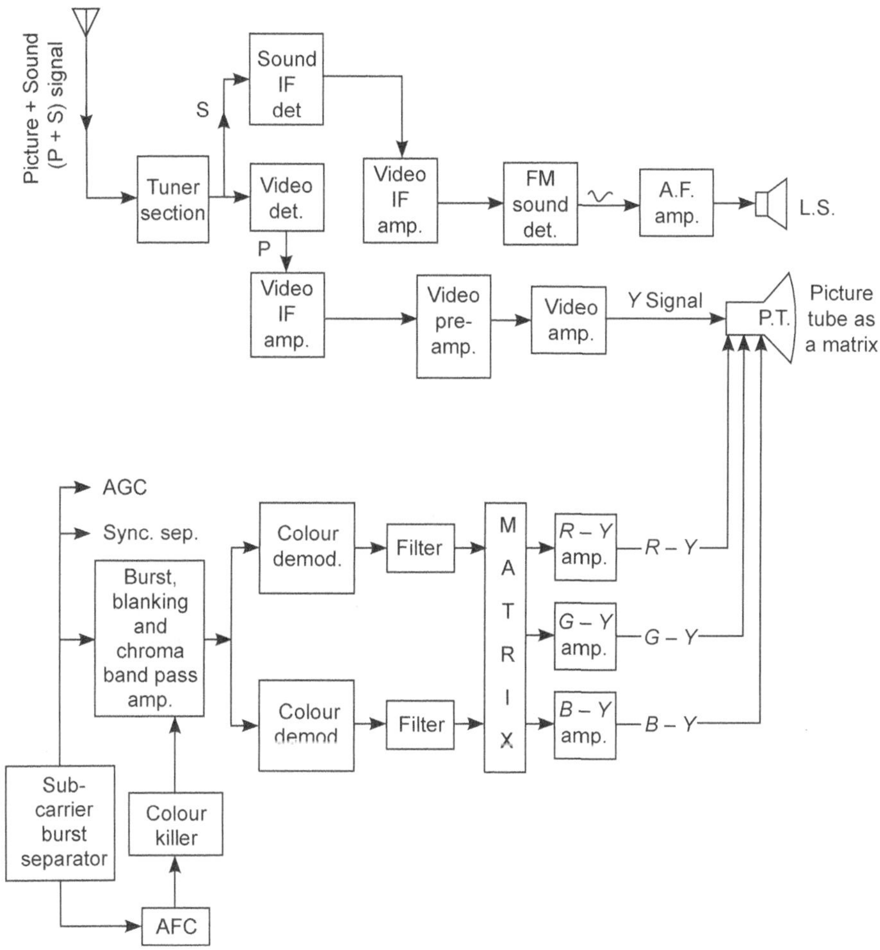

Fig. 10.22

(*ii*) *Colour Demodulators*: The colour demodulators detect the original signal. They are a combination of phase and amplitude detectors, because the output is dependent on phase as well as on amplitude of the chroma signal. Each demodulator has two input signals—the *chroma* which is to the demodulated and a constant output from the local oscillator.

(*iii*) *Burst Separator*: This circuit has to separate colour burst which is transmitted on the back porch of every horizontal sync. pulse. The circuit is tuned to the subcarrier frequency. The burst output is fed to AFC circuit.

(*iv*) *Colour Killer Circuit*: As the name indicates, this circuit becomes ON and disables the band pass amplifier during monochrome reception. In other words, it prevents any spurious signal from getting through the demodulators which may cause any colour interference on the screen. When the colour killer circuit is off, the chroma band pass amplifier is ON for colour information.

(*v*) *Matrix*: This circuit combines signal in specific proportions. At the receiver, the picture tube acts as matrix for inputs of Y signal and $R–Y$, $G–Y$, $B–Y$ signals to produce R, G and B lights on the screen.

Table 10.1 Trouble Shooting of TV Receivers

Troubles	Faults and Remedy
1. No sound, no picture but normal raster.	Loose antenna or lead connections. Check RF tuner and I.F. Sections.
2. Sound distorted, picture normal.	Defective speaker, defective capacitors in Audio Amplifiers.
3. Sound normal, no picture but normal raster.	Check "picture tube circuit".
4. Insufficient vertical height.	Check vertical sweep circuit.
5. Insufficient width of raster.	Weak output transistor.
6. Fold over in picture, sound normal.	Check horizontal deflection circuit.
7. Picture not centered but sound normal.	Check centering controls of the picture tube.
8. Weak picture and sound but of good quality.	Check tuner and I.F. sections.
9. Hum.	Check loudspeaker and Audio Section.
10. Snow in the picture.	Insufficient signal strength. The lead from antenna to receiver may be open.
11. Bars in the picture.	Horizontal bars are produced when frequency of signal is less than 15625 Hz. Vertical bars are produced when frequency is more than 15625 Hz.
12. Ghosts in the picture.	The antenna receives also the reflected signal from neighbouring buildings, mountains, hills, etc. This produces the duplicate of the picture called *ghost*. A high directional antenna is the solution.
13. White vertical bars.	The reason is the production of oscillations in the horizontal scanning currents.
14. Black vertical bars.	The reason is the radiation of harmonics of the horizontal scanning currents, RF chokes. "Ferrite heads" may be used in amplifier.

Troubles	Faults and Remedy
15. White spots on the screen after making the set off.	A short circuit between control grid and cathode of the PT.
16. Tilted picture.	Wrong adjustment of Deflection coils of P.T.
17. Black shadow on the border of the screen.	Wrong position of deflection coils on neck of the P.T.
18. Sound OK, outline of the picture looks like gears (*Gear tooth effect*).	AFC circuit is faulty.
19. Sound OK, but a tree shape is seen on the screen (*chrismas tree effect*).	Wrong adjustment of the horizontal oscillator core.
20. Size of the picture not OK (*Trapezoidal effect*).	Short circuit in the vertical deflection coils.
21. The loud Speaker gives a motor boat noise (motor boat effect).	EHT stage is faulty.

Problem 10.1. *Find time for scanning one horizontal line for frames repeated at 60 Hz and 525 lines per frame. If scanning is progressive and without interlacing.*

Solution. Time for scanning one line

$$\frac{1}{60 \times 525} = 31.75 \ \mu s \quad \textbf{Ans.}$$

Problem 10.2. *In a TV picture tube, the cathode is at +230 V and control grid is at 190 V, find grid bias.*

Solution. Grid bias = 190 – 230 = – 40 V **Ans.**

Problem 10.3. *The period of an equalizing pulse is 31.77 μs. What is the frequency repetitive rate.*

Solution. Frequency repetive rate (FRR)

$$= \frac{1}{31.77 \times 10^{-6}} = 31476.23 \text{ Hz} \quad \textbf{Ans.}$$

Problem 10.4. *The time required for horizontal blanking is 16% of each horizontal time. If horizontal time is 63.5 μs, fmd horizontal blanking time for each line.*

Solution. Horizontal blanking time for each line

$$= \frac{63.5 \times 16}{100} = 10.2 \ \mu s \quad \textbf{Ans.}$$

Problem 10.5. *If 100 picture elements are scanned in 50 μs. Find time needed to scan 4 picture elements.*

Solution. Time to scan 4 elements

$$= \frac{50}{400} \times 4 = 0.5\ \mu s \quad \textbf{Ans.}$$

Problem 10.6. *A picture has 625 vertical and 50 horizontal elements. Find total number of elements.*

Solution. Total number of picture elements = $625 \times 500 = 31250$ **Ans.**

Problem 10.7. *Find % modulation for the following frequency deviation in FM sound signal.*

(*a*) *5 kHz* (*b*) *10 kHz* (*c*) *25 kHz*

Solution. (*a*) $\frac{5}{25} \times 100 = 20\%$ **Ans.**

(*b*) $\frac{10}{25} \times 100 = 40\%$ **Ans.**

(*c*) $\frac{25}{25} \times 100 = 100\%$ **Ans.**

Problem 10.8. *The high tension supply to a colour TV is 24.5 kV and the total beam current from the three guns is 2000 μA. Find power of the picture tube.*

Solution.

$$\text{Power} = \text{Current} \times \text{Voltage}$$
$$= (2000 \times 10^{-6})\,(24.5 \times 10^{3})$$
$$= 49\ \text{W} \quad \textbf{Ans.}$$

Problem 10.9. *A colour TV has a 53 cm screen (Fig. 10.23). The aspect ratio is 4/3. Find height and width of TV screen.*

Solution.

$$(3x)^2 + (4x)^2 = (53)^2$$
$$x = 10.6$$
$$3x = 3 \times 10.6$$
$$= 31.8.\ \text{cm} \quad \textbf{Ans.}$$
$$4x = 4 \times 10.6$$
$$= 42.4\ \text{cm} \quad \textbf{Ans.}$$
$$\text{Width} = 42.4 \times 31.8\ \text{cm} \quad \textbf{Ans.}$$

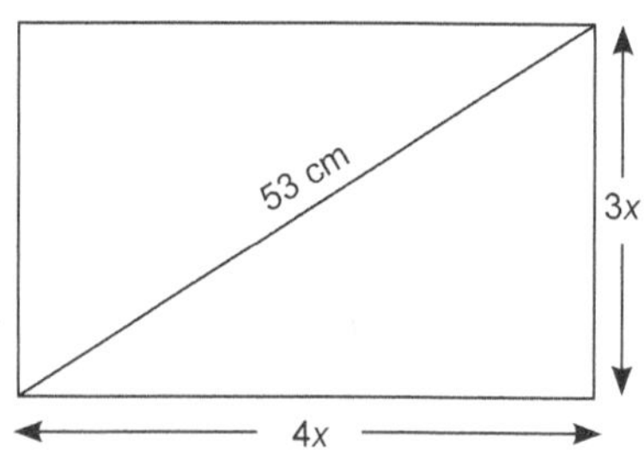

Fig. 10.23

Problem 10.10. *Find the number of possible colours with a 4 line digital input monitor.*

Solution. Number of possible colours

$$= 2^4 = 16 \quad \textbf{Ans.}$$

Problem 10.11. *(Fig. 10.24) The audio carrier for a TV channel is 529.75 MHz and IF is 33.4 MHz. Find frequency of the local oscillator.* (*L.O.*)

Solution. Frequency of L.O.

$$f_0 = f_c + \text{IF}$$

$$f_0 = 529.75 + 33.4$$

$$= 563.15 \text{ MHz} \quad \textbf{Ans.}$$

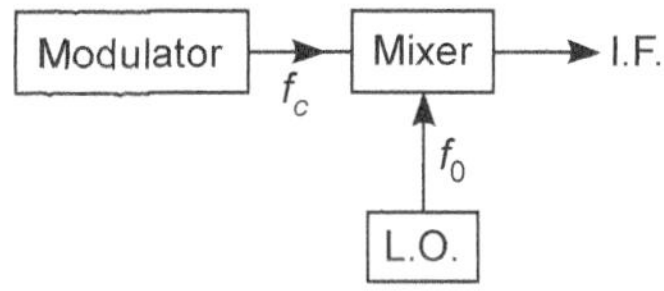

Fig. 10.24

Problem 10.12. *Find vertical and horizontal resolution for TV. Assume K = 0.7, total no. of lines = 625, 20 lines are lost per field and aspect ratio = 4/3.*

Solution.

$$\text{Total lines} = 625$$

$$\text{Lines lost } 2 \times 20 = 40 \text{ (For 2 fields)}$$

$$\text{Lines} = 625 - 40 = 585.$$

$$\text{Vertical resolution} = 585 \times 0.7 = 409.5 \quad \textbf{Ans.}$$

$$\text{Horizontal resolution} = 409.5 \times \frac{4}{3} = 546 \quad \textbf{Ans.}$$

10.13 SPECIAL TVs

Here, we shall discuss the following TVs

1. Projection TV
2. Closed circuit TV
3. Flat Panel TV
4. Digital TV
5. Three-dimension TV
6. HDTV
7. LCD/LED TV
8. Plasma TV.
9. Home Theatre TV
10. Satellite TV
11. Cable TV
12. DTH

10.13.1 Projection TV

The projection TV is the outcome of the quest of generating a picture whose size considerably exceeds the dimensions of the TV picture tube. Such a large sized picture (1 m × 1 m or even larger) may not be required for home but it certainly has a no. of professional applications. Today, a projection TV is used in a university where the students will follow the lectures on otherwise dull subjects with greater interest. It is used in disco theatres, where the young ones will flock to see their favourites.

It is used in hospitals and medical colleges where the medical researchers will be able to follow more closely the achievements of their colleagues in other advanced countries where the video filming of new techniques is very common. Similarly, there are many more similar applications.

A projection TV set up comprises of a projector and screen [Fig. 10.25]. The projector houses the three projection (picture) tubes for *R*, *G* and *B* (red, green and blue) signals (in case of a colour TV projector) and the associated optics. Both floor mounted and ceiling mounted projectors are commercially available. These projectors in addition to showing the normal TV programme can also be hooked up to a video cassette recorder (VCR) or a personal computer (P.C.) for a variety of applications.

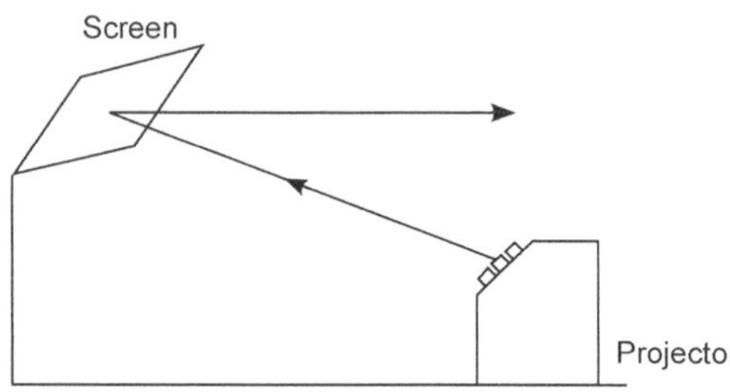

Fig. 10.25

10.13.2 Closed Circuit TV

In a Closed Circuit TV (CCTV) setup, the video signal at the camera output is directly fed to a TV monitor at a remote location by means of a cable. Since the video signal is not modulated and is routed through the cable, the TV monitor is nothing but a TV receiver without RF and IF sections. In a typical CCTV set-up, one video camera may feed more than one monitors [Fig. 17.26]. Projection TVs with large screens can also be used in place of TV monitors.

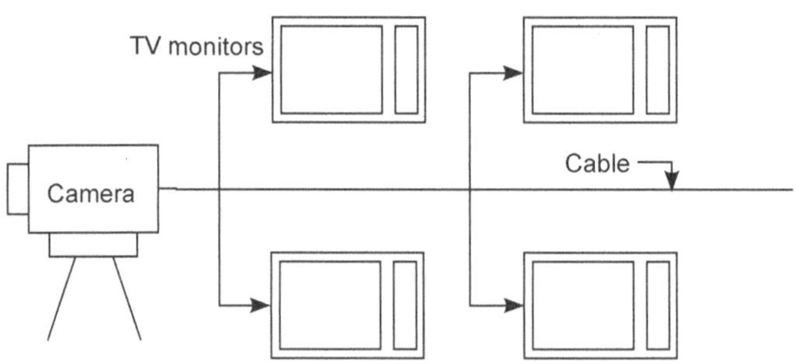

Fig. 10.26

CCTV finds extensive applications in a large number of areas like education, industry, medicine, traffic control and so on. One such application is in the concept of one teacher for many class rooms. A lecture delivered before a camera could be viewed in a number of class rooms on different monitors or projection TVs. The CCTV also finds wide use at airports, railway stations, harbours etc., for making public announcements like arrival and departure schedules and also for silent paging of passengers and personnels. In hospitals, CCTV network ensures remote observation of patients in a ward at a central location. In another application in surgery, all phases of an operation can be transmitted live to a large group of medical students and researchers outside the operation theatre. The CCTV can also be used for observation of blasts from a distance and also for observation and subsequent control of crushers, hoists, etc., in the mining industry. The problems created by today's vehicular traffic can only be tackled through a high speed information system, of which a CCTV forms an integral part.

10.13.3 Flat Panel Television

A flat panel television receiver is the one in which the conventional CRT type picture tube is replaced by a flat panel display. The dimensions of a CRT picture tube put a limit on the degree of compactness achievable in a television receiver. Flat panel TVs are ultracompact due to small sized flat panel displays. The initial initiative and thrust to develop flat panel displays was due to their extensive applications in watches, pocket calculators and so on. Later on, applications of these displays to pocket size TVs has really hastened the pace of development.

There are two distinct approaches being followed in the development of flat panel displays for TV applications. The first and the earlier one is that of using a simple dot matrix with tricolour film dots. The second approach is that of using an active dot matrix. The LCD displays being used in pocket sized television receivers follow the simple dot matrix approach.

10.13.4 Digital Television

When we talk about Digital television, we not only imply use of digital television receivers that have been successfully commercialized, we also mean an analog television studio being replaced by a digital television studio. Though analog video cameras are being replaced by digital video cameras and analog viodeo tape recorders (VTRs) are being replaced by their digital counterparts, a complete changeover can take place only after universal acceptance of a common digital TV standard.

Digital technology has a large number of advantages. Digital equipment produces predictable results and the resulting pictures are stable and immune to noise. The signal if in the digital form can be regenerated as often as desired without any signal degradation so that long transmission lines and multiple tape generation are no longer problems. The digital signals can be stored in a memory for any length of time and can be read from the memory and written into the memory at different speeds. It can be time stretched, delayed, compressed and read in a form different from the one in which it was written. It has made possible the special visual effects that we see on television.

10.13.5 Three Dimensional Television (3 DTV)

In comparison to a two dimensional picture which appears flat and appears to lie in the plane of the screen, in three dimensional television viewing, all the three dimensions-length, breadth and depth are depicted on the screen. The picture appears to have all the qualities of a live scene as seen with natural vision.

We, the human beings are able to see in three dimensions as the left eye and right eye see a slightly different image of the same object while viewing it. These images are transmitted to the brain where they are processed and interpreted as a single composite image in 3-D instead of two overlapping flat images. This in fact is the starting point for achieving a 3-D television broadcast and reception.

The stereo video signals can be obtained from a pair of video cameras suitably located to take a binocular view of the scene to be televised. These video signals are kept separate throughout the transmission and reception processes. Various techniques have been demonstrated for the stereo video display. One such technique is to use a television receiver with a shadow mask picture tube. All the phosphor dots of one colour (say Green) are excited by one of the electron beams intensity modulated by the video signal corresponding to one of the cameras. Similarly, all phosphor dots of another colour (say Red) are excited by another electron beam intensity modulated by the video signal corresponding to the second video camera. The third beam is kept off and the third colour phosphors (Blue in this case) are not excited. A viewer wearing *green* filter glasses in one eye and *red* filter glasses in the other eye will view the scene in stereo in yellow. In the improved version of the same principle, the picture tube has only two types of phosphor dots, blue and yellow. In such a case, a viewer wearing blue and yellow filters in respective eyes will see a 3-D view of the scene in monochrome, as yellow and blue are complementary colours.

The 3D television though already demonstrated by various companies using different techniques is not yet commercialised.

10.13.6 HDTV (High Definition Television)

It is a new generation TV. Displaying images with much greater resolution on screens having an *aspect ratio* of 16 : 9 (instead of 4 : 3 at present) will enable the viewers to experience the visual quality comparable to 350 mm film and a sound quality of a CD system. The HDTV system demands complete transformation to the new system. Today there are 3 HDTV systems: The

Japanese MUSE system, European EUREKA system and American DIGITAL System.

The Fig. 10.27 shows the Japanese muse system, which is a commonly used system.

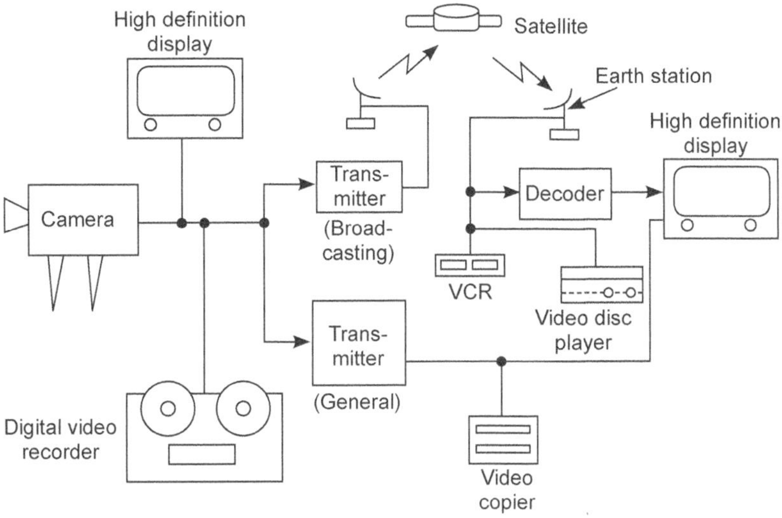

Fig. 10.27

10.13.7 LCD/LED TV

We today witness slow death of CRT (Cathode Ray Tube) TVs. Now we have LCD/LED TVs, which replace CRT completely. The LCD TV uses a LCD (Liquid Crystal Display), which gets activated, when an electric current is applied to it.

The LED TV works on the same principle with the difference that the CRT has been replaced by an LED (Light Emitting Diode). The LCD TVs use LCD as back light. So difference between LCD and LED TVs is only the back light they use. Moreover, the LED TV has less power consumption but costly than the LCD TV. Here we shall describe LCD TV.

(a) LCD TV: An LCD is in liquid state but has crystalline structure which when a voltage is applied, changes the arrangement of liquid crystal molecules. This characteristic is used to activate a shutter which alternately shifts off and passes the light.

An LCD TV produces picture by the above principle, but picture elements equivalent to scanning lines of ordinary TV are minute liquid crystal moleculars. The key to clear image is, how fast the liquid crystal molecules can be changed in response to the slight fluctuation of voltage.

The Fig. 10.28 shows LCD-TV circuit.

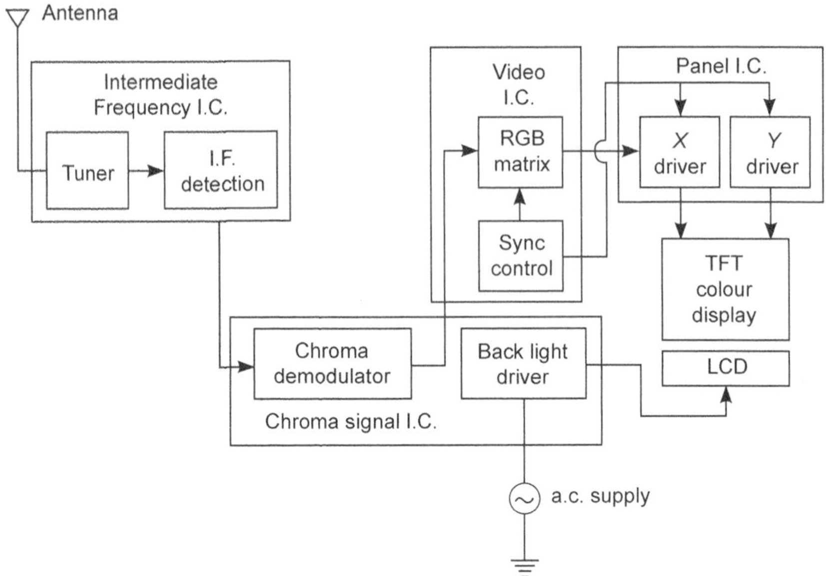

Fig. 10.28

In uses following (advanced) ICs

(*i*) The **IF (Intermediate frequency) signal IC** contains circuits for video and sound, synchronising circuit, sync. circuit, AFC and AGC. It has a tuner which can receive VHF (Very High Frequency) as well as UHF (Ultra High Frequency) signals.

(*ii*) *The chroma signal IC* contains colour demodulator, and RGB (Red, Green, Blue) generating circuits. As the liquid crystal is driven by ac supply, the polarities of video signals for red, green and blue colour are reversed by "ac RGB video generating circuit".

(*iii*) The **video IC** contain RGB matrix and sync. control circuits. It is necessary to change the video signals to red, green and blue for each period of horizontal scanning.

(*iv*) The **panel IC** Board contains *X* and *Y* driver circuits. The panel has two ICs for *X* driver and two ICs for *Y* driver.

The outputs of *X* and *Y* drives goes to TFT (Thin Film Transistor) colour display.

(*v*) Video I.C. (RGB matrix) may be used two ways:

(*a*) Transparent electrodes are placed over the entire screen. The electric current flows in both the directions to drive LCD molecules to create picture. In this method, the picture quality is not so good.

(*b*) The each tiny picture element has a thin film transistor (TFT), which applies voltage on LCD to create images. In this method the picture quality is good but the cost of the TV is higher.

10.13.8 Plasma TV

In a CRT television, a gun fires a beam of electrons (negatively-charged particles) inside a large glass tube. The electrons excite phosphors along the screen of P.T. which causes the phosphors to light up. The television image is produced by lighting up different areas of the phosphor coating with different colours at different intensities.

Cathode ray tubes produce crisp, vibrant images, but they do have a serious drawback: They are bulky. In order to increase the screen width in a CRT set, you also have to increase the length of the tube (to give the scanning electron gun room to reach all parts of the screen). Consequently, any big-screen CRT television is going to be heavy and take up a sizable part of a room.

Recently, a new alternative has popped up: the plasma flat panel display. These televisions have wide screens, comparable to the largest CRT sets, but they are only about 6 inches (15 cm) thick. Based on the information in a video signal, the television lights up thousands of tiny dots (called pixels) with a high-energy beam of electrons. In most of the systems, there are three pixel colours– red, green and blue – which are evenly distributed on the screen. By combining these colours in different proportions, the television can produce the entire colour spectrum.

The basic idea of a plasma display is to illuminate tiny coloured fluorescent lights to form an image. Each pixel is made up of three fluorescent lights — a red light, a green light and a blue light. Just like a CRT television, the plasma display varies the intensities of the different lights to produce a full range of colours.

The central element in a fluorescent *light is a plasma, a gas made up of free-flowing ions* (electrically charged atoms) and electrons (negatively charged particles). Under normal conditions, a gas is mainly made up of uncharged particles. That is, the individual gas atoms include equal number of protons (positively charged particles in the atom's nucleus) and electrons. The negatively charged electrons perfectly balance the positively charged protons, so the atom has a net charge of zero.

If you introduce many free electrons into the gas by establishing an electrical voltage across it, the situation changes very quickly. The free electrons collide with the atoms, knocking loose other electrons. With a missing electron, an atom loses its balance. It has a net positive charge, making it an ion.

In a plasma T.V., with an electric current running through it, negatively charged particles are rushing towards the positively charged area of the plasma, and positively charged particles are rushing towards the negatively charged area. In this mad rush, particles are constantly bumping into each other. These collisions excite the gas atoms in the plasma, causing them to release photons of energy.

The Fig. 10.29 explains the plasma technology *i.e.*, how atoms emit light.

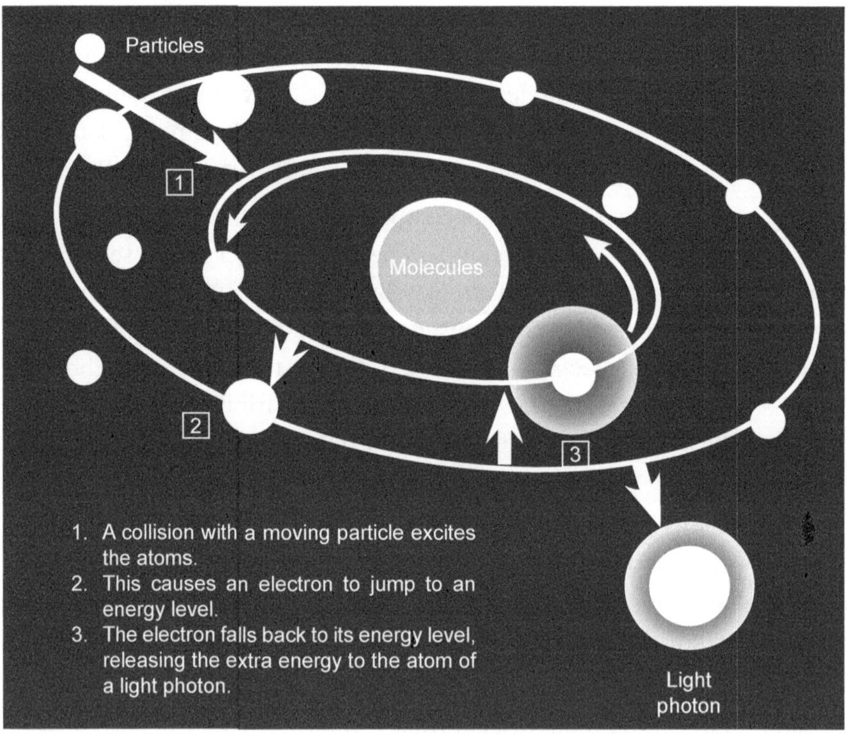

Fig. 10.29

The *Xenon and neon atoms*, used in plasma screens, release light photons when they are excited. Mostly, these atoms release ultraviolet light photons, which are invisible to the human eye. But ultraviolet photons *can be used to excite visible light* photons. The xenon and neon gas in a plasma television is contained in hundreds of thousands of tiny cells positioned between two plates of glass.

The main advantage of display plasma technology is that you can produce a very wide screen using extremely thin materials and because each pixel is lit individually, the image is very bright and looks good from almost every angle. The image *quality isn't quite up to the standards* of the best cathode ray tube sets, but it is fairly good.

The biggest drawback of this technology is its very high price.

10.13.9 Home Theatre TV

This is a system which provides effects of a THEATRE in a home. We get in our living room a big sound and big picture giving best audio video experience comparable to a fine theatre.

Components/Accessories of a Home Theatre TV System alongwith typical specifications are given below. (Fig. 10.30)

1. A large colour TV (CTV) 53 cm (21″)/63 cm (25″)/73 cm (29″) with full and flat square tube, menu driven remote control.

   NTSC (National TV System Committee) or **PAL** (Phase Alternation by Line) playback 4.43/3.58 MHz. Compatible system 100 program memory, S band and hyperband tuner (cable ready), "on screen" display and a separate enclosure speaker system.

2. *Speaker System*. A *sub woofer* with a powerful "built in" amplifier, alongwith 5 sleek and compact speakers (tweeters, and squawkers (2 front + 1 rear 2 side speakers) together giving a combined output of 1000 Watt. PMPO (Peak Music Power Output)— such that if we are watching a helicopter on the screen, we will feel that it is hovering in the room, we are sitting.

Fig. 10.30

3. *ACD/VCD-Player*. Audio and Video CD player—3 discs playing continuously with remote control and playback control.

   In addition any audio system can be connected and a PMPO of 1000 W can be obtained.

   The stereophonic effects are obtained, even if a monophony signal is transmitted by the cable operator.

10.13.10 Satellite Television

The TV programs from a telecasting station cannot be viewed beyond a certain distance. Using the communication satellites launched in space, the TV programs from a telecasting station can be viewed even for away from the place. The *Geo stationary satellites* placed in space around the earth have enabled us to see national as well as international TV programs. If three satellites are placed in space at a certain distance above the equator at 120° from each other, they can cover almost the entire world (See Fig. 10.31).

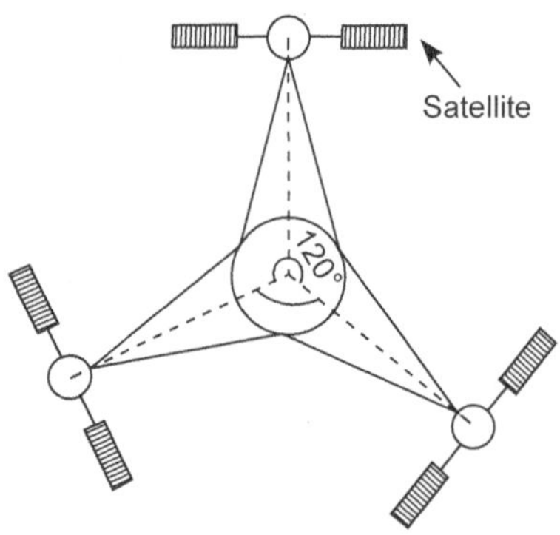

Fig. 10.31

(a) The Process: The programs are prepared in a TV studio and are fed to the earth station in the form of electric signals. The earth station converts these signals into assigned uplink microwave frequency (5.9 to 6.4 GHz) and feeds it to the transmitting dish antenna. The dish antenna beams the signals in the direction of the satellite. The satellite is equipped with its own antenna which receives the uplink signal and feeds it to the receiver in the *transponder*. The receiver amplifies the signal, changes it into another frequency known as *downlink frequency* (3.7 to 4.2 GHz). This is to be mentioned that these frequencies allocations are made by International Telecommunication Union (I.T.U.) in Geneva. Here we are to remind the readers that a *transponder* is an equipment on the satellite which contains a receiver, amplifier, frequency converter and the transmitter.

A satellite has a number of such transponders for handling different channels' signals. These channels are used for telecommunication, radio and TV transmissions and for many other functions. The commercial broadcasters lease the transponders with assigned specific frequencies for transmitting the signals from the ground station.

The transmitter of the transponder retransmits the signals back to earth through its transmitting antenna. These signals are sent to earth at a down link frequency (between 3.7 to 4.2 GHz).

Note. The values of 'up' and 'down' frequencies for satellite TV are different from that required for other satellite communications.

(b) Collection and Use of Signals: The receiving earth station collects these down link signals using a parabolic dish antenna (Fig. 10.32). The. downlink signals are weak, so a high gain, large diameter antenna is required to collect the sufficient amount of the signal. The signals falling on the antenna are reflected to its "focal point" (*F*) and collected by its " feed horn".

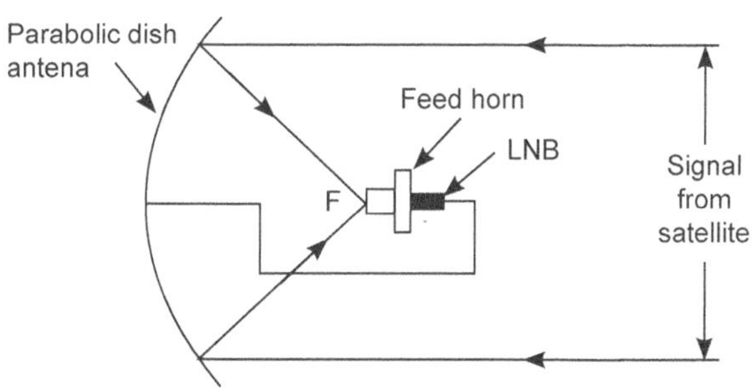

Fig. 10.32

The signals are then fed to a "low noise (high gain) block (LNB) amplifier and a down converter", which amplifies the signal and converts the down link frequency to a lower frequency range (950 – 1450 MHz) and the output is connected to a receiver through a co-axial cable.

(c) Distribution: The signals from the receiver are further processed and distributed through an other system called "cable TV network". For this purpose the signal is fed into a "distribution amplifier", which boosts the signal sufficiently high so that it may reach to individual TV receiver.

The output from the amplifier is fed to "splitters" through co-axial cables and then sent to individual TV sets. The Fig. 10.33 shows receiving system, signal processing and distribution of the signal.

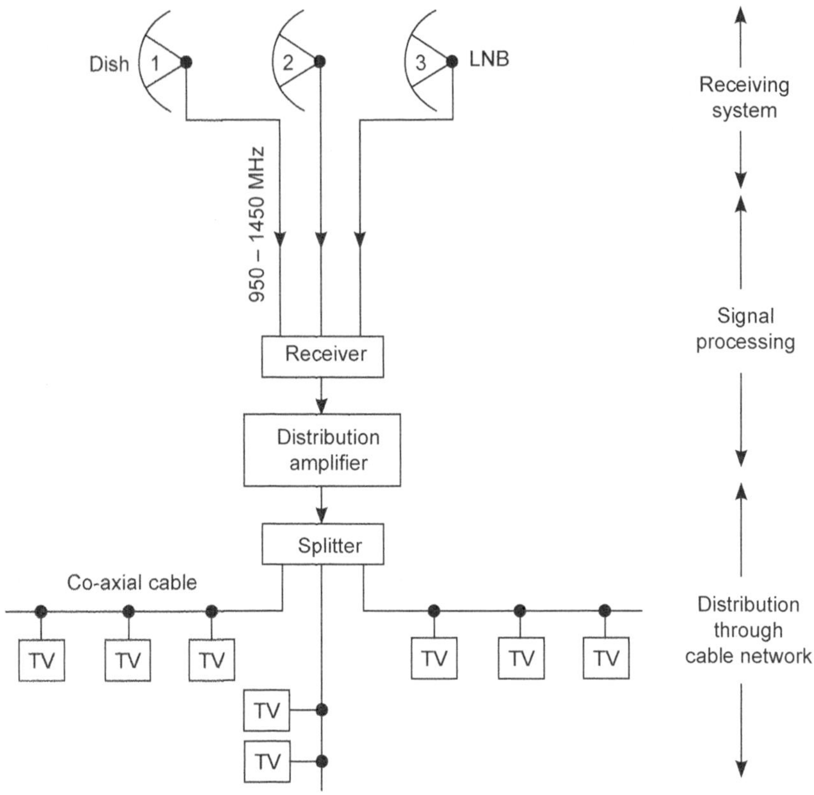

Fig. 10.33

10.13.11 CABLE TV (CATV)

The feeding of many TV receivers from a single antenna is called cable T.V. A single antenna can feed a complete locality. The antenna is installed at the top of a building and the signal is fed to the houses through cable. The antenna is generally dish antenna which receives international signals through satellite and this signal is sent to houses through cable. The signal should be sufficiently strong so that the picture is clear at the TV sets at each home. Also there should be no interference between the signal received and the signal sent to the viewers. The weak signal is amplified before sending to the subscribers. The signals from VCR etc. may also be sent to the subscribers through cable. Generally modulation is not required and the cable used is a coaxial cable.

The CATV system needs miles of cables. They suffer interference in the way moreover, with distance, the signal goes on weaker and when it reaches a TV receiver, it is distorted. To solve this problem, amplifiers are installed at regular distances to keep the signal strong. This also poses problem, if there is a breakdown in any of the amplifiers, in that case, there would be no signal at all, so regular check up of the whole system is must. The Fig. 10.34 shows block diagram of cable TV network.

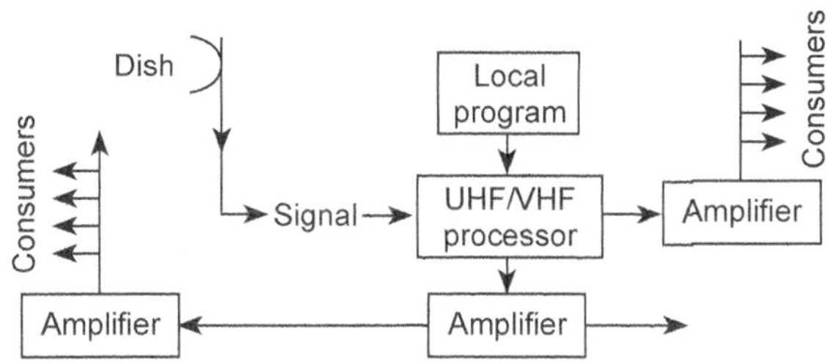

Fig. 10.34

The connection to consumers may be direct or through a converter i.e. Set Top Box: The signals suffer an attenuation @ 15 dB per 500 m cable; so use of repeaters/amplifiers is necessary. An UHF/VHF signal processor is used which is positioned at a central place.

It is necessary to convert UHF channels to VHF channels, because signal loss at VHF is lesser. Moreover the processor can accept signals from dish as well as the locally made programs. The local made programs include announcements, local advertisement etc.

Note:

(*i*) The signals can also be sent through cable directly from VCR (Video Cassette Recorder) without any modulation.

(*ii*) If there is a breakdown in any of the amplifiers, there is a total breakdown of the signal ahead so the system needs regular checkup.

CATV Through Internet

The CATV through Internet is possible using telephone lines as a medium of transfer and using a dial up **modem** (modulator-demodulator).

For cable internet access on a PC (personal computer), a cable MODEM is required at the user's end. A cable modem is an external device that is connected to the computer to provide high speed data access via CATV network. A cable modem sends and receives data from the internet, using the cable network. The modem translates signals in the same way, a telephone modem does from the telephone line. It translates radio frequency signals to (and from) the cable into internet. The modem is connected to the computer through *N.I.C.* (network interface card). This provides connectivity between cable and computer and translates the signals from the modem so that the computer software displays these signals.

The CATV network has high bandwidth about 500 to 720 MHz. It handles more than 100 channels. The CATV through internet is a two way network.

This is done by upgrading the amplifiers. The internet signals are digital signals and they are to be interfaced with analog CATV network. This interface is known as "Cable modem termination service" (CMTC), which consists of input interface, cable modem and microprocessor.

About 5 to 10 PCs (personal computers) may be connected with the internet through Digital Subscriber Line (DSL). The feature films and other programs can be down loaded over the internet and can be put on the cable network; so that the consumer can enjoy them on their PC at their home.

In near future, the cable connection will be made through optical fibre cable (OFC) replacing the coaxial cable. This will improve the reception further. The data rate may be as high as 100 megabits per second, so the 3 hour movie can be down loaded in few minutes.

The internet service providers offer telephony, internet access and television services together, resulting in the economics to the consumers.

The Fig. 10.35 shows block diagram of CATV through internet.

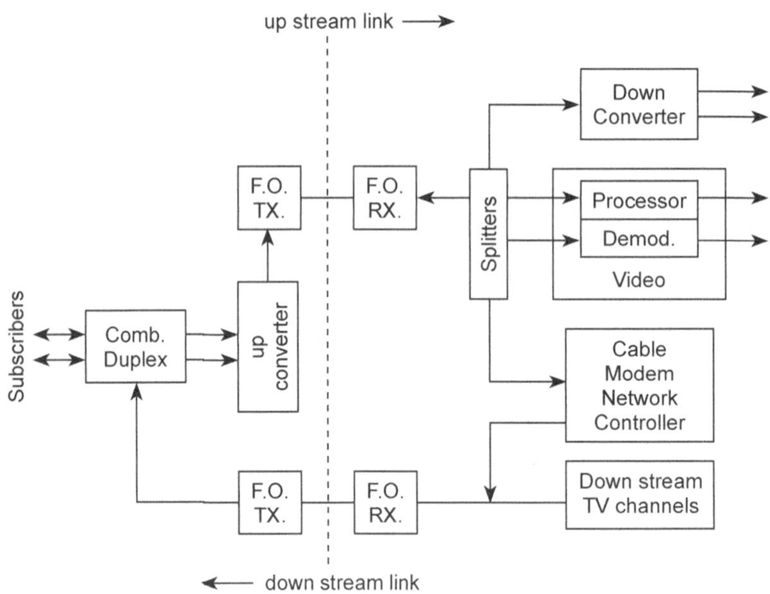

Fig. 10.35

10.13.12 DTH (Direct To Home) TV (Fig. 10.36)

The DTH refers to the reception of satellite signals direct on a TV with a personal dish in an individual home.

The DTH service is a satellite transmission technology. It has numerous merits for the sender as well as for the receiver. Due to digital technology, it can handle many channels over a single delivery platform as compared to the analog technology.

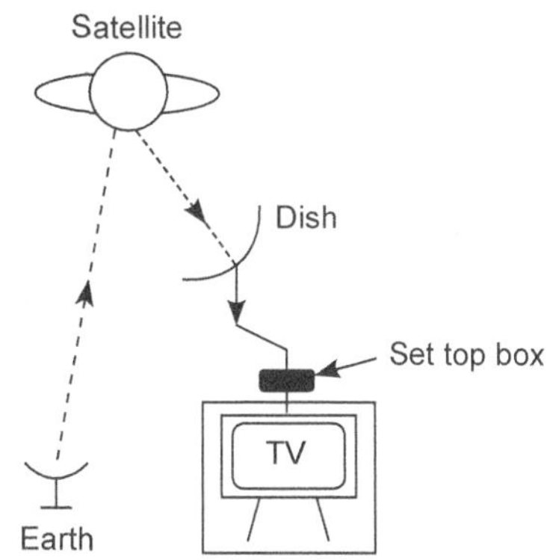

Fig. 10.36

The DTH service is that in which large number of channels are digitally compressed and are sent through high power satellite. The programs can be directly received at home. The consumer is to install a small dish antenna at the roof of his house. The transmission eliminates cable operators completely as the consumer receives signals directly form the satellite. The consumer also needs a digital receiver. To receive the multiplexed signals as compared to CATV, this has easy control and clear picture.

Transmissions in DTH is done in **Ku (or *C*) band** which is most appropriate. As the signals are digital, this band provides high resolution picture and also better sound. All merits of digital transmission are also applicable.

The Fig. 10.37 shows the DTH dish set with LNB, Set Top Box with its remote.

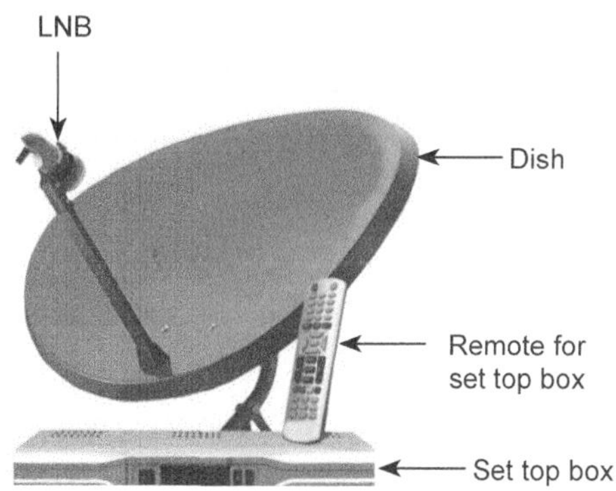

Fig. 10.37

The LNB in the dish is "Low noise Block". It receives the very low level microwave signals from the satellite, amplifies them, changes the signal to a lower frequency band and sends them down the cable to the individual receiver.

The set up box is installed near TV, and selects the desired signal improves reception etc.

Merits of DTH Communication

As DTH is a digital technology, all merits of a digital communications are applicable. It has numerous merits to the subscribers few are listed below:

1. Digital quality picture
2. Stereophonic sound
3. Uninterrupted viewing
4. Multi audio feed.
5. Capacity upto 400 channels
6. Movie on demand
7. Interactive TV

10.14 TV STUDIO

The important TV studio equipment is listed below:

1. Microphones
2. Loudspeakers
3. VCRs/VCPs
4. TV cameras
5. Picture tubes
6. Video Display Units—monitors, CRO etc.
7. Film Projectors/cinema to graph
8. Studio lighting.

All above equipment has already been discussed except the last two.

(*i*) *Film Projector*. The important parts of a film projector are shown in the Fig. 10.38.

As the lenses produce an inverted image, the film is placed *up side down* in the projector to get an erect image on the screen.

As the *shutter* of the projector rotates, the gate opens and closes alternatively. The film is placed slightly away from the focal length of the projecting lens, which produces an enlarged and erect image on the screen. The screen is placed at a distance greater than twice of focal length of the lens. During operation of the projector, the film reel unwinds from the upper reel and after passing through the gate, it rewinds on lower reel of the projector.

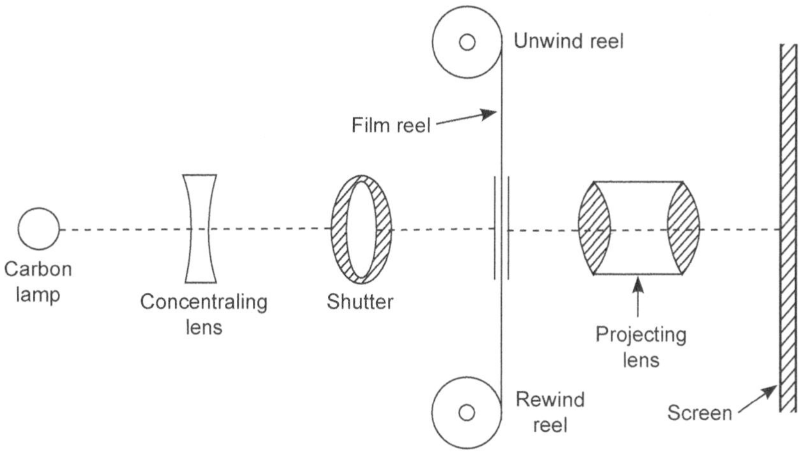

Fig. 10.38

The projector projects at the rate of 16 films per second on the screen. Due to *persistence of vision* we cannot visualise this small difference of time and illusion of " image continuity" is maintained.

(*ii*) *Studio lighting.* In a TV studio, lighting is necessary in addition to the illumination required at the scene to be televised. The lighting scheme:

1. should give equal brightness throughout the whole studio
2. should not produce shadows and glare.

About 50 to 100 light fittings are provided in the main studio of wattage between 500 to 5000 W. Generally pendent lights are preferred.

Plan of a TV Studio

The Fig. 10.39 shows plan of a TV studio. Here T = Tube light, F = Fan.

TV studio may cover a total of 2 to 3 hectares of land but its "programing hall" measures about 40 × 20 × 15 m. About 3 to 4 stages can be installed at a time. Lights are mostly pendent type but adjustable lights are also there. At a time, 3 to 4 microphones and cameras are at work. One stage is used to transmit the program while the others for rehearsals. All conversations among the staff is through "signals by hand", speaking is avoided or head phones can also be used. Director of the program can visualise all activities going on in the hall from his room.

The control room of the studio has the following officials to control the program.

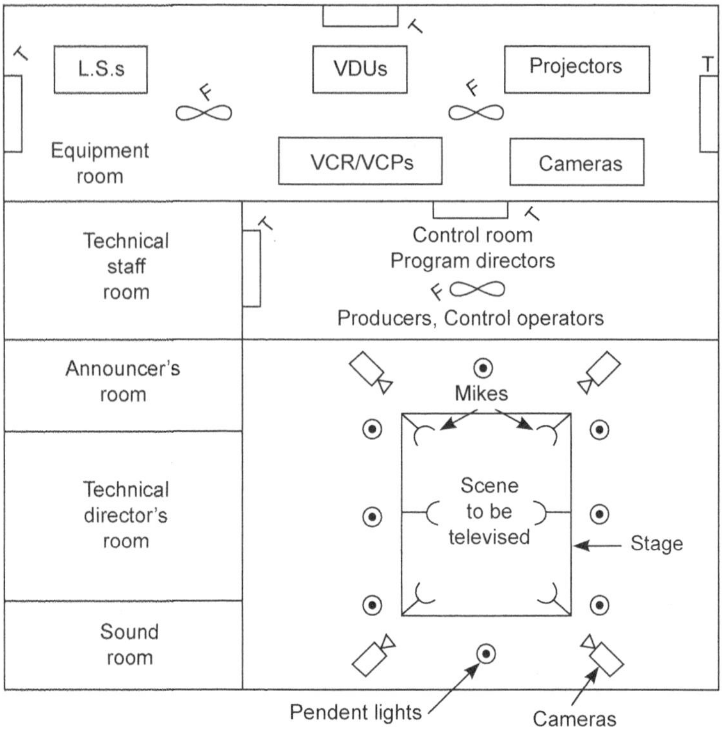

Fig. 10.39

1. *Control operator*. He controls the camera's output. Facing him, there is a control panel. By various controls he tries to obtain best results from the camera.
2. *Technical Director*. A master control panel is facing the technical director who is fully responsible for the program. All camera control operators send their camera output to the control panel of the director and the director selects the best output for transmission.
3. *Sound man*. The sound man receives output of all the microphones working at the stage. He selects best sound for transmission. The Announcer's sound also first of all reaches to the soundman.
4. *Monitors room*. In the control room or by its side, there are 2 monitors:
   (*i*) *Air monitor*. The camera output selected by the director is monitored on this monitor.
   (*ii*) *Preview monitor*. This monitor shows the final picture going to be transmitted to TV receivers.

SUMMARY

1. The red, green and blue are called primary colours. Any colour can be prepared by mixing primary colours in right proportions.
2. Colour triangle gives the percentage of primary colours to be mixed to obtain a particular secondary colour *e.g.*, 30% red + 59% green and 11% yellow gives white colour.
3. Various TV systems used are: PAL., NTSC and SECAM. In India, PAL (Phase Alternate Line) system is popularly Used.
4. A colour camera basically comprises of three Black and White Cameras.
5. A colour picture tube basically comprises of three monochrome picture tubes (guns) of red, green and blue colours in a prescribed configuration.
6. The colour picture tubes are of three types: Delta tube, gun in line and single gun tube.
7. The TV signal is a composite video signal, which contains audio, video and other signals.
8. There are many types of colour TVs such as LCD, LED, Plasma TV, cable TV, DTH etc.
9. Home Theatre TV system gives theatre effect in home.
10. The TV studio equipment are : Microphones, Loudspeakers, TV, Camera etc.

SELF TEST QUESTIONS

1. What is the difference between monochrome and colour TV?
2. What are primary and secondary colours? Explain preparation of secondary colours with the help of colour triangle.
3. Draw and explain "Visibility curve".
4. Explain the terms—White, Hue, Saturation, Luminance, Chrominance, Subcarrier and multiplexing.
5. Describe various TV systems you know, which one is used in India and why.
6. Write a note on colour TV camera.
7. Enlist types of colour picture tubes. Describe Delta tube.
8. How a colour picture tube is different from a monochrome tube?
9. Describe various colour TV screens.
10. Describe a PAL TV transmitter and receiver.
11. Prepare a trouble shooting chart for TV receivers.
12. Write a note on LCD TV, plasma TV, CATV, DTH studio.

OBJECTIVE TYPE QUESTIONS

I. Fill in the blanks:

1. The three primary colours are______ .
2. The secondary colours are prepared by mixing______ colours.
3. Yellow colour can be prepared by mixing______ .
4. While colour can be obtained by mixing _____ .
5. Expand PAL.
6. Expand NTSC.
7. Expand SECAM.
8. In India______ system is used.
9. The colour signal contains two signals namely _____ .
10. Expand V.S.B.

ANSWERS

1. Red, Blue, Green
2. Primary
3. Red and Green
4. Red and Cyan
5. Phase alternate line
6. National Television System Committee
7. Sequential colour and memory
8. PAL
9. Luminance and chroma
10. Vestigial Side Band

❑❑❑

11

Facsimile (FAX)

Facsimile telegraphy is a technique to send printed, typed or hand written messages from one station to another. As compared with the ordinary telegraphic system, the following points are to be noted:

(*i*) By this system, the messages written in the language not known can also be sent, whereas it is not possible in the ordinary system.

(*ii*) A bandwidth of 120 Hz is needed in the facsimile system, where as the ordinary telegraph system needs a very high BW of the order of 2200 Hz.

(*iii*) For facsimile, trained operator is required, whereas in ordinary system this is not essential.

(*iv*) The equipment needed in FAX is costly as compared to the conventional telegraphy.

In wider sense, the picture telegraphy and facsimile telegraphy systems are same. Now-a-days these systems are used over *lines* as well as over *radio* (through space). In line facsimile, *AM signals* are used where as in radio facsimile, *FM signals* are invariably used. Our modern transmitters and receivers are capable to handle AM as well as FM signals.

The word "Facsimile" means "exact reproduction" . In facsimile transmission, an exact reproduction of the picture or document is provided at the receiving end.

11.1 FACSIMILE (FAX)

Fax, short for facsimile transmission is a way of sending text of pictures over telephone lines.

A device called a facsimile (FAX) machine is used for sending and receiving the messages or pictures. The machine is connected to a telephone. To send a document, the sender inserts it into the machine and dials the telephone number of the receiving fax machine. When the connection is made, an electronic scanner in the transmitting machine converts the images of the document into a set of electric signals. The signals travel over the telephone lines to the receiving machine where the signals are reconverted into a copy of the original document.

The first fax machine began operations in France between Paris and Lyons in 1866. The signals were sent through telegraph wires. It could fax documents and drawings. The first photograph was faxed from Paris to London in 1907. The rate is the same as for an ordinary telephone call.

11.1.1 FAX vs T.V. Transmission

A fax is a machine that can transmit photographs, written documents, etc. The same is done in television transmission. The following is the difference between the two:

1. In TV, the scene includes "movement" whereas, such is not possible in fax transmission. The TV can send moving pictures. The FAX can transmit only the still pictures.
2. The transmission in TV is much faster than in FAX.
3. Due to the above, TV transmission needs much larger bandwidth than the FAX. This is the reason that FAX messages can be *sent over ordinary telephone lines.*

11.1.2 Applications of FAX

Important applications of FAX are given below:

1. *Business*: The business documents like purchase order, invoice, agreement etc., available at one place can be made available at other place in no time through fax without relying on postal department. The most common example is news paper printing. News, photographs, reports available at one place can be sent to other place to be published in the next morning edition. The photographic plates of a complete news paper can be sent in no time to any corner of the world through fax. It has become very easy to print a news paper from many places simultaneously.
2. *Weather*: The weather reports, maps from a place can be sent to TV stations for broadcasting. Similarly, they can be sent to airports, seaports and to any place through fax for their use.
3. *Legal applications*: The finger prints, photographs, signatures, and documents of a criminal can be sent from one place to another through FAX for quick verification.

4. *Personal applications*: Personal documents such as certificates, photographs and other documents urgently needed by a person at a distance from his residence can be obtained through FAX in no time. No postal service/courier can be too fast.

11.2 TYPES OF FAX MACHINES

Mainly two types of faxes are used.

1. Thermal Paper Fax
2. Plain Paper Fax.

1. *Thermal Paper Fax*: Thermal printing works by firing minute dots of heat at a special heat sensitive paper. This triggers a chemical reaction so that where the heat hits the paper, it turns black forming the required image.
2. *Plain Paper Fax.*

   The plain paper faxes are of the following types:

   (*a*) *Thermal Transfer fax*: Thermal transfer fax adopts a similar process where minute dots of heat are fired at a heat sensitive ribbon, which in turn transfers carbon like material on to plain paper in the form of an image.

   (*b*) *Laser Fax*: Laser faxes use fundamentally the same technology as laser printer which is, similar to that used in *photocopier.* The only difference is that instead of receiving digital pulses from the computer the laser fax receives them over a telephone line.

   (*c*) *LED Fax*: Light Emitting Diode (LED) printing is very similar to laser printing. Instead of firing a laser beam at a drum, LED uses a row of infrared lights to produce a charge which then goes on to produce the image.

   (*d*) *Inkjet Faxes*: Inkjet printing creates an image by firing jets of ink from a matrix of nossels onto a page.

11.3 BASIC FAX SYSTEM

FAX machines provide an easy way to send documents to any phone number equipped with another fax machine. The process of sending the document is called *facsimile.* It is a method by which a page (printed, written or photographic) is converted into electrical signals, sent quickly through the *public switching telephone network* (PSTN) and recorded as a copy in the remote fax machine. In most of the cases, faxes are a cost-effective-alternative to overnight delivery and a faster option than regular mail.

The Facsimile process involves three basic steps:

1. **Reading** or *converting the documents into electrical signals.*
2. **Sending** *or transmitting the signals through a telephone system to another fax unit.*
3. **Converting** *a received signals into a "facsimile" (true copy) of the transmitted documents."*

A fax machine works by scanning each outgoing page, converting the image into a series of light and dark dots. This pattern is then translated into audio tones, and sent over regular phone lines. The receiving fax hears the tones and prints the total compilation of dots. The resulting document is black and white like the original page.

11.4 PRINCIPLE OF FAX

Fax involves the following steps of operation which can be very well understood from the block diagram shown in Fig. 11.1 (a):

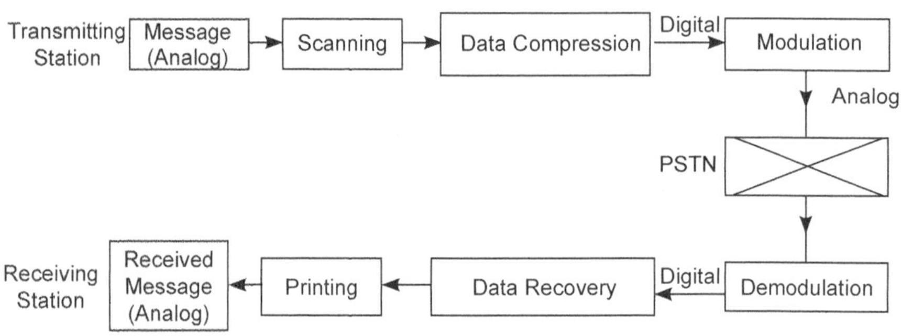

Fig. 11.1 (a)

1. *Scanning*: **(Reading and converting the document into electrical signals)**. The transmitted document originally consists of two-dimensional structure. We have to make it one-dimensional structure in order to consider document as the concept of line and then, we convert the line into the dot structure, we regard it as "SCAN".

   The scanner is the part of a fax unit that converts the marks on the page being sent into electrical signals. A photosensor looks at a very small "picture element (Pixel)", determines whether it is black or white, and generates a strong or weak electrical pulse. To read a page electronically, all of the spots in a thin strip 1/100th inch high across the top of the page are read, one at a time, starting with the upper left-hand corner. A *stepper motor* moves the page down and the photosensor reads the next strip below. Successive strips are read until the whole page is converted into electrical pulses.

2. *Modem* (*Modulation and Demodulation*): A modem is a modulation-demodulation device that accepts the digital information and modulates it into an analog signal to be sent over a voice telephone line. At the receiver, the analog signal is converted back into the digital.

   We have to modulate the scanned picture element to convert digital signal into analog signal at transmitting station in order to transfer through the PSTN (*Public Switching Telephone Network*).

3. *Printing i.e., converting the received signals into a facsimile (copy) of the transmitted documents. Normally we use* the Print Head Operation. The Print head has a row of very small resistor-element spots across the recording paper with very-very small width. The thermal recording paper touches this row of resistors. A pulse of current through a resistor causes it to become hot enough to mark the paper in a spot of about 0.005 inch in diameter. The spot temperature must .be changed from non marking temperature to marking temperature and return to non marking temperature before the paper steps to the next recording line. Marking temperature is about 200° F. The *thermal paper* is a special paper that turns black on heating. Either *thin-film* or *thick-film* technology is used for making the *print heads*.

Note: The signal obtained from scanning cannot be directly transmitted, as it is difficult to amplify low frequencies. Hence process of modulation is carried out. The Fig. 11.1 (b) shows the transmission and reception process. The signal is "amplitude modulated" with a carrier. After this, we use AM/FM converter, as frequency

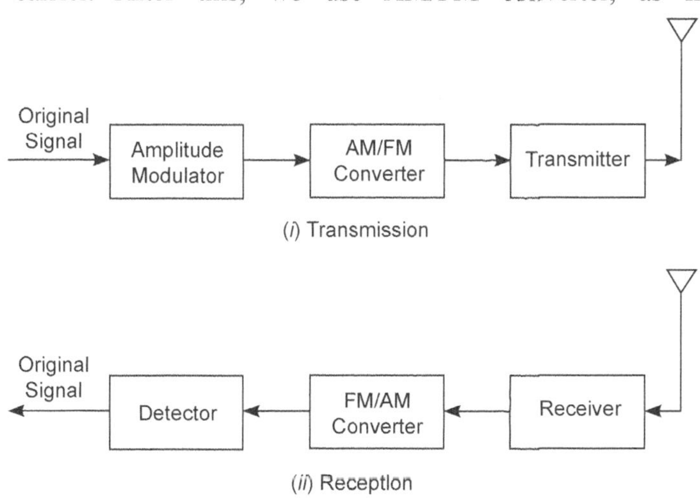

Fig. 11.1 (b)

modulation is superior to that of the amplitude modulation and the signal is transmitted. At **reception side** we again use an FM/AM converter and the detecter, which detects the original signal and the carrier is separated. The carrier frequencies used for both the modulations are between 1200 Hz to 1800 Hz.

11.5 CONVERSION OF OPTICAL SIGNAL INTO ELECTRICAL SIGNAL

This is the basic process which occurs in FAX. Following devices are used for this purpose:

1. *Photodiode* (Fig. 11.2): This is the simplest device which can be used to convert light into electrical signal. When light from picture or printed page falls on a photodiode, an electric current flows through the diode. This current is made to flow through a resistive circuit to produce a voltage which is later on amplified. The output of the amplifier is proportional to the light intensity falling on the diode.
2. *Photo Multiplier tube* (Fig. 11.3): The photo multiplier tube has so many "photo electrodes" . The light is made to fall on the first electrode which emits electrons. These emitted electrons strike the next electrode and more no. of electrons are emitted from the later. The process is continued and thus multiplied output of electrons *i.e.*, a large photo current reaches at the collecting electrode (anode). The amplification in this way takes place within the device and very little amplification is further needed. The device rather needs much higher voltage to operate and thus it is costly.

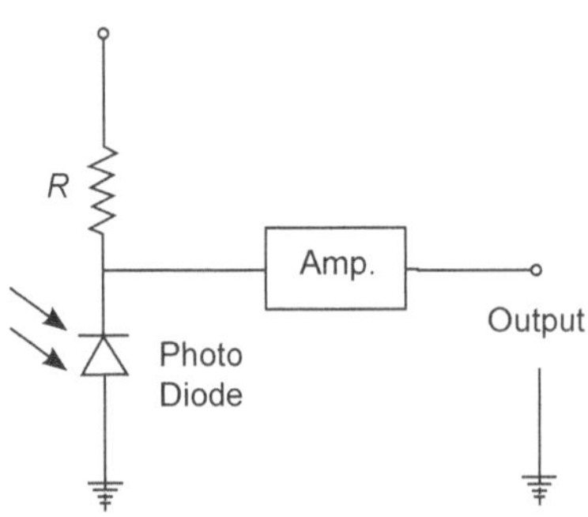

Fig. 11.2

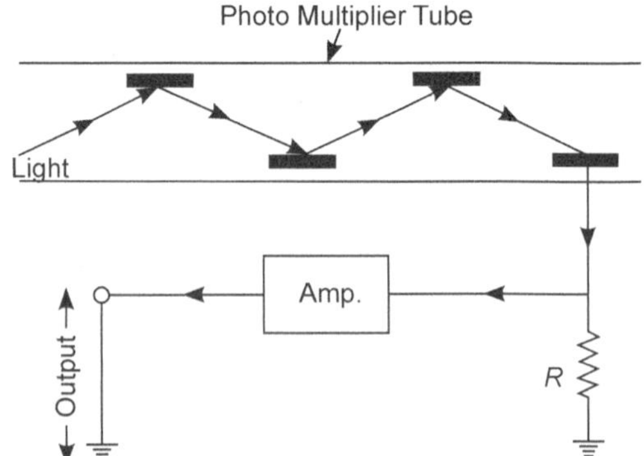

Fig. 11.3

Note: Whatever the type of photodetector (photodiode or photo multiplier tube) is used to convert light into the electrical signal, the light should be restricted on the particular part of the picture or document to be transmitted. There may be following methods.

(*a*) In the first approach, the light is focussed through a lens on the part which is to be transmitted. The reflected light from the part reaches the photo diode, which converts the lightness (or darkness) of the part into voltages. A pure white part generates a high voltage and a dark part generates a low voltage. (See Fig. 11.4(a)

(*b*) In another approach, though the whole surface of the document is illuminated, but reflected light from a restricted area (which is to be transmitted) is allowed to reach the photo diode. The light out of this area is not detected by the photodetector. (See Fig. 11.4(b)

The first method is superior as it can be achieved by using a LASER beam which gives a better control on the light intensity.

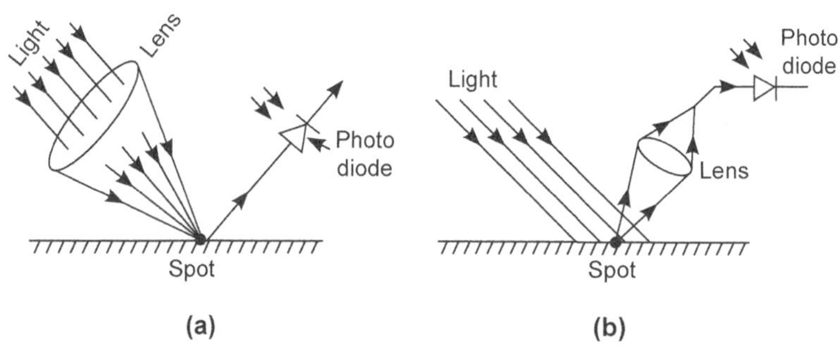

Fig. 11.4

11.6 FAX TRANSMITTER AND RECEIVER

11.6.1 FAX Transmitter

The message is "scanned", usually by *optical scanning* in the transmitter. Two methods are adopted for optical scanning:

1. Cylindrical Scanning. 2. Tape Scanning.

1. *Cylindrical Scanning* (Fig. 11.5): The message is wrapped around a cylindrical drum to allow complete scanning. The cylinder (drum) is rotated and the scanning spot is kept fixed. The light reflected from the drum is fed to a photo diode, which converts the picture signal (light) into electrical signal. The electrical output from the photo diode

is amplified and transmitted. The drum is rotated at about 60 RPM and scanning rate is kept at 5 lines/minute.

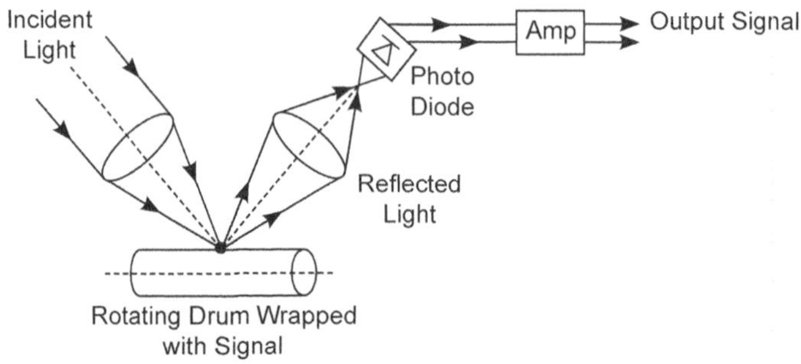

Fig. 11.5

2. *Tape Scanning*: This is also an optical scanning method. In this method message is taken directly from the printed tape. Due to width limitations, the method has lesser in use. The Fig. 11.6 shows the method.

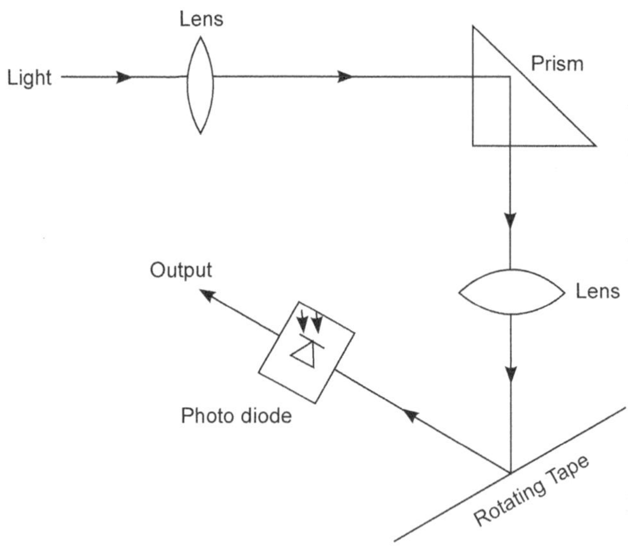

Fig. 11.6

Scanning Spot

The shape of scanning spot in optical scanning may be rectangular [Fig. 11.7 (*a*) or a trapezoidal (*b*)]. The shape of scanning spot decides the size of the wave shape of the output signal.

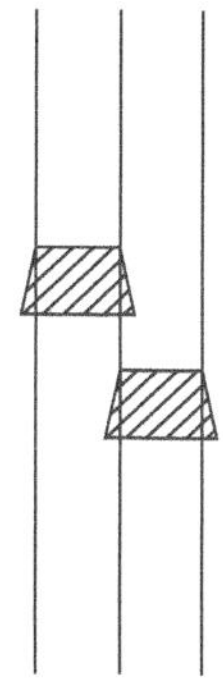

(a) Rectangular Spot (b) Trapezoidal Spot

Fig. 11.7

Usually a rectangular spot is used. The size of the spot is very small and covers an area of constant illumination on the message. The change in the average illumination gives the signal output.

11.6.2 Fax Receiver

In the receiver, "reverse" is done, of what happened in the transmitter. While transmission, the picture message was converted into an electrical signal, in the receiver, the electrical signal is converted back into picture signal. The equipment (drum etc.) in the receiver is however, almost same as used in the transmitter. We may get an output free of distortion, for this it is necessary that the signal is properly **"synchronised"**.

Synchronisation of the Signal

(*a*) When the signal is a written message (*e.g.*, a page of a book), synchronous motors which rotate at constant speeds are used both at transmission and receiving ends. These motors are operated under controlled frequency supply.

If f is the frequency of the supply fed to a synchronous motor with no. of poles P, the synchronous speed of the motor will be

$$N_s = \frac{120 \times f}{P}$$

The synchronous speed N_s will remain constant so long frequency of the supply f is kept constant by some means.

(*b*) If it is a picture (photograph) signal, a synchronising signal of 1 kHz is sent for synchronization.

When the signal modulates a carrier e.g., in the case of F.D.M. (Frequency Division Multiplexing), the carrier is sent along with upper side band (U.S.B.). The carrier helps in the total recovery of the synchronising signal later on.

SUMMARY

1. The fax telegraphy is a technique to send printed message from one station to another.
2. In comparison to TV transmission, the fax needs a much shorter bandwidth, this is the reason that fax messages can be sent over ordinary telephone lines.
3. The fax finds application in business, weather calculations and personal use etc.
4. The fax process involves three basic steps: converting the document into electrical signal, transmitting the electrical signal through telephone wires to another FAX machine and there converting the received signal into the original document.
5. The picture (document) is converted into electrical pulses by a photodiode or photo multiplier.
6. In transmission, AM-FM converter is used. In receiver, FM-AM converter is used.

SELF TEST QUESTIONS

1. What is a FAX? What are its various applications?
2. Explain the basic principle of FAX.
3. What are the various types of FAX machines.
4. Give two methods employed in FAX to convert the picture (document) signal into electrical signal.
5. Explain FAX transmitter and receiver.

OBJECTIVE TYPE QUESTIONS

I. Fill in the blanks:

1. A bandwidth of ______ Hz is needed in FAX system.
2. Facsimile means exact ______ .

3. The first photograph was faxed in _______.
4. The FAX can send _______ messages.
5. A _______ can convert picture signal into electrical signal.
6. The FAX message can be sent on ordinary ______ wires.

ANSWERS

1. 120	2. reproduction
3. 1907	4. written
5. photodiode	6. telephone or telegraph

II. State True/False:

1. The ordinary telegraphy needs larger bandwidth than the FAX.
2. The TV transmission is faster than FAX.
3. FAX can be used to transmit message only in Hindi and English.
4. Modulation is not needed in FAX transmission.
5. Photodiode can be used to convert optical information into electrical pulses.
6. FAX signals are synchronised on both sides by using synchronous motors.
7. FAX stands for facsimile.

ANSWERS

1. True	2. True	3. False
4. False	5. True	6. True
7. True		

❑❑❑

12

Radar Systems

Radar is basically a means of gathering information about distant objects or targets (aeroplane etc.). This is done by sending electro-magnetic waves at them and analyzing the *echo* received.

12.1 THE RADARS

The term "RADAR" means "Radio Detection and Ranging". At first it was used as a method to detect an approaching aircraft for many other purposes.

The radar has multifarious jobs to perform. It can be used as a navigational aid. In military, radar is used for aiming guns at ships and aircrafts, directing missiles, searching submarines, attacking enemy crafts etc.

In an aircraft, a radar can provide information of the path in case of darkness or poor visibility; similarly in a ship it can give information about other ships, land masses etc.

Radars can also be used for safe landing of aircrafts. On approaching the airfield, the pilot is guided by the radar. Radar helps the *meteorologists* in forecasting weather. As the rain drops are able to reflect radar signals, height of the clouds can be measured. Radar can be used for discovering ores, oil fields and buried metals.

Pilots sometimes fly at very low levels to avoid radar. When they do this, radar reflection given off by the plane become indistinguishable from those given off by buildings and other tall objects, and the plane can go undetected. Some planes are made of materials that do not reflect radar waves but instead, absorb them. As no radar waves are reflected back to the radar ground station, the plane becomes invisible to radar.

12.2 BASIC PRINCIPLES OF RADAR

Basically a radar consists of a *transmitter* and a *receiver*, each connected to a *directional antenna.* The transmitter sends VHF (*i.e.*, microwave) power through the antenna. The receiver collects as much energy as possible from the *echo* reflected from the target and analyses the information received and later on displays it on a CRO screen. The receiving antenna is same as the transmitting antenna. All this is accomplished through TDM (Time Division Multiplexing).

The Fig. 12.1 shows the fundamental block diagram of a radar to explain its functioning. For each transmitted pulse, the cycle of events is as follows: In response to an internally generated trigger signal, the transmitter generates a short *rectangular* pulse. As soon as a small fraction of the pulse power is fed to the *duplexer*, the receiver is disconnected and the transmitter is connected to the antenna. The antenna moves in a predetermined pattern called *scanning pattern.* Either way, the antenna is highly directional and sends out the generated pulses towards the target. The *scanning speed* is high but it is small compared to the time taken by the pulses to return from the normal range of the targets. Thus when echoes are received from the target the antenna still points in the right directions to *receive* them. Now the duplexer disconnects the transmitter from the antenna and reconnects the receiver with the antenna. The received echo is *processed* in the receiver *i.e.*, amplified and demodulated. The receiver is a *super-hetrodyne type*. The output pulses (along the noise of course) are then fed to the displaying device (such as CRO). Now one cycle is complete and the set is once again ready for transmission of the next cycle, while the antenna goes on scanning along its path.

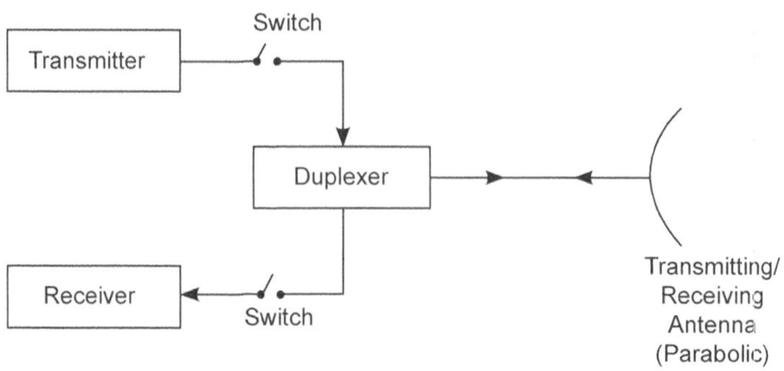

Fig. 12.1

The radar is able to show position of a target because the information about the horizontal and vertical direction of the antenna is available. In addition, distance of the target can also be calculated from the total time taken by the pulse on its forward and return journey. We know that speed of electro magnetic pulses is 3×10^8 m/s, and if a pulse returns in 1 μs, *i.e.*, it covers *go and return* distance in 1 μs, the distance of the target can be calculated from the following formulae:

$$\text{Speed} = \frac{\text{Distance}}{\text{Times}}$$

$$3 \times 10^8 \text{ m/s} = \frac{\text{Distance}}{1\ \mu\text{s}}$$

$$\text{Distance} = 3 \times 10^8 \times 1 \times 10^{-6} (1\ \mu\text{s} = 1 \times 10^{-6}\text{s})$$

$$= \mathbf{300\ meters}$$

As the pulse has covered go + return distances, the distance of the target is

$$= \frac{300}{2} = 150 \text{ meters.}$$

However, a limit exists on the smallest number of pulses that can be sent per second. Practically the system sends out several hundred pulses per second, with a duration of 1 μs.

Frequencies and Powers used in Radars

The radars mostly operate in the frequency range of 100 MHz to 250 GHz. Long range radars operate between 1 to 4 GHz. The radars designed for high *resolution* operate above 20 GHz. This permits to limit the size of antenna within practical length.

Table 12.1: Frequencies and Powers used in Radars

Band	Frequency Range (GHz) 1 GHz = 10^9 Hz	Max. power available in the transmitter (in MW) (1 MW = 10^6 W)
UHF	0.3 – 1.0	5.0
L	1.0 – 1.5	30.0
S	1.5 – 3.9	25.0
C	3.9 – 8.0	15.0
X	8.0 – 12.5	10.0
Ku	12.5 – 18.0	2.0
K	18.0 – 26.5	0.6
Ka	26.5 – 40.0	0.25
The output power of a radar transmitter is given in terms of maximum power. Generally radar power lies within the range of 100 to 500 kW.		

12.3 FUNCTIONS (APPLICATIONS) OF A RADAR SYSTEM

A radar system has to perform three basic functions: (*i*) Searching (*ii*) Tracking (*iii*) Measurement of azimuth and range.

A radar system *searches* a target and once target is found, it follows or *tracks* the target to ascertain its position and range (*i.e.*, distance) etc. Sometimes the same radar performs both the functions or two radar systems may be employed for the above two tasks.

12.3.1 Search Radar Systems

For searching a target, the radar system should be capable of *scanning* its region rapidly. Narrow beam antenna pattern takes long time for scanning therefore *broad beam scanning* is a better solution. Once the approximate position of the target is obtained the information may be passed on to the tracking radar which will quickly follow it. Another solution to this problem is to use two "fan shaped" beams, so that one is in the direction of *azimuth* and other in the direction of *elevation*. The two antennas rotate together: one *searches* and other acts as a *height finder*. *Air traffic* control (ARC) radars are its example.

12.3.2 Tracking Radar Systems

Once a target is found, it may be then *followed* or *tracked* to find its actual position in the space. For this, *conical scanning* may be employed. Tracking may be of two types:

(*a*) *Tracking in angle* means to find the angular position of the target.

(*b*) *Tracking in range* means to find the range in meters (as well as the angle) of the target.

- *Airborn* radar is a system, in which a single radar can perform both the functions. First the system works in *search mode,* once the target searched, the system may be switched on to the *track mode*. The difficulty with this radar is that when it is "tracking" it cannot "search" and the radar is *blind* in all directions expect the one.
- The *Track While* Scan (TWS) is a radar which can perform both functions simultaneously but this radar can be used only for small areas.
- Radars are also available which can track satellites, spacecrafts, and space rockets. They use special way of tracking. In this system only *searching* is needed as the approximate *position* of the target (*e.g.* moon) is known. For this purpose, a very high transmitting power, sensitive receivers and antennas are required.
- It should be mentioned here that the radars for deep space tracking are *monostatic* (transmitting and receiving antenna located at the same point). In exceptional cases, they are *bistatic i.e.,* transmitter and receiver are separate.

12.3.3 Measurement of Azimuth (Bearing or Position) and Range

Since a radar works on the principle of wave reflection (*i.e.*, echo) it uses *quasi optical microwaves* in the range from 1000 to 10,000 MHz. These can be shaped by dish (parabolic) reflector into sharply focussed beams like those of search lights. This beam is made to scan the sky through rotation of the radar antenna and the reflector. When the beam is reflected from a target, the direction of the antenna at that moment gives azimuth or bearing (*i.e.*, position) of the target at that moment. The azimuth or bearing in degrees is usually shown on the indicator.

The distance of the target can be measured directly from the CRT. In Fig. 12.2(*a*), the pulse is seen leaving the antenna simultaneously with indication on CRT. The pulse is known as "*main bang*". After reflection from the target, the '*echo*' is also recorded on the CRT. The distance of the target can be calculated from the gap between the *main bang* and *echo* obtained on the CRT, for which CRT can be calibrated in a suitable scale.

The Fig. (*b*) shows that range of several targets may be indicated on the face of a CRT simultaneously.

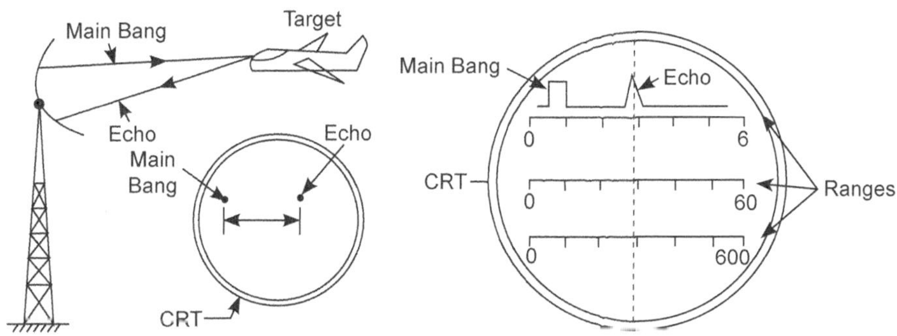

(*a*) Measurement of distance of the target

(*b*) Simultaneous measurement of range of several targets on a CRT

Fig. 12.2

12.4 ANTENNA SCANNING PATTERNS

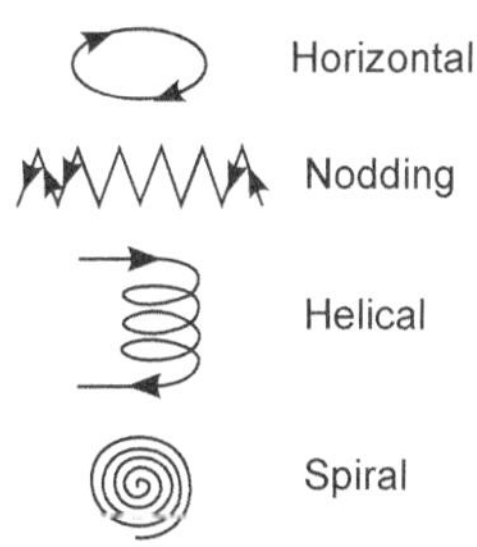

Fig. 12.3

As already mentioned radar antennas are often made to *scan* a given area in the space. For different applications, different scanning patterns are popular. Few are briefly described below (Fig. 12.3):

(*i*) *Horizontal scanning:* This is simplest, in which antenna is made to scan in the horizontal plane only. This pattern is used in *ship to ship* radars.

(*ii*) *Nodding scanning:* In this, antenna scans in both *i.e.*, horizontal as well as in vertical planes. This method can be used to scan a limited area of the space.

(*iii*) *Helical scanning:* In this system also, antenna scans in both the planes and through a large area in the space. The antenna returns to its starting point at the end of each cycle. The typical speed in this pattern is 6 r.p.m.

(*iv*) *Spiral scanning*: This method is used when a smaller area in circular shape is to be scanned. In radars "scanning antenna" is used. The purpose of using scanning antenna is to find direction of the target w.r.t. to the target. The direction of antenna, at the instant when echo is received gives direction of the target.

12.5 TRANSMISSION AND RECEPTION IN RADAR SYSTEM

The process of sending out wave energy in short pulses is known as *pulsing* and the same is used in radars. The essential parts of a simple radar system is shown in Fig. 12.4. The *indicator* unit consists of a cathode ray tube (CRT) similar to that used in a TV set. The CRT displays the original transmitted

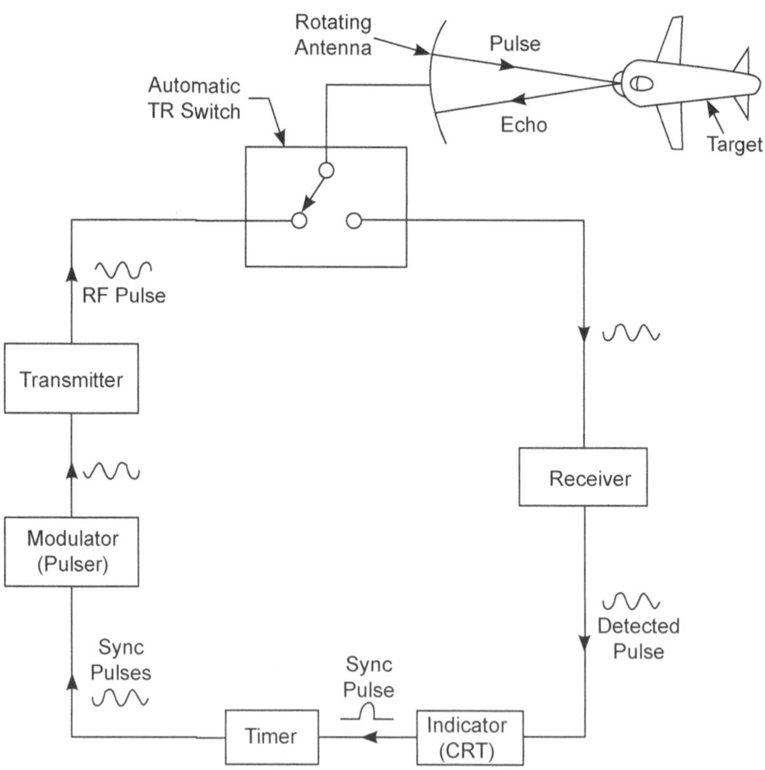

Fig. 12.4

pulse as well as the return echo pulse along a scale calibrated in terms of distance. There may be other indicators for drawing a map of the area searched by the radar antenna, but the system shown here is only to give information about the *range and direction of the target.*

The *Timer* in the block diagram assures that a single pulse from the *pulser* turns on the transmitter exactly at the same time as the indicator begins to draw the pulses along its scale. The *transmitter* (high power magnetron) then sends out the short pulses of very high frequencies through a *Transmit-Receiver* (TR) switch. At this moment, switch is connecting transmitter and terminating the receiver off the circuit. These pulses are sent to the rotating antenna, which radiates them into space. At the same time the transmitted pulse is displayed on the indicator scale. If the transmitted pulse strikes a target some where in the space, an echo will be received by the antenna. The echo will be maximum when the target is exactly in the centre of the transmitted pulses. In practice, this is achieved by turning the antenna till a maximum echo is obtained.

In the mean time, the fast acting automatic TR switch has already turned towards the *receiver* where echo pulses are amplified and demodulated by a *superheterodyne* circuit. The output of the receiver is given to the indicator (say a CRT) for display. The distance between the *original* pulse and the *echo* can be read out directly (say in kilometers) on the CRT, thus giving *range* of the target. The direction of the antenna at this time gives direction of the target.

12.6 CLASSIFICATION AND DESCRIPTION OF RADARS

The various radar systems can be classified as below:

1. Pulse radars
2. Moving target indication (MTI) radar
3. Beacons radar
4. CW radars
5. Tracking radar
6. Laser radar

12.6.1 Pulse Radar (Fig. 12.5)

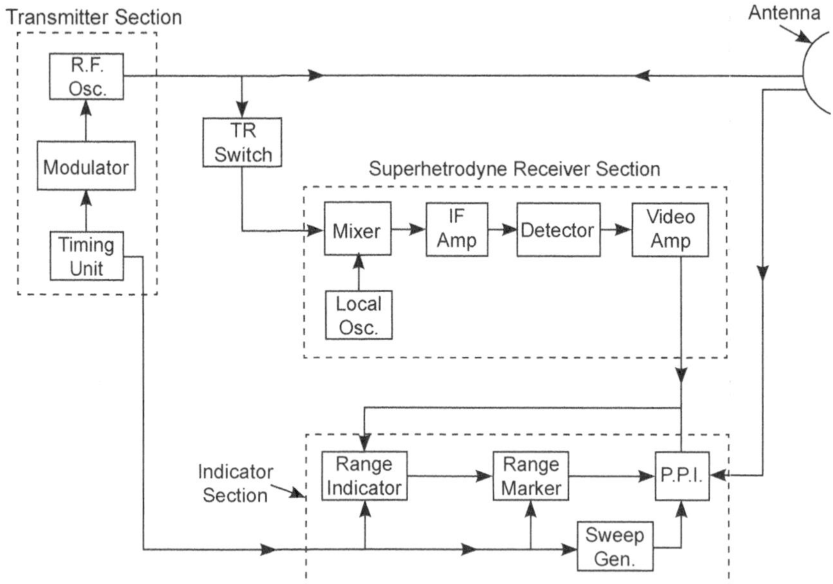

Fig. 12.5

1. *Transmitter*: In this section, a high power *oscillator* (usually a magnetron) generates RF oscillations. The voltage pulses are formed in a modulator which are applied in its plate cathode circuit. The pulse operation is controlled by a *timer* unit, which is basically a multivibrator. The trigger pulses obtained from the timer unit are also applied to the indicator section for synchronisation.

   The RF energy of the oscillator is transmitted through an *antenna* which is a specially shaped and designed parabolic reflector.

2. *Transmits Receive* (*TR*) *switch*:

   The TR switch which periodically makes contact with transmitter and receiver employs a gas discharge tube. The high power pulses from oscillator (magnetron) break down the gap in the tube and short circuits the line to the receiver.

3. *Receiver*: The radar receiver is a superheterodyne receiver. The intermediate frequency (IF) is kept usually 30 to 60 MHz. A diode is used as a *detector*. The video amplifier raises the signal amplitude to the required level to be applied to the indicators.

4. *Indicators*: The indicator panel of the radar is provided with the various measuring instruments and indicators *e.g.* CRO, pulse position

Indicator (PP1) etc. It is provided also with a sweep generator for range measurements.

5. *T.R. Switch* (*Duplexer*) (*Fig. 12.6*): ATR switch or duplexer is a circuit which enables the same antenna to be used for both *transmission* as well as *reception* without any interference. In radars, generally *branch type* duplexer is used. It has two switches (TR and anti TR) which alternately connect transmitter and receiver with the antenna. Cold cathode tubes are generally used for these switches.

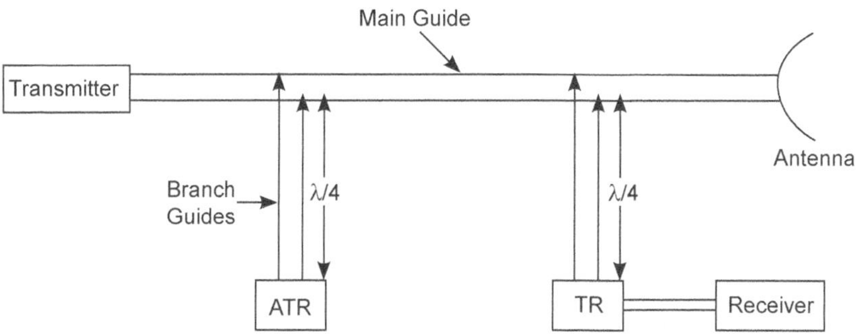

Fig. 12.6

Operation: A *main guide* connects the transmitter to the antenna. With this main guide, TR and ATR (Anti TR) switches are connected through the *branch guides* of quarter wave lengths.

When the transmitter produces RF pulses, both switches get short circuited. The short circuiting of TR switch prevents RF power from entering the receiver. At the termination of the transmission, both switches are open circuited. The ATR switch throws a *short circuit* thus preventing any signal from entering into transmitter. The guide leading through TR switch to the receiver becomes continuous, thus the echo reflected back from the antenna can go to the receiver.

6. *Modulator*: Modulator is a component of radar transmitter. The function of modulator is to switch *on* and *off* the output tube as and when required. There are two types of modulators in use:

   - **Line pulsing modulator:** In this, anode of the output tube is modulated directly by a system that generates and provides large pulses of supply voltage. This is achieved by slowly charging and discharging a transmission line. But practically a *Pulse Forming Net work* (PFNW) replaces the line. As shown in the Fig. 12.7 the PFNW behaves as a transmission line for frequencies below $f = 1/2\pi\sqrt{LC}$ where L and C are inductance and capacitance respectively. For switch(S), a *thyratron or a SCR may be employed.*

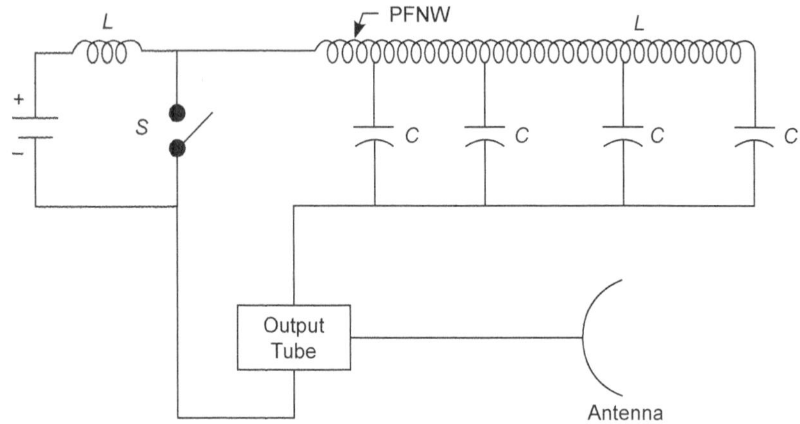

Fig. 12.7

- **Active switch modulator:** The line pulsing modulator cannot give variable pulse length, therefore active switch modulator may be used. In this modulator, pulses are generated at low power level and later are amplified by an amplifier, but this modulator is less efficient, more complex and bulky, so rarely used.

7. *Antenna*: The majority of radars use *dipole* or *horn type* parabolic *reflector* as antenna. The beam width in the vertical direction is worst than in the horizontal direction, but this is immaterial in *ground to ground* or even *air to ground* radars. It has the advantage of having small size and weight, reduced wind load. Moreover they need small motors.

   The two shapes used for radar antenna are shown in the Fig. 12.8 (a) and (b).

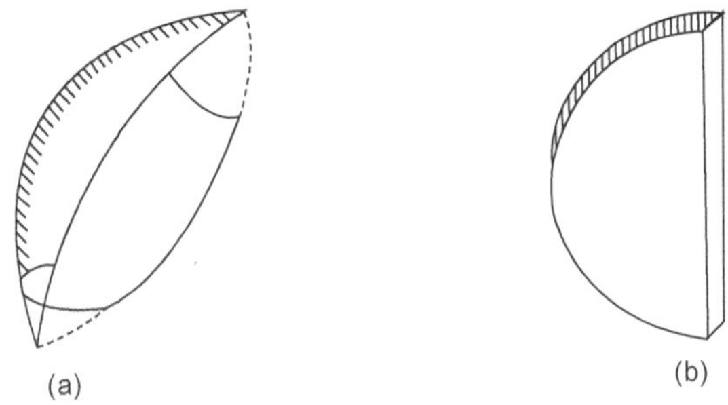

Fig. 12.8

12.6.2 MTI Radar

In Moving target Indicating (MTI) radar system, only moving targets are displayed on the screen, and the echo obtained from the stationary targets is not displayed at all. The MTI radar employs *Doppler's effect* explained below.

(a) Doppler Effect

Doppler in 1842 gave a postulate known as "Doppler Effect", which says:

"The apparent frequency of an electromagnetic (*i.e.*, light, sound) wave depends on the relative motion of *source* and the *observer*."

If source and the observer are moving away from each other, the apparent frequency will decrease and if they are moving towards each other, the apparent frequency will increase.

The Fig. 12.9 shows block diagram of MTI radar

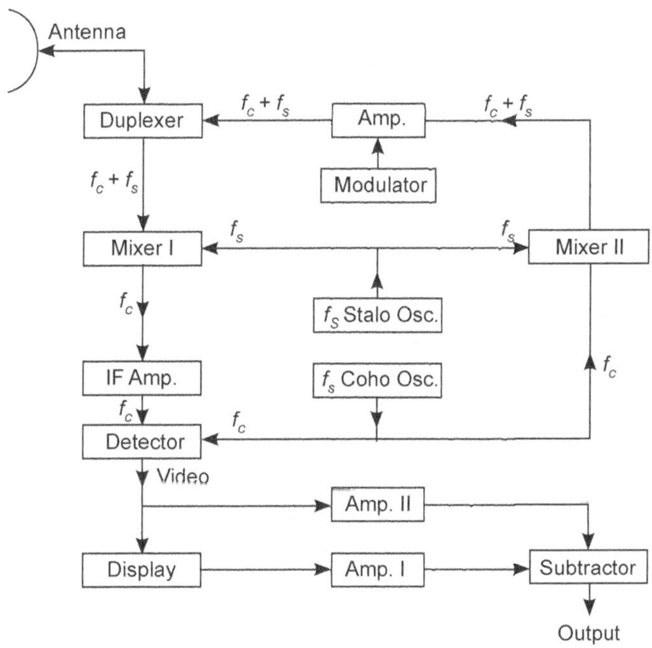

Fig. 12.9

In case of radar, involving a moving target, the signal undergoes a change due to doppler effect. The target acts as a *source* of the reflected waves. Now we have a moving source (target) and a stationary observer (radar). If the target is moving towards the radar, the overall effect is doubled. The doppler effect is observed only in *radial* motions and not in the *tangential* motions. If target (source) is rotating, the radar can distinguish its *leading* edge from its *trailing* edge.

Basically, a MTI radar system compares a set of received echoes with those received during previous sweep. The echoes whose phase remains constant are cancelled out as they are obtained from stationary targets, but the echoes obtained from moving targets show a *phase difference* due to *doppler* effect and hence not cancelled. It also helps in the detection of moving targets whose echoes are very small due to long distance than those of nearby stationary targets.

The transmitted frequency is the sum of the outputs of two oscillators $(f_c + f_s)$. The first oscillator is called *Stalo* (stable) and other a *coho* (coherent). (Fig. 12.9) The stalo is a local oscillator and coho operates at intermediate frequency (I.F.) and provides *coherent signal*. The mixers I and II are identical and both use the *stalo* thus phase relations of their inputs are preserved in their outputs also. The output of the I.F. amplifier and signal of coho oscillator is fed to a detector (phase discriminator) circuit.

The coho oscillator is used to generate RF signal as well as the signal for the detector. Since the output of the detector is phase sensitive, an output will be obtained for all fixed and moving targets. The phase difference between *transmitted* and *received* signals will be constant for fixed targets and it will be changing for moving targets as per the doppler effect.

(b) Blind speeds in MTI radars

In an MTI radar, the phase difference between transmitted pulse and the echo remains constant from pulse to pulse in case of stationary targets. The phase difference, however, varies in case of moving targets. That is, why a MTI radar can distinguish between stationary and moving targets.

If the target happens to be moving with a velocity, such that its velocity component along the radar axis results in a phase difference of 2π radians or its integral multiple between successive sets of pulses to the radar, the targets appear as stationary. These target velocities are said to be **Blind speeds**.

If the target velocity is such that it moves a distance equal to $n \cdot \lambda/2$, where n is an integer and λ the transmitted wave length along the radar axis, the phase difference will be $2\pi \cdot n$ radians and the blind speed is expressed by

$$V_b = \frac{n\lambda/2}{1/fr} = \frac{n\lambda fr}{2}$$

when $n = 1$ the lowest blind speed $= \lambda\, fr/2$.

Where, n is an integer, λ = wave length transmitted and fr = pulse repetition frequency (PRF).

12.6.3 Beacon Radar (Responders)

The *beacon* is a small radar set consisting of a receiver, transmitter and an *all directional* antenna. When another radar transmits a coded signal to a beacon *i.e.*, when a beacon is *interrogated,* the beacon *responds* by sending back its own coded pulses. A beacon thus is called a *responder* and it may respond in the same frequency as interrogated by the *interrogating* radar or in a special beacon frequency in which case a separate receiver is required by the interrogating radar. It is to be noted that a *beacon* does not transmit pulses continuously but it only *responds*, when *interrogated* by a radar.

One of the functions of a beacon is to *identify*. For example, a beacon may be installed on target such as an aircraft and it will transmit a coded pulse when interrogated. These pulses appear on the plan position indicator (P.P.I.) of the *Interrogating radar* and inform it about identity of the target. When used as such it is called *Identification Friend or Foe* (IFF). An IFF operates satisfactorily with naval *targets.* It however, fails if large number of fast planes, some friends and some foes (enemies) are involved.

Another application of a beacon is that it can work as a "light house". A beacon light house can operate over much larger distances. A lost ship can calculate its position by interrogating a beacon lighthouse.

A Beacon is generally used to identify whether an approaching target (*e.g.* plane) is a friend or a foe.

12.6.4 CW (Continuous Waves) Radars

A CW radar sends *continuous waves* (sine waves) rather than pulses. It also uses the *Doppler effect* to detect the frequency change caused by the moving target and displays it as a *velocity* of the target in kilometer per hour.

Two types of continous wave (CW) radars are described below:

(a) **Unmodulated Continuous Wave (CW) Radars:** An un-modulated CW radar has been shown in Fig. 12.10. Since the transmission is continuous, a *duplexer* is not needed. Separate antennas for transmission and reception have been shown. The arrangement increases isolation between transmitter and the receiver.

As shown, a small portion of the transmitter output is mixed with output of a local oscillator and the sum is fed to the mixer. This also receives the *shifted signal* from the receiving antenna and produces an output difference frequency (plus or minus) called *doppler frequency*. The output of this mixer is amplified, and signal from the detector gives the doppler frequency. After amplification the output is given to the indicator.

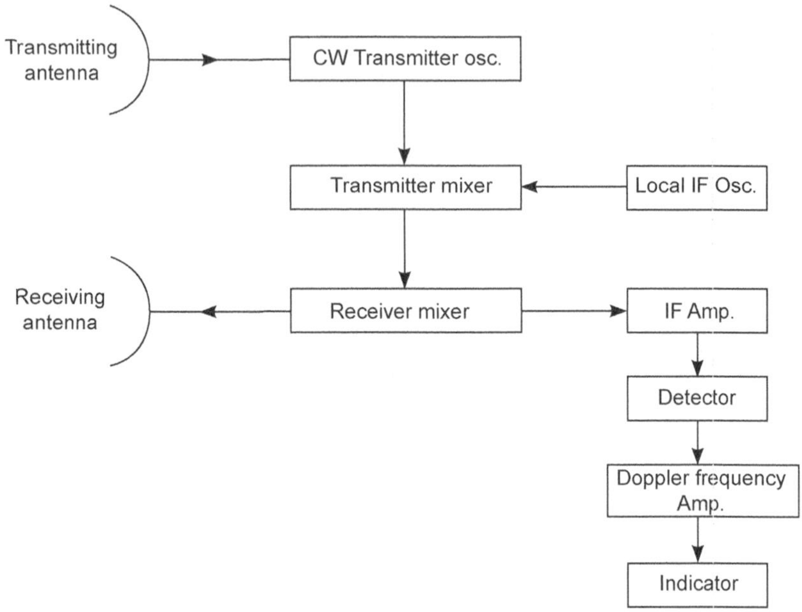

Fig. 12.10

Advantages

(*i*) Un-modulated CW radar gives accurate measurement of relative velocities, using low transmitter power and low power consumption.

(*ii*) This radar is comparatively of lesser size than the widely used pulse radars.

(*iii*) It is uneffected by stationary targets.

(*iv*) The receiver of the radar is ON all the time, it can be operated even at zero range.

(*v*) It can find the direction of the target in addition to its speed.

Disadvantages

(*i*) It has limitations of its maximum range, which is very smaller than of pulse radar.

(*ii*) It is incapable to indicate the range of the target, it can show only its velocity.

Applications

(*i*) It is used in aircraft navigation.

(*ii*) As it is unaffected by stationary objects, it is able to detect presence of aircrafts.

(*iii*) It can measure speed of missiles, automobiles etc.

(b) Frequency Modulated CW (FMCW) Radar: The greatest limitation of unmodulated CW radars is that they can not measure the range. If the transmitted signal is frequency modulated, the same radar will be able to measure range, because then it will be able to distinguish one cycle from the other. Using FM will require an increase in BW of the system.

In this way, a FMCW (Frequency modulated continuous wave) radar is an improved form of a CW unmodulated radar. As the main application of a FMCW radar is to measure height of an aircraft or plane, it is generally known as an *Altimeter*. The Fig. 12.11 shows the block diagram of a FMCW radar being used as an altimeter.

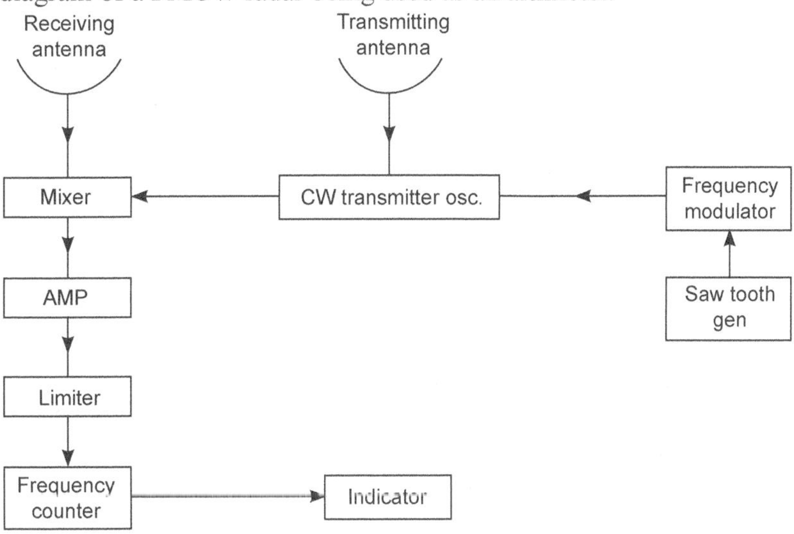

Fig. 12.11

As an Altimeter, as it has to measure height, in this case the target is earth, which is stationary w.r.t. to a plane whose height is to be measured. A frequency difference proportional to the height of the plane will exist between transmitted and the received signals, because the signal now being received was sent at a time, when the frequency was different. If the *rate of change of frequency* w.r.t. time due to FM process is known, the difference in the time between *sent* and the received signals may be found and thus height of the plane above earth can be easily calculated. Thus output of the mixer which produces the frequency difference can be *amplified* and fed to the *frequency* counter and then to the indicator, whose output is calibrated in kilometers.

As an altimeter, FMCW radar is preferred to the pulse radar, as the FMCW radar has no limit on the minimum range, as short heights are involved in this measurement. Further simple, low power and smaller equipment may be used with small antenna. A typical FMCW radar uses a 1-2 W transmitter power easily obtained from a diode and has a range of about 10 km or more.

12.6.5 Tracking Radar

The tracking radar does the job of continuously tracking a moving target. It is usually a ground base system used to track the air born targets. The tracking radar antenna sends out a very narrow beam (called pencil beam), whose beam width could be anywhere between fraction of a degree to one degree both in azimuth and elevation to get the requisite resolution for tracking purpose. One can visualize that it is imperative to use a search radar and acquire the target with comparatively much larger bandwidth before a track action is intiated.

12.6.6 Laser Radar

A laser radar uses an optical beam instead of microwaves. In other words, the e.m. energy transmitted lies in the optical spectrum in laser radars where as in microwave radars, it lies in the microwave region. In laser radar, the frequencies involved are very high. It is possible to generate laser pulses as narrow as a fraction of picosecond.

12.7 RADAR DISPLAYS

The output of a radar may be displayed in any of the following devices

(*i*) A scope

(*ii*) Plan position indicator

(*iii*) Monitor

12.7.1 A scope (See Fig. 12.12)

The operation of this system is more or less similar to that of a CRO. A sweep wave form is supplied to the horizontal deflection plates of the CRT and moves the beam slowly left to right across the face of the tube. The flyback period is rapid just as in a TV and the beam rapidly returns to its starting point. In the absence of a signal, the display is a horizontal straight line.

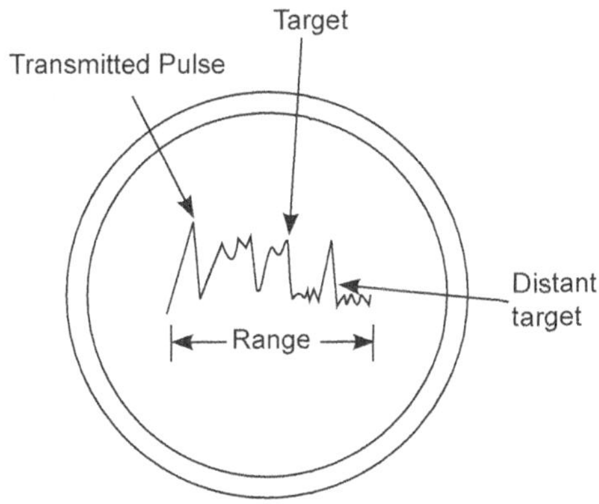

Fig. 12.12

12.7.2 Plan Position Indicator (PPI)

[Fig. 12.13 (*a*), (*b*)] The PPI displays a map of the targetted area. The signal obtained from the receiver after demodulation is applied to the grid of the CRT. Long persistence phosphors are used at the face of PPI screen, so that it does not flicker. The PPI display is used with search radars.

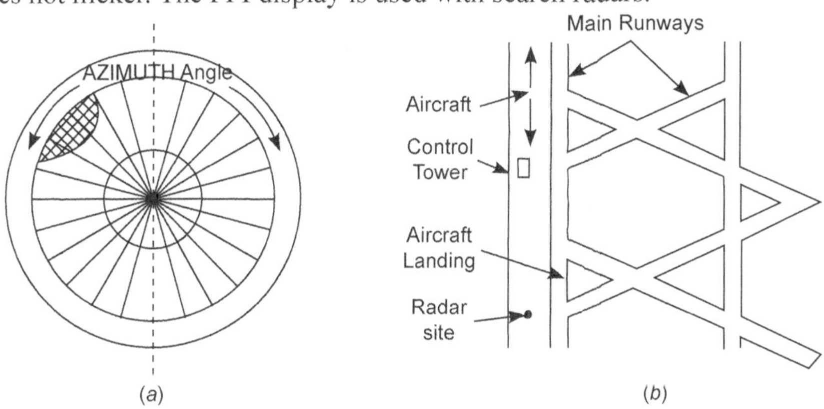

Fig. 12.13

12.7.3 Monitors

They are just like TV receivers with a screen to display and a picture tube and all other components are missing. The output is given to the picture tube. The panel has all control of a TV receiver like brightness, contrast, focus, etc.

12.8 TERMS RELATED TO RADAR SYSTEMS

The important terms related to radar systems are given below:

1. *Pulse Repetition Rate* (*PRR*): The PRR stands for no. of pulses transmitted per second. The PRR is determined by the time difference between two pulses which should be large enough, so that the echo of

a particular pulse lies within the time interval between the transmitted and succeeding pulses. But if the time difference is much larger, there is every possibility that the radar fails to detect the target, *i.e.*, the radar is not transmitting any pulse, when the antenna points in the direction of the target.

2. *Range*: The range of a radar is the time lapse between the transmitted pulse and received echo. The maximum possible range of radar is given by

$$[R_{max}]^4 = \frac{P.G.A.\sigma}{(4\pi)^2.S\,\text{min}}$$

Where P = Transmitted power in Watt.

G = Gain of the transmitting antenna

A = Effective aperature area of receiving antenna in m^2

σ = Target cross-reaction area in m^2

$\pi = 3.14$

S_{min} = Minimum detectable signal

3. *Resolution:* This is the ability of a radar to distinguish between two closely spaced targets.

4. *Range ambiguity:* The inability of a radar to distinguish between two targets at different ranges is called the range ambiguity (See Figure 12.14). The radar can not distinguish between targets represented by the positions A and B.

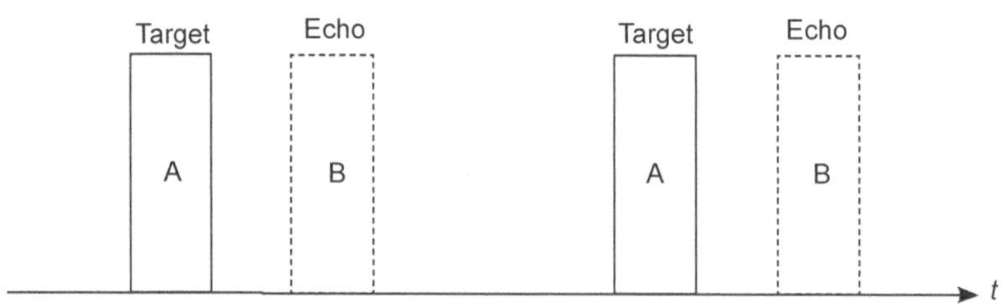

Fig. 12.14

5. *Clutter*: All those objects, which reflect the radar signal, and which are other than the target are termed as clutter. This is a type of noise.

   The echo corresponding to the stationary target is also called clutter.

6. *Blind range*: When a target lies at such a distance, that the radar receives the echo while it is transmitting, it fails to detect the target. In such a case the target is called to be at the blind range.

7. *Precision*: The standard deviation in the estimate of certain parameters such as dopplers shift, range etc amidst noise is termed as precision. The precision is a function of single noise ratio and the wave form.

8. *Monostatic and Bistatic radar systems*: In monostatic radar system (Fig. 12.15) single antenna is used for both transmission and reception. A duplexer is used to separate the two functions, both in time and power amplitude domains.

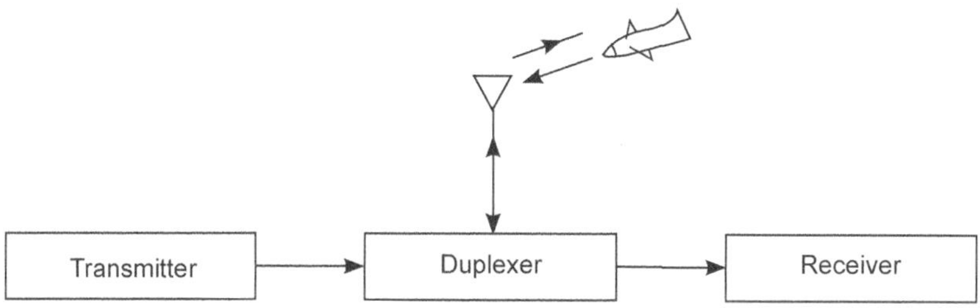

Fig. 12.15

In Bi stable radar system, separate antennas are used for transmission and reception. These antennas may be sometimes miles apart, see Fig. 12.16.

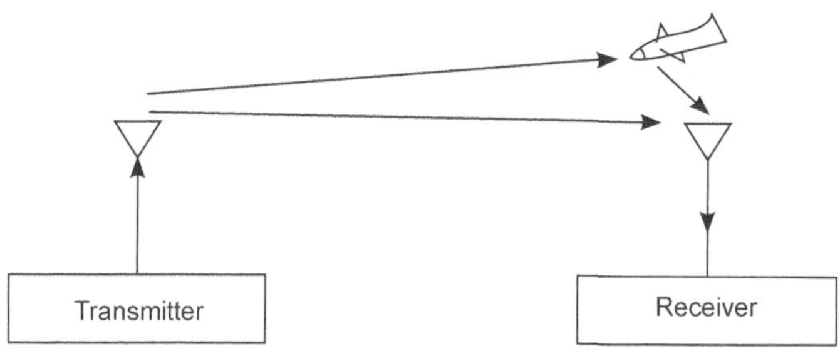

Fig. 12.16 Bi-stable radar system

9. *Duplexer*: A device which disconnects the receiver from the antenna and connects it with the transmitter and vice versa.

10. *Nautical miles* (*Nmi*): a unit of distance and one *nautical mile* (Nmi) is equal to 1852 m, this unit is used in radar systems.

11. *TWT* (*Travelling Wave Tube*): This is a microwave amplifier.

12. *CFA* (*Cross Field Amplifier*): This is also a micro wave power amplifier similar to a magnetron.

13. *Grass*: In radar system, the noise which is not constant in amplitude and position is termed as grass.

SUMMARY

1. Radar stands for "Radio detection and ranging". It detects and gives range (distance) of a plane. It can also follow the enemy planes.
2. If a plane flies too low, it cannot be detected.
3. The radar has a transmitting as well as receiving antenna. The transmitting antenna sends VHF signal which is reflected by the enemy plane. The radar receiver antenna receives the reflected signal and displays the same on the monitor. The time between transmission and reception of signal helps in the calculation of distance (range) of the enemy plane.
4. The various radars are: pulse radar, MTI radar, Beacon radar CW radar etc.
5. The MTI radar is based on the doppler effect which says that the apparent frequency of an e.m. wave depends on the relative motion of source and the observer.
6. The most popular video display unit used for radars is "plane position indicator".

SELF TEST QUESTIONS

1. What do you mean by a "radar"? Give its applications.
2. Explain the principle of its working.
3. How it "detects" and "ranges" a target?
4. Draw block diagram of a simple radar system and explain its important blocks.
5. Classify radars.
6. Explain a "pulse radar"?
7. Explain scanning of a radar antenna.
8. Draw block diagram of a MTI radar system. What is its special significance?
9. Explain a *Beacon*. How it differs from a MTI radar system?
10. Describe an *Altimeter*. What are its limitations?
11. Write a note on "blind speeds" in MTI radars.
12. What are: (*i*) Tracking radar (*ii*) laser radar.

13. Explain the terms: PRR, Blind range, Precision, Mono and Bistatic radars.
14. What are the display devices used for radars?

OBJECTIVE TYPE QUESTIONS

I. Fill in the blanks:

1. Radar stands for _________ .
2. Very high frequency waves are called _________.
3. TDM stands for _________ .
4. VDU means a _________ .
5. Examples of VDU are _________ .
6. PPI means _________ .
7. The MTI radar uses _________ .
8. IFF means _________ .
9. A radar receives an echo 2 ms after the transmission of pulse. The target is at a distance of _________ km.
10. In a radar, rotating antenna is used to determine _________ of the target.
11. The device used to detect objects under water is called _________.
12. The PPI can display range as well as _________ of the target.
13. The radar can determine velocity of the target.
14. The radar that does not use the echo signal is _________ radar.
15. The mono pulse radar is a _________ radar.

ANSWERS

1. Radio detection and ranging
2. microwaves
3. Time Division Multiplexing
4. Video display unit
5. C.R.O., monitor etc.
6. Plane position indicator
7. doppler effect
8. Identification friend or foe (radar)
9. 300
10. bearing
11. Sonar
12. direction
13. MTI
14. surveillence
15. tracking.

II. State True/False:

1. The radars can be avoided.
2. The antenna for transmitter and the receiver may be separate.
3. A duplexer is a two way switch.
4. The radars operate at very low frequencies.
5. All radars employ doppler effect.
6. The radar can identify whether the approaching radar is a friend or enemy.
7. In radar system, the noise is called clutter.
8. In radar systems, rotating antenna is used.
9. The radar cannot determine velocity of the target.
10. All radars use echo for their operation.

ANSWERS

1. True	2. True
3. True	4. False
5. False	6. True
7. True	8. True
9. False	10. False.

❑❑❑

13

Telegraphy and Telephony

Telegraphy and Telephony are very important communication systems. Telegraphy is a system of sending written messages, while Telephony is a system of sending verbal messages.

13.1 TELEGRAPHY

Telegraphy is a communication system of sending a written information/ signal/massage from one place to another.

The massages are expressed with the help of words while the words are made of letters or *characters*. Further the characters are made up of combination of *"signal elements"*.

This representation of characters by elements is known as Telegraphic *codes*.

At the transmitter, the massage is *"coded"* and sent through lines. At the receiver the massage is *"decoded"* to reobtain the original massage.

In *"manual"* telegraphy, a telegraphic *"key"* is operated to code the massage and in *"speed"* telegraphy, codes are formed on perforated paper.

13.1.1 Telegraphy Codes

In telegraphy the codes which are used are two namely:

(*i*) Morse code (*ii*) Five Unit Code.

(a) Morse Code: In Morse code, the letters are represented by "dot and dash". The number of "dot and dash" is different for different letters. The maximum and minimum number of signal elements present in a letter is 4 and 1 respectively and the "duration" of dot or dash is *inverse* of the speed. The morse code is given in the Fig. 13.1

The important points in using the code are:

(*i*) The duration of a dash (—) is *three* times the duration of a dot (.) signal.

(*ii*) The duration of space between signal elements is equal to the time period of a *dot* element. The duration of space between two letters is equal to a *dash* and the duration of the space between two words is four times of a *dot*.

(A) Letters

a · —	n —·
b —· · ·	o — — —
c —· —·	p · — — ·
d —· ·	q — — · —
e ·	r · — ·
f · ·	s · · ·
g — —·	t —
h · · · ·	u · · —
i · ·	v · · · —
j · — — —	w · — —
k —· —	x — · · —
l · —· ·	y — · — —
m — —	z — — · ·

(B) Digits

1. · — — — —	6. — · · · ·
2. · · — — —	7. — — · · ·
3. · · · — —	8. — — — · ·
4. · · · · —	9. — — — — ·
5. · · · · · —	10. — — — —

(C) Punctuation

Full stop	· — · — · —
Comma	— — · · — —

Fig. 13.1

The transmission may be by

(*i*) Single current method

(*ii*) Double current method.

In *single current method*, the current flows through the line in the same direction during the *signal period* only and no current flows during the *space periods*.

In *double current method* the current always flows through the line, and during space periods it flows in opposite direction.

These methods are also called "Two conditions method" as only two conditions are possible on the line either *"current or on current"* or *"positive or negative current"*.

Illustration: If we have to send the word "dc" and we know that

d is formed by dash-dot-dot (— · ·)

and *c* is formed by dash-dot-dash-dot (— · — ·)

(*i*) In *single current method*, duration of pulse for dash is about 3 times of the dot and no current is to flow between *d* and *c*. [See Fig. 13.2 (*a*)]

(*ii*) In *double current method*, the reverse current is to flow between the period of *d* and *c*. [See Fig. 13.2 (*b*)]

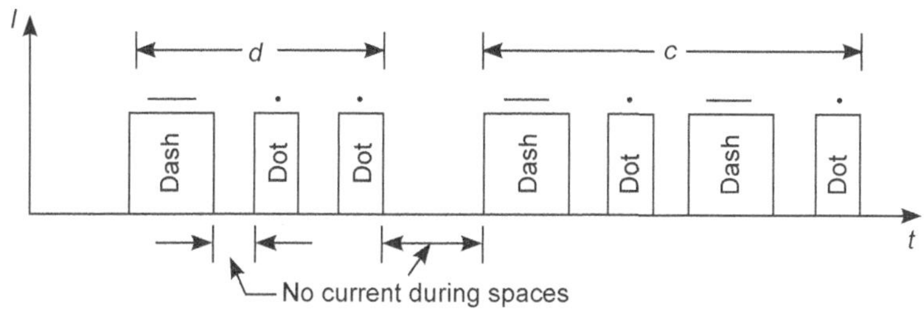

(*a*) Single Current Method

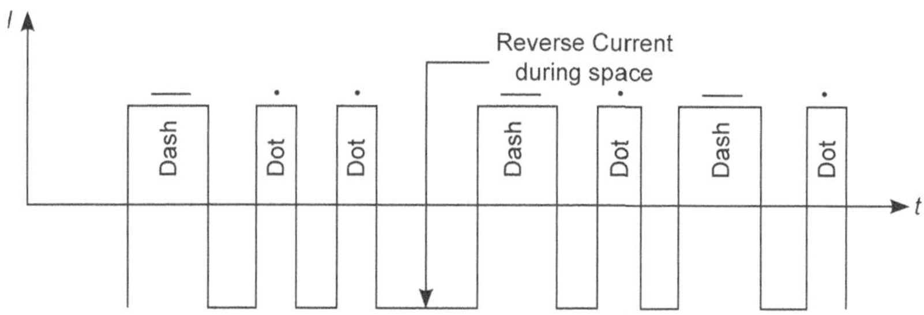

(*b*) Double Current Method

Fig. 13.2

Note: Morse code is now almost absolete and has a limited use.

(b) Five Unit Code: This is used in teleprinters and will be discussed in coming pages.

13.1.2 Punched Paper Transmission

For high speed transmission, the signals are obtained in the form of "holes" or "punches" on a paper. This punched paper is rapidly passed in the transmitter and sent through the lines. For this also *Morse code* is used.

(*i*) *For dot.* two bigger holes on the sides of a small hole.

(*ii*) *For dash.* Two smaller holes on the sides of two bigger holes.

(*iii*) *For space.* A small hole in the centre.

The Fig. 13.3 (*a* & *b*) shows the paper tape punched in the morse code.

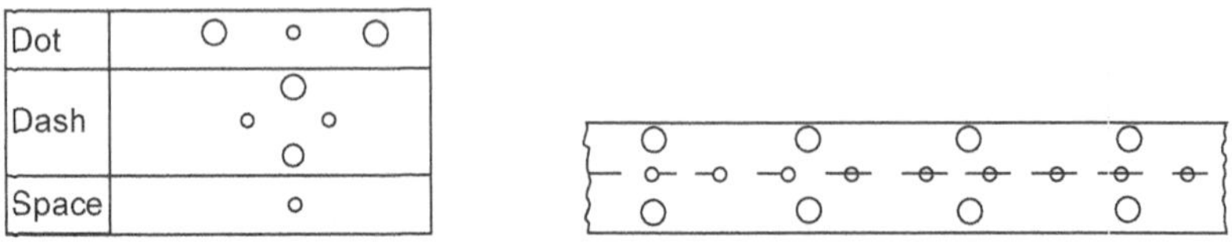

(*a*) Morse code on paper (*b*) Punched paper with morse code

Fig. 13.3

Note: The punched paper transmission is now absolete and has a limited use.

13.1.3 Ordinary Telegraphic System

In a telegraphic communication, following important equipment is used:

1. *Transmitter.* Here signals are formed from the message. In other words, message or information is converted into signal for transmission.

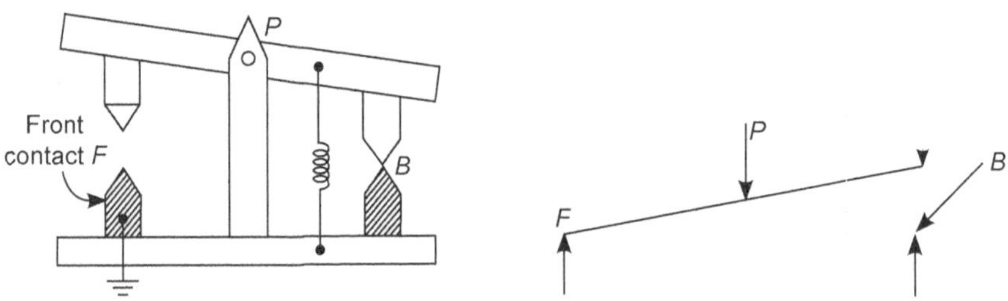

(*a*) Pictorial view (*b*) Symbolic representation

Fig. 13.4

In ordinary telegraphic system, a simple *telegraphic key* (Fig. 13.4) is used as a transmitter. By pressing or releasing the key according to the morse code, any message may be converted into the electric signal. The key may be a *single current* key or a *double current* key.

2. *Transmitting medium.* The lines or wires on poles are the carriers of the signal, which is in the form of electric currents. The wires of galvanised iron are erected on iron poles.

3. *Receiver.* At the receiving stations the signals are received and reconverted into the original massage.

Now: The manual telegraphy is now absolete and has a limited use.

13.1.4 Components of Telecommunication Lines

The tele messages are sent by overhead wires. It has the following components: (Fig. 13.5)

(*i*) *Conductors or wires.* Generally galvanised iron or steel wires are used to carry the telegraph signals. The conductivity of iron is poor but as currents are smaller, which can be easily carried. Also the mechanical strength of iron is sufficient. For ordinary frequencies, one conductor is required as the *earth* acts as the return conductor. At high frequencies, two conductors are required, the second conductor acts as return conductor.

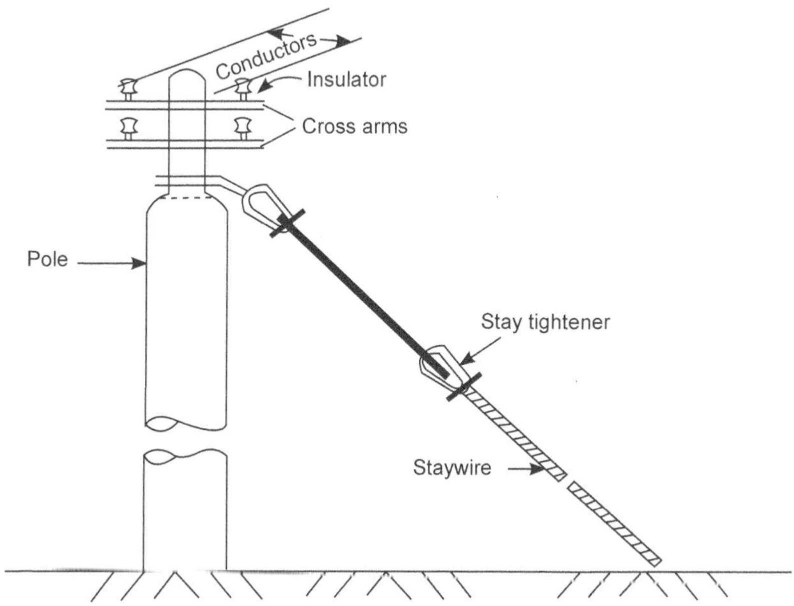

Fig. 13.5

(*ii*) *Poles.* In rural areas, wooden poles may be used. These poles are made of well seasoned teak wood. Further, they are treated against dampness and insects.

Otherwise galvanised iron pole may be used. They are of approximate 8 meters in length. Between poles, a span of about 100 m is used.

(*iii*) *Cross arms.* The poles are fitted with cross arms which are to carry insulators. The cross arms used are of wood or a channel section for wooden and iron pole respectively.

(*iv*) *Stay.* Due to weight of cross arms, and wires etc. the pole may turn down. As a balance force, a stay wire is tightened with the pole in the opposite direction to keep it "erect".

(*v*) *Insulators* (*See Fig.* 13.6). The insulators are fitted on the crossarms to carry conductors. The insulators are made of some insulating material like porcelain and prevent the direct contact of conductor with the cross arms. If wire or conductor comes in direct contact with the pole, the signal currents will also flow through the pole and it will create problems. They are generally pin type insulators, with the help of the pin they are fitted with the cross arms. They are *galzed* to prevent accumulation of water on its surface.

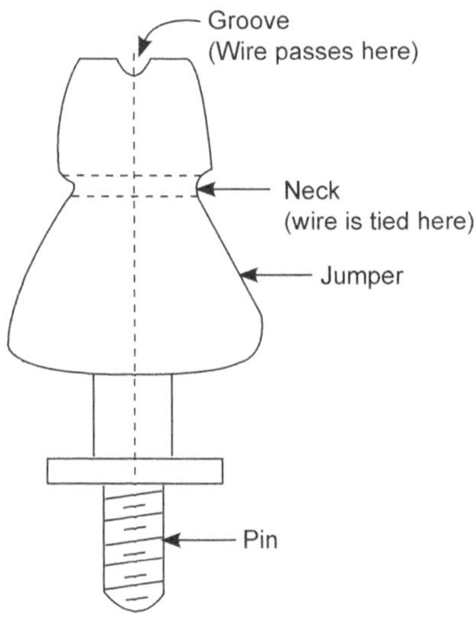

Fig. 13.6

13.1.5 VARIOUS TELEGRAPHY SYSTEMS

For short distances, telegraph messages are transmitted over the line as direct current (dc) pulses. For this purpose, a battery may be used. For long distances however, the dc signals are converted into ac signals at transmitting ends and converted back into dc signals at receiving ends. For sending and receiving telegraphic messages, the following systems are used:

(*i*) *Simplex Working*. This is the system of sending one message at a time through a line *i.e.*, one station "sends" and other station "receives" the message.

(*ii*) *Duplex Working*. This is the system of sending two messages at a time through the same line but in the opposite directions, *i.e.*, both the stations at a time send one message and receive the other. It is further of two types:

(*a*) ***Full duplex*** is a simultaneous two way communication in two direction. [See Figure 13.7 (*a*)].

(*b*) ***Half duplex*** is a two way communication but in only one direction at a time. [See Fig. (*b*)].

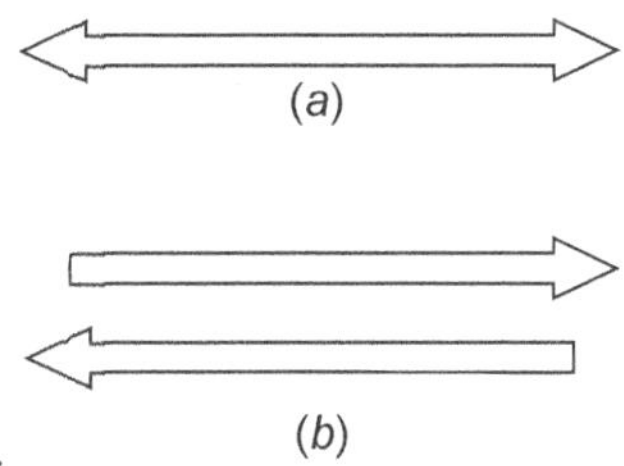

Fig. 13.7

(*iii*) *Diplex Working*. This is the process of sending two messages in the same direction through a line.

(*iv*) *Multiplex Working*. This is the process of sending many messages through the same line. There are however, two important multiplex systems:

(*a*) *Time Division Multiplexing* (*TDM*). In this system, same circuit is used for sending more than one messages on time distribution basis. "Boudot multiplexing" is the example. Several operators share same transmission on a time distribution basis.

(*b*) *Frequency Division Multiplexing* (*FDM*). In this system, same circuit is used for sending more than one messages on frequency distribution basis. The "Voice Frequency Telegraphy" with 24 channels is example. Each signal modulates a particular carrier frequency allotted to the channel.

13.1.6 Carrier Telegraphic System

Now, the old *Duplex* and *Quadruplex* telegraph systems are being replaced by **carrier telegraphy system**. It enables us to send many messages through many number of *channels* simultaneously through the same line. The system is popular as "Multichannel Voice Frequency" (MCVF) system. Generally 6, 12, 18 or 24 channels are simultaneously used. This is worked with *teleprinter* on both sides. The system is more or less similar to radio communication.

(a) Modulation Systems Used: Two types of modulation systems are used in *carrier* telegraphy.

(*i*) Amplitude Modulation (AM) (See Fig. 13.8)

(*ii*) Frequency Modulation (FM) (See Fig. 13.9)

As described earlier, in AM, the amplitude of the carrier is varied according to the amplitude of the telegraph signal and in FM, the frequency of the carrier is varied according to the amplitude of the telegraph signal to be sent. In modulated wave, we get the carrier and sidebands.

If different carriers having different frequencies are used for different telegraph signals; different side bands of the different frequencies would be obtained and thus number of signals can be simultaneously sent over the same telegraph lines. Similarly at the receiver these different signals can be demodulated and detected. This principle is more or less same as of sending *radio signals*.

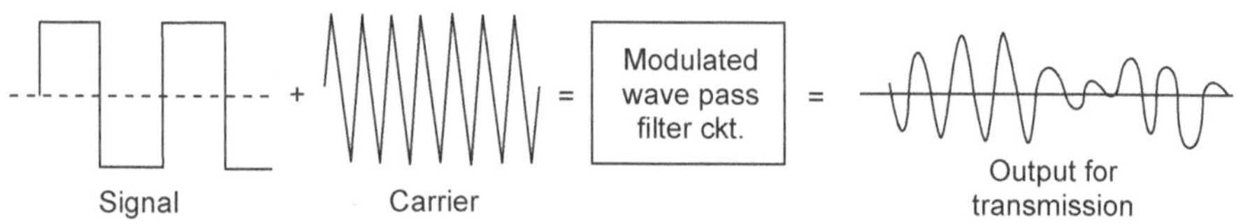

Fig. 13.8

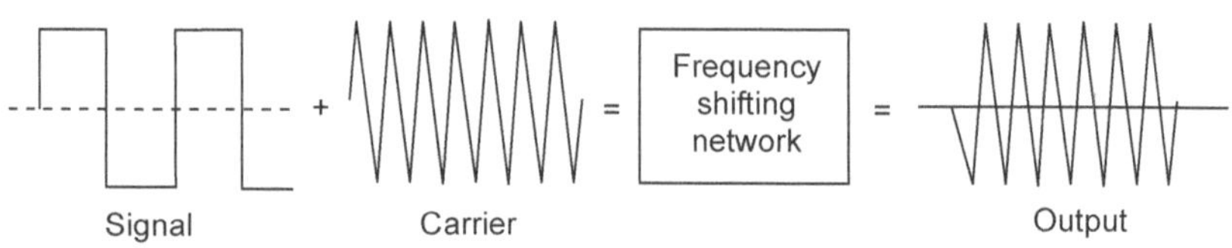

Fig. 13.9

(b) Square Wave Telegraphy: Here it is important that in telegraphy the signal is a *square* wave which besides the fundamental frequencies contains *harmonics also*; therefore a modulated wave should be filtered out of the higher frequency of sidebands and this is what is actually done in telegraphy.

As all the harmonics are filtered out except fundamental frequency, as a result a bandwidth of 60 Hz is required for transmission of the signal and of 120 Hz for side bands. Generally 180 Hz-BW is used per channel transmission.

(c) Multichannel System: The multichannel system uses carrier frequency between 400 Hz to 3 kHz. For higher channel, the carrier frequency also increases, *e.g.* For channel No. 1, this value is 400 Hz, for channel 12 this is 1720 Hz and for channel No. 24, it is 3 kHz.

13.1.7 Picture Telegraphy

The FAX is an example of Picture telegraphy. This is a technique to send pictures from one place to another. For this purpose "photo telegraph machines" are used. The *photo electric* cell is an indispensible component of all these machines.

The principle is more or less same as used in *Television*. The picture is divided into small elements or lines and they are scanned (Fig. 13.10). The scanning is same as in sending TV signals. A picture of normal size is divided into 2×10^5 elements. A photo electric cell generates a "video signal" according to the intensity of the light.

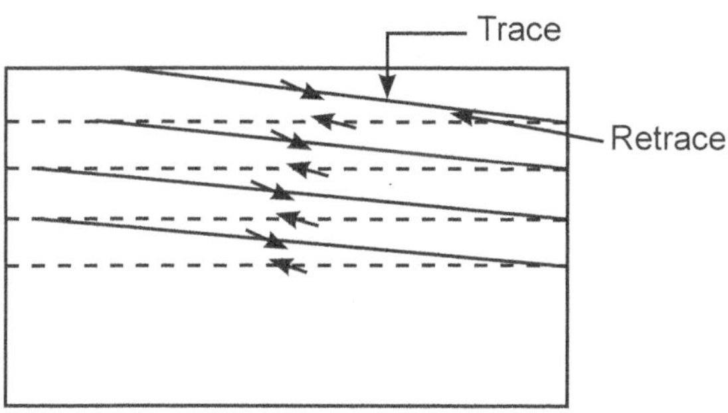

Fig. 13.10

The picture to be scanned is rolled over a drum which is rotated at a determined speed by a motor (Fig. 13.11). The speed of the scanning motor should be same at the transmitting and receiving ends. In other words, the two scanning should be properly synchronised.

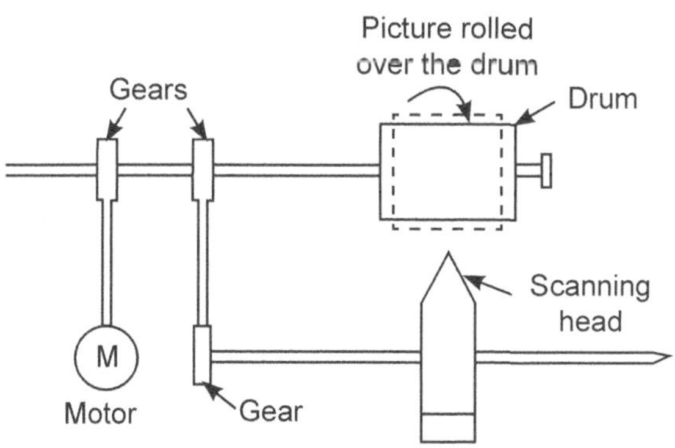

Fig. 13.11

(a) **Picture Telegraph Transmitter (Fig. 13.12):** The picture is rolled over the drum and the scanning head carries the photo electric cell. As a result of scanning, the output of the cell is a *video signal* proportional to the intensity of the light.

The video signal output of the photo cell *modulates* a carrier of suitable frequency and the obtained modulated signal is then amplified.

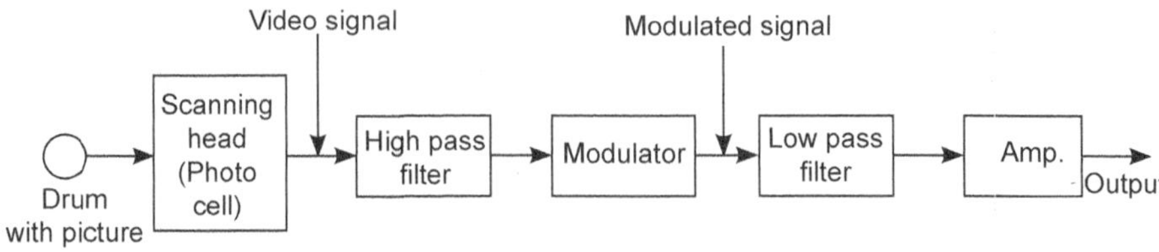

Fig. 13.12

The normal telephone circuits deal with a frequency band of 300–2200 Hz and the carrier frequency used is of 1200 Hz. This carrier is modulated by video signal *frequency* produced as a result of the scanning.

In case of radio telegraphy, the transmission is done by "*Sub Carrier Frequency Modulation*" to keep off the noise. For modulation, 7 kHz frequencies are utilised and the output obtained is of 1400 Hz (for white light) and 2400 Hz (for black light).

(b) **Picture Telegraph Receiver:** The AM signal at the receiving end is first passed through a high pass filter circuit. Its output is amplified and then demodulated. The output of the demodulator is fed to a low pass filter circuit. The output is used to move the oscillograph, which records the message on a photographic film rolled on a drum.

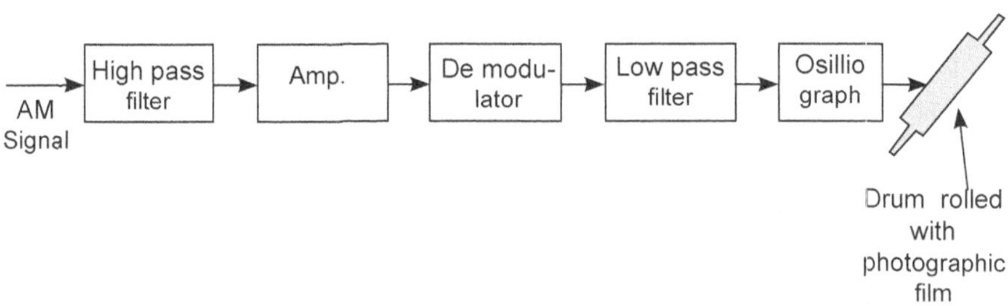

Fig. 13.13

13.1.8 Teleprinter

So long, we have studied manual telegraphy systems. The *teleprinters* are automatic machines for sending and receiving messages. **The "morse key" and**

other manual equipment have been replaced by automatic transmitters and receivers. In this way, we can handle the increasing telegraph "traffic". The teleprinters use "start/stop" telegraphy. The essential components of teleprinters are: Key board, Transmitter, Receiver and Printer.

(a) Five Unit Teleprinter Code (Murray Code): For teleprinter, a (5/7.5) unit code also known as CCITT (Consultative Committee for International Telephone and Telegraph) code is used. In this, each "character" consists of 5 elements.

Where as "Morse" code is uneven and of non uniform length, the teleprinter code is even and uniform. But it is a *binary* code (Binary means two). We have 25 alphabets, 10 numerals and various punctuation marks in English language, therefore on teleprinter key board (like that of a typewriter and computer), we have "letter shift" which when pressed, the "printer" prints only "letters". We have also "figure shift" on the keyboard, which when pressed, the printer prints only the figures (numericals). Though we may have only 32 combinations ($2^5 = 32$) but by the technique explained above, we may have the capacity of the code as almost doubled.

When this code is used in teleprinter, the five units of the code is preceeded by a 'start' (space) and followed by a 'stop' (mark). This is therefore also called as a "start/stop" code. Usually the duration of 'start' and 'stop' is same as that of the 5 code elements thus the total length of the code becomes 7 units. A part of the code is given below.

Letter Shift	Figure shift	Start	Code Element					Stop
			1	2	3	4	5	
A	–	0	1	1	0	0	0	1
B	?	0	1	0	0	1	1	1
C	:	0	0	1	1	1	0	1

(b) Block Diagram of a Teleprinter: The Fig. 13.14 shows various functional blocks involved in a teleprinter circuit connecting two *keyboard send receiver* (KSR) units which are used for direct communication between two operators. The key board and the printer both are mechanically driven from a common motor through a set of shafts.

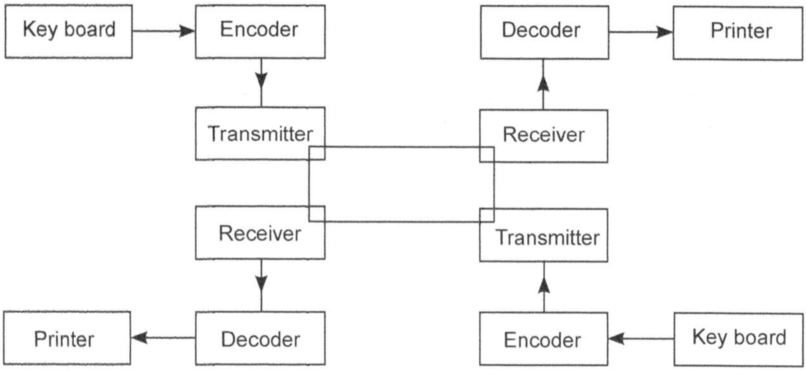

Fig. 13.14

(c) Telegraphy Terms: The important terms related to telegraphy are:

1. *Speed.* This is equal to the reciprocal of the time in seconds used up in the transmission of unit element. It is measured in "BAUDS".

   e.g., A teleprinter operates at 600 letters per minute for a 7.5 unit code.

$$\text{time for one letter} = \frac{1 \text{ minute}}{600} = \frac{60 \text{ seconds}}{600} = 0.1 \text{ sec.}$$

$$\text{time for unit} = \frac{0.1}{7.5} = 0.013 \text{ sec.}$$

   Speed is the reciprocal of 0.013 = 1/0.013=77 Bauds

   or

   The speed in bauds may also be defined as the " inverse" of the period of the shortest information unit (which corresponds to 1 bit) used in the code transmission. The BAUD is thus the maximum information rate of the code transmission *i.e.*,

$$\text{speed} = \frac{1}{T} \text{ Baud.}$$

   where T is the duration of shortest information unit in minutes.

   The digit 0 or 1 is called a "bit". The digital communication is carried out through bits.

2. *Distortion.* Ideally the received signal should be the "exact replica" of the sent signal. When there is a difference between the two, distortion is said to have occured.

   Distortion is defined as the maximum difference between any two instants of modulation. Distortion is always expressed in *percentage*.

3. *Margin*. The margin of a teleprinter is defined as the percentage of distortion which is tolerable. This should not be more than 30%.

13.2 TELEPHONY

Alexander Graham Bell

The word "telephony" is derived from the greek word "tele" meaning distant and "phone" means moving sound.

Telephony is a technique of sending verbal massages from one place to other.

Telephone was invented by Alexander Grahm Bell.

13.2.1 A Telephone Hand Set

A telephone hand set consists of two major parts – *transmitter* and the *receiver.* Both are contained in a handy bakelite piece, which can be taken in hand for use.

Its transmitter is nothing but a *microphone* before which the person speaks. The microphone converts his speech into electrical signal. The receiver is a *loud speaker* and converts the electrical signal back into original speech, therefore, it is kept near the ear. The human voice has a frequency of about 3.5 kHz, therefore, the mike and speaker of the telephone should be able to handle these frequencies.

It has following components:

(*i*) *Transmitter or microphone* (Fig. 13.15). The transmitter is an improved form of the carbon microphone. It is *immersed* type as the carbon electrodes are totally immersed in carbon granules from all sides. As different persons handle it in different

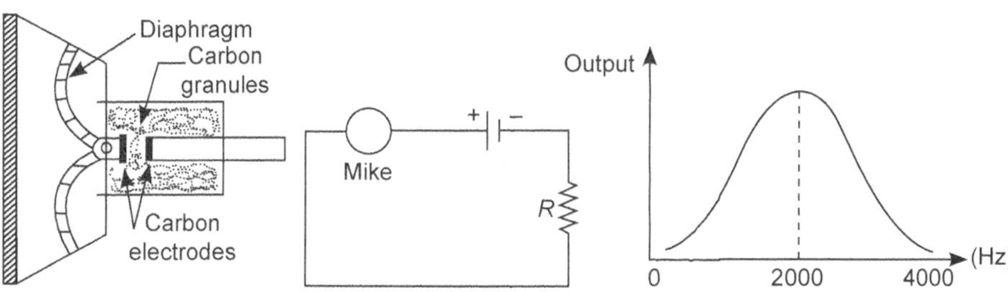

(*a*) Construction of mike (*b*) Connection (*c*) Frequency response

Fig. 13.15

way while using, this ensures uniform pressure on the electrodes in all positions thus variation of dc resistance is also negligible.

The electrodes are also made of carbon and the diaphragm is of light aluminium alloy called *duralumin* (Al + Cu + Mg), thus weight of the moving parts is very less and it increases sensitivity of the instrument.

The frequency response curve of this microphone is also better.

(*ii*) *Telephone receiver*: The telephone receiver used in a hand telephone is shown in the Fig. 13.16. It consists of a permanent magnet with two pole pieces on which coils are put with large number of turns. The electrical signal currents received from the transmitter flow through these coils. A diaphragm is put in front of these coils. The diaphragm moves to and fro according to the signal strength and thus original sound can be heard. The whole arrangement is contained in a bakelite body which is covered with an ear cap with holes so that the sound may come out.

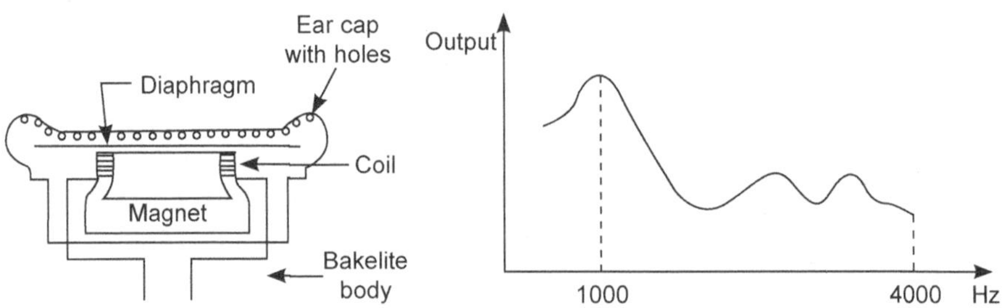

(*a*) Construction (*b*) Frequency response curve

Fig. 13.16

The frequency response curve for the receiver is shown in Fig. (*b*). It has a peak value for the output at about 1 kHz.

(*iii*) *Connection with lines* (Fig. 13.17). The telephone transmitter and receiver are connected with the lines through transformers.

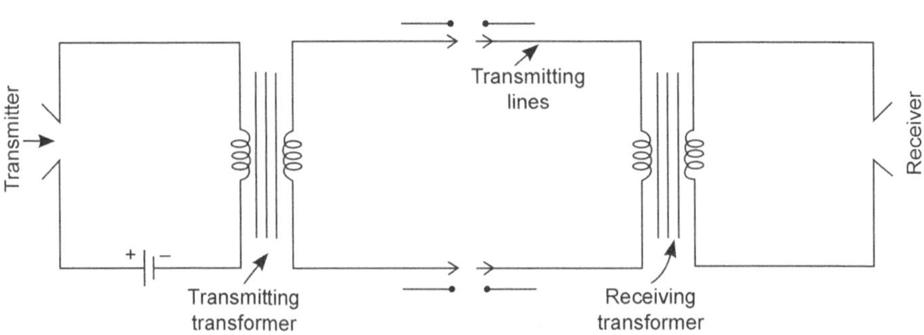

Fig. 13.17

Impedence matching is to be kept in mind for good results. Thus *one* impedence matching is needed between transmitter and the lines and the *second* in between line and the receiver.

(*iv*) *Dial* (Fig. 13.18). A dial on the telephone's face is an important device as this produces the *train of impulses.* There is a *Finger plate* with 10 holes, fixed on the main spindle. At the backside is a *number ring* with numbers, 1 to 10 (1, 2, 3, 4, 5, 6, 7, 8, 9 and 0). When a number is to be dialled, the finger is inserted into a particular hole of that number and rotated upto the *finger stop* and released. When the dial returns, it sends required number *of pulses.* Pulses of any number between 0 to 10 can be produced, the frequency being 10 pulse per second *e.g.,* when we dial "7", on its return journey a "train of 7 pulses", is produced by the ring.

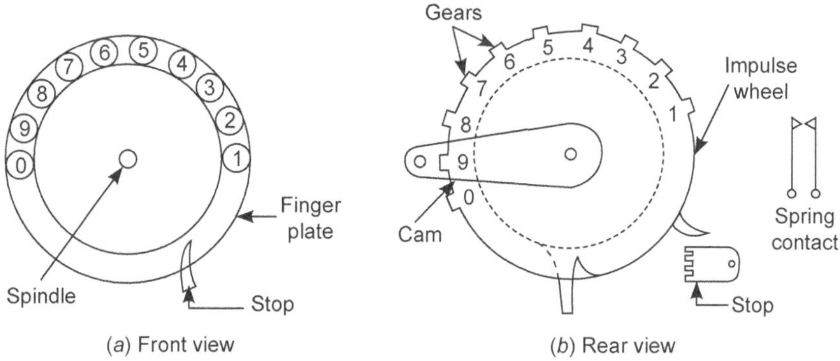

Fig. 13.18

At the rearside of the dial, there is also an *impulse wheel* with teeth. If we dial '5', It means 5 slots will pass beyond the *spring* contact and during return of the dial, these slots will operate the contact *five* times. Thus *make and break* of the contacts will produce a train of 5 pulses per second.

The nature of impulse currents produced by the dial is also shown (See Fig. 13.19). If t_1 is the time for which no current flows and t_2 is the time for which the flow of current takes place, the frequency will be equal to

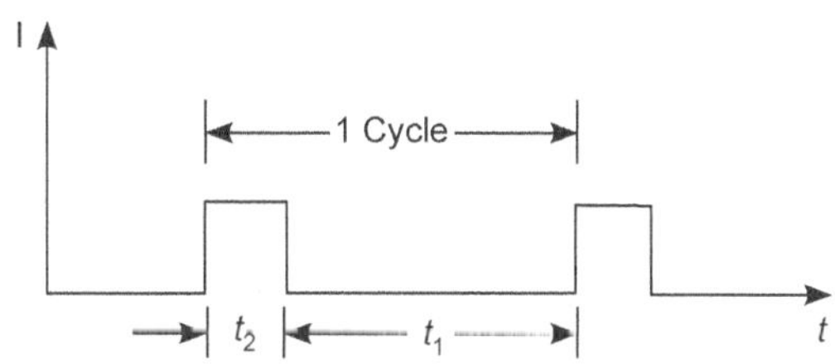

Fig. 13.19

$$f = \frac{1}{t_1 + t_2}$$

Generally, $t_1 = 90$ ms and $t_2 = 10$ ms.

Note: Now, this type of dial is disolute and a dial with digits is used and particular digit is pressed by a finger.

13.2.2 Telephone Lines

The Telephone lines are of three types:

(*a*) For telephony, same lines are used as for the telegraphy.

The power lines (50 Hz) going near by, can create an interference in telephone wires but this can be eliminated by *transposing* the telephone wires. The Fig. 13.20 (*a*) shows how interference can be picked up by telephone lines from power lines through inductive (L) and capacitive (C) couplings. The Fig. (*b*) shows *transposition* of the telephone wires. By transposing the telephone wires, the constants (resistance, inductance and capacitance) of the wires become uniformly distributed over the entire length of the line and therefore, interference is reduced.

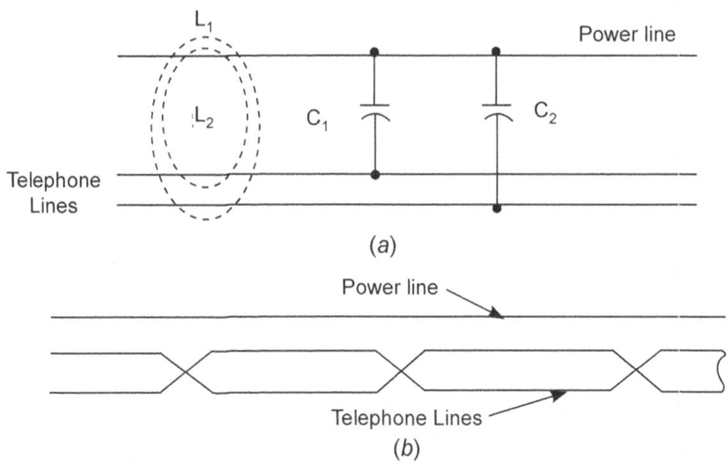

Fig. 13.20

(*b*) Where open lines are not feasible, wires are formed into multicore cable assembly which is usually carried *underground* in ducts. These wires are made of annealed (soft) copper, which is flexible, and each strand is insulated with high grade paper tape. A lead sheath is then given as an outermost covering.

Now, optical fibres are preferred.

(*c*) For high frequency signals (*e.g.*, carrier telephony and television), co-axial cables are used.

13.2.3 Electronic Telephone

The hand telephone set described above makes use of electro magnetic and electro-mechanical devices. They suffer from the following drawbacks:

1. They are delayed devices and take time to actuate and operation.
2. They are heavy and equipment become even heavier. They take more floor space, whereas the world is advancing towards *minitiarisation.*
3. Electromagnetic and electromechanical devices need periodic maintenance.
4. Due to wear and tear, their life is short.

Therefore, we are nowadays using electronic telephone set employing *electronic switches.*

(*i*) Electronic switches can make ON and OFF a circuit very fast. Their frequency of operation may be 10^9 per second.

(*ii*) They are light in weight, free of maintenance and very accurate in operation.

(*iii*) They have long life with no wear and tear and are more efficient.

The important electronic switches used in electronic telephones are briefly described here

1. *Vaccum diode or diode (Fig. 13.21).* When the cathode is more negative to anode, the vacuum diode conducts current. However when the anode is made more negative w.r.t. cathode, the diode does not conduct. In other words, in the first condition, it is a closed switch with negligible resistance and in the later condition, it is an open switch offering a very high resistance (in megaohms).

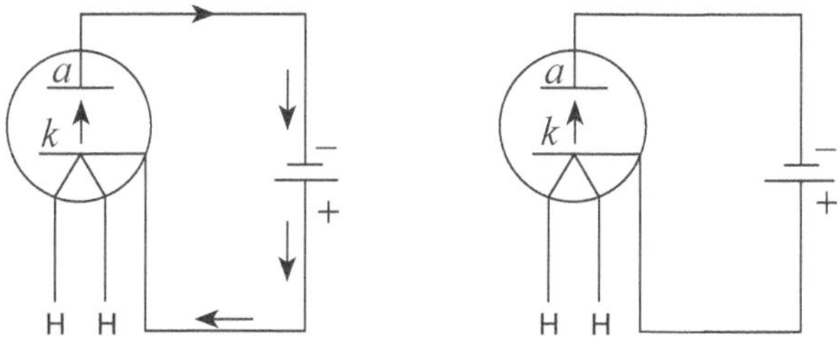

(a) Tube is on (b) Tube is off (No current)

Fig. 13.21

2. *Semiconductor diode* (Fig. 13.22) A semi conductor diode allows current in Forward bias (F.B.) state only i.e. diode is ON. In reverse bias (R.B.) condition, the diode is off.

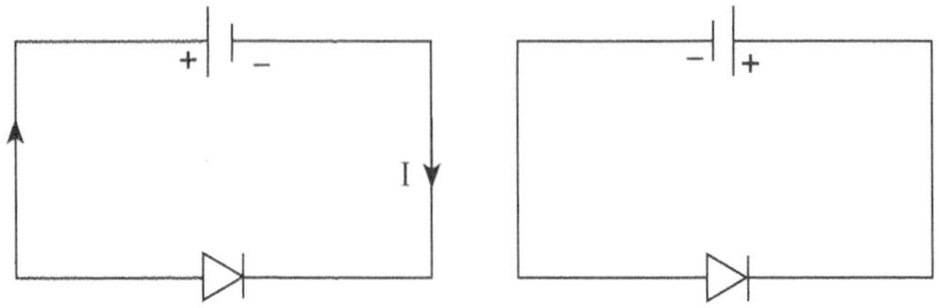

(a) Diode is on (*b*) Diode is off (No current)

Fig. 13.22

The merit point in favour of semi conductor diode is that it does not need any filament supply. It has smaller size and smaller biasing is required by them as compared to the tubes.

3. *Transistors* (Fig. 13.23). When a transistor (NPN or PNP) is FB at its input junction and RB at its output junction it acts as a very good switch. If the input junction is made R.B., the transistor becomes off. (Fig. a & b)

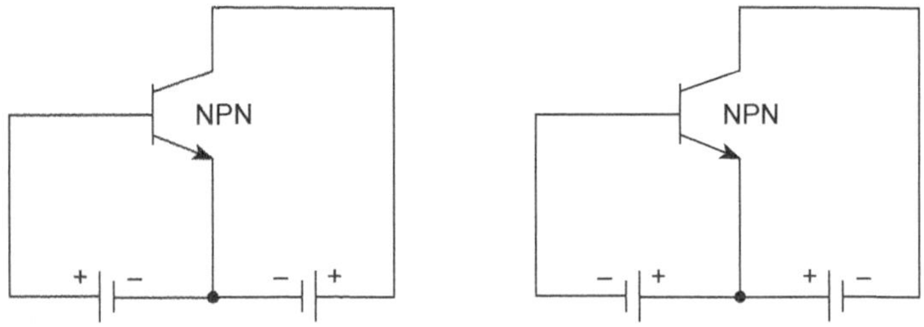

(*a*) Transistor is on (*b*) Transistor is off

Fig. 13.23

There are many other electronic switches like SCR etc.

13.2.4 Electronic Exchange

The electronic telephone exchanges generally use the following systems:

(a) Time Division Multiplexing (TDM) (Fig. 13.24): If a speech wave is *sampled* at a frequency greater than the highest frequency present in the speech, we get a sample wave as shown in which the speech signal is present and same may be obtained at the receiver.

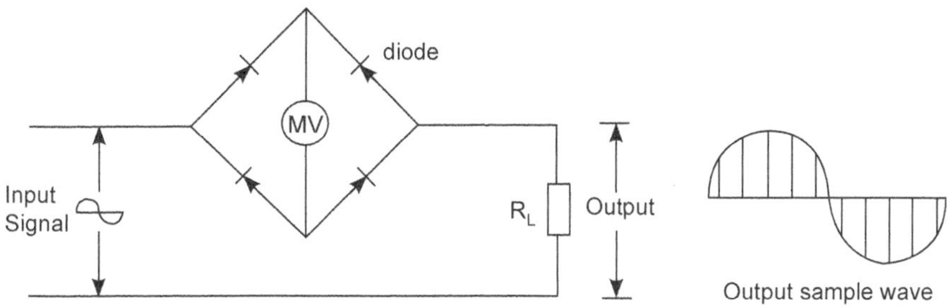

Fig. 13.24

For generating such a wave at the transmitter, a square wave generator (Multi-vibrator MV) producing 10 kHz carrier wave with a bridge rectifier circuit is used. Across the load R_L, the sampling wave is obtained. At the receiver the same circuit detects the sample wave across its load, provided both the circuits (at transmitter as well as at receiver) work in *synchronism*. If two persons are provided with such a circuit, they can hear each other. Moreover, a number of signals can pass through the same transmission line.

(b) **Frequency Division Multiplexing (FDM):** This is also known as **carrier** telephone system. In this system, a carrier frequency is modulated by the speech signal and as a *modulated wave*, we get original carrier along with sidebands. Only the side bands which carry the signal are sent. At the receiver the original speech signal is detected or **demodulated**. By using different carrier frequencies, we get different side bands of different frequencies and, therefore, any number of signals can be transmitted simultaneously through the same lines. Similarly, the signals can be received at the receivers by using different filter circuits.

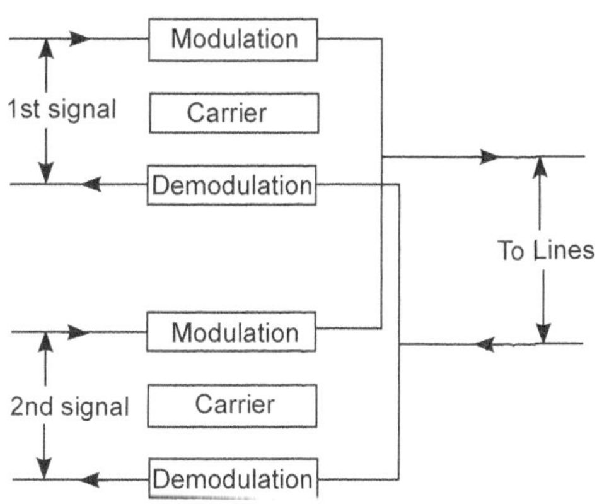

Fig. 13.25

The Fig. 13.25. shows simplified diagram of an electronic telephone exchange using FDM, only two signals have been shown but any number of signals can be sent through a pair of lines.

(c) Telephone Modulation System (Fig. 13.26): Speech produces much more complex modulation pattern as compared to other signals but the principle of *amplitude modulation* in telephony remains the same.

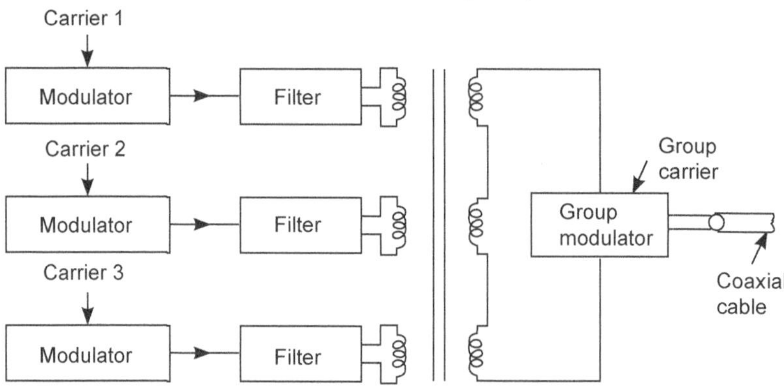

Fig. 13.26

By having a number of carrier frequencies, telephonic conversations can be "stacked" one above the other in terms of frequencies and, therefore, takes place without interfering with one another. Upto 300 simultaneous conversations may be carried out in this way over a single pair of cable.

13.2.5 PLCC (Power Line Carrier Communication)

Whenever some fault or a breakdown occurs in a power line, the message to rectify the fault is to be sent to the substation. The matter being of utmost importance needs quickness in sending the message. The ordinary telephone system remains almost engaged, therefore, that cannot be used. So, the power lines are used for this purpose. The message is modulated by IF carriers and sent as radio waves through the power lines. This is the reason that the system is known as "power line carrier communication" (PLCC) system.

The carriers to modulate the message are of 50 to 500 kHz. This modulated signal is amplified and transmitted over the power lines of above 66 kV. At the receiving end, the signal is separated. It is to be noted that the signal received has exactly the same frequency as the original signal transmitted.

(a) Advantages & Disadvantages of PLCC:

Advantages

1. It is cheaper, as the same power lines are used for telephone purposes and separate telephone lines are not to be laid.
2. It is more reliable, as the power lines are mechanically stronger than the telephone lines. The power lines are made of high grade copper, where as telephone lines are made of iron, which may be corroded in unfavourable weather conditions.
3. It has short length, as the power lines provide shortest route between the power stations. Because they can run on high towers and can cross fields, rivers, railway lines etc.
4. It causes less attenuation of the signal, as power lines have very low resistivity.
5. It causes less leakage of power between conductors and ground.
6. The large spacing between the power lines reduce cross talk.

Disadvantages

1. The power lines carry high voltages and currents, proper protection to the PLCC equipment is must.
2. The power communication suffers from noise, as problems of corona etc., exist on the lines.

   In fine weather, the phenomenon of production of arc/flash along with the hissing sound between high voltage power lines is called *Corona*.

(b) Components of a PLCC System: The important components of a PLCC systems are:

1. Coupling capacitor
2. Wave trap
3. Carrier set.
4. Coupling filter circuit/LTU

1. *Coupling Capacitor*. It is of very large size and therefore, mounted on a pedestal outdoor foundation. It is filled with oil which offers low impedance to carrier frequencies. It is protected against high voltage surges and also against grounding. The value of capacitor is in between 2000 pF to 10,000 pF and voltage rating from 33 kV to 400 kV.

The protective devices like spark gap, ground switch etc., are provided in the base of the coupling capacitor.

2. *Wave Trap*. (Fig. 13.27). They have coils of inductance of 2 mH to 2H which carry full line currents. The wave traps are connected in series with the power line and have high impedance to carrier frequencies. In this way, they "trap" the carrier but allow power frequencies (50 Hz) to flow through them. They have inductance and one or more capacitors in parallel and hence can be "resonated" to a particular frequency. The tuning (resonant) circuit makes it suitable to be used on more than one channel. They also contain lightening arrester as a protection against surges. The wave traps are generally hanging from the line.

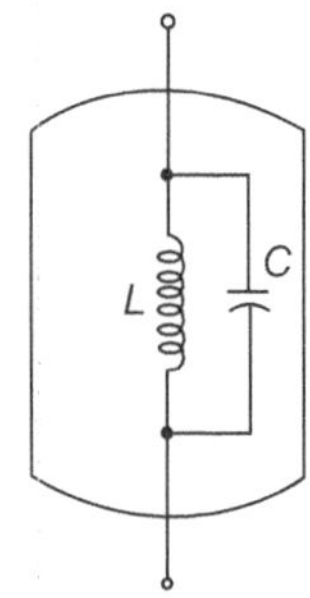

Fig. 13.27

3. *The Carrier Set* (*Terminal*). This is the "terminal", where speech signal (message) is modulated with the carrier. The demodulation is also carried out in this terminal. In other words, it contains transmitter and receiver of the system.

4. *Coupling Filter*. This facilitates feeding of modulated output to the coupling capacitor. It is also called as "line tuning unit" (LTU). The Fig. 13.28 shows one channel being fed to the power line. The variable inductance L in series with coupling capacitor C forms a resonant circuit which passes carrier frequency. The matching transformer is used for impedance matching between co-axial cable and the coupling filter circuit.

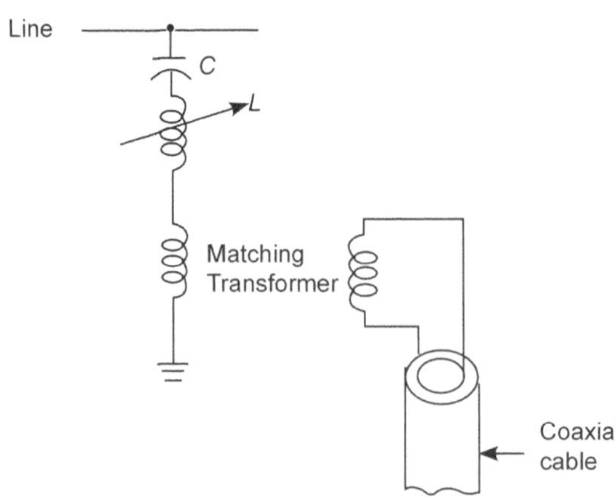

Fig. 13.28

The Fig. 13.29 shows PLCC system.

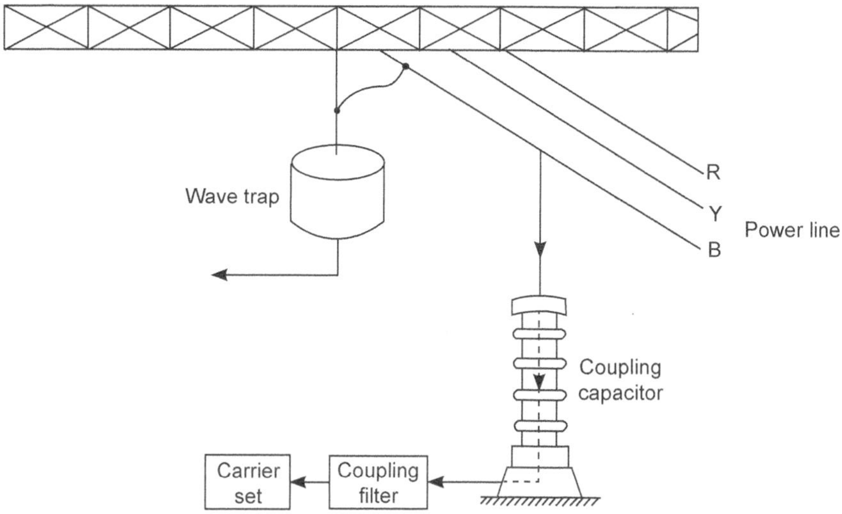

(*b*)

Fig. 13.29

Note:

1. The telegraph and telephone lines basically are the medium frequency lines. The voice massages (conversation) are transmitted between 250 Hz to 2800 Hz and radio massages over telephone lines are transmitted at between 50 to 7200 Hz frequencies.
2. The telegraph and telephone lines may be "open wires" running on poles with insulators or cables run overhead, underground or laid in the ocean (ocean cables) for overseas communications. The cables have low inductance but high capacitance. The cables can handle frequencies upto 2800 MHz.
3. Some telegraph systems use earth as one side of the circuit, while other use a complete metallic path (pair of wires) for the purpose.
4. By carrier communication, many telephone massages can be sent through telephone lines at a time. Usually amplitude modulation is used and the carrier frequency used for this purpose is between 5 kHz to 150 kHz.
5. As mentioned above, under ground cables can handle frequencies in MHz. Now co-axial cables are developed which enable telecommunication in MHz and that too can handle no. of channels over one line. However "repeaters" (amplifiers) are to be used at short intervals along the route.

SUMMARY

1. The telegraphy is a communication system of sending massage/ document from one place to other. The line telegraphy uses two codes : Morse code and Five unit code.
2. A line consists of poles, insulators galvanised steel wires etc.
3. The carrier telegraphy employs "amplitude modulation" or "frequency modulation".
4. Picture telegraphy works almost on the principle of Television.
5. Teleprinters are automatic machines for sending and receiving massages.
6. Fax or Fascimile is a technique of sending printed massages or documents from one place to other.
7. Telephony is a technique to send and receive audio signals.
8. Today, we have electronic telephony. The electronic telephone exchanges employ T.D.M (Time division multiplexing) and FDM (Frequency division multiplexing) for sending messages.
9. The PLCC (Power Line Carrier Communication) is a technique of sending telephonic messages through high tension (66 KV and above) power lines.

SELF TEST QUESTIONS

1. What is telegraphy and telephony. Describe "Morse code" .
2. Briefly describe: telegraphic line, supports and insulators.
3. Describe carrier telegraphy and telephony.
4. Write short notes on Teleprinter.
5. Write working of a hand telephone set.
6. Describe few electronic components and their use in electronic telephony.
7. What do you mean by TDM and FDM ?
8. Describe a PLCC system.

OBJECTIVE TYPE QUESTIONS

I. Fill in the blanks:

1. Ordinary telegraphy and telephony are examples of _________ communication.
2. The popular code used for telegraphy is a _________ code.
3. M.C.V.F. stands for _________.
4. For telephony _________ cable is used.
5. Picture telephony and telegraphy are examples of _________ communication.
6. FAX stands for _________.
7. On a telephone dial, if we deal '5', it will produce _________.
8. F.D.M. means _________.
9. PLCC stands for _________.

ANSWERS

1. wire
2. morse
3. multiplex carrier on voice frequency
4. co-axial
5. wireless
6. fascimile
7. a train of 5 pulses
8. frequency division multiplexing
9. Power line carrier communication.

❑❑❑

14

Satellite Communication

A satellite is a station in space. The "communication" satellites, receive signals from earth, amplify and return back the same to earth. The communication satellite is stationed at a distance of 35800 km from earth and completes one revolution in 24 hrs (as the earth does), such a satellite appears "Geo stationary" or "stationary'.

Many countries have stationed their satellites in space. We have also set up a series of Satellites and are using them for transmission of telephone, TV and fax signals.

A satellite is basically a repeater (amplifier) station, as long distance signals cannot reach to the users, unless they are amplified at regular intervals. This is, what actually a satellite does, *i.e.* it receives a signal, amplifies and after processing, retransmits it.

14.1 SATELLITES

A satellite as already pointed out is an "off surface" earth repeater station. The function to be performed by a satellite varies from type to type, but almost all satellites have the capability to receive a signal, process and transmit it again. However, from operational point of view, satellites can be of the following categories:

1. *Observatory Satellites*. These satellites are launched into a nearby elliptical earth orbit. They are meant for gathering information about the earth's crust for hidden minerals, melting of glaciers, onset of rains, cyclones, earthquakes etc. The *Indian remote sensing (IRS) satellite* is an example of such a satellite.

Other important function of these satellites include "remote sensing". Satellites are used by some advance countries for their defence as well.

2. *Communication Satellites*. A communication satellite does not have the capacity of generating its own signal. It receives a signal from earth, processes it and transmits again to the earth. It may be mentioned that the signal in "up link" (upward) journey carries the same information as the one in "down link" (downward) journey. These satellites are generally launched into an orbit at a distance of 35800 km from the surface of earth. This satellite is also called "Geostationary" because, it completes one revolution around the earth in the same time as the earth itself *i.e.*, in 24 hrs. Because of the "speed synchronism", the satellite appears to be stationary at all times to an observer on earth. This gives it the name as "Geostationary" (stationary w.r.t. an observer on earth).

   Such satellites work on solar energy and are controlled through computers at the controlling centres on the earth.

   As we know that moon is a natural satellite of earth. This paved the way for the man made satellites. This is based on Newton's theory about gravitational forces towards the earth and motion of moon around earth. The Geostationary satellite rotates in *unison* with the earth, due to which it always remains at the same place above earth, where it was placed.

   A communication satellite carries microwave receiving and transmitting equipment in order to relay signals from one point to the other on the earth. Satellites are placed above *Ionosphere*, therefore, microwave frequencies must be used to penetrate the ionosphere. Similarly, microwave frequencies are also required to handle the wideband signals used in present day communication.

14.1.1 Satellite Communication Orbits

A communication satellite orbit may be of the following types:

1. Equatorial
2. Polar
3. Inclined

The Fig. 14.1 shows (*a*) polar and (*b*) "inclined" and (c) equatorial orbit.

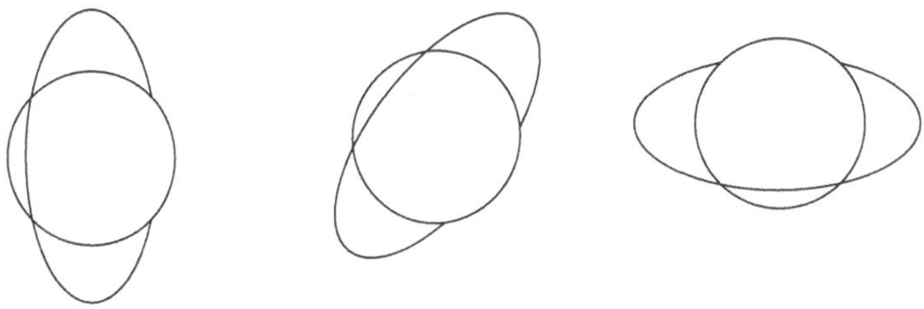

(*a*) (*b*) (*c*)

Fig. 14.1

The *'polar'* orbits are not used because majority of the world population would then remain uncovered, similarly " Inclined" orbits are also not popular for the same reason.

'Equatorial' orbits are most commonly used. All *INTELSATS* (International telecommunication satellites) use these orbits due to following advantages:

(*i*) they give uninterrupted communication.

(*ii*) Only 3 satellites (displaced at 120°) are needed for full global coverage.

(*iii*) Very limited earth station tracking is required.

However, they,

(*i*) need greater launching capability.

(*ii*) give no coverage to polar regions.

(*iii*) have a one way delay of about 300 milliseconds.

Note : Other orbits in use are

1. Elliptical
2. Circular

14.2 SATELLITE COMMUNICATION TERMS

We can define "satellite communication" as a process to send a signal from "earth" to a "communication satellite" located in space and receive back the same signal to earth to utilise for various purposes.

The important terms related to the satellite communication are following:

1. *Apogee*. See Fig. 14.2 This is the *highest* point attained by a satellite moving round the earth.

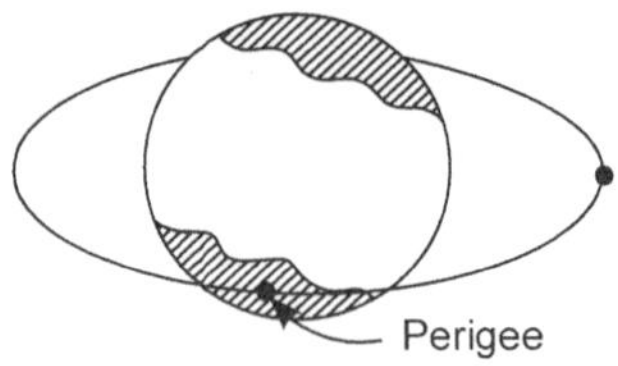

Fig. 14.2

2. *Perigee*. See Fig. 14.2. This is the *lowest* point attained by a satellite moving round the earth.

   The APOGEE and PERIGEE have relevance only, if the satellite has an elliptical path. In case of circular orbits, both are equal.

3. *Geosynchronous*. By this term, we mean that the satellite is moving round the *earth* in synchronism *i.e.*, both complete one rotation in 24 hrs. A geosynchronous satellite is always *geostationary*.

4. *Orbital Velocity* (v). It is the velocity with which the satellite moves in its assigned orbit. It is given by

   $$v = R\sqrt{g/R + H}$$

   where
   R = radius of earth = 6400 km
   H = height of satellite = 35800 km
   g = acceleration due to gravity = 9.81 m/s.

5. *Escape Velocity*. If we throw a stone vertically upwards, it falls back to the surface of the earth after attaining a height depending upon the velocity of projection. If the stone is given a sufficiently large velocity it may escape the gravitational pull of the earth and may never return to the earth.

   The minimum velocity with which a body must be projected in the atmosphere so as to enable it to just overcome the gravitational pull, is known as escape velocity.

   The escape velocity is given by

   $$v = \sqrt{2\,gR}$$

   where
   R = radius of earth,
   g = 9.81 m/s.

   Like orbital velocity, escape velocity is also independent of the mass of the body to be projected.

The escape velocity at earth is 11.2 km/s (taking radius of earth = 6400 km).

6. *Payload:* This is another name for the satellite. In fact, the complete system comprises of a rocket (SLV) and a satellite. The satellite which is later separated from the SLV is called "Payload". The SLV stands for "Satellite Launch Vehicle" *i.e.,* a rocket.
7. *Satellite Launch Vehicle* (SLV). This is the rocket, that carries the satellites. After it is fired, it goes up and puts the satellite into orbit.
8. *Figure of Merit* (FOM). The "Figure of merit" of on earth station is expressed by 10 log (G/T), where *G* is the gain of the receiving antenna of the earth station and *T* is the noise temperature of the receiving system.

   The FOM is the direct indicator of the overall performance of a satellite system.
9. *EIRP:* The EIRP stands for *effective isotropic* radiated power. This is the actual power radiated by the antenna of a satellite multiplied by its " Isotropic gain".

   Note that an *"Isotropic"* antenna radiates energy uniformly in all directions. Alternately, gain of an isotropic antenna is called *"Isotropic gain".*
10. *C/N Ratio:* This is meant by carrier/noise ratio. It is equal to the ratio of the carrier signal power of the receiving antenna to the noise power.
11. *Noise Figure or (Noise Factor):* The "noise figure or factor"(NF) is defined as the ratio of the signal/noise power supplied to the input terminals to the signal/noise power received at the output terminals thus,

$$\text{NF} = \frac{\text{S/N input}}{\text{S/N output}}$$

   The typical value of noise figure is around 15 dB.
12. *Noise Temperature:* The noise power produced by a resistor is directly proportional to the temperature and is independent of the resistance value. The power is specified in terms of noise temperature and denoted by Kelvin, the typical value being 175 K.
13. *Gain/Noise Temperature Ratio:* The gain/noise temperature ratio is the ratio of the antenna gain at 6 GHz to the receiving system noise temperature.

14.3 SATELLITE ANTENNA

The gain of satellite antenna has to be compatible with the earth coverage required by satellite transmission. See table 14.1

Table 14.1 Beamwidth and Size (dia) of Antenna

Coverage	Beamwidth	Size (Diameter) of Antenna dish in cm (at 6 GHz)
1. Global coverage	18°	23
2. Continental coverage	4°	100
3. National coverage	1°	415

14.4 DOMSAT

The DOMSAT is an acronym for "Domestic Satellites (system). In this system, the geostationary satellites are used for domestic communication. In India, **INSAT** series satellites are a part of domsat. It has helped in achieving nation wide TV broadcast in the country.

The Fig. 14.3 shows the operation of DOMSAT. Like other satellite communication setups, in DOMSAT also, the signal from the transmitting station is beamed up at the satellite which in turn retransmits it to a number

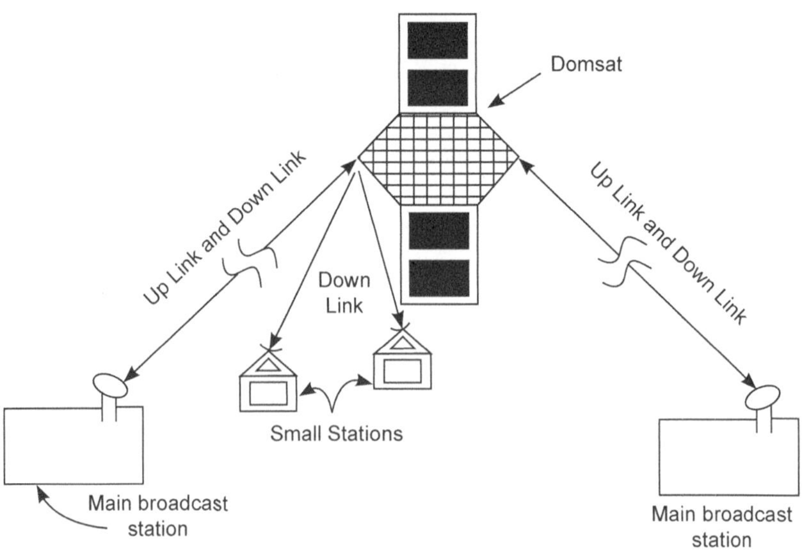

Fig. 14.3

of small stations lying within its coverage. In a set up like this, there may be as many as 100 small stations and only a few main broadcasting stations. It may also be mentioned that the main broadcast stations have both transmitting as well as receiving antenna, whereas the small stations have got only the receiving facilities.

A domsat has a much smaller coverage, while "Intelset" must be accessible roughly to l/3rd of the world with its antenna having a coverage of about 170 million square kilometers. On the other hand, a *circular antenna* having a radius of 1490 km would cover entire India.

14.5 INTELSAT

The *"Intelsat"* stands for "International tele-communication Satellite". These satellites are meant for global communication. The Fig. 14.4 shows a typical "set up" for demonstrating communication from one country to the another. The signal from the main broadcast station, where the program is generated reaches the earth station nearer to the broadcast station and also the one lying within the satellite coverage *via* a chain of microwave links. It may be mentioned that the broadcast station may or may not lie within the satellite coverage. The signal from the earth station is beamed upto the satellite. The satellite retransmits the signal down towards the earth. It is then picked up by the interested earth station on the other side. The signal then finds its way through a similar chain of microwave links to the main broadcast station from where it is retransmitted either through cables or on a "wireless" media for the interested subscribers.

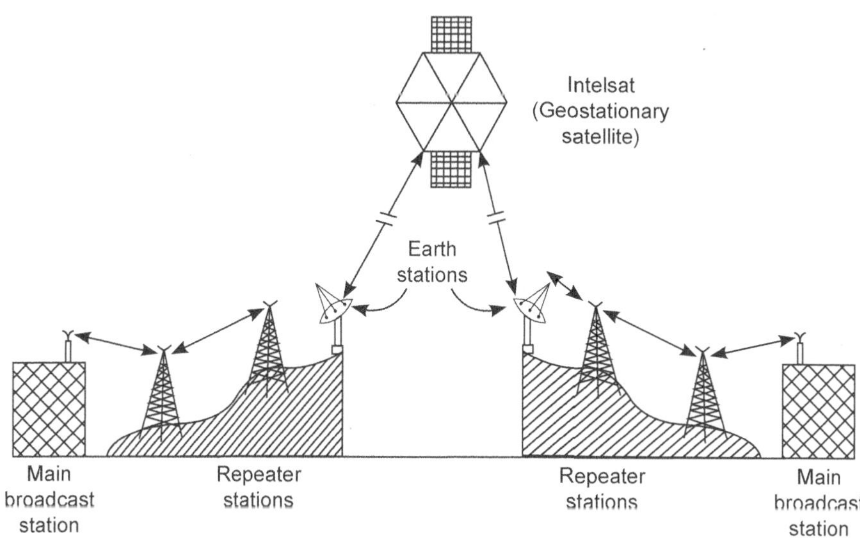

Fig. 14.4

It may be pointed out here, that both the earth stations should be within the coverage of the satellite. There may, however, be as many as 30 to 40 earth stations within the coverage of an INTELSAT satellite.

The Fig. 14.5 shows photograph of INTELSAT V, the most advanced satellite in current use. It is 15.9 m long and its overall height is 6.4 m. When it is in the orbit, all its antennas point towards earth. It uses solar panels. Like all, this satellite also uses 5.925 to 6.425 GHz up link frequency and 3.7 to 4.2 GHz as down link frequency. It has 11 low noise 6.6 kHz receivers consisting of 8 transistor amplifiers and a low noise mixer. Five of these amplifiers operate at a time and others remain "stand by". This satellite was launched in 1980 and now its capacity is about 15000 circuits.

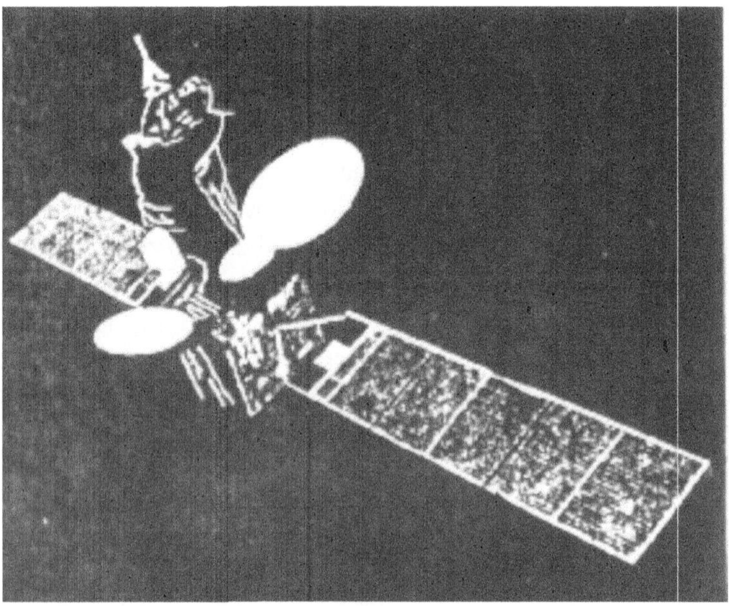

Fig. 14.5

14.6 SATELLITE REPEATERS (AMPLIFIERS)

Present day satellites use non modulating type of repeaters (amplifiers), which are further classified as

1. Double conversion repeaters
2. Single conversion repeaters.

(*i*) In double conversion system shown in Fig. 14.6 (a), the received signal is first *converted down* to an Intermediate Frequency (I.F.), where most of the amplification occurs. It is then *converted up to* the transmitting frequency and then again amplified before it is retransmitted. This type of repeater were used in early satellites.

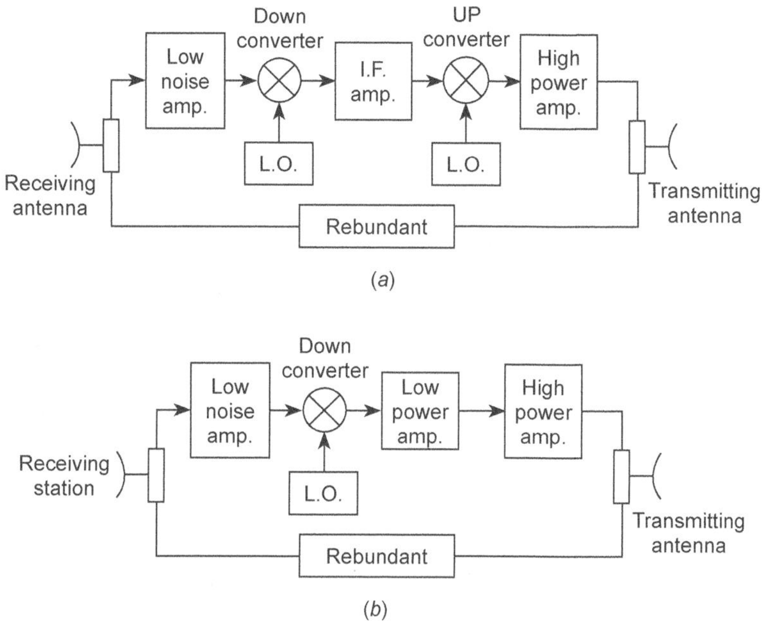

Fig. 14.6

(*ii*) The *single conversion* repeater system is shown in Fig. 14.6(*b*). The received signal is converted down to the final frequency in one stage. The amplification of the signal both prior to and after conversion is necessary. It is highly suitable for broad band satellite systems. The present day satellites use single conversion technique.

14.7 THE SATELLITE COMMUNICATION SYSTEM (FIG. 14.7)

The basic components of a satellite communication system are:

(***a***) Earth station

(*b*) Satellite station

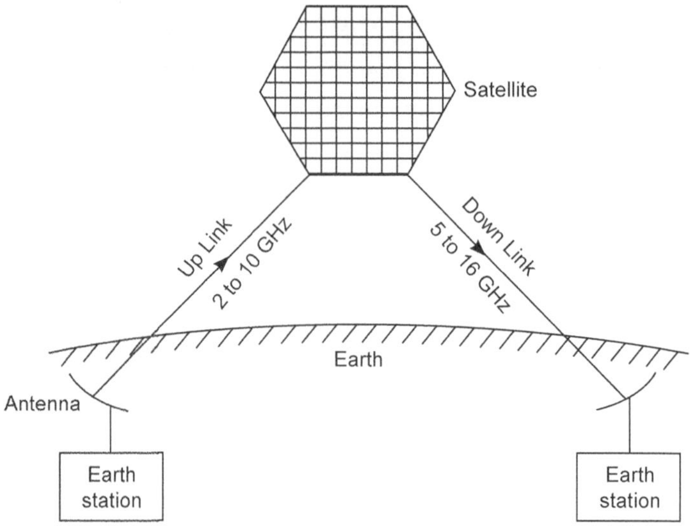

Fig. 14.7

The earth station sends signals to the satellite and receives back the signals from the satellite. The frequency range is 2 GHz to 10 GHz for "uplink" and 5 GHz to 16 GHz for "down link". The two frequencies are kept different to avoid any interference between the coming and going signals.

14.7.1 Earth Station

Few important Earth station systems are briefly described below:

(*i*) *Power Supply (Fig. 14.8)*: An uninterrupted power supply (**UPS**) is needed to be supplied to the earth station. In emergency, batteries provide the necessary supply which are kept on charging during normal supply hours.

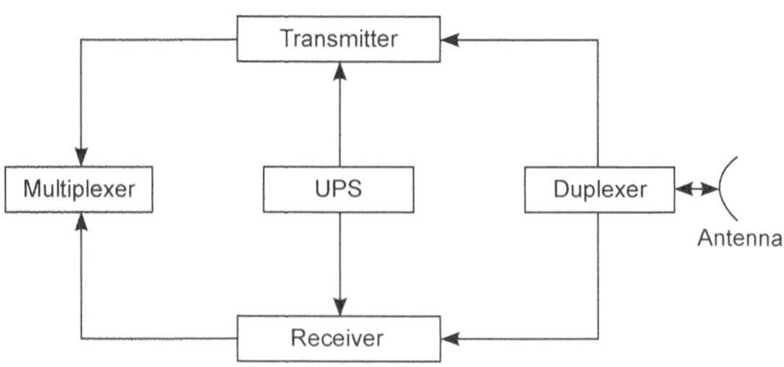

Fig. 14.8

(*ii*) *Transmitter*: The transmission system has modulators, amplifiers (repeaters which may be Double Conversion type or Single Conversion type) and other accessories. The system also has frequency converters which convert 65 MHz (IF) to 5 GHz for transmission.

(*iii*) *Receiver System*: The receiving system has demodulators, amplifiers (which amplify in stages) and other accessories. The system also has frequency converters which converts the incoming frequency again to standard intermediate frequency (65 MHz).

(*iv*) *Antenna*: The earth station has "transmitting" antenna as well as a "receiving" antenna. The transmitting antenna should throw all power towards the satellite and the receiving antenna should receive all the power sent by the satellite to the earth station. They should perform their job without any or little power loss, and also they should not pick up any noise, in other words both the antennas should be *"directional"*. Generally parabolic or horn antenna are used. The earth station antennas are 25–35 m in diameter and may be of hundred tons of weight.

(*v*) *Duplexer and Multiplexer System:* These are the systems to send/receive more than one message simultaneously through the same line. The Fig. 14.9 shows a complete block diagram of a typical satellite earth station.

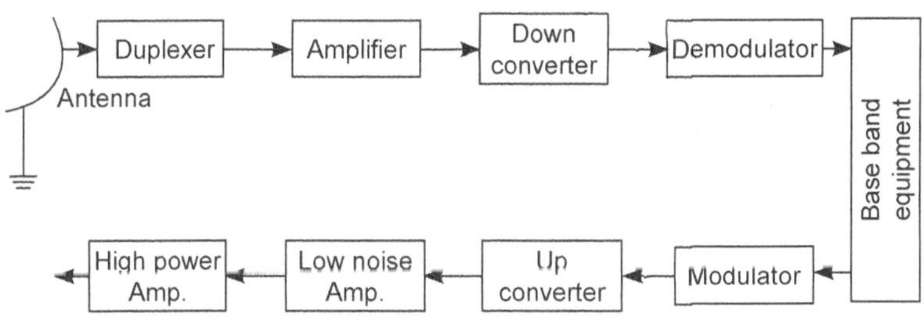

Fig. 14.9

14.7.2 Satellite Station (Fig. 14.10)

The few important systems at satellite station are described below:

(*i*) *Power Supply:* The power supply for the satellite is obtained by using a panel of *solar cells* which convert sunlight into electrical energy for working of the satellite. The solar cells are connected in series/parallel combination to get the required voltage and current for the satellite. It is estimated that to generate 100 W of electrical energy, we need about 2200 solar cells.

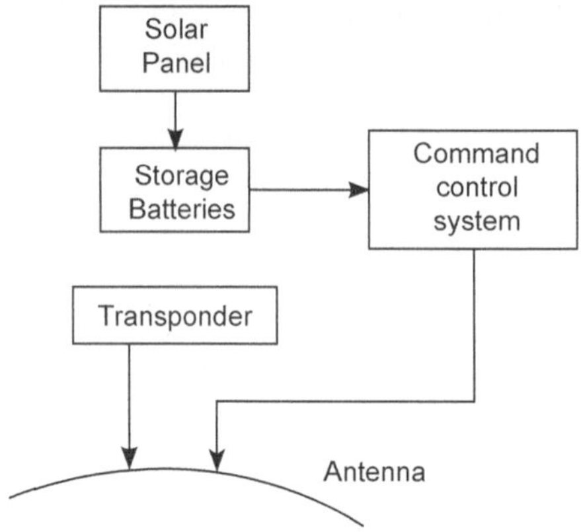

Fig. 14.10

(*ii*) *Transponder:* A *transponder* is an equipment on the satellite which receives a signal, amplifies it, changes its frequency and retransmits it to the earth station. The satellite may have more than one transponder. They function at a bandwidth of 35 MHz.

(*iii*) *Command and Control System:* This system is for receiving commands from the earth station and to adjust the position of the satellite accordingly.

(*iv*) *Antenna:* The antenna should be highly directional so that it can receive and transmit the signals.

SUMMARY

1. Satellite Communication is a technique to send and receive massages directly from a man made satellite placed in the space.
2. The satellites may be "observatory" type or "communication" type.
3. The INTELSAT is an International communication Satellite.
4. A satellite system consists of two earth stations and a satellite. One earth station sends signals to the satellite at "up link frequency" the satellite reflects back the signal to the other earth station (at a down link frequency), where signal is required. The uplink frequency is 2 - 10 GHz and down link frequency is 5 - 16 GHz.
5. The satellites are used to obtain TV signals Cable TV signals, DTH (direct to home) TV signals etc.
6. Satellites are also used to get information about weather, cyclones, rains, thunder etc.

SELF TEST QUESTIONS

1. What do you mean by a satellite?
2. Give brief specifications of a satellite.
3. Enlist communication satellites orbits.
4. Define: Apogee, Perigee, Escape velocity and FOM.
5. Write short notes: DOMSAT, INTELSAT.
6. With a sketch, describe a satellite system.

OBJECTIVE TYPE QUESTIONS

I. Fill in the blanks:

1. Expand IRS __________ .
2. The highest point attained by a satellite is called __________ .
3. The lowest point attained by a satellite is called __________ .
4. Expand EIRP __________ .
5. The noise factor is the ratio of S/N Input and __________ .
6. Expand Intelsat __________ .
7. In satellite communication, the up link frequency is __________ and down link frequency is __________ .

ANSWERS

1. Indian remote sensing (satellite)
2. Apogee
3. Perigee
4. Effective Isotropic Radiated Power
5. S/N output
6. International Tele-communication Satellite
7. 2–10 GHz, 5–16 GHz.

❑❑❑

15

Transmission Lines, Cables and Waveguides

Energy can be transmitted either through electromagnetic waves (such as in radio, TV) or it can be transmitted through conductors called **transmission lines**, cables and waveguides.

A **transmission** line may be defined as a conductive method of sending electrical energy from one place to another. In electronic communication, these lines are used as a link between antenna and a transmitter (or receiver). A **cable** is a wire which is covered by an insulation and, may be laid underground. The **waveguides** are hollow pipes used for transmission at UHF. The telegraph and telephone signals are sent through over head wires and also through underground cables.

15.1 TRANSMISSION LINES: TYPES

Transmission lines are of following types:

(*i*) *Parallel (open) wire lines*: This is a very common form of transmission line. The ordinary telephone lines running on overhead poles are its examples. Fig. 15.1 (*a*) shows cross section of parallel wire lines.

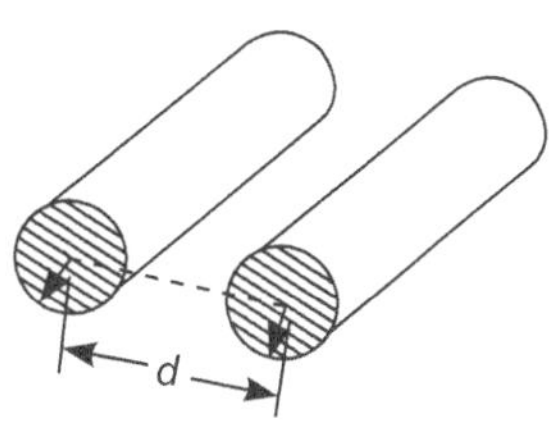

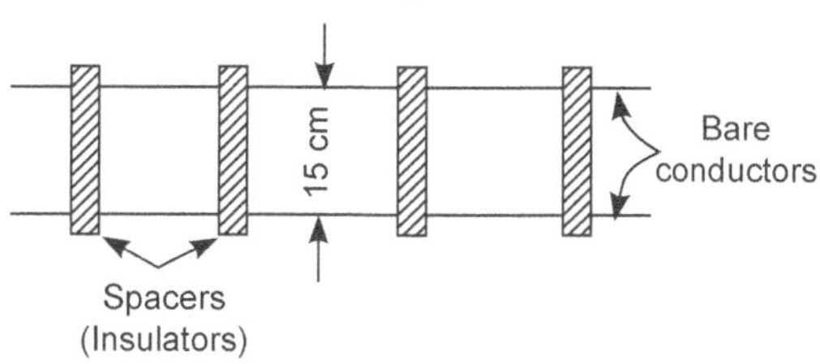

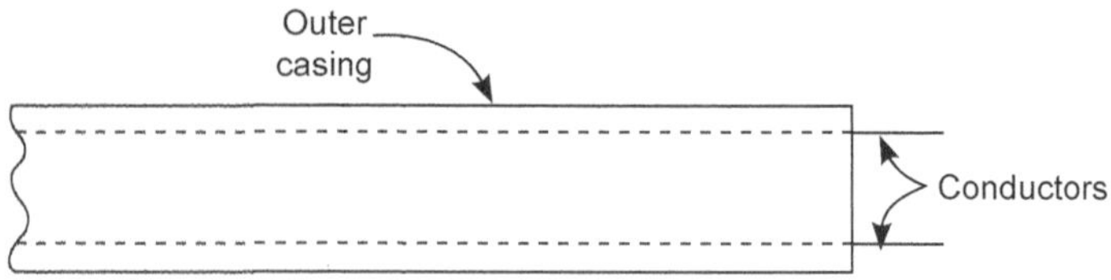

(*c*)

Fig. 15.1

These lines are constructed by keeping insulator spacers between the parallel bare conductors with a spacing of about 15 cm [See Fig. 15.1 (*b*)]. Impedence of this line is about 400 ohm. The Fig. 15.1 (*c*) shows a parallel line in its simple construction.

(*ii*) *Flat twin lead* [Fig. 15.2 (*a*) and (*b*)]: This is also an open wire line. The flat lead is one of the most popular transmission lines in use for very high frequency (VHF). The wires are enclosed in a flat plastic cover and, therefore, are unaffected by weather conditions. Its impedence ranges from 75 ohms to 300 ohms. This line is balanced. It is unshielded and used for noisy locations.

It should not run close to power lines so that it does not pick up hum. As most of the receivers have an input impedence of 300 ohms, a twin lead of 300 ohms is ideal for impedence matching.

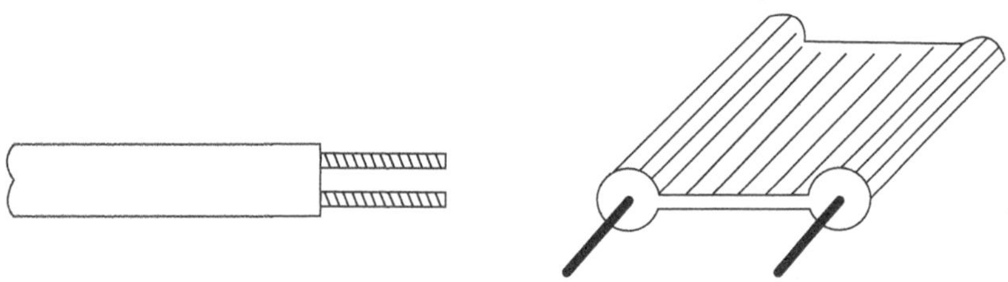

(*a*) (*b*)

Fig. 15.2

(*iii*) *Tubular twin lead* (*Fig.* 15.3): This is also an open wire line, it is similar to the flat twin lead except that in this, the twin conductors are enclosed in a tubular plastic tubing with air as dielectric. It has low losses and suited for VHF bands. The plastic tubing is strong, flexible and protects the lead against bad weather conditions.

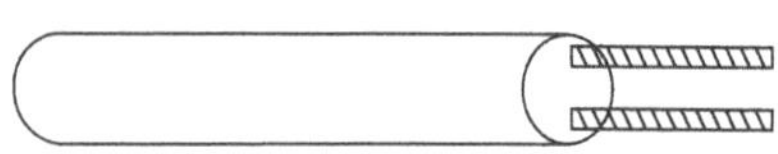

Fig. 15.3

15.1.1 Special Transmission Lines

So-long we have studied open wire lines and co-axial cables. For microwave transmission, we have modern transmission lines to replace the conventional lines. These are high power lines used to feed radio frequency power to antenna in kW range. The *solid dielectric is omitted* to keep losses at minimum. Few special lines are described below:

(*i*) *Microstrip line*. This line consists of a conductor mounted over a metallic ground and separated by a dielectric. However, above the conductor, the dielectric is air. See Fig. 15.4 (*i*). The Fig. 15.4 (*ii*) shows its cross-section and (*iii*) its equivalent LC circuit.

Microstrip is similar to open wires, consisting of the top strip and its image below the ground. The dielectric used are teflon ($\varepsilon_r = 4$), alumina ($\varepsilon_r = 8$) or silicon ($\varepsilon_r = 10$), where ε_r is the relative permittivity.

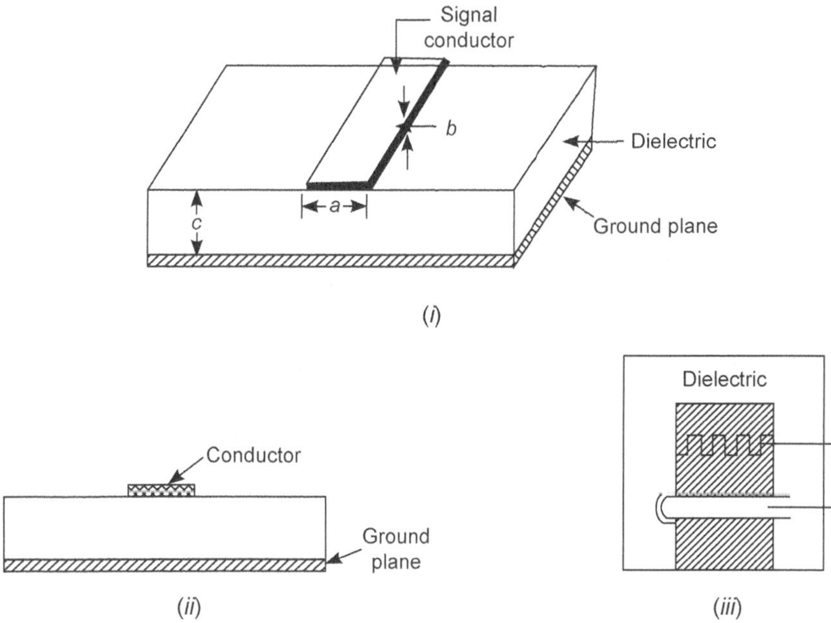

Fig. 15.4

The microstrip lines are constructed by the photographic process used for I.Cs.

Owing to the open type structure of microstrip lines, the electromagnetic field is not confined to the dielectric but partly in the surrounding air. The Fig. 15.5 shows electrical fields (EF) and magnetic fields (MF) around the dielectric.

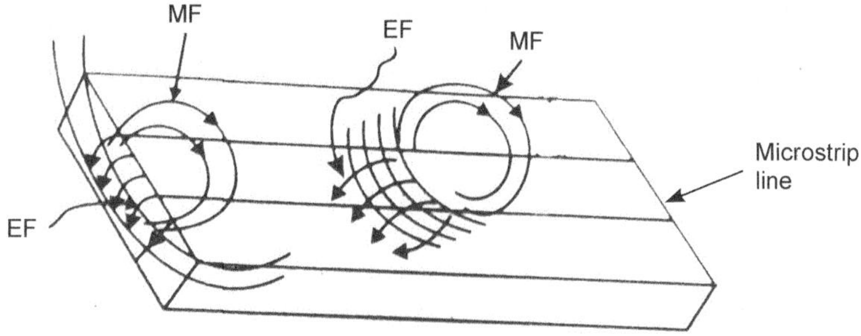

Fig. 15.5

The microstrip line suffers from radiation losses from its irregular surfaces and sharp corners. The power handling capacity of this line is also low. The lines are used upto 10 GHz.

The microstrip lines belong to the group of parallel plate lines. These lines are widely used in electronics. Apart from being used as transmission lines for microwave I.C.s., the microstrips are also used for circuit components such as filters, couplers, resonators and antennas. As compared to co-axial cables, microstrip offers more flexibility and compactness.

(*ii*) *Strip Line*. This line consists of a conductor placed between two ground planes as shown in Fig. 15.6 (*i*), the Fig. 15.6 (*ii*) shows its cross section. The strip line is a balanced line whereas the microstrip line discussed above is an unbalanced line.

Note that, a balanced line is that in which electric field distribution between the two is symmetrical, whereas in an unbalanced line it is unsymmetrical.

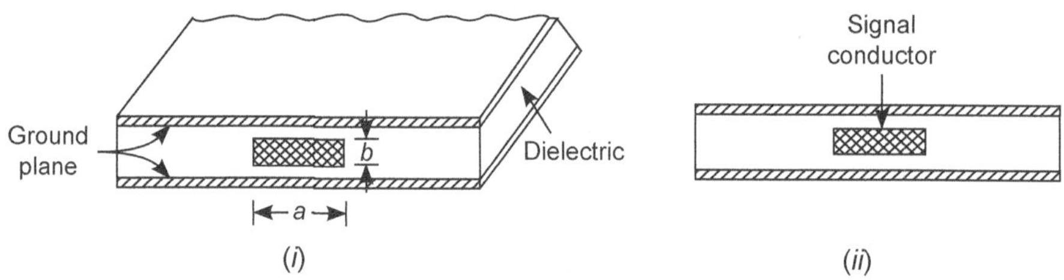

Fig. 15.6

The strip lines are used upto 12 GHz (1 GHz = 10^9 Hz).

The strip line circuits are manufactured using printed circuit board (PCB) methods. The special board materials are used which have low losses at microwave frequencies, examples are "woven fibre glass" and "polyoteflin". The boards may be copper cladded on one or both sides. Two PCBs are used each having a ground plane. The Fig. 15.7 (*a*) shows method of strip line assembly and (*b*) shows its assembled view.

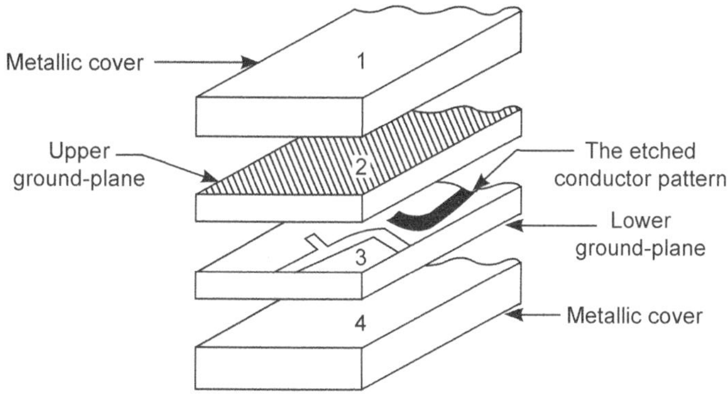

(*a*) Method of Assembly

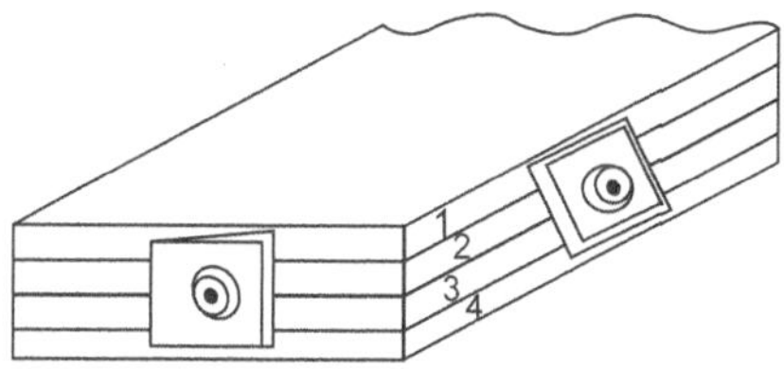

(*b*) Assembled view

Fig. 15.7

15.1.2 Characteristic Impedence of A Transmission Line

The input impedence of an infinite length of a transmission line is known as its "Characteristic Impedence". Its unit is **ohm** and it is also sometimes referred as **Surge impedence**.

The characteristic impedance (or surge impedance), Z_0 of a line may be defined as the ratio of square root of series impedance and shunt admittance per unit length. It is equal to:

$$Z_0 = \sqrt{\frac{R + j\omega L}{G + j\omega C}}$$

where R = resistance, G = Conductance, ωL = Inductive reactance and ωC = capacitive reactance of the transmission line.

It should be kept in mind that the characteristic impedence Z_0 does not involve length of the line, but is determined by characteristic of the line for unit length. It is not the impedence, that a line itself possesses.

(*i*) For an open wire line

$$Z_0 = 276 \log_{10} \frac{d}{r}$$

where d is the spacing between the wires and r is the radius of each wire (See Fig. 15.8 a).

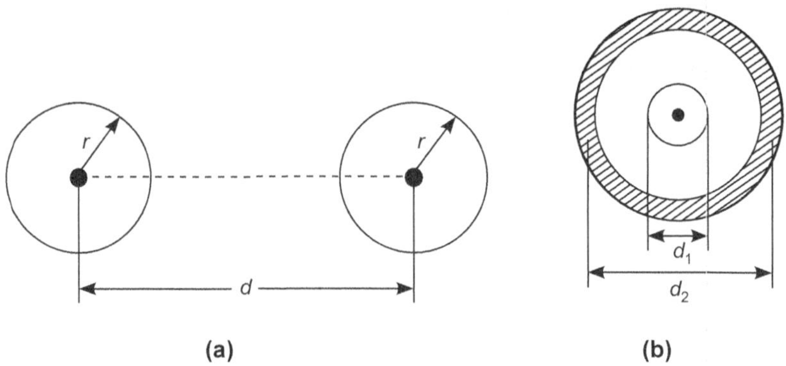

Fig. 15.8

(*ii*) For co-axial cables (Fig. 15.8 b)

$$Z_0 = 138 \log_{10} \frac{d_2}{d_1}$$

where d_2 is diameter of the outer conductor and d_1 is the diameter of the inner conductor.

15.1.3 The Requirements of Transmission Lines

The desirable requirements of transmission lines are:

(*i*) The losses along the line should be minimum.

(*ii*) There should be no reflection of the signal on the lines.

(*iii*) The line itself should not pick up any stray signals, to prevent this, the line should be *shielded.*

(*iv*) There should be a balanced matching between antenna and the receiver.

The Fig. 15.9 shows a balanced match between antenna and the receiver.

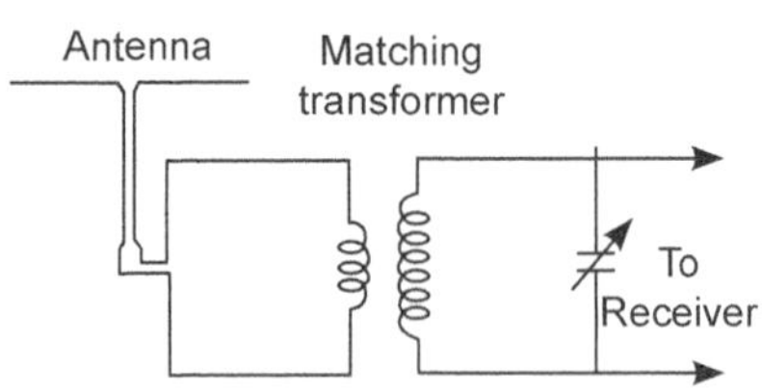

Fig. 15.9

15.1.4 Standing waves

When a transmission line is not correctly (properly) terminated, the travelling waves at the receiving end is reflected completely or partially at the termination. There will be forward as well as *reflected* travelling waves. The combination of incident and reflected waves give rise to sinusoidal (or non-sinusoidal distribution) of power, these are called *standing* or non progressive waves of current and voltage. These have definite maxima and minima of current and voltage along the line (See Fig. 15.10).

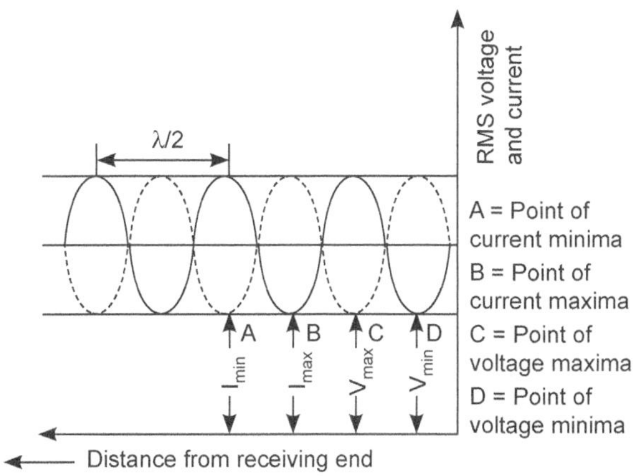

Fig. 15.10

15.1.5 Standing Wave Ratio (SWR)

The ratio of maximum and minimum magnitude of currents or voltage on a line, having standing waves is called *standing wave ratio* (SWR).

$$\text{SWR} = \frac{V_{max}}{V_{min}} = \frac{I_{max}}{I_{min}}$$

So there will be two SWRs.

(*i*) *Voltage SWR (VSWR)*

The ratio of maximum and minimum magnitudes of voltage on a line having standing waves is called voltage SWR.

$$\text{VSWR} = \frac{|V_{max}|}{|V_{min}|}$$

The value of VSWR is always greater than 1. When it is equal to 1, the line is correctly terminated and there is no reflection.

(*ii*) *Current SWR (CSWR)*

The ratio of maximum and minimum magnitude of current on a line having standing waves is called CSWR.

$$\text{CSWR} = \frac{I_{max}}{I_{min}}$$

15.1.6 Impedance Maching of Lines

The transmission lines are used for transmission of power and information. For radio frequency power transmission, it is desirable that max power is transmitted to the load with minimum losses. This requires that the load must be "matched" to the "characteristic impedance" of the line, so that the standing wave ratio (SWR) on the line is close to unity as possible. The mis-matching produces echoes and *distortions* in the signal.

The following methods are used for impedance matching of the lines:

(*i*) By quarter wave transformer

The **disadvantage** of the quarter wave transformer is that it is a "narrow band device".

(*ii*) By stub matching.

Though the single lumped inductors or capacitors can match the line. But it is more common to use the suceptive properties of short circuited pieces of lines. A *stub* is a piece of transmission line. It is possible to connect sections of open or short circuited lines called **stub (Tuner)** with the main line, at some points to obtain impedance matching.

The stubs have the following advantages:

- The length and characteristic impedance of the line remains unchanged.
- Adjustable susceptance can be added in shunt with the line.

15.2 CABLES

Open wire lines create following problems at higher frequencies:

(*a*) Radiation losses from the line become excessive high.

(*b*) An external shield becomes necessary to protect the lines from external interference.

15.2.1 Co-Axial Cables

These problems have been removed in co-axial cables. At high frequencies, co-axial pair of wires has half the inductance and twice the capacitance of the open parallel wires of the same diameter, hence radiation losses are reduced. Moreover, the outer cover protects the line against external interference.

It consists of two conductors arranged co-axially (both have same axis), one conductor is hollow and the other being located inside it, co-axially. The dielectric between the two may be a solid, air or a gaseous medium. A plastic cover over the line provides it protection against bad weather. Over the plastic cover, a metallic shield is provided which is grounded so that the cable does not pick up any stray signals or noise. With outside diameter of about 1 cm, the impedance of the cable is about 75 ohm. It is costly but suitable for noisy areas and placed where no. of cables are to be laid close to each other. Sometimes, ***foam*** is also used as dielectric,

The Fig. 15.11(a) shows dimensions and Fig. 15.11(b) shows cross section of a typical co-axial cable. The **Interstices Quads** (I.Q.) are provided to give the cable a perfect circular shape. The Fig. 15.11(c) shows simple layout of a co-axial cable.

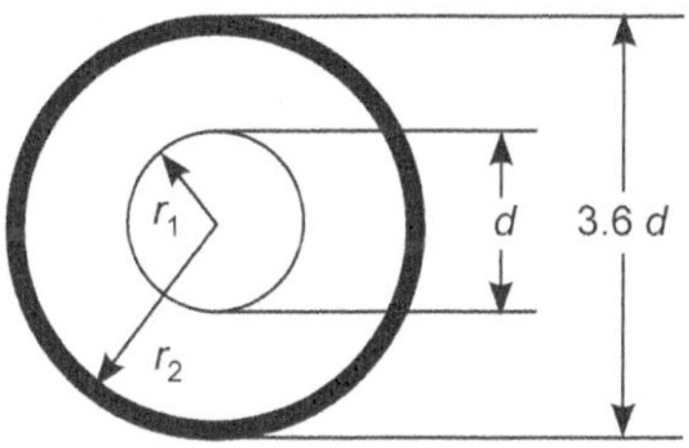

Fig. 15.11 (a)

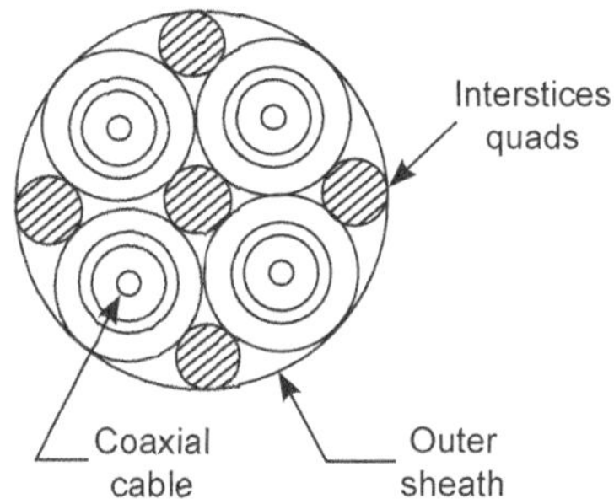

Fig. 15.11 (b)

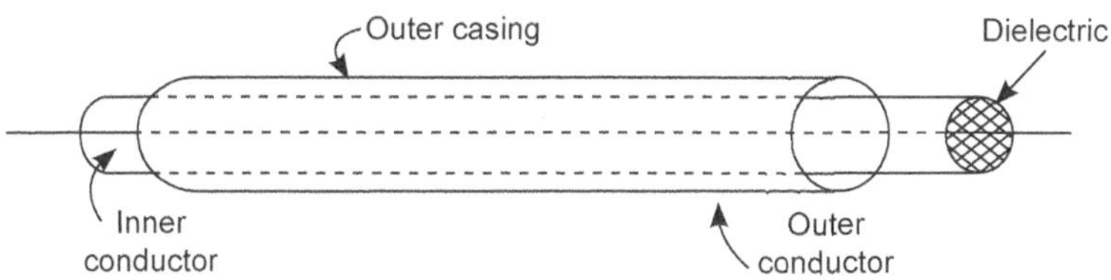

Fig. 15.11 (c)

A co-axial cable is a two conductor transmission line with a centre conductor surrounded by a braided shield. The inner conductor is supported by some form of dielectric insulation.

The most significant feature of a co-axial cable is its shielded structure. In a shielded structure, the electromagnetic fields associated with the two conductors remain confined within the space between the inner conductor and the outer shield. As the frequency of the signal carried by the cable increases, the current concentrates itself inside of the other conductor. As a result, a co-axial structure is a self-shielding with the shielding characteristics improving with increase in frequency. The major application of co-axial cable is in transmission of high frequency broadband signals. These are rarely used at lower frequencies, as their shielding properties become poor. They tend to be more expensive than the open wire lines for the same transmission loss.

15.2.2 Distortion Less Condition in Cables

For distortion less condition, $\frac{R}{L} = \frac{G}{C}$, should be satisfied. The methods are:

(*i*) Increase L (inductance)

(*ii*) Decrease C (Capacitance)

(*iii*) Reduce R (Resistance)

(*iv*) Increase G (Conductance)

15.3 WAVEGUIDES

Waveguides are basically hollow metallic (rectangular, circular or elliptical) tubes or pipes. They transmit electromagnetic waves by repeatedly reflecting the signal from their inner walls. The waveguides are used as transmission lines at UHF. Important features of waveguides are given below:

(*i*) The waveguides are most suitable at **UHF** (Ultra High Frequencies), where these are even superior than the co-axial cables. At low frequencies the dimensions of the waveguides will be very large and thus will be impractical.

(*ii*) The input Impedance (or *characteristic impedance*) of a waveguide is directly proportional to the frequency, where as in conventional transmission lines, it is dependent upon the geometry of the line.

(*iii*) Waveguides are usually air filled and can also be *coupled* with co-axial cables.

(*iv*) A waveguide acts like a high pass filter. Hence it has a *cut off frequency* and a corresponding *cut off wave* length. Any wave length greater than this does not propagate through the guide and is reflected back. It is determined by the dimensions of the guide. This also makes guides suitable only for UHF.

15.3.1 Shape of Waveguides

The waveguides can be of any shapes as shown in Fig. 15.12.

(*i*) The rectangular guides are used for general purpose.

(*ii*) The circular guides are preferred for rotating antennas (*e.g.*, radar).

(*iii*) Elliptical shapes are popular for flexible waveguides. The a and b are called its dimensions.

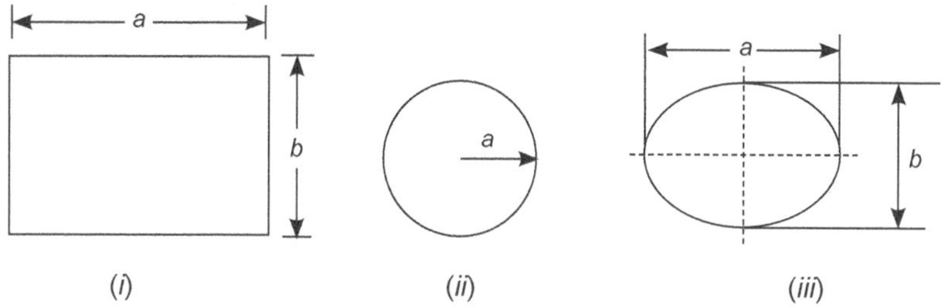

Fig. 15.12

15.3.2 Types of Waveguides

There are two important types

(*a*) *Co-planer Waveguide*. The shape of this guide is similar to the microstrip line. In this line also, there is a ground plane and a dielectric. On the dielectric, a conductor is deposited in the centre. In addition, there are two more conductors deposited on both sides of the "centre conductor" but with some separation as shown in Fig. 15.13.

This line is not much used.

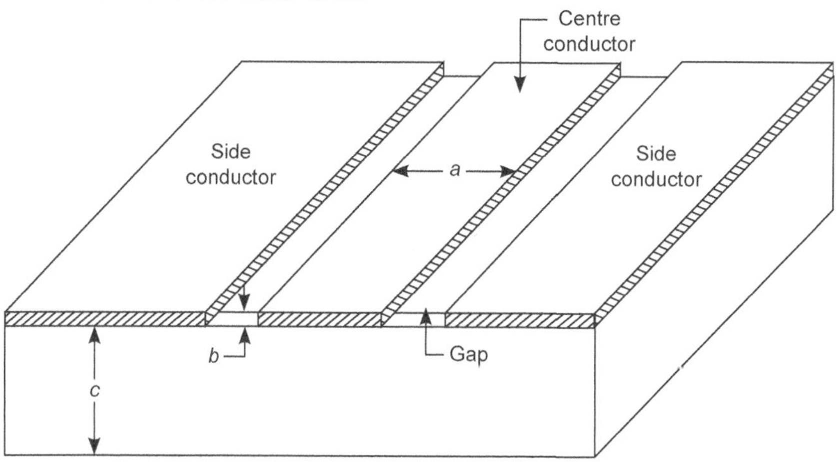

Fig. 15.13

(*b*) *Optical Waveguide* (*Optical Fiber Cable*). The optical waveguides are the advanced technique waveguides operated optically. They have a wider scope for communication.

In construction, it is a solid tube of glass or silica of maximum purity. The diameter of the tube is about 0.02 mm, and any length can be obtained by joining the pieces through highly advanced technique. There is a cladding over the glass or silica core. The refractive index of cladding is lower than that of the core but it is also of the same material as the core. The optical waveguide works on the principle of total internal reflection (T.I.R.). See Fig. 15.14, where n_1 refraction index of core, n_2 refraction index of cladding ($n_1 > n_2$). An optical waveguide differs from the ordinary waveguides in the following ways:

(*i*) In optical waveguides, refraction is obtained through total internal refraction, whereas in case of ordinary waveguide, it was only from the walls.

(*ii*) The loss of light energy due to dispersion is very less in optical waveguides.

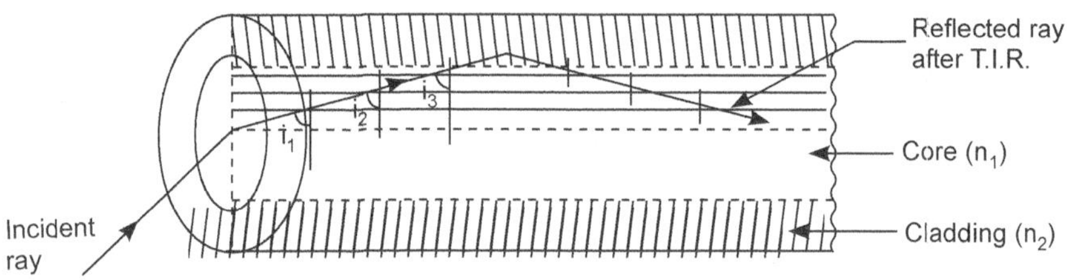

Fig. 15.14

(*iii*) The optical waveguide proves to be cheaper as it does not need *repeaters* (amplifiers) upto 50 km line, because the *attenuation* is negligible.

(*iv*) It has low *crosstalk* losses.

(*v*) The cable is free from *electro-magnetic interference* (EMI).

(*vi*) The light wave used is only of wavelength of 0.75 μm and therefore, losses are very low. Moreover, the frequency handling capacity of guide at this wavelength comes to be very high to the tune of 30 GHz (1 GHz = 10^9 Hz).

15.3.3 Waveguide Modes

A waveguide does the same job at microwave frequencies (in GHz) which the transmission lines usually do at relatively lower frequencies. At microwaves frequencies, it is more convenient to talk in terms of electric and magnetic fields propagating in the transmission medium rather than voltages and currents which is in the case of transmission lines. A waveguide is nothing but a conducting tube through which energy is transmitted in the form of electromagnetic waves. The waveguide can be considered to be a boundary which confines the waves in the space enclosed by boundary walls. The waveguide can assume any shape theoretically but the analysis of irregularly shaped guides becomes very difficult. Two popular types are the *rectangular* waveguides and *circular* waveguides and again out of the two, former is more extensively used.

Here we will discuss rectangular wave guide modes.

A rectangular waveguide is characterized by its wide dimension (a) and narrow dimension (b). There will be infinite number of possible electric and magnetic field configurations inside the waveguide if there was no upper limit for the frequency of the signal to be transmitted. Each of these field configurations in which a wave may propagate in a wave-guide is called a **MODE**. There are two types of modes; **TM** (Transverse Magnetic) and TE (Transverse Electric). In TM mode, magnetic lines are entirely transverse to the direction of propagation of e.m. wave whereas, the electric field has a component in that direction. In TE mode (Transverse Electric), the electric field lines are entirely transverse to the direction of propagation whereas magnetic field has a component along the direction of propagation.

The propagation modes, both TM and TE are designated by two subscripts. The first subscript indicates the number of half wave variations of the electric field in the wide dimension (a) of the waveguide whereas the second subscript indicates the number of half-wave variations of electric field along the narrow dimension (b) of the waveguide. For instance, in TE_{10} mode, which is the simplest mode, there is only one half-wave variation of electric field along the wide dimension and there is no electric field variation along the narrow dimension. The Fig. 15.15 shows rectangular wave guide in TE_7 mode. It may be mentioned that this subscript notation is only for rectangular waveguides. In circular waveguides, the subscripts are there but they do not carry the same meaning as they do in case of rectangular waveguides.

Accordingly there are TE_{10}, TE_{11}, TM_{10}, TM_{11} modes.

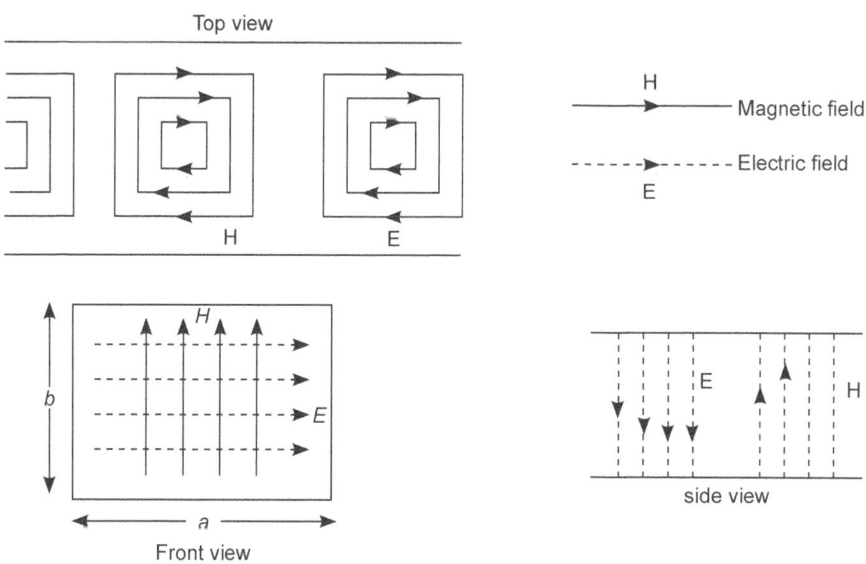

Fig. 15.15

15.3.4 Propagation Modes in Wave Guides

Following are the wave propagation modes in wave guides (Fig. 15.16):

(*i*) **TE mode:** In *transverse electric* (TE) wave propagation mode, there exists a component of $\overline{H}$ (Magnetic field) and no component of $\overline{E}$ (electric field) exists in the direction of propagation (z), *i.e.*

$$E_z = 0 H_z \neq 0$$

(*ii*) **TM mode:** In *transverse magnetic* (TM) wave propagation, there exists a component of $\overline{E}$, but no component of $\overline{H}$ exists in the direction of propagation (z), *i.e.*

$$E_z \neq 0 H_z \neq 0$$

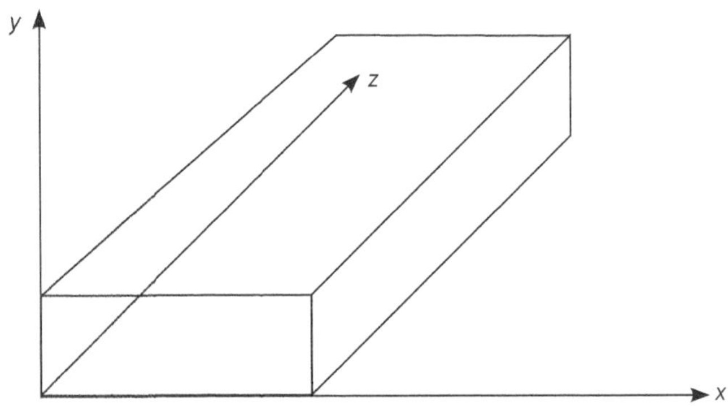

Fig. 15.16

15.3.5 Coupling a Waveguide with Co-axial Cable

Sometimes a direct coupling is needed between a waveguide and a co-axial cable. The method is shown in Fig. 15.17.

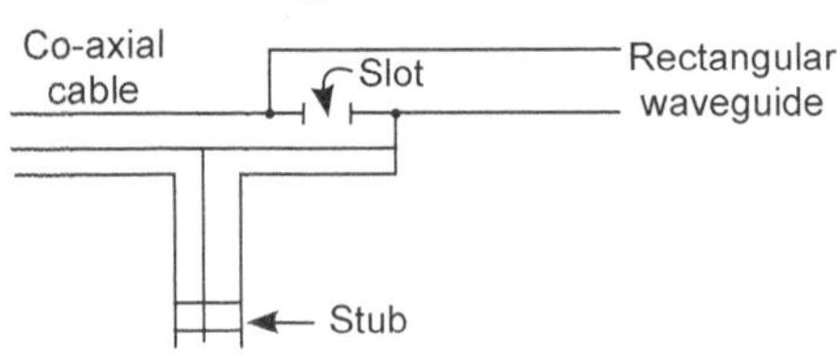

Fig. 15.17

The Fig. 15.17 shows a slot in a common wall where energy from co-axial cable is coupled into the waveguide.

15.3.6 Cavity Resonators

A simple cavity resonator consists of a waveguide closed at both ends with metallic planes. If the propagation is in longitudinal direction *standing waves* will be set up in the closed structure and the closed unit will start resonating.

The cavity resonator is analogous to *LC* turned circuits. They have frequency coverage equal to that of waveguides. They can be used as tuned circuits of amplifiers, oscillators etc. Their most important application is as cavity wave meter used for measuring microwave frequencies. The Fig. 15.18 shows a simple cavity resonator.

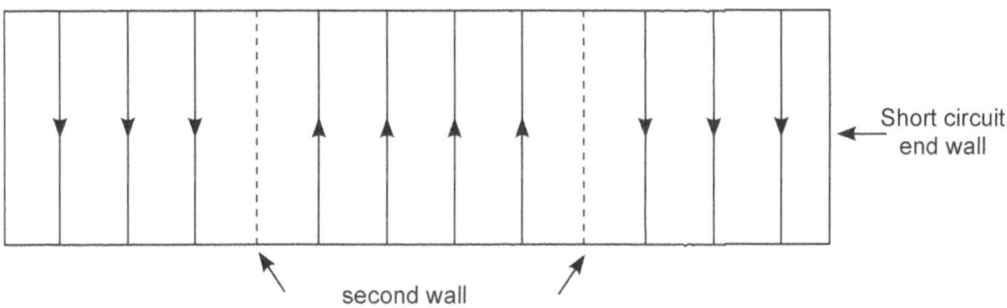

Fig. 15.18

SUMMARY

1. The transmission lines may be open parallel wires or co-axial cables.
2. The lines used at radio frequencies are microstrip and strip lines. These can be used at 10–12 GHz.
3. The impedence of an open line = 276 log d/r and of a co-axial cable = 138 log d_2/d_1. There should be a balanced impedence match between antenna and the receiver.
4. Waveguides are basically hollow pipes used for transmission at ultra high frequency.
5. The waveguide may be of various shapes: rectangular, circular or elliptical.
6. The two important types of wave guides are coplanar waveguides and optical waveguides.
7. The electric and magnetic field configurations inside waveguide is called *mode*.
8. There are two types of modes: transverse magnetic (TM) mode and transverse electric (TE) mode.

SELF TEST QUESTIONS

1. Define transmission line.
2. Write short note on types of lines.
3. What are co-axial cables? Explain their use.
4. Define characteristic impedence of a cable. Give its value.
5. What are waveguides? Explain their construction and working.
6. What are RF lines? Give their types and construction.
7. What are desirable requirements of a line?

OBJECTIVE TYPE QUESTIONS

I. Fill in the blanks:

1. Co-axial cable consists of two conductors arranged _______ .
2. Characteristic impedance of a line is the input impedance of an _______ line.
3. Waveguides are _______ metallic pipes.
4. Expand UHF _______ .
5. The far and near end coupling losses between two wires going close to each other are called _______ .
6. Expand EMI _______ .

ANSWERS

1. Co-axially
2. Infinite
3. hollow
4. ultra high frequency
5. cross talk
6. Electromagnetic interference.

II. State True/False:

1. The overhead transmission lines used are usually naked.
2. The co-axial cable consists of two conductors with different axis.
3. The strip lines are used at Gigahertz.
4. Impedance match is not required between a line and the receiver.
5. The waveguides are parallel solid tubes.

6. The waveguide and a co-axial cable cannot be coupled together.
7. The input impedance of an infinite length of a line is called characteristic impedance.
8. There should be no cross talk between the lines.

ANSWERS

1. True	2. False
3. True	4. False
5. False	6. False
7. True	8. True.

❑❑❑

16

Optical Fibre Cables

These are basically *optical wave guides*. The optical fibre cables have brought a revolution in digital communication. At present all mobile/computer/Internet signals being sent through underground optic cables.

16.1 FIBRE OPTICS

The basic medium of fibre optics is a *hair-thin* fibre that is sometimes made of plastic, but more often of glass. A typical glass optical fibre has a diameter of 125 micrometers or 0.125 mm (0.005 inch). This is actually the diameter of the *cladding* or outer reflecting layer. The core or inner transmitting cylinder, may have a diameter as small as 0.10 mm. Through a process called *total internal reflection,* light rays beamed into the fibre can propagate within the core for great distances with remarkably little attenuation or reduction in intensity. The degree of attenuation over distance varies according to the wavelength of the light and composition of the fibre.

16.2 OPTIC FIBRE COMMUNICATION

It is the science of transmitting data, voice and images by the passage of light through thin, transparent fibres. It is also spelt as *fibre optics*. In telecommunication, fibre optic technology has virtually replaced the copper wire in long distance telephone lines.

Optical telecommunication is usually conducted with infrared light in the wavelength range of 0.8–0.9 m, that are efficiently generated by light-emitting diodes (LED) or semiconductor lasers and suffer least attenuation in glass fibrcs.

16.2.1 Advantages of Optical Fibre Communication

In addition of having extra bandwidth using light as carrier signal, the optical fibre communication systems have several other advantages over the conventional systems:

(*a*) Extra advantages of being light in weight and small in size, particularly in underwater cables.

(*b*) No possibility of internal noise and crosstalk generation along with the immunity to ambient electrical noise, echoes or electromagnetic interference.

(*c*) No hazards of short circuits as in metal wires.

(*d*) No problems when used in explosive environments.

(*e*) Immunity to adverse temperature and moisture conditions.

(*f*) Lower cost of cables per unit length compared to that of metal counterpart.

(*g*) No need for additional equipment to protect against grounding and voltage problems.

(*h*) Very nominal handling and installation costs, and

(*i*) Lesser problems in space radiation shielding and line to line data isolation.

16.2.2 Applications

Because of these advantages, fibre optic communication is being currently utilised in telephones; computers, cable television, space vehicles, ships, submarine cable security & alarm systems, electronic instrumentation systems, medical systems, satellite ground stations and industrial automation & process controls.

16.3 PRINCIPLES OF OPTICAL COMMUNICATION

The basic principle and law, governing optical communications are:

(*i*) Total Internal reflection.

(*ii*) Snell Law

16.3.1 Total Internal Reflection (T.I.R)

If the angle of incidence is greater than the critical angle, the light is reflected back into the original dielectric medium; this is called *Total Internal Refleraction* (T.I.R.)

Note that T.I.R. occurs at the interface between dielectrics of different (refractive) index from higher to lower, when ray of light is incident on the dielectric of lower index from the dielectric of higher index and the angle of incidence is greater than the critical angle. The ray in this case is also called as **meridional** ray; as that passes through the axis of the fibre core.

The Fig. 16.1 shows, total internal reflection in a multimode G.I.F. (Graded Index Fiber) cable with a parabolic profile core.

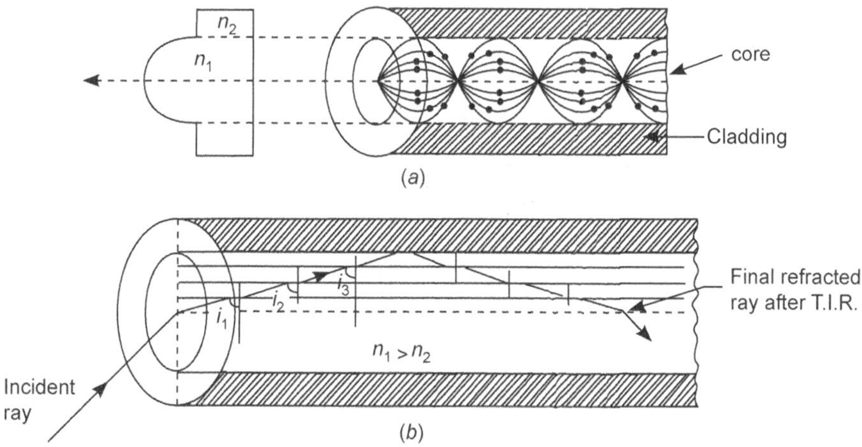

Fig. 16.1

The gradual decrease in refractive index from centre to boundary causes many refractions. The ray goes through increasing angle of incidence (i_1, i_2, i_3) through layers of the core till total internal reflection is achieved and the ray returns towards the axis. [See Fig. 16.1 (*b*)].

The value of refractive index of the core (n_1) decreases as the radial distance from the centre of the core increases. The value of n_1 decreases till it approaches the value of refractive index of cladding (n_2). For proper mode propagation, the n_1 should be higher than n_2.

16.3.2 Snell's law (Figure 16.2)

The Snell's law gives relationship at the interface of two mediums, having different refractive index. Snell's law states, how a ray of light reacts when it meets the boundary between two mediums. When the ray of light encounters the boundary, a part of the light is *refracted* to the other media, while other part is *reflected* back to the first (same) media.

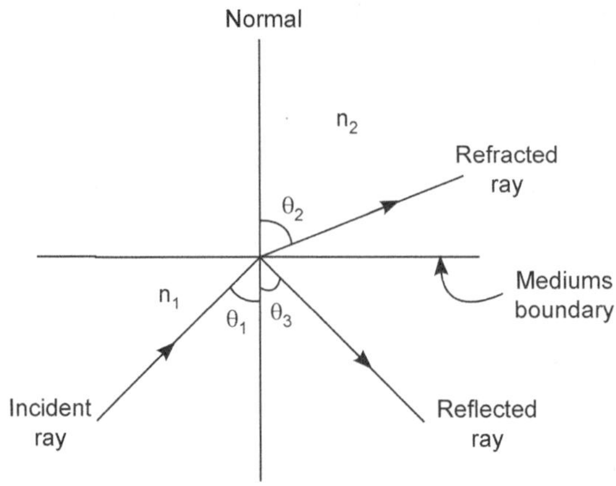

Fig. 16.2

16.4 THE OPTICAL FIBRE

The fibre is the core material used for making optical cables.

An optical fibre is a piece of very thin, highly pure glass with outside cladding that is similar but of slightly different composition and also with different refractive index. The fibres may be of two types:

(*i*) *Step Index Fibre* (SIF): A Step Index fibre has core diameter of 2 to 200 μm. The Fig. 16.3 shows two views of a Step Index Fibre.

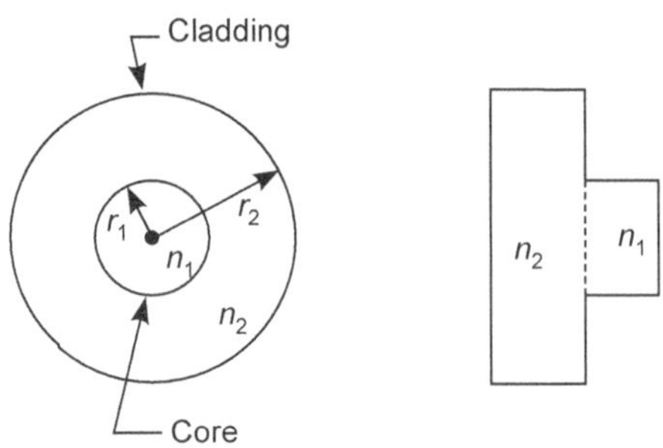

Fig. 16.3

(*ii*) *Graded Index Fibre* (GIF): The optical fibre, with refractive index reducing gradually from centre to outside of the fibre is called a **Graded Index Fibre**. The Fig. 16.4 shows its two views.

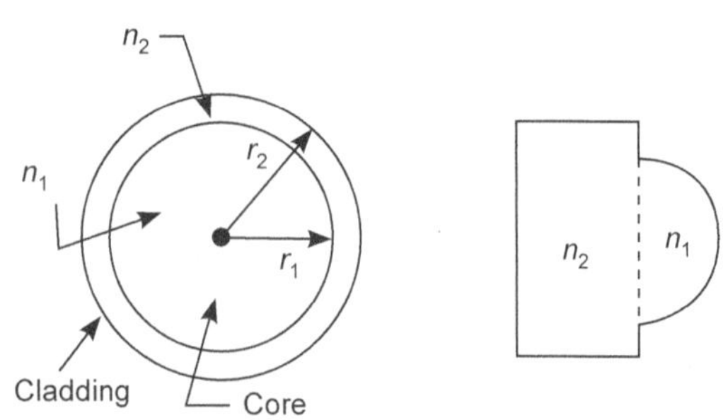

Fig. 16.4

However, Step Index Fibres are more in use because of their higher *attenuation*. This can also be used for multimode or singlemode transmission, which yields best performance. The core diameter required for single mode transmission is 2 to 10 μm and for multimode transmission it is 50 to 200 μm. This can also be successfully used for digital transmission.

Remember that in metal wires, the signal is carried by **electrons** but in optic fibre cables, the signal is carried by **Photons**.

It may be mentioned again, that the core and the cladding both of an optical fibre are made of glass but with different composition and refractive index. The glass should be of highest purity and quality. The optical fibre covered with protective layers gives an *optical fibre cable*. The protective layers provide the cable extra mechanical strength and protection against environment.

16.5 OPTIC FIBRE CABLE (OFC)

In the case of telephone, trunk and trans-oceanic communication systems instead of fibre bundles, fibre optic cables are required which must have some competitive advantages over the properties of their metallic counterparts.

16.5.1 Requirements

Basically these *fibre optic cables* must be such that (*a*) their handling might be done in the same way as most ordinary communication wire and coaxial cables, (*b*) their mechanical and electrical properties are compatible for their specific use and surrounding conditions, including delivered powers to repeaters and resistance to applied static and dynamic stresses, (*c*) permanent splicing and connecting in the field may be done with ease and in reasonable time, (*d*) these are competitive economically with existing wire and coaxial communication cables, (*e*) their attenuation is quite low up to to the order of the individual fibre's and (*f*) fibres within the cables must be distinguishable.

16.5.2 Structure

Based on these particular requirements, a fibre optic cable can be designed. The constructional details of fibre optic cables used in telephone and trunk systems are shown in Fig. 16.5. As evident from this figure, there are seven major components in it.

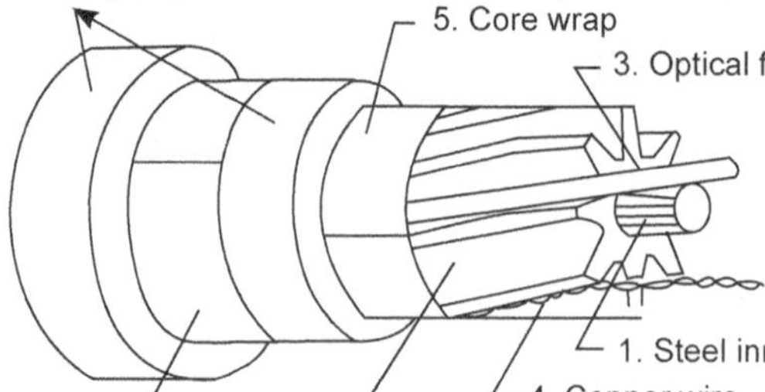

Fig. 16.5

1. *Steel Inner Core:* This core is made of aircraft steel and runs through the centre of the optical cable to absorb tension applied during cable installation in ducts.
2. *Grooved Polyethylene Core:* Polyethylene pellets are melted and forced through a mould, forming grooves around the inner core. The core is twisted so the grooves spiral around the steel core first in one direction and then in the other. Thus the inner core wire gets surrounded by an extruded plastic core with six slots running in a reverse helix pattern.
3. *Optical Fibres:* Colour coded strands of fibre are laid in the spiral grooves. The structure is designed to *prevent pinching and squeezing* which could break the fibres or affect their performance. The colour aids in identification of each fibre during installation.

   The optical fibre is made by depositing many thin layers of glass from gaseous raw materials on the inside of a pure glass tube. The hollow form is then collapsed into a solid rod heated to high and precisely controlled temperature in a special furnace and 'drawn' into a fibre as thin *as a human hair* (approximately 0.125 millimetre in diameter) and several kilometers in length.
4. *Copper Wires:* A pair of plastic insulated copper wires are added for testing and to allow installers to "tap in" and talk to each other during installation.
5. *Core Wrap:* The bound cable is further wrapped in a protective film.
6. *Aluminium Tape:* A thin layer of aluminium is applied to further protect the cable from moisture and potential damage.
7. *Polyethylene Jackets:* The cable is coated with a final thick jacket of polythylene to add strength, keep water out and to prevent damage.

16.5.3 Types of Optical Fibre Cables

According to the construction, the cables may be

(*i*) Single fibre cable (*ii*) Double fibre cable

(*iii*) Multiple fibre cable.

(*i*) *Single Fibre Cable*. This is shown in the Fig. 16.6(a). It has a single buffer, while the fibre itself acts as "strengthener". This cable is more suitable for indoor applications.

(*ii*) *Double Fibre Cable*. The cross section of a double fibre cable is shown in Fig. 16.6(b). This is also provided with a steel wire fencing for better mechanical strength and environmental protection.

(*iii*) *Multiple Fibre Cable*. The cable has multiple buffer as shown in Fig. 16.6(c).

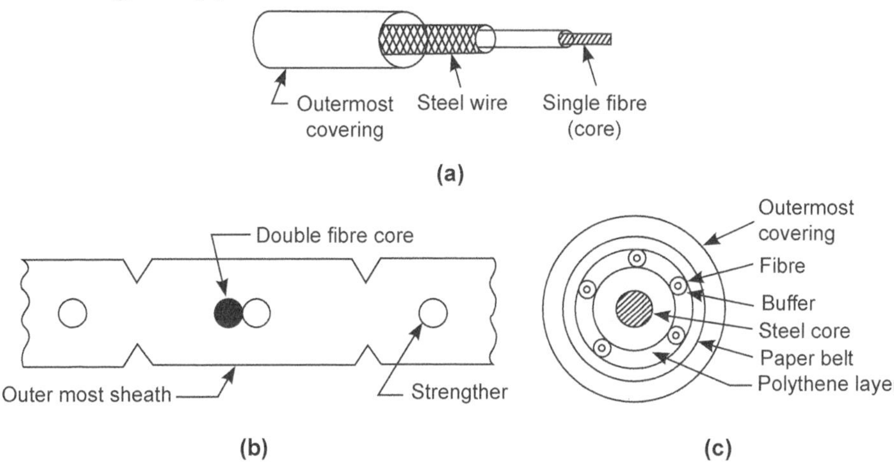

Fig. 16.6

16.5.4 Losses in Optical Fibre Cables

Following losses occur in fibre cables

(*i*) *Scattering Losses*. The glass used for optical fibres is an amorphous (non crystalline) solid, which is formed by melting the glass and then cooling it. While in plastic form, it is drawn into fibres. This is the reason of scattering losses occurring in the fibres. This loss can be reduced by adopting careful manufacturing techniques.

(*ii*) *Absorption Losses*. These may be ultraviolet or infrared absorptions. The absorptions occur in the walls of the fibre cables.

(*iii*) *Bending Losses*. The losses may be due to "micro" bending or "constant radius" bending.

16.6 LIGHT SOURCES FOR FIBRE CABLES

The light sources used for fibre cables act as " light transmitters" . They should meet the following requirements:

(*i*) Their light should be "monochromatic" (*i.e.*, of single frequency). The gas ionisation lamps, LEDs and laser diodes have lights of comparatively narrow band of frequencies though not pure monochromatic.

(*ii*) They should have light of high intensity to overcome and compensate the various losses (scattering loss, bending loss, etc.) during transmission of the signal.

(*iii*) The light source should be capable to be "modulated". Preferably due to lesser noisy reception *pulse code modulation* (PCM) is carried out.

(*iv*) They should be small, compact and can be easily coupled to the fibre.

Usually LED and diode laser are used, as they meet the above requirements. (See Table 16.1)

Table 16.1 Performance of some diode lasers as Light Sources

Diode	Material	Wavelength Range (μm)	Power (dbm)	Maximum bandwidth (GHz)	Light emitted	Spectral width
LED	GaAlAs	0.75–0.90	0.7	0.1	Converging	Bigger
	InGaAsP	1.3	0	0.2		
Diode Laser	GaAlAs	0.8–0.9	1	1	Diverging	Smaller
	InGaAsP	1.3	1	1		
In the Table : Ga: Gallium, Al: Aluminium, As: Arsenic, P: Phosphorus, In: Indium,						

16.7 BLOCK DIAGRAM OF FIBRE OPTIC COMMUNICATION

Fibre optic communication systems would involve a series of components which are analogous to those in radio frequency communication, but are different in design. A basic block diagram of fibre optic communication system is shown in Fig. 16.7. It is apparent that the main blocks of the system are transmitter, repeater and receives. The Transmitter includes modulators, switches and drive circuits to lasers. It also has couplers to couple the transmitted light signals with the fibre. Fibres also may require connectors to increase the length of fibre medium. Repeaters detect the transmitted signals, regenerate, amplify and then retransmit them. Receivers amplify the signal and then convert them into the desired output.

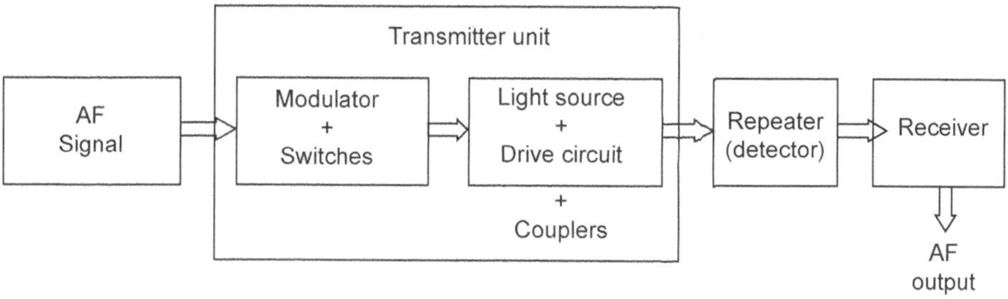

Fig. 16.7

Miniaturised version of above optical components like modulators, couplers, switches, detectors, etc., have also been developed. These involve guided wave optics into thin film waveguides. The principle is called "*integrated optics*" in analogy with the integrated semiconductor electronics. The main importance of such integrated optical components is that these are rugged, require less space and consume very little power. In addition, these are quite efficient and fast.

16.7.1 Brief Description of Components/Blocks of Optical Communication

Below is given brief description of components:

(*i*) *Electronic Transmitter*: The information to be conveyed enters into Electronic Transmitter, where it is converted into electrical form, modulated and multiplexed. (Fig. 16.8(a))

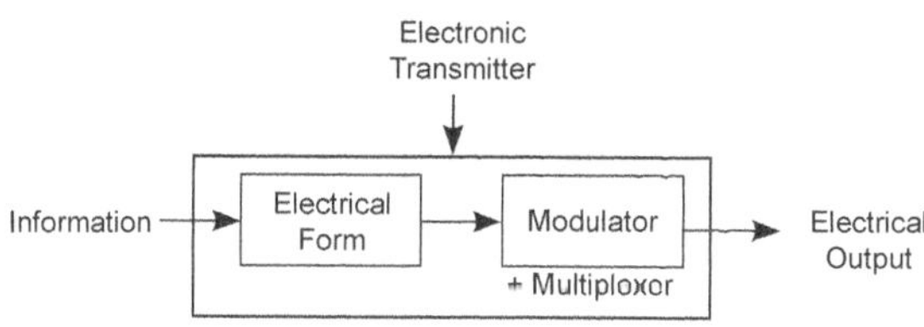

Fig. 16.8(a)

(*ii*) *Optical Transmitter* [Fig. 16.8(b)]: The electrical signal then goes into *optical transmitter*, where it is converted into optical form *i.e.*, light. The optical transmitter consists of drive circuit *optical source*, *modulator* and *a channel coupler*.

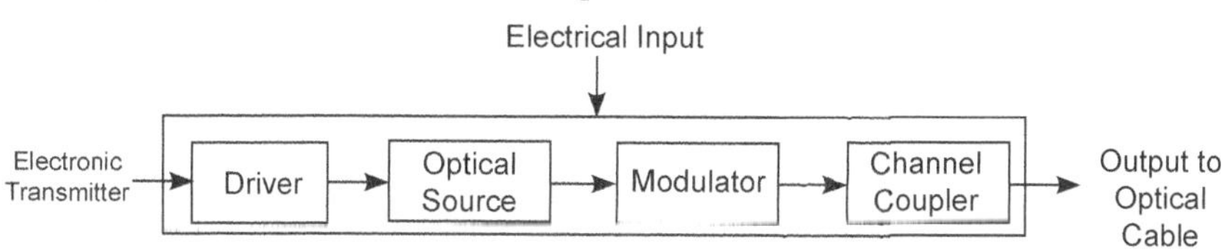

Fig. 16.8(b)

Either LED or a Laser diode is used as optical source. The input signal **modulates** the intensity of light (called intensity modulation) coming from the optical source. The intensity modulation can be carried out electrically or optically. Analog or digital modulation is possible. Now-a-days **Electro-optic modulators** are used, which modulate the light by changing its refractive index through the electrical signal. The **channel coupler** couples the modulated optical signal to the optical cable.

(*iii*) *Optical Cable*: Through the optical cable, the signal travels **by total internal reflection.**

(*iv*) *Optical Receiver*: The main function of optical receiver is to convert the received soptical signal to original electrical signal. The optical receiver consists of a channel coupler, optical photo detector, and a demodulator [Fig. 16.8(c)].

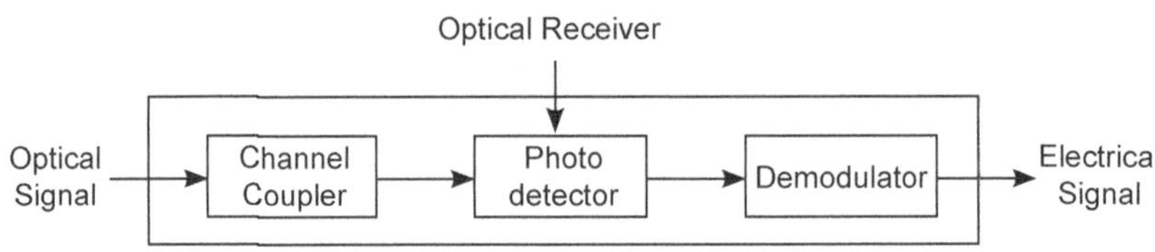

Fig. 16.8(c)

The coupler focusses the received optical signal on to the photodetector. The photo-detector converts the optical pulses into electrical pulses. The demodulator demodulates the received electrical signal.

(*v*) *Electronic Receiver* [Fig. 16.8(d)]: The electrical signal is then processed in the electronic receiver to get the exact original information.

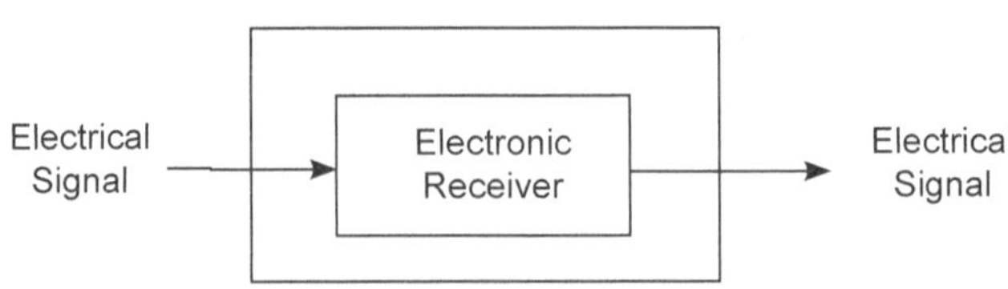

Fig. 16.8(d)

16.8 DIGITAL MODULATION SCHEMES USED IN OPTICAL COMMUNICATION

In optical communication the various digital modulation schemes used are (Fig. 16.9):

(*i*) Amplitude shift key (ASK)

(*ii*) Frequency shift key (FSK)

(*iii*) Phase shift key (PSK)

The Fig. 16.9 (*a*) shows binary ASK, the Fig. 16.19 (*b*) shows FSK, in which the binary 1 is transmitted at higher optical frequency than the 0 bit. The Fig. 16.9 (*c*) shows PSK, which shows 180° phase shift between Binary 1 and 0 bits.

It should be noted that in ASK, the amplitude of the optical carrier is effectively switched ON or OFF whereas, the amplitude of the optical carrier remains constant in other two modulation (FSK, PSK) schemes. This is the reason that **ASK modulation is preferred**.

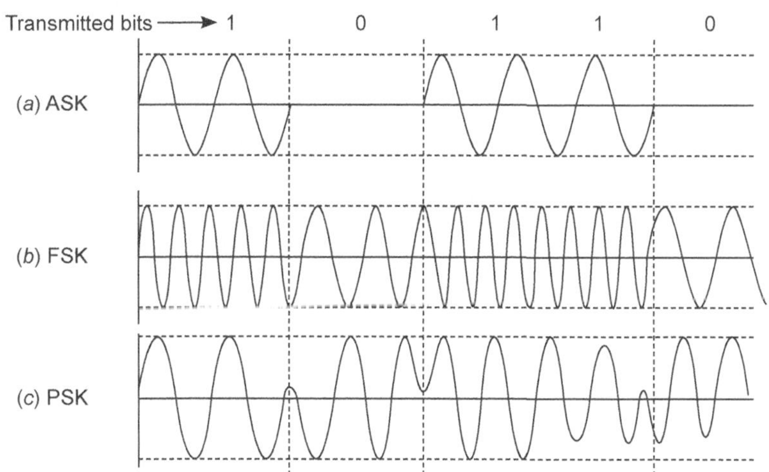

Fig. 16.9

Note: "Amplitude Shift Key (ASK) modulation" is analogous to "amplitude modulation, "Frequency Shift Key (FSK) modulation" is analogous to "Frequency modulation" and "Phase Shift Key (PSK) modulation" is analogons to "Phase modulation". Note that ASK, FSK and PSK are digital modulation techniques.

16.8.1 Digital Transmission and Reception through Optical Cable

The Fig. 16.10 shows digital transmission through optical fibre cable.

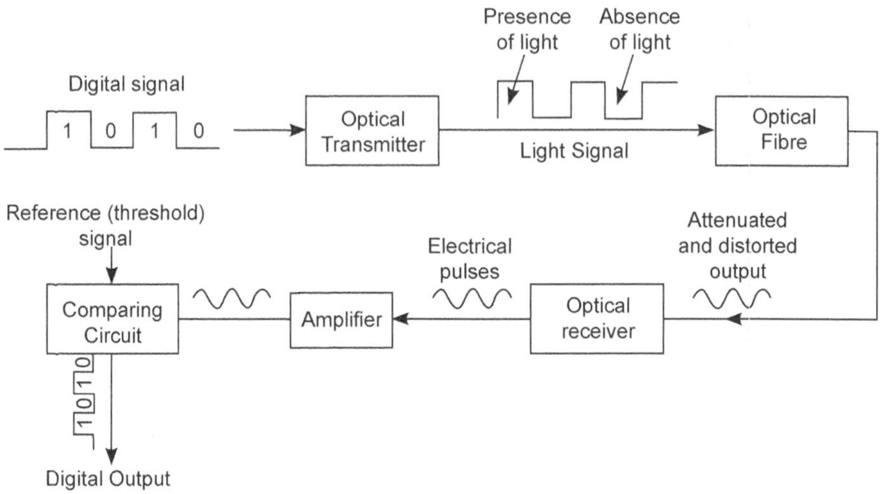

Fig. 16.10

(*i*) The digital data is sent with the help of "amplitude shift key" (ASK) technique.

(*ii*) The optical transmitter converts the electrical signal into optical (light) signal. Any time-varying current can be used to **modulate** an optical *source* to produce optical (light) signal.

(*iii*) The optical signal is coupled to optical fibre. The optical signal gets attenuated and distorted, as it propagates through a fibre.

(*iv*) The optical signal, now is coupled to optical receiver which uses a PIN diode or Avalanche photodiode as a detector. It converts the optical signal back to original electrical signal (pulses).

(*v*) A "comparer circuit" then compares this amplified signal in each "time slot" with a certain "reference" signal known as "threshold level."

If the received signal is more than the threshold level, an output binary '1' is obtained. If the received signal is less than the threshold level, a binary '0' is obtained.

SUMMARY

1. The optical fibre communication has higher bandwidth and uses light as carrier signal. The fibre cables have light weight, small size and have no risk of short circuit.
2. The optical fibre is the core material for making optical fibre cables. The fibre may be step index fibre or graded index fibre.
3. The materials used for making fibre are silica glass and plastic.
4. The optical fibre cable may be single, double or multiple fibre type.
5. The propagation in the optical fibre cable is obtained through *total internal reflection*.
6. The important fibre optical sources are: LED and LASER diodes.

SELF TEST QUESTIONS

1. What do you know about fibre optic communication? What are its advantages over conventional communication systems?
2. Explain the basics of fibre optic communication.
3. Enlist the types of fibres. Describe the structure of a fibre.
4. Write note on:

   (*i*) Step Index Fibre (*ii*) Graded Index Fibre
5. What are the materials used for making fibres?
6. Give construction of an optical fibre cable.
7. Write note on single fibre cable and double fibre cable.
8. Explain how propagation occurs through an optical cable. Enlist the losses occuring in the process.
9. What are the devices used as fibre optical sources for fibre cables? What are their desirable requirements?
10. Describe the following fibre optical sources in brief:

    (*i*) LED (*ii*) LASER diode
11. Draw block diagram of fibre optical communication system and explain function of each block.

OBJECTIVE TYPE QUESTIONS

I. Fill in the blanks:

1. The fibres may be active and ________________ type.
2. For fibre optic communication, the ________________ part of electro magnetic radiation is used.
3. For few applications, fibre________________are used.
4. For ______________, the use of fibre cable is must.
5. The copper wires are embedded in the fibre cables for __________ during the installation.
6. The last layer on the optical cable is ________________ sheath.
7. The frequency range of operation of optical cable is in _________ Hz.
8. The optical cable works on the principle of total _______________ reflection.
9. Now-a-days the optical sources used for optical cables are _______ diodes.
10. The optical source should be capable to ___________ modulation.
11. Give five semiconducting materials to manufacture optical sources___________ .

ANSWERS

1. passive	2. visible
3. bundles	4. telephony
5. talking each other	6. polyethylene
7. gega	8. internal
9. laser	10. pulse code

11. Gallium, Arsenic, Aluminium, Indium and Phosphorus

II. State True/False:

1. The carrier signal used in optical fibre communication is light energy.
2. The fibre communication provides very low bandwidth.
3. Silica glass and plastic are the materials used for making optical cables.

4. The frequency of light radiations in Fibre communications is in GHz.
5. There is always a high risk of short circuits in fibre cables.
6. The propagation through fibre cables is by refraction of light.
7. Fibre cables for under water use are also available.
8. The light sources used for fibre cables should have monochromatic light.
9. Pulse code modulation (PCM) is used in fibre optic communication
10. Fibre communication finds use in telephony.

ANSWERS

1. True	2. False
3. True	4. True
5. False	6. True
7. True	8. True
9. True	10. True.

❑❑❑

17

Consumer Devices

In this chapter we shall discuss few commonly used consumer electronic devices.

17.1 SEVEN SEGMENT DISPLAY

This is used almost in all electronic devices such as watch, calculator, multimeter etc. The display uses 7 light emitting diodes (LEDs).

For displaying certain digit, the concerned diodes are illuminated. For example for displaying '9' as shown 6 LEDs become ON and others remains off. [Fig. 17.1(*a*)]. The Fig. (*b*) shows the circuit. The LCDs can also be used in place of LEDs.

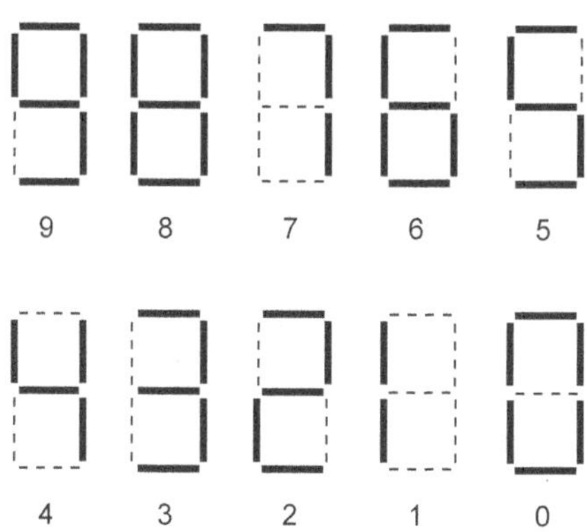

Fig. 17.1(*a*)

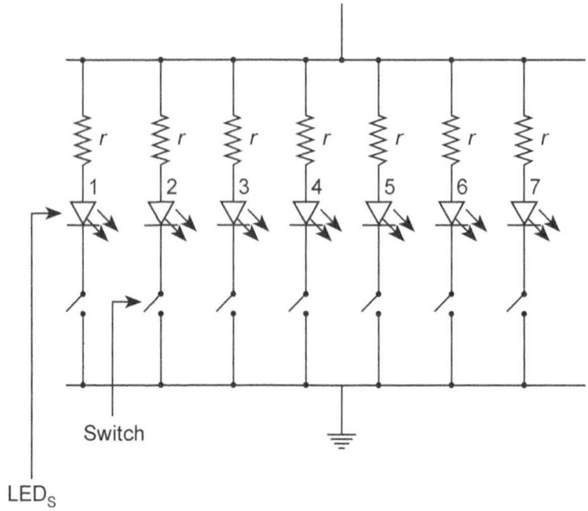

Fig. 17.1 (*b*)

17.2 DIGITAL WATCH/CLOCK

The digital watches/clocks are now common. The mechanical watches here almost been replaced due to following reasons:

(*i*) The digital watches have no moving part, it increases their life.

(*ii*) They are accurate and more reliable.

(*iii*) Time upto seconds can be read quickly and correctly.

Construction

The Fig. 17.2 shows block diagram of a digital watch. The main blocks are—crystal oscillator, Logic circuit, Digit driver, Segment driver and the LED display.

A frequency signal of 32 kHz is generated by the crystal oscillator (usually a quartz crystal is used). This signal is processed and is decoded by circuits to generate the drive signal in the form of digits.

For the display, LEDs (light emitting diodes) or LCDs (liquid crystal diodes) are used. Usually Indoor watches/clocks employ LEDs, where as the outdoor watches/clocks use LCDs for their display.

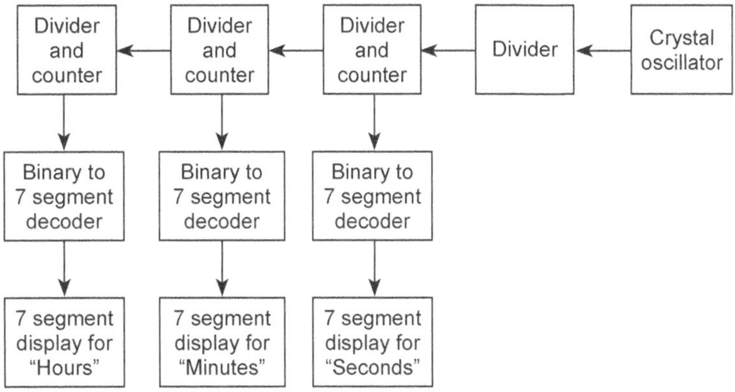

Fig. 17.2

These watches/clocks employ I.Cs. for the circuitary and therefore consume very little power. Usually *Button Cells* are used in these watches. The button cells are mercury cells which use "Potassium hydroxide plus zinc oxide" as electrolyte, the negative electrode is made of (Zn + Cu) alloy, while the positive electrode of mercurous oxide. Their emf is 1.35 V, size 12.5 mm (dia) × 3 mm thickness and life about one year.

The modem digital watches have the facilities like : stop watch, alarm, date reminder, a built in calculator, temperature display etc.

17.3 DIGITAL CALCULATOR

The digital calculator is a very simple device for making mathematical calculations. These are available in various varieties/ types described below :

1. *General purpose calculator.* On this calculator, functions available are arithmetic functions (+, –, ×, ÷), percentage (%), square and square root, add to memory (M^+), subtract from memory (M^-), recall memory (RM), clear memory (CM) and reciprocal (l/x).
2. *Scientific calculator.* On this, all functions of general purpose calculators are available. In addition, we have trigonometrical functions (sine, cos, tan and their inverse), log, antilog (at the base of 10 and at the base of e), Rectangular/polar, degree/ radian, hyperbolic/inverse hyperbolic, pie calculations, brackets, etc. etc. Today we have calculators having many other functions.
3. *Commercial/desk calculators.* These calculators have display printer, sub/grand total facilities, floating/fixed decimal, cost register facilities. Now a days.

   Today programmable calculators are also available.

17.3.1 Principle of Working

The Fig. 17.3 shows block diagram of a calculator. It has three main blocks:

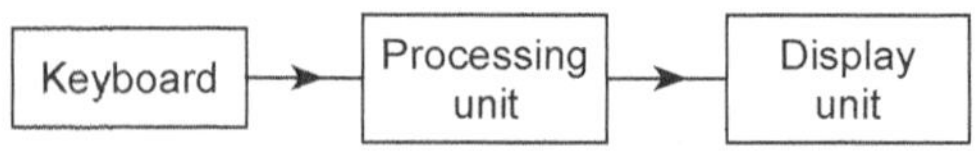

Fig. 17.3

1. *Keyboard (Input device).* This has many key. When we press a key, a particular signal is generated, which is fed into the calculator. By pressing key, we feed in the figures which are to be used in calculations.
2. *The processing unit.* This is a L.S.I. (Large Scale Integrated I.C.) circuit. This is the heart of the calculator and all mathematical operations are performed in this unit.
3. *Display unit.* This is a LED/LCD display, which gives output of the calculator.

   The calculators are also low power devices. *Pencil cells* (number of cells depends upon the size of the calculator) are used. External supply (adopter) may also be used, for this, an arrangement is provided in the calculator. Solar cell calculators are also available.

The Fig. 17.4 shows a complete block diagram of the digital calculator.

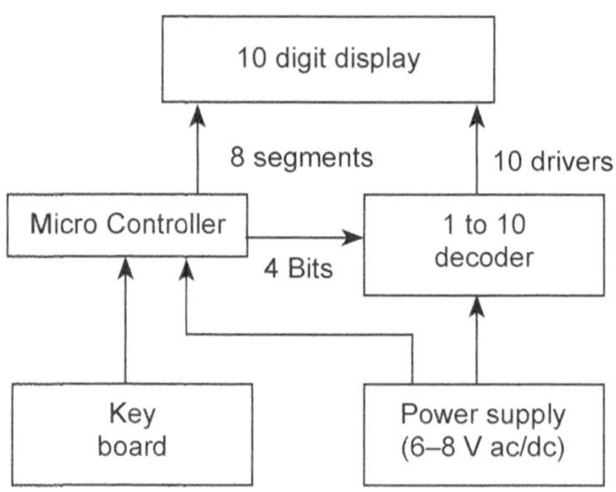

Fig. 17.4

17.4 MICROWAVE (μ-WAVE) OVEN

The working of a microwave oven is based on the principle of dielectric (high frequency/microwave frequency) heating. For industry applications, usually 915-2500 MHz frequencies are used.

(*a*) Advantages of Microwave Heating

The microwave heating has many advantages over other conventional heating methods :

(*i*) It provides uniform heating of the substance.

(*ii*) The system is neat and clean.

(*iii*) The system provides quick heating.

(*iv*) The depth of penetration of heat into the material is much more.

(*v*) The heat is generated directly within the material, it gives much faster temperature rise.

(*b*) Construction of μ-Wave Oven

The μ-wave oven basically converts the μ-wave energy into thermal energy and supplies to the substance to be heated. The industrial ovens may be of several kilowatts depending upon the size of the oven, whereas the domestic ovens are of about 1 kW rating.

A micro-wave oven basically consists of three blocks: (Fig. 17.5)

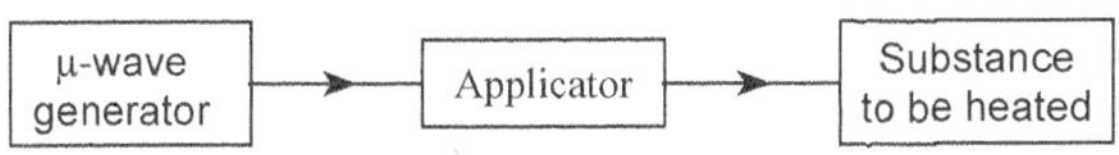

Fig. 17.5

(*i*) *Microwave generator*: This generates micro-wave energy to be supplied to the material to be heated.

(*ii*) *Applicator*: An applicator is a link between micro-wave generator and the material to be heated. There should be proper impedence matching between the two for maximum power transfer to the material and that too without any leakage.

The applicator may be of the following types :

(*i*) monomode cavity type

(*ii*) multimode cavity type

(*iii*) waveguide type.

The Fig. 17.6(a) shows a wave guide of a (micro-wave oven. The wave guide is a hollow tube, which carries micro-wave power from the (micro-wave generator to substance to be heated. The wave guide has holes (perforations) on its surface, through these holes micro-wave power comes out. The material to be heated flows over a conveyor belt, which moves very close to the wave guide. The amount of power coming out of the wave guide and the temperature of substance is controlled by the size and number of holes on the wave guide. The Fig. 17.6(b) shows circuit of a programmable micro-wave oven.

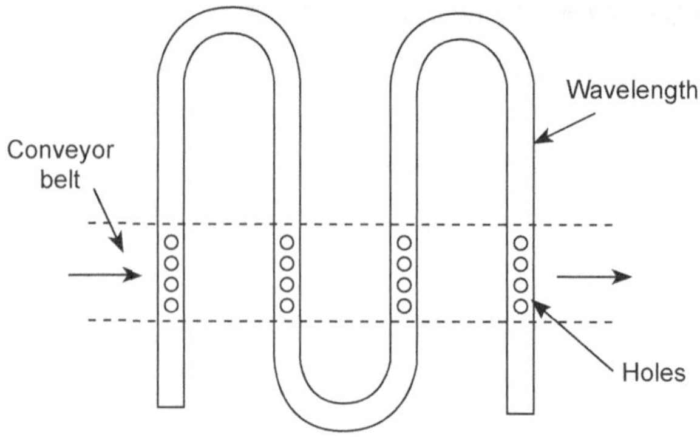

Fig. 17.6 (a)

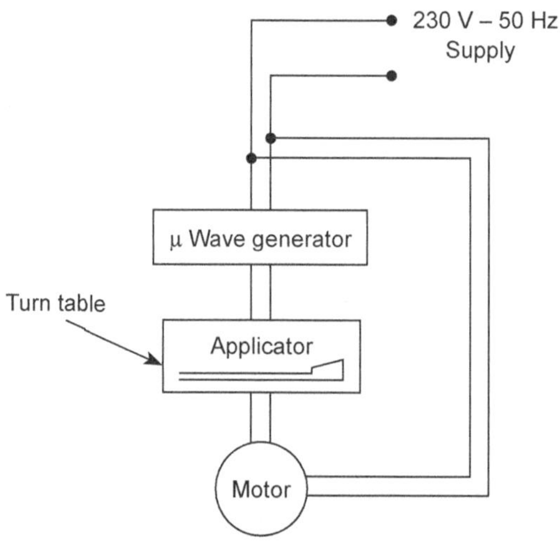

Fig. 17.6 (b)

The modern μ-wave oven is a "feather touch" oven using a *microprocessor* that provides a wide variety of cooking programs which are impossible by conventional control methods. The operation of oven is controlled by touching the appropriate "pads" arranged on the surface of the control panel. Each time we touch the control panel to make an entry, a "signal tone" is heard. At the end of the cooking cycle, again an audible signal sounds and also a display takes place which gives visual indication that cooking is complete.

It has the following facilities :

1. *Instant minute.* By simply touching the "Instant minute" pad, you can cook high for one minute.

2. *Power level.* There are five power levels:

Power level	% Power
High	100%
Medium high	80%
Medium	60%
Medium low	40%
Low	25%

3. *Auto defrost.* We can auto defrost a frozen food by inserting a correct programme.

(*d*) Technical Specifications of a Typical Micro-Wave Oven

(*i*)	Power	220 V 50 ~ AC
(*ii*)	Power consumption	1 kW
(*iii*)	μ-wave output power	600 W
(*iv*)	μ-wave frequency	2500 MHz
(*v*)	Cavity dimensions	300 × 290 × 210 mm.

(e) μ-WAVE COOKING MECHANISM

In a μ-wave oven, electricity is converted into electromagnetic waves, these waves generate energy, that is used to cook the food.

These waves are nothing but high frequency radio waves similar to those used by radio or TV. These waves have very short wavelength (and very high frequency) hence known as "microwaves".

Microwaves vibrate millions of times per second (2400-2500 MHz). In μ-wave oven these waves are confined inside the oven cavity and are reflected from its walls and doors, once the door is opened all μ-waves are automatically switched off.

When μ-wave energy comes into contact with a substance, it is reflected, transmitted or absorbed. These waves are *reflected* by metals, *transmitted* through paper, glass, plastic etc. and *absorbed* by water/moisture present in the food. When this energy is absorbed, heat is produced and cooking takes place.

The μ-waves are attracted to water, fat and sugar molecules. They cause these molecules to vibrate @ 2500 MHz/second leading to a friction within the food, which generates heat and begins the cooking process.

(*f*) Block Diagram of Microwave Oven

The block diagram of microwave oven is shown in Fig. 17.7

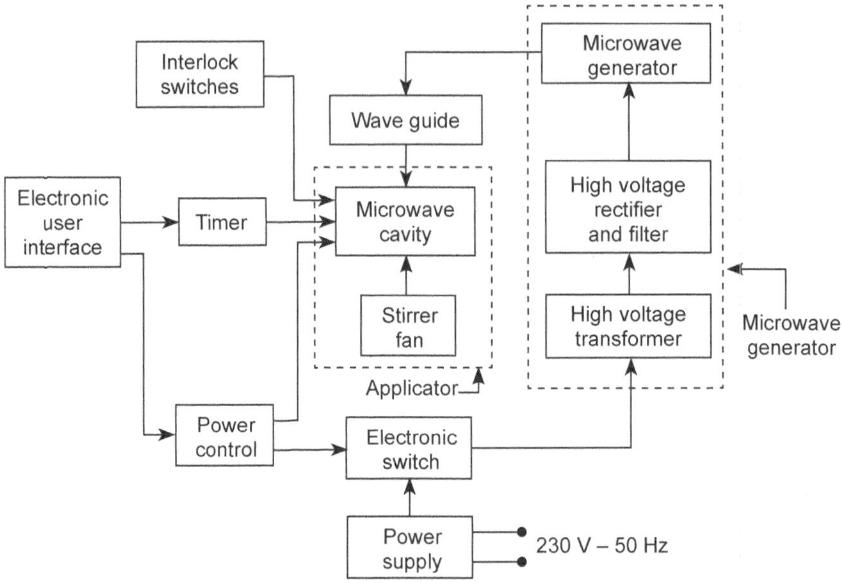

Fig. 17.7

A microwave oven mainly consists of:

- Microwave generator (Magnetron tube)
- Applicator
- Wave guide
- Electronic Switch
- Timer

(*g*) Working of μv-Oven

Microwave energy is more easily directed, controlled and concentrated than low frequency e.m. waves. Also various atomic and molecular resonances occur at μw frequencies; creating divesse application areas in basic science, remote sensing and heating methods. The heating properties of μw power are useful in wide variety of commercial and industrial applications. The Fig. 17.8 shows a cross section of typical μw-oven.

When the *magnetron* oscillates, μw energy is extracted from the *resonant cavities*. The reflection from the stationary walls and motion of the *stirring fan* causes the μw energy to be well distributed, thus the μw energy enables the cooking process to be fast and even. The μw heating properties are also used in *physical diathermy* and in drying potato chips, paper, cloth and so on.

A μw oven contains components such as source, transmission line, wave guide, attenuators, resonantors, circulators and filters etc.

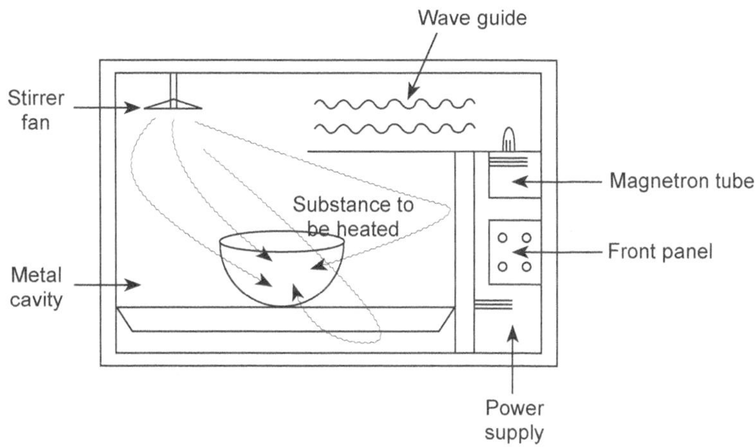

Fig. 17.8

(h) Other Applications of μ-wave Heating

The μ-wave heating is a versatile process having following applications :

1. Food processing/kitchen work
2. Baking: manufacture of bread etc.
3. Drying of paper and textiles
4. Treatment of cancer and other diseases
5. Manufacture of plastics
6. Processing of timber, cement etc.

17.5 ELECTRONIC WASHING MACHINE

It is an ordinary/conventional washing machine plus electronic control and programmable. The Fig. 17.9 shows outlook of a domestic electronic programmable washing machine. The typical technical data are given below :

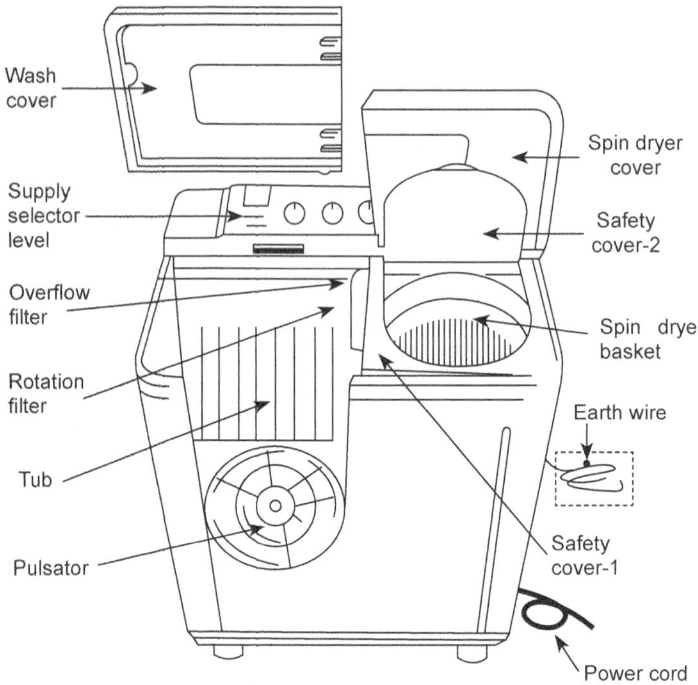

Fig. 17.9

(*a*) Technical data of a typical washing machine.

1.	Programmes	18 wash care programms
2.	Selector Buttons	1/2 load, spin cancel, speed selector
3.	Motor rating	310 W (wash), 510 W (Spin)
4.	Spin speed	800 and 1100 RPM
5.	Heater rating	1.9 kW
6.	Temperature selection	30-95°C
7.	Drain pump rating	30 W
8.	Water consumption	20-100 litre as per programm selected
9.	Power consumption	0.09-1.50 kwh
10.	Connected load	2.2 kW
11.	Cycle time	12.5-92.5 min
12.	Power supply	Single phase 220-50 Hz
13.	Water pressure	1.03 kg/cm^2-10.3 kg/cm^2

The Fig. 17.10 (*a*) shows control panel of a typical washing machine and Fig. (*b*) shows its few accessories.

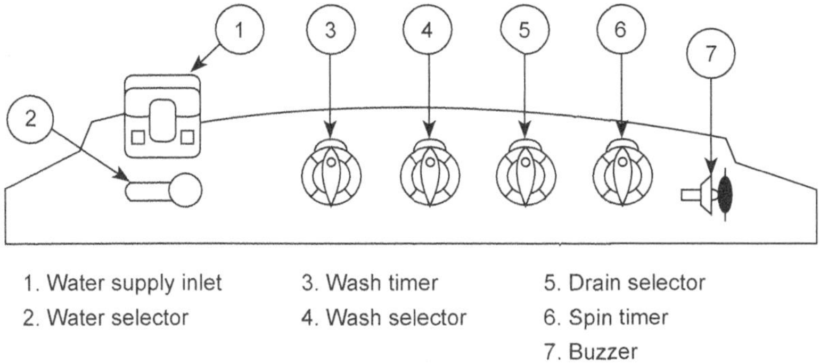

(*a*)

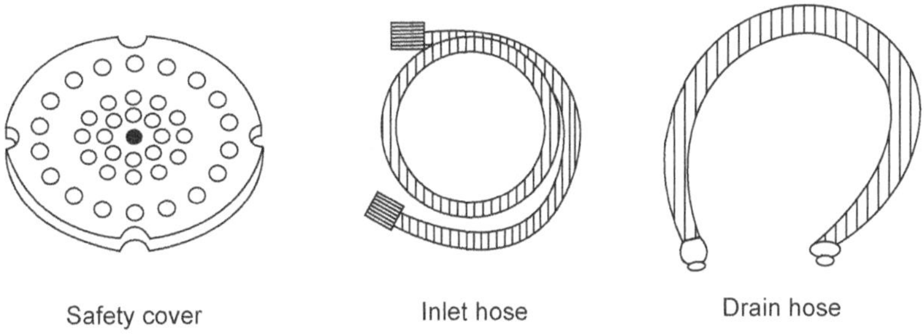

(*b*)

Fig. 17.10

The Fig. 17.11 (*a*) shows input and output devices and (*b*) shows Time control of an automatic washing machine.

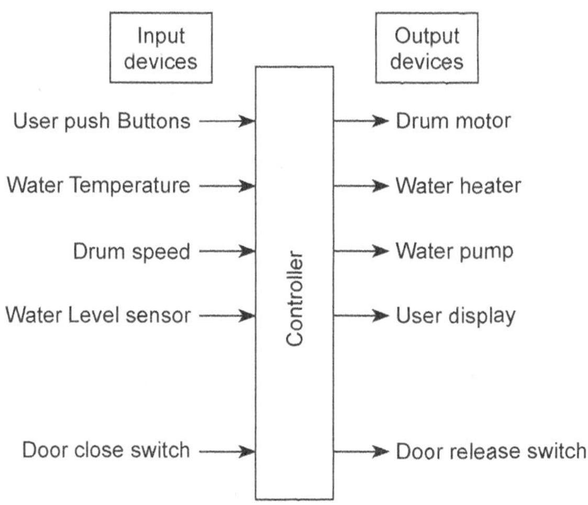

Fig. 17.11 (*a*)

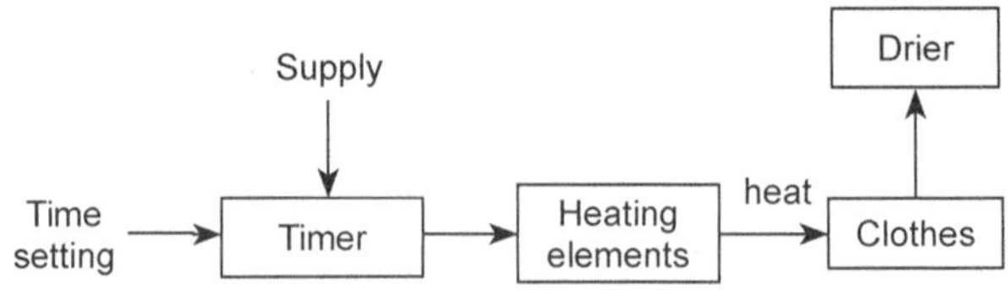

Fig. 17.11 (*b*)

The Fig. 17.12 shows complete block diagram of an automatic washing machine.

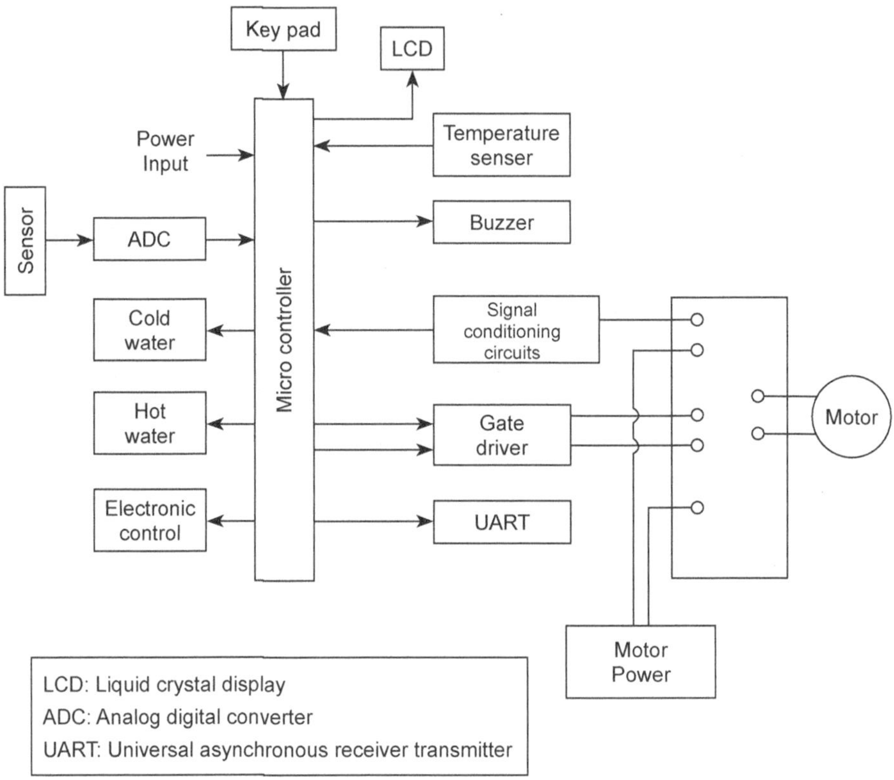

Fig. 17.12

(*b*) Precautions While Using an Electronic Washing Machine

1. The machine should be used at 220 V 50 Hz, single phase supply.
2. The machine must be properly earthed before operation. For this, 3 pin plug and properly earthed socket of 15 A rating should be used. The power supply plug head is generally provided with an inbuilt fuse of 12 Amp. rating. The same can be replaced (in case it is blown off) by a good fuse of proper rating.
3. Use of extension board should be avoided.

4. It is advisable to run the machine empty at 60°C for the first time after installation.
5. The program knob should be turned only clockwise.
6. Program selection must be done prior to switching ON the machine. For resetting a program, switch off the machine, select the programm and restart.
7. Ensure that the electrical line on which machine is being used can take the required load.

Table 17.1 Trouble Shooting in Washing Machines

S.No.	Fault	Reason/Remedy
1.	Pilot lamp does not glow.	(*i*) Check if the machine is ON. (*ii*) Check fuse. (*iii*) Check for loose connections.
2.	Water does not enter machine/ detergent not being flushed into drum.	(*i*) Check water supply. (*ii*) Water tap not turned ON. (*iii*) Port hole door not firmly closed.
3.	Water leakage from machine.	(*i*) Inlet/outlet hose connection loose. (*ii*) Inlet hose leaking. (*iii*) Fluff-filter not screwed proper.
4.	Machine vibrates during spinning.	(*i*) Machine not levelled properly. (*ii*) Clothes not loaded properly.
5.	Machine does not start.	(*i*) Check position of programm selector knob. (*ii*) Check water supply. (*iii*) Check if the portable door is closed.
6.	Machine does not spin.	Check, if spin button is pressed.

17.6 ELECTROSTATE/PHOTOSTATE MACHINE/ PHOTOCOPIER

The electrostate machine is used to transfer an image/figure/message on a paper from the original. The process involves the following steps:

(*i*) The transfer of image on a photo conductor surface from the original.

(*ii*) Then fixing the image on the paper.

The automatic electrostate machine is similar to a manual machine except, that a *selenium coated drum* replaces the conventional selenium plate. Note that selenium is a photo conductor material.

(*a*) Schematic diagram

The Fig. 17.13(a) shows schematic arrangement of an electrostate machine. It uses fibre optics, which reduces' the size and weight of the machine. It also reduces loss of light intensity. Halogen lamps are used in the machine, which further reduce exposure time of the selenium drum.

The Fig. 17.13(b) shows of the machine with important parts

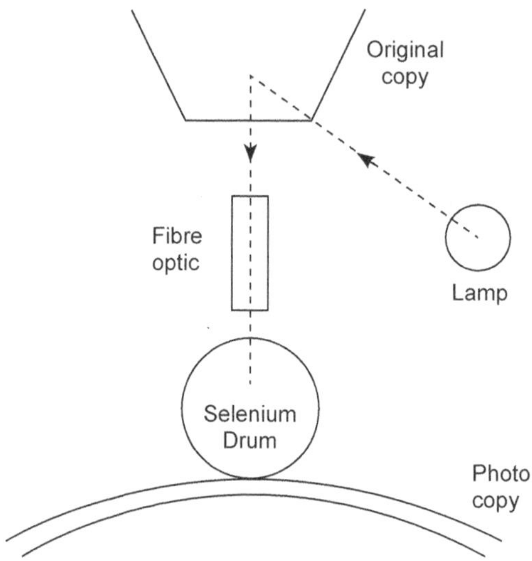

Fig. 17.13(a)

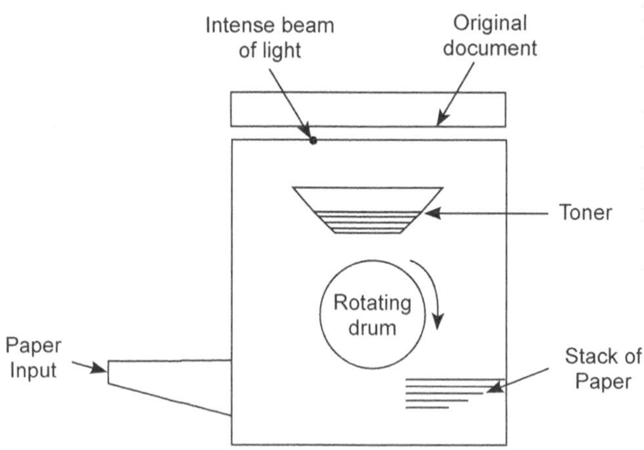

Fig. 17.13(b)

(*b*) Principle of Operation

The photostate (xerographic) copying machine is an important application of electrostatics. The surface of the photo-conductor is initially charged uniformly as shown in Fig. 17.14(*a*). When light from the document to be copied is focussed on the photo conductor, the charges on the lower surface combine with those on the upper surface to neutralise each other. (Fig. *b*) The image is now developed by pouring a charged black powder over the surface of the photo conductor. The electric field attracts the charged powder which is lateron transferred to the paper and melted to form a permanent image.

Note the permittivity of the upper surface is ε_1 and of the lower surface is ε_2.

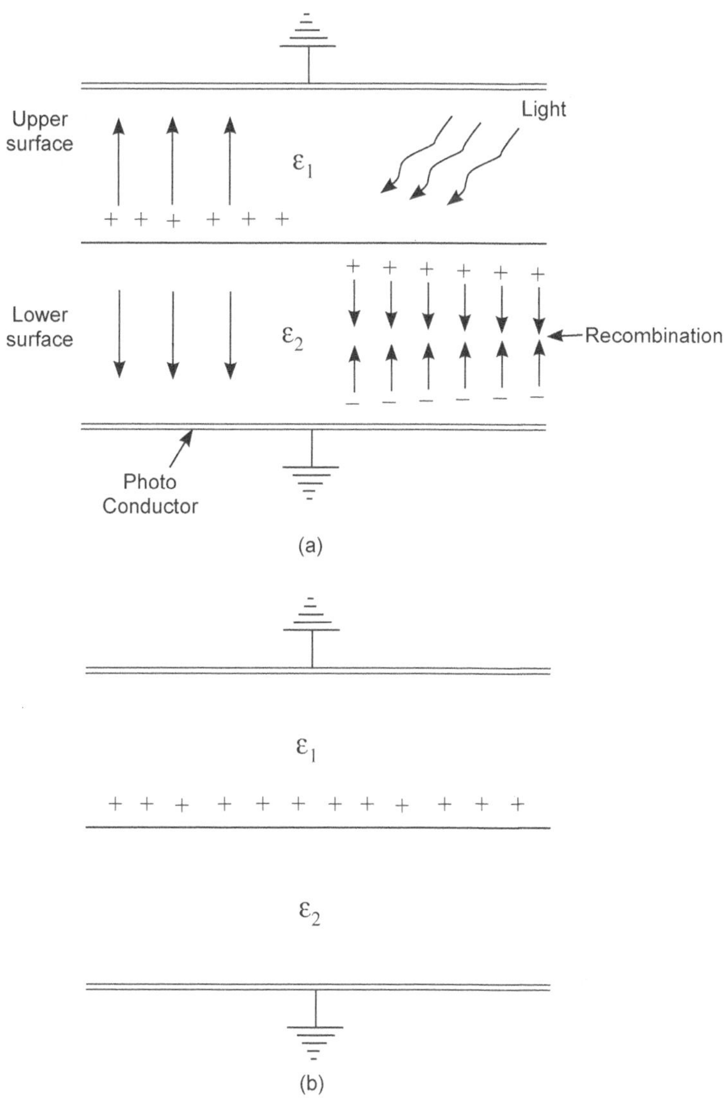

Fig. 17.14

17.7 ELECTRONIC IGNITION SYSTEM FOR AUTOMOBILES

The Fig. 17.15 shows an electronic ignition system for automobiles. The circuit uses a NPN transistor in common emitter configuration.

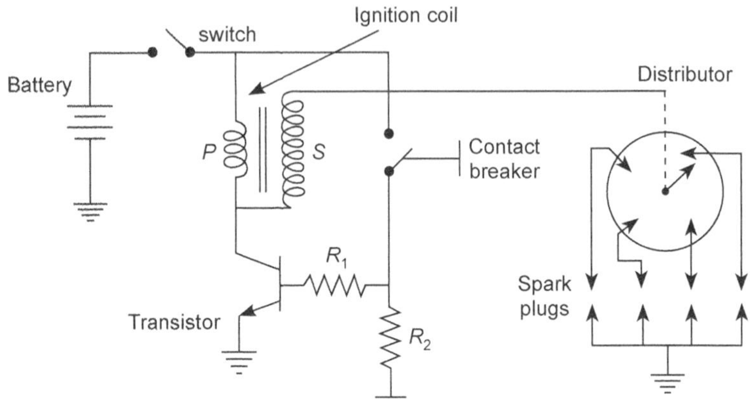

Fig. 17.15

When the contact beaker is closed, a small current (= 0.2 Amp) flows at the base of the transistor, as a result the transistor becomes ON and a collector current flows in the primary winding (P) of ignition coil. Generally the primary and secondary windings of the ignition coil are connected in an auto-transformer configuration. The secondary winding (S) has large number of turns about 50 times of the primary.

When a spark is needed, the cam opens the contact breaker. This makes the transistor OFF. As a result, the primary is open and a high voltage is induced in the secondary of the ignition coil. This high voltage is applied to the spark plugs turn by turn of the multi cylinder engine through the distributor, which results in the ignition.

The contact breaker may be altogether replaced by a "pulse generator", thus problem of erosion of contacts can be totally eliminated.

(*a*) Block Diagram of Electronic Ignition of Automobiles

The simplified block diagram of an electronic ignition system is shown is Fig. 17.6. The voltage pulses produced by the contact breaker are reshaped by the schmitt trigger and applied to monostable multiviberator (MV-1). This stage has been arranged such, that it provides pulses of about 1.8 ms at its output. This corresponds exactly to the required spark duration. In a four-stroke/four cylinder patrol engine, each cylinder is ignited at every second revolution of the engine. Since the four cylinders catch fire at regular intervals, two sparks are needed for each engine revolution, at an engine speed of 6000 rev./min (rpm) and interval between two sparks in 5 ms. As the spark duration is 1.8 ms, the coil has 3.6 ms to restore its energy.

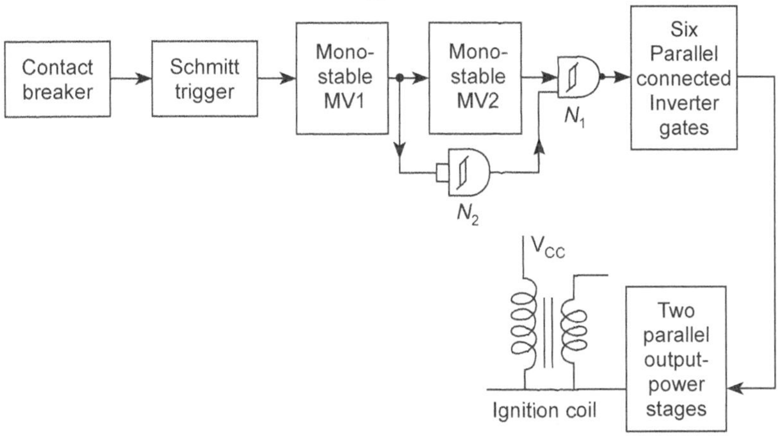

Fig. 17.16

Monostable multiviberator MV 1 is triggered at its input every time, it gets a pulse at its output. This pulse is applied via NAND (N_2 and N_1) gates to parallel connected inverter gates in order to drive the power output stages. For safety, the two parallel output power stages are used. Monostable multi vibrator (MV2) is triggered by the output pulses of MV1 and generates for each of these, a pulse of about 0.5 s at its output. The duration of this pulse is determined by the *time constant*. The pulse ensures that NAND gate N_1 remains open to accept control pulses. When the engine stops, the contact breaker no longer provides control pulses and the NAND gate closes after 0.5 s. This ensures that the ignition coil does not burn out, when the engine is not running. In that condition, parallel-connected resistors allow a current of about 250 mA to prevent the contact breaker points from corroding.

(*b*) Advantages of Electronic Ignition System Over Mechanical Ignition System

(*i*) The problem of erosion of contacts is eliminated.

(*ii*) The ignition is as per the speed variations.

(*iii*) The system needs little servicing.

17.8 RECTIFIERS/BATTERY ELIMINATORS

We know that generation, transmission and distribution of A.C. is economical but at few places, we need D.C. supply, *e.g.,* electronic devices and circuits. Here, we convert the available A.C. supply into D.C. supply. The process is called 'rectification' and the devices used for rectification are called 'rectifiers'.

Diodes are used as rectifiers. Battery eliminators are basically rectifiers. They eliminate battery, hence the name.

The rectifiers (battery eliminators) can be classified as:

1. Half-wave rectifiers
2. Full-wave rectifiers
   (*a*) Centre tap rectifiers
   (*b*) Bridge rectifiers

The readers should recall that a diode acts only in one direction, *i.e.,* when it is forward biased. Due to this property, diodes are used as rectifier.

17.8.1 Half-Wave (H.W.) Rectifiers/Battery Eliminators

These circuits rectify only the half (positive) cycle of the A.C. supply and hence the name. The circuit uses only one diode.

Circuit

The A.C. supply to the diode is given through a stepdown transformer which steps down the voltage to be supplied to the diode. The output D.C. voltage is obtained across a load (See Fig. 17.17 (*a*))

Operation

The stepdown voltage appears across secondary MN of the transformer. This becomes the A.C. input to the diode:

(*i*) When positive cycle of the A.C. input appears across the diode, *i.e.,* the end *M* is positive and end *N* is negative, the diode becomes forward biased and is short circuited. As a result, the whole A.C. input of the positive cycle appears across the load.

(*ii*) When negative cycle of the supply appears, *i.e.,* end *M* becomes negative and N positive. The diode is reverse biased and is open circuited. As a result, the whole A.C. input of the negative cycle appears across the diode and output across the load is zero.

Figure 17.17 (*b*) shows waveform for A.C. input and D.C. output.

Note:

(*i*) Only positive cycle is rectified and negative cycle is dropped across the diode, hence efficiency is low (about 40%).

(*ii*) The D.C. output is not pure and such an output is called 'pulsating' D.C. output, *i.e.,* it contains some A.C. ripples (components) also. By using filter circuit, we get pure DC.

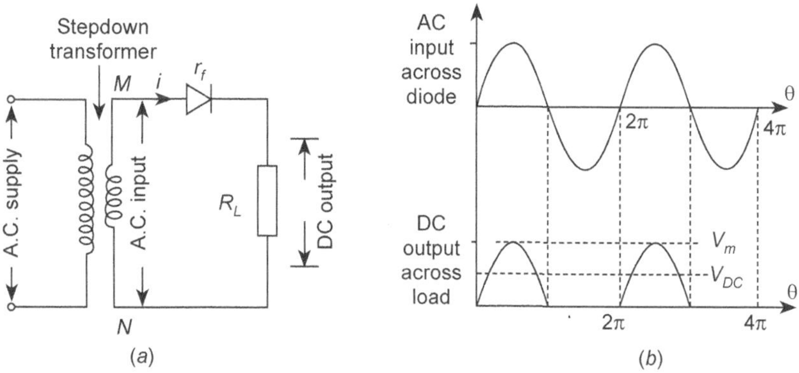

Fig. 17.17

(*iii*) During negative cycle, the diode is reverse biased. The whole negative cycle appears across the diode; hence, the diode used for half-wave rectification should have high value of peak inverse voltage (PIV).

17.8.2 Full-Wave (F.W.) Rectifiers/Battery Eliminators

In these circuits more than one diode is used. This enables the circuit to process both cycles of the A.C. supply. During both the cycles, current flows through the load in the same direction.

They are of two types:

(a) Centre tap F.W. rectifier:

Circuit

The circuit uses a stepdown transformer with the centre tapped (C.T.) secondary. In addition to this, two diodes are connected as shown in Fig. 17.18 (*a*). One diode processes positive half cycle and the other diode processes negative half cycle.

Operation

(*i*) During positive half cycle of the A.C. supply, diode D_1 is forward biased as the end A is positive and B negative. This makes diode D_1 forward biased and D_2 reverse biased. As a result current flows through the diode D_1, and through the load R_L as shown by arrows. The diode D_2 does not conduct during this period.

(*ii*) During negative cycle, end B becomes positive and A negative. This makes D_2 forward biased and D_1 reverse biased. The D_2 conducts and current flows through D_2 and load R_L as shown by arrows. The diode D_1 does not conduct during this period. It can be seen that the current in both the cases flows through R_L, from S to K, hence we get a unidirectional (direct) current.

Figure 17.18(*b*) shows the wave forms for A.C. input and D.C. output.

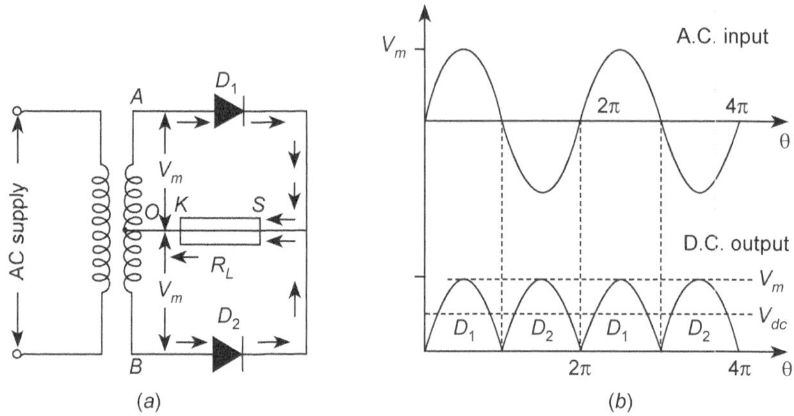

Fig. 17.18

Note:

(*i*) Both the cycles of A.C. input are rectified. One cycle appears across *AO* and the second cycle across BO. (Fig. 17.18). Both are rectified by different diodes. The efficiency of this circuit is double of that of the H.W. circuit.

(*ii*) The D.C. output in this case is also not pure, i.e., we get pulsating D.C. which contains A.C. components, but their number is less than in H.W. rectified output. By use of filter circuit, we obtain pure DC.

(*iii*) The diodes must be capable of bearing high value of PIV The value of PIV across each diode in this case is twice the maximum voltage across the half secondary winding (PIV = 2 V_m).

(*iv*) The D.C. output is small, because each diode processes one half of the voltage. Moreover, it is difficult to locate the centre of the secondary of the transformer correctly. Due to these disadvantages bridge rectifier is preferred.

(b) F.W. bridge rectifers/battery eliminators: As already mentioned, in the case of centre tap (C.T.) rectifier circuit, it is difficult to locate the centre point on the secondary of the transformer correctly and this may give distortion in the rectified D.C. output. Moreover, PIV value is also high in C.T. rectifier circuit. These drawbacks have been eliminated in bridge rectifier which employs 4 diodes. By using 4 diodes, its output is twice that of the C.T. circuit for the same secondary voltage. As shown in Fig. 17.19, the load R_L is connected between the two ends P and R, the arrangement gives it the shape of a bridge hence the name.

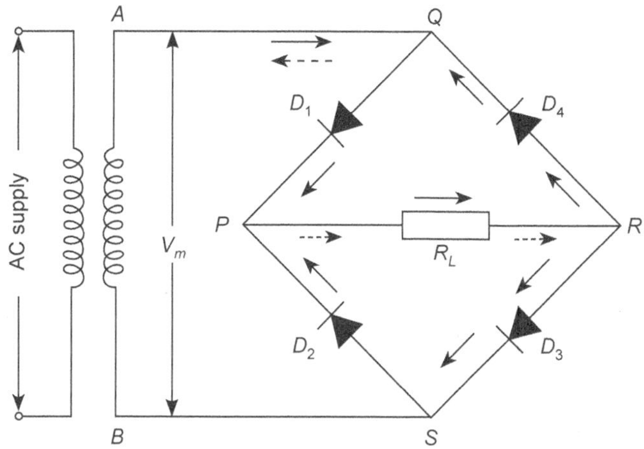

Fig. 17.19

Operation

During positive half cycle of the secondary voltage, the end A of the secondary becomes positive and B becomes negative. As a result the diodes D_1 and D_3 become forward biased which conduct. The direction of current is shown by solid arrows, *i.e.*, through $A\ Q\ P\ R\ S\ B\ A$.

During the negative cycle, the end A becomes negative and B becomes positive. The diodes D_2 and D_4 become forward biased and conduct in the direction shown by dotted arrows, *i.e.*, $B\ S\ P\ R\ Q\ A\ B$.

It can be seen that current in both the cases flow through R_L from P to R, hence we get a unidirectional (direct) current.

Figure 17.20 (a and b) shows waveform for A.C. input and D.C. output, respectively.

Note:

(*i*) Each of the cycle is rectified by two diodes simultaneously; hence, the output is more than the C.T. rectifier circuit.

(*ii*) The D.C. output in this case is also not pure, *i.e.*, we get pulsating D.C. however, in this circuit. A.C.

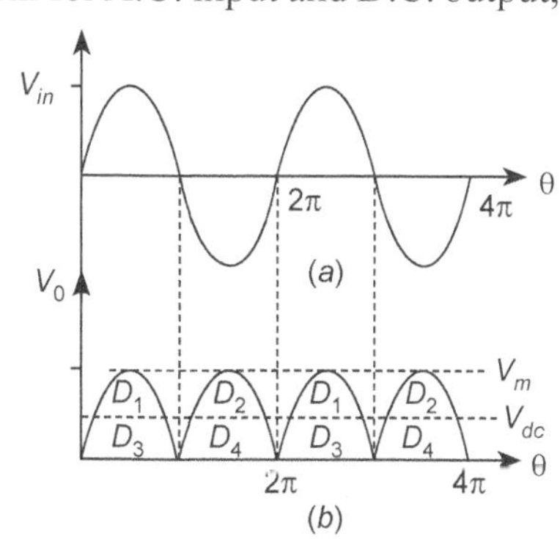

Fig. 17.20

components (supplies) are further reduced. We get pure DC through filter circuit.

(*iii*) The diodes required are of lesser PIV value than in C.T. circuit. The value of PIV across each diode in this case is only V_m, the max. voltage across the secondary (*i.e.*, PIV = V_m). Remember that the value of P.I.V. is same as in the case of HW circuit.

(*iv*) The two diodes simultaneously conduct in series, hence voltage drop in them increases.

17.9 REGULATED POWER SUPPLY

A power supply which maintains a constant D.C. output irrespective of fluctuations in the A.C. input or in the load, is called a regulated power supply. It needs an additional device called regulator after the filter output (See Fig. 17.21).

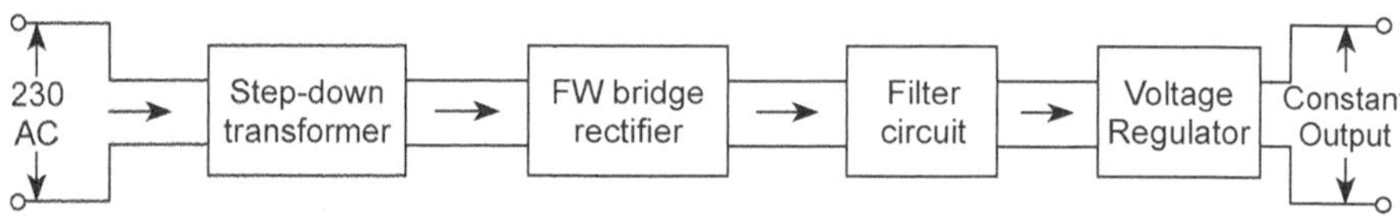

Fig. 17.21

1. Voltage (or line) regulation

The change in the D.C. output voltage from no load to full load with respect to full load voltage of a power supply is called its *voltage regulation*. It is given in percentage.

$$\% \text{ Regulation} = \frac{V_O - V_L}{V_L} \times 100$$

where

V_O = no load D.C. output voltage

V_L = full load D.C. output voltage

The regulation of unregulated (ordinary) power supply is poor. An ideal regulated power supply should have a regulation not more than 1%.

For example, if no load voltage of a supply is 10 V, and full load voltage is 9.9 V, the % voltage regulation (V.R.) of the supply would be

$$\text{V.R.} = \frac{10\ V - 9.9\ V}{10} \times 100 = 1\%$$

17.9.1 Type of Regulated Power Supplies

Regulator Supplies are of two types

(1) Linear supply

(1) Switching mode power supply (SMPS)

The conventional AC/DC supply comprising a transformer, rectifier, filter and a regulator (series or shunt) constitute the *linear supply*. The active device used as regulator is always operated in the active region of its characteristics.

On the other hand, the regulator in Switching Mode Power Supply (SMPS) is always operated in the *cut off or saturation region* of the characteristics. The input DC is chopped at high frequency using an active device and a converter transformer. The chopped wavform is rectified and filtered. A sample of output is used as feedback to adjust the duty cycle to achieve regulation.

S.No.	Linear supply	Switching Mode Power Supply (SMPS)
1.	Voltage regulation of a linear supply is between 0.02 – 0.05 percent.	Voltage regulation of a switching supply is between 0.05 to 0.1 percent.
2.	The peak to peak ripple in this case is 5 mV.	The peak to peak ripple in this case is 100 mV.
3.	The efficiency of a linear supply may be upto 50%.	The efficiency of switching supply may be upto 90%.
4.	Power density of a linear supply is 0.5 watts per sq. inch.	Power density of a switching supply is 3 watts per square inch.
Note: In the first 2 parameters, linear supply is superior to switching supply. In the last two, switching supply is superior to linear supply.		

17.9.2 Regulators/Stabiliziers

The device that used as to regulate/stabilize the supply are called regulators these are classified as:

1. *Linear regulators:* upto 50V, zener regulators and above transistorized regulators are used.

   The transistorized regulators control the output voltage by continously adjusting the voltage across a series pass transistor. This is simple but its efficiency is poor due to power dissipation in the transistor.

2. *Switching (SMPS) regulators:* These operate the series pass transistor as a switch. This reduces power dissipation across the series pass transistor, thus efficiency of the circuit is improved. These are also called SMPS (Switching mode power supply) regulators.

   (a) Linear zener diode regulator: The Fig. 17.22 shows circuit for a *Zener regulator*. The Zener diode is in reverse biased connection. The resistor R_S is connected to limit the (reverse) current (I_z) through Zener to a safer value. The R_S absorbs output voltage fluctuations so as to maintain voltage across the load as constant. The input voltage to the zener (output voltage of the filter circuit) V_{in} and the R_S are so selected that the Zener operates in the *breakdown region*. The voltage across Zener (V_Z) in this region is also the voltage across load (V_O).

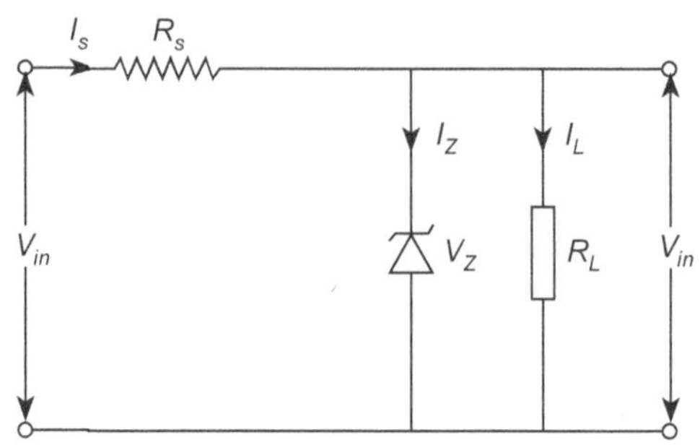

Fig. 17.22

So long the voltage is less than breakdown voltage V_z, the Zener does not conduct. As the V_{in} increases the voltage across Zener exceeds its breakdown voltage, the Zener conducts in its breakdown region. In this situation, the voltage across the load remains constant even current through Zener (I_z) may vary.

(b) SMPS Regulator: The term 'SMPS' stands for switched mode power supplies. The examples of SMP supplies are DC to DC converters, and AC to DC converters (Inverters), U.P.S. etc.

A switched mode power supply (SMPS) employs electronic switching device (BJT, MOSFET) to obtain dc power supply, that has negligible ac ripples and can also be easily controlled. The SMPS is very useful for low and medium power ranges.

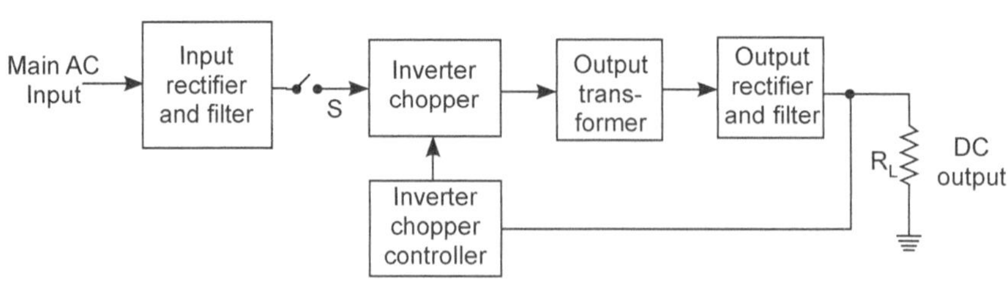

Fig. 17.23

The Fig. 17.23 shows a block diagram of a SMPS regulator. The switching device (S) may be a pair of transistor or thyristor. An electronic device chops the output wavefrom and LC filter does the final filtering. If the switching rate is very high, the filter components will be small.

The rate of chopping should be limited to a frequency of 40 kHz for transistors and upto 250 kHz for thysistors. A higher frequency reduces the size of the components.

The advantage of SMPS is that the electronic switching device is either ON or OFF and thus its efficiency is maximum.

The operation of the switching device is governed by a feedback loop.

17.10 AUTOMATIC VOLTAGE REGULATOR OR STABILIZER

To stabilize the voltage, a voltage regulator or voltage stabilizer is used with domestic refrigerators and air conditioners (A.Cs.). The voltage regulator may be manual or automatic. In manual voltage regulator, one has to check the voltmeter reading and adjust it. Automatic voltage regulator employs relays and other electronic circuitry which automatically keeps the output voltage within very narrow range. The automatic voltage stabilizers are of following ranges:

S. No.	Input voltage	Output voltage	Capacity
1.	165 to 280 volts	220 to 240 volts	0.5 KVA
2.	135 to 280 volts	220 to 240 volts	0.6 KVA

The automatic regulator is costly as compared to manual but is more advantageous because it automatically takes care of the refrigerator and air conditioner. The voltmeter of such regulators, need not to be switched on all the time.

The Fig. 17.24 (*a*) shows the outlook of a automatic voltage stabilizer and Fig. (*b*) shows the block diagram.

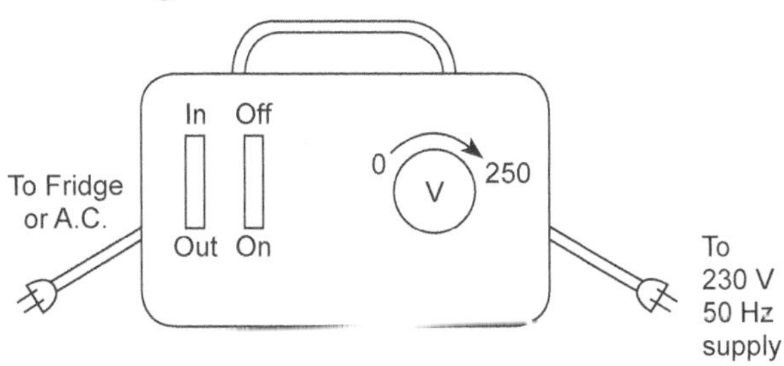

Fig. 17.24(*a*)

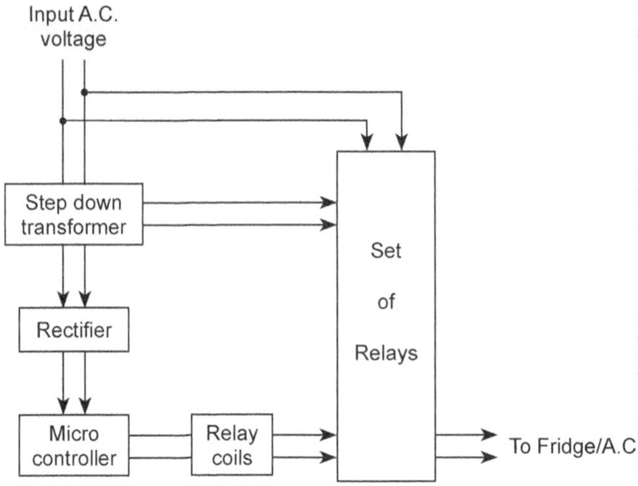

Fig. 17.24 (b)

The Fig. 17.25 shows circuit of a typical automatic voltage stabilizer.

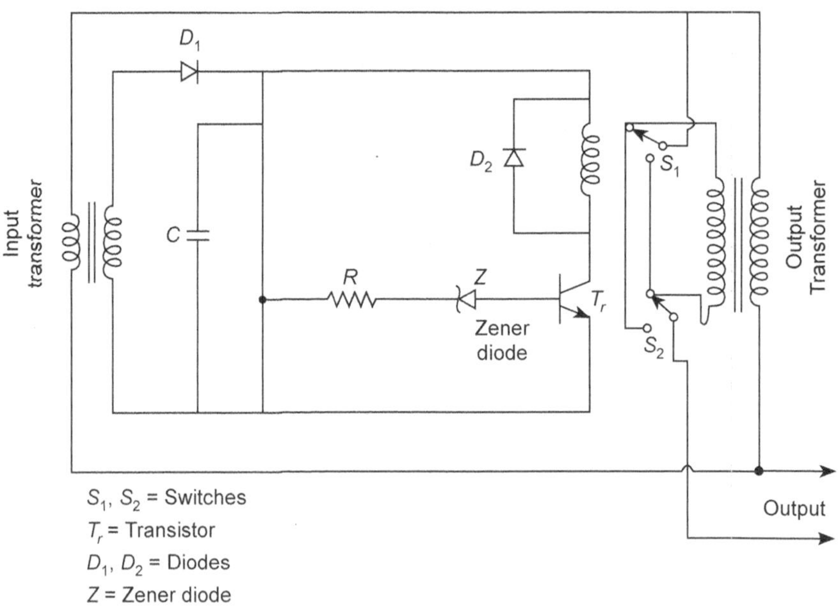

Fig. 17.25

Table 17.2 Trouble shooting in voltage regulator (for Fridge/A.C.)

S.No.	Problem/Symptoms	Possible cause	Solution
1.	The LED not glowing	Now power supply	Wait for the supply
2.	Load is working but voltage display not coming	Faulty display indicator	Contact the manufacturer
3.	Display showing the power but load does not work.	Connect the load directly without the stabilizer if load is not working.	Contact the manufacturer.
4.	Mild shock	Poor earthing	Check the house wiring
5.	Noise/vibration	The relay may be producing noise	Change the relay

17.11 DOMESTIC INVERTER

An inverter converts dc supply to ac supply and is an example of switched mode power supply.

In case of failure of supply, an inverter converts dc supply of the battery into ac and supplies to a house or to an office in emergency.

The output wave from of an ideal inverter should be sinusoidal. However the wave form of practical inverters are non-sinusoidal and contain undesirable harmonics. For low and medium power applications square, quasi square or stepped wave forms may be accepted. However for high power applications pure sinusoidal wave forms are required.

(*a*) Working principle

The Fig. 17.26 explains principle of an inverter.

Switch S_1 is normally closed so that a.c. mains feeds the load directly. The rectifier converts ac to dc and keeps the battery in working condition. The switch S_2 is an overload relay which operates in case of overload and switches off the inverter. The switch S_3 operates in case of under voltage from the supply and connects inverter with the load when main supply is stored.

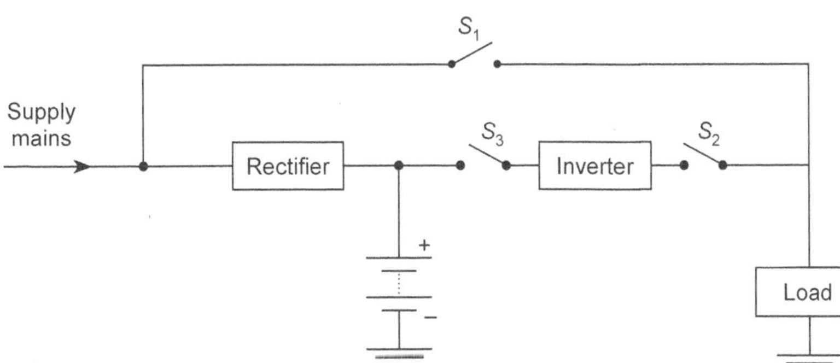

Fig. 17.26

(*b*) Typical Features

Important features of a domestic inverter are:

• Input voltage	220 V ± 40 %
• Output voltage on inverter mode	220 V ± 20 %
• Output voltage on mains mode	Same as input
• Input Frequency	43 Hz to 65 Hz
• Output frequency on inverter mode	50 Hz ± 1 %
• Output wave form on mains mode	Same as Input
• Output wave form on inverter mode	Square wave
• Battery charging current	Constant charging, Approx, 6% of rated battery current
• Efficiency	More than 92%
• Switching from Mains to inverter or inverter to mains	Automatic

(*c*) Suggested load distribution

Rating of Inverter	Suggested load on the Inverter
500 VA	3 fans + 2 tubes or 2 fans + 2 tubes + 1 T.V.
625 VA	4 fans + 2 tubes OR 3 fans + 2 tubes + 1 T.V
750 VA	4 fans + 4 tubes + 1 T.V. or 2 fans + 2 tubes + 1 Room Cooler
1250 VA	6 fans + 8 tubes + 1 T.V. or 2 fans + 2 tubes + 1 T.V. + 1 Desert Cooler

Non sinusoidal output inverter can be used for house and office and sinusoidal output inverter for TV, VCR etc. These inverters use pulse width modulation (PWM) to give sine wave output. See Fig. 17.27.

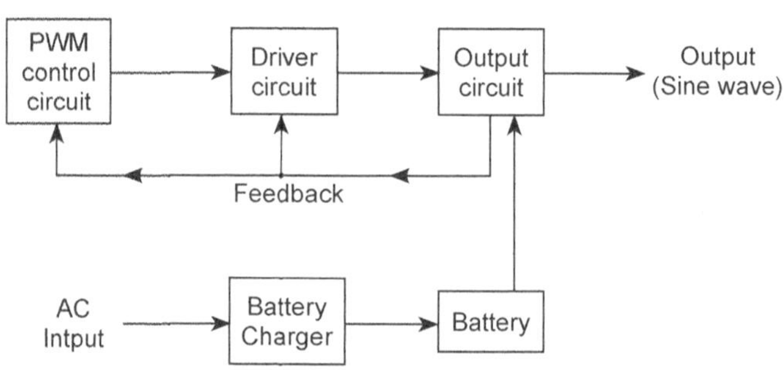

Fig. 17.27

Inverter may be fabricated of transistors, mosfets or of thyristors (SCRs), however *thyristor inverters* are costly.

Fig. 17.28 shows a typical inverter circuit using SCRs. The input to this circuit will be D.C., while the output across load will be A.C. The magnitude of the output will depend upon the turn ratio of the transformer and frequency of the output will depend upon the frequency of the triggering circuit for the SCRs.

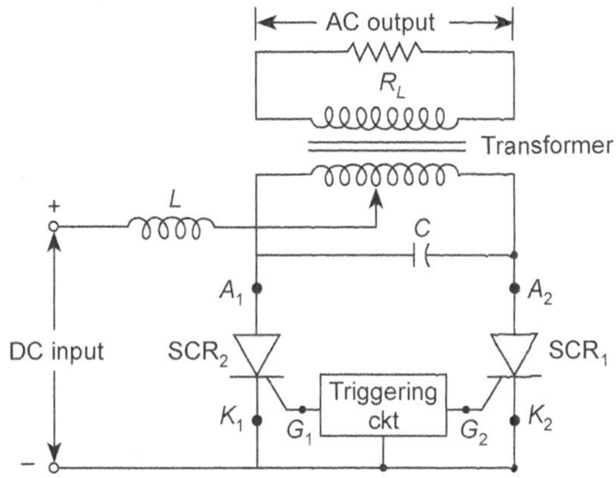

Fig. 17.28

17.12 UNINTERRUPTED POWER SUPPLY (UPS)

This is also a switched mode power supply.

Where as an inverter takes few seconds for switching to the battery supply, a UPS takes milli seconds for this changeover, so the supply is without any interruptions this is, why this is called "uninterrupted" power supply.

The UPS is used for critical loads such as computers. This also provides complete protection against over voltage and under voltages.

17.12.1 Principle of Working

The Fig. 17.29 shows block diagram to explain working principle of a UPS.

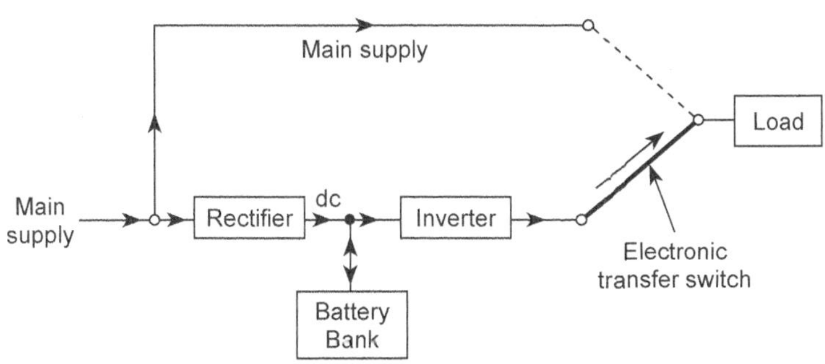

Fig. 17.29

The function of various blocks is given below:

1. *Rectifier:* A rectifier is used to supply power to inverter. Generally a Bridge rectifier is used.
2. *Batteries Bank:* A bank of lead acid batteries is used. When the main supply is off, the batteries supply power to the load. When the main supply is restored, the batteries are put to charging.
3. *Inverter:* This converts D.C. power of the batteries into A.C. power.
4. *Transfer Switch:* When the normal supply is off, the transfer switch transfer the load to the UPS. When the supply is restored, it transfers the load onto the mains. This is an electronic switch (*e.g.,* SCR).

17.12.2 Types of UPS Systems

These are 3 types of UPS systems

1. On line UPS
2. Off line UPS
3. Electronic generator UPS

1. **On line UPS:** In this case, whether the main supply is ON or OFF, the battery operated inverter is ON all the time and supplies the ac output voltage. When the main supply goes off, the UPS supplies power only till the batteries get discharged.

   However, once the main power restors, the bateries get charged again. The switching time of this system is almost zero. Usually maintenance free batteries are used and running time of the inverter is less than 30 minutes. The Fig. 17.30 shows its block diagram. At the input, an *electromagnetic interference* (EMI) filter is used.

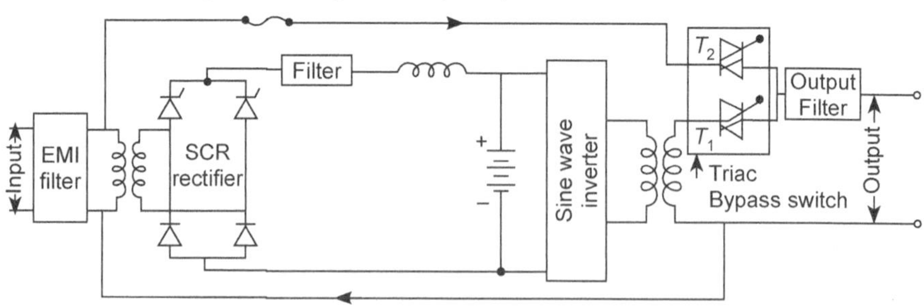

Fig. 17.30

2. **OFF line UPS:** In this case when the main power is ON the inverter is OFF, and the output voltage is obtained directly from the main. The inverter turns ON only, when the main supply goes OFF. Its switching time is less than 5 ms. These are employed with computers, where power interruption for 5 ms does not harm Usually lead acid batteries are employed. The running time of the inverter is almost same as for ON line UPS i.e., upto 30 minutes. The Fig 17.31 shows its block diagram.

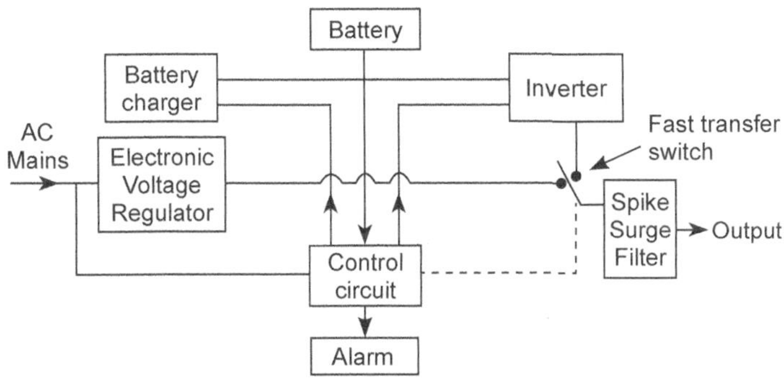

Fig. 17.31

3. **Electronic generator (Home UPS, Fig. 17.32):** Similar to OFF line UPS, in this case also when the main supply is ON, the Inverter is OFF and the output voltage is obtained directly from the mains. Its switching time is 30-100 ms. This UPS is used in houses to run lights, fans, TV etc. The running time of this inverter is from 1 to 4 hours. In this UPS, Pulse width modulation PWM switching and switch mode power supplies (SMPS) control have been used.

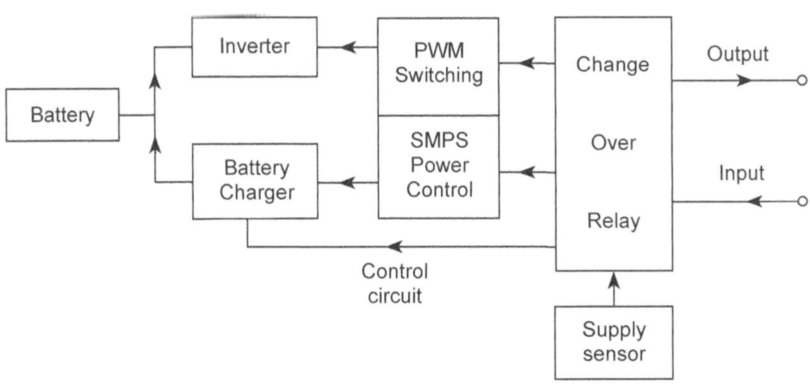

Fig. 17.32

Table 17.3 Comparison of three UPS systems

S. No.	ON-LINE UPS	OFF-LINE UPS	Electronic Generator (Home UPS)
1.	Inverter is ON all the time and supplies output power.	Inverter is ON only when mains is OFF.	Inverter is ON only when mains is OFF.
2.	Switching time=0s	Switching time < 5ms	Switching time 30-100 ms
3.	Used for mainframe or workstation computers or where uninterrupted supply is no problem	Used with PCs or computers or other appliances where a power interruption for 5 ms or less does not matter, when mains supply fails.	Used in houses and offices to run fans, lights, TV, VCR, medical equipment PC where a switching time of 100 ms does not matter
4.	Generally sinewave inverter used.	Sinewave inverters/square wave inverters used.	Generally, squarewave inverters used
5.	Cost is the highest	Cost is medium	Cost is the lowest
6.	High quality sealed maintenance-free batteries used	Sealed maintenance free batteries used.	Generally lead-acid batteries used.
7.	Inverter gives supply for 10–30 minutes	Running time of the inverter is generally less (10 min to 30 min.)	Running time of inverters is high 1 to 4 hours
8.	Built-in batteries used in low capacity models	Built-in batteries used in low capacity models.	The batteries are used separately.
9.	Another alternate power back-up like diesel gen-era-tor will be used as the inverter runs only for 30 minutes.	Another alternate power back-up like diesel generator will be used, as the inverter runs only for 30 minutes.	No alternate power backup is required as the electronic generator will run for a long time.

17.13 IC VOLTAGE REGULATORS

The digital ICs require a $V_{CC} = +5V$ with a variation of ± 5%. When an extremely stable power supply is needed, IC voltage regulator is used. The IC regulators can supply load current beween 100 mA to 3 Amp.

These have three terminals:

The I.C. regulators may be classified as Positive regulators and Negative regulators.

17.13.1 Three Terminal Positive Voltage Regulators

In their basic operational mode, they do not need any external components. They are available both in fixed voltage type and adjustable voltage type, with current ratings of 100 mA, 500 mA, 1.5A and 3.0A. They have positive output.

The examples of these regulators are : LM 78XX and LM 340XX series. These have been designed primarily as Fixed voltage regulators, yet can be used with external components to obtain adjustable (variable) voltages and currents.

The Fig. 17.33 shows the basic operational mode of these regulators. The capacitor C_1 is a ceramic disc capacitor of 0.22 μF capacity and is connected when regulator is located away from the power supply filter. The capacitor C_2 is of 0.1 μF is also of ceramic disc type and is used to improve the transient response.

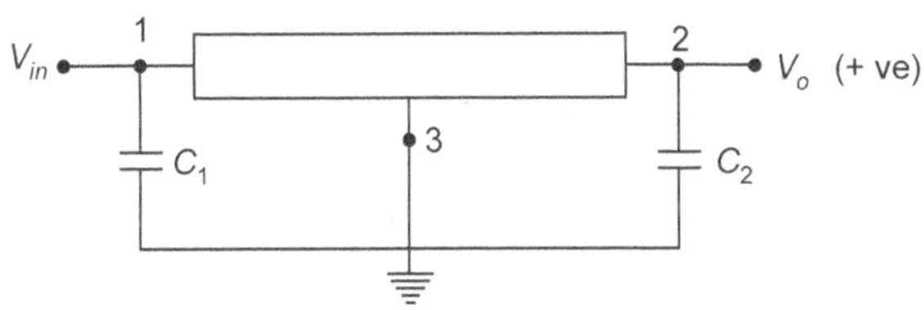

Fig. 17.33

The Fig 17.34 shows views of positive regulators.

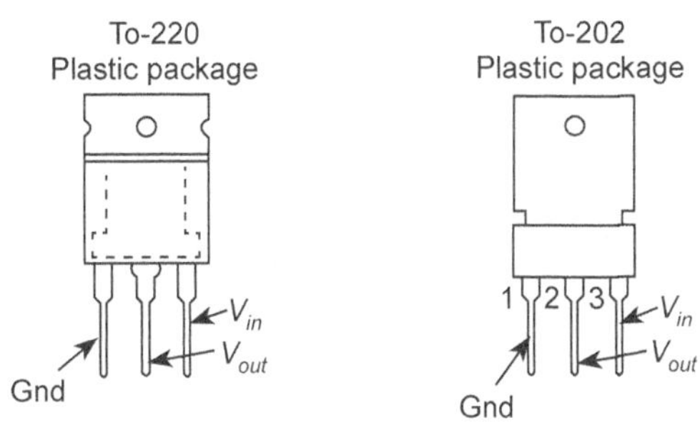

Fig. 17.34

17.13.2 Three Terminal Negative Voltage Regulators

They are similar to the positive regulators with a difference that they have a negative output. (See Fig. 17.35)

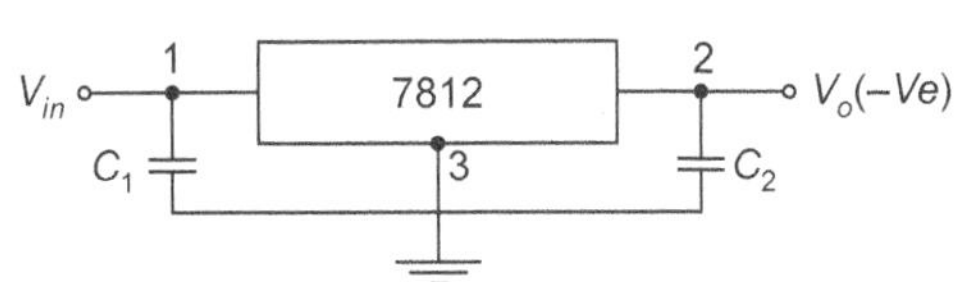

Fig. 17.35

The Fig. 17.36 shows views of negative regulators.

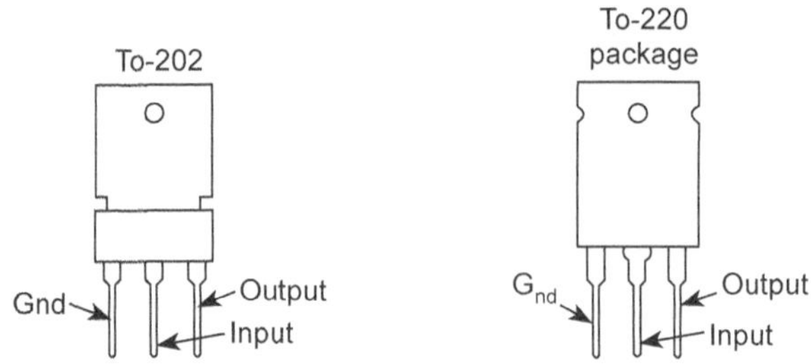

(a) (b)

Fig. 17.36

The IC regulators have the following inbuilt protections:

(*i*) Short circuit protection

(*ii*) Current limiting protection

(*iii*) Thermal overload protection.

17.14 LIGHT EMITTING DIODE (LED)

A LED is a special *P-N* junction (See Fig. 17.37 (*a*). We know, when a *P-N* junction is forward biased the phenomenon of recombination of electrons and holes (charge carriers) takes place. Under this process, with increasing forward bias and decreasing depletion layer at the junction, electrons cross the barrier and go other side, where they fall into a 'hole' and recombination takes place. Each recombination radiates energy.

In ordinary diodes, it is in the form of heat, whereas in the LED it is in the form of light.

Figure (*b*) shows a symbol for LED. The arrows coming out of a crystal diode shows the emission of light. Figure (*c*) shows the process of electro-luminence in a LED.

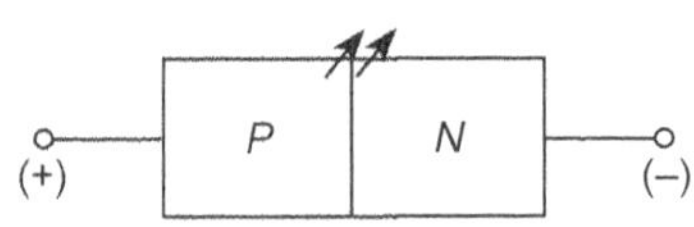

17.37(*a*)

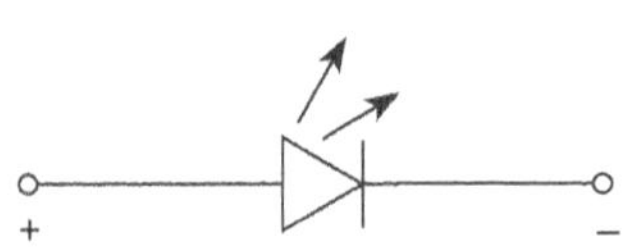

Fig. 17.37 (*b*)

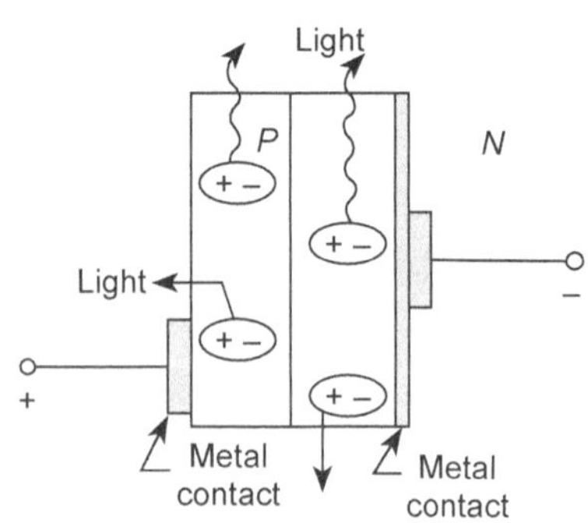

Fig. 17.37 (*c*)

(*a*) Applications

The various applications of LED are:

(*i*) They can be used in burglar alarm/indicating circuits.

(*ii*) They are used as indicator light on panel boards.

(*iii*) In watches, calculators, multimeters, etc., these devices are used as 'Seven-segment display' made of seven LEDs.

(*iv*) We have LED televisions.

(*b*) LED Versus Incandescent Lamps

The following are the advantages of LEDs over conventional incandescent lamps:

(*i*) LED works on 2 V and 10 mA currents, whereas incandescent lamps work on 230 V, 5 A current; hence LED consumes very little power.

(*ii*) LED is fast in action, whereas lamps take 'warm up' time before giving light.

(*iii*) LED is small in size and weight.

(*iv*) LED is shock proof and can be carried anywhere, whereas lamps are fragile and there is every risk of breaking of their filament wire or glass envelope during transportation.

(*v*) LED has very long life. Lamps have an average life of 1000 hours whereas LED goes for years.

But, it should be remembered that LED has a very small power output.

(*c*) LED Versus Ordinary Diode

A LED is different from an ordinary P-N junction in the materials used. A P-N junction is made of germanium or silicon, which emits heat radiations, whereas a LED is made of galium arsenide (GaAs) which emits light radiations during the recombination of charges.

17.15 LIQUID CRYSTAL DIODE (LCD)

They are like L.E.D.s with an advantage that they need a very low power in μW (microwatt) for the display, whereas L.E.D.s need mW (milliwatts) power for the purpose. On the other side, L.C.D.s need a light source (external or internal), whereas L.E.D.s do not need such a source; also, L.C.D.s are degradable with time.

(*a*) Construction

A liquid crystal is a solid organic material which flows like a liquid. The solid organic material is a rigid polymer. Chemically it is pentyl-bi-phenyl

($C_{18}H_{19}N$). The voltage (5 to 24 V) is applied across the crystal, which disturbs the inside molecular arrangement. The incident light when enters the crystal (through the transparent indium oxide layer) the light is reflected and scattered in all directions (See Fig. 17.38).

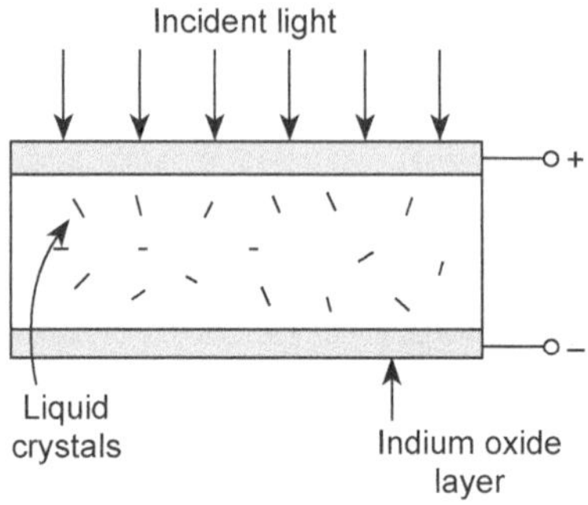

Fig. 17.38

(*b*) Types of LCDs

The LCDs are of two types:

1. *Dynamic Scattering LCD:* It needs an external light source. The liquid crystal fluid in a dynamic scattering LCD is normally clear. Under the application of electric field, ion activity causes turbulence which scatters the external light, yielding a lighted display segment on dark background.
2. *Field Effect LCD:* Field effect LCD has a light source within the LCD. It also contains front and back polarizers at right angles to each other. Without electrical excitation, the light coming through the front polarizers is rotated by 90° in the fluid, passes through the rear polarizers, and is reflected by the mirror. When an electrostatic field is applied, the LCD fluid molecules rotate 90° so that the light is not rotated 90° and is therefore, absorbed by the rear polarizer. This causes the appearance of dark digits on a light background. Most LCDs are prone to premature failure if operated with dc potential, hence an ac square wave drive is usually required.

(*c*) Applications

The L.C.D.s (like L.E.D.s) are also used for seven-segment display in calculators, watches, radiograms. etc.

We have LCD television also.

(*d*) LCD Vs LED

1. The L.C.D.s are slower than L.E.D.s. The response time of L.C.D.s is in milliseconds, while that of L.E.D.s is in nanoseconds.
2. The L.C.D.s have lower power consumption than L.E.D.s
3. The life of L.C.D.s is also more than that of L.E.D.s.
4. There is more colour choice in L.C.D.s.
5. L.C.D.s need an external or internal light source, while L.E.D.s generate light of their own.
6. LCDs suffer from lack of established reliability, limited temperature range, poor visibility, low ambient lighting, slow speed, and need for an ac drive.

17.16 SOLAR CELLS

The solar cell is an example of photovoltaic cell.

The solar cells convert the solar energy into electrical energy. The solar energy is the cheapest and easily available. It is a non-conventional energy source.

(*a*) Construction (Fig. 17.39)

A solar cell is basically a semiconductor diode (*PN* junction). The incident sun light enters through the *P*-type layer and the photons carrying light energy collides with valence electrons in *P* and *N*-type regions and free *electron-hole pairs are generated*, this consititutes a current. The solar cell needs no external bias.

The selenium and silicon are mostly used materials for solar cells but gallium arsenide, indium arsenide, cadmium sulphide, etc., are also used. The spectral response (2000 Å – 7500 Å) of selenium closely matches with that of the human eye.

The solar cells are popularly used as 'solar array (panel)' for the space shuttles and satellites, where they convert solar energy into electrical energy for the shuttle/satellite. In solar panel, large banks of solar cells are used in series-parallel combinations to form solar batteries.

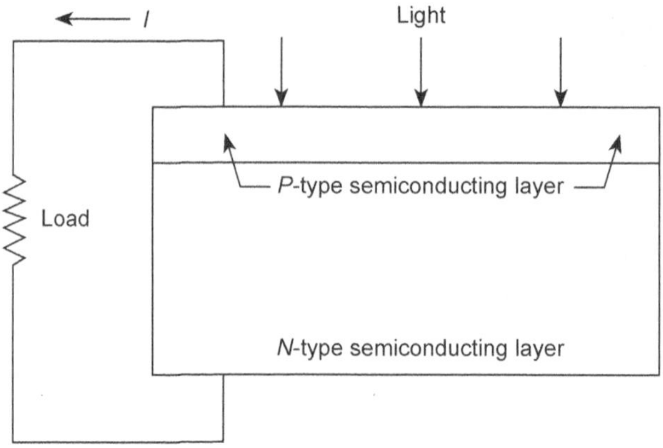

Fig. 17.39

(*b*) Efficiency

The efficiency of solar cell may be defined as the ratio of electrical power developed by the cell to the solar power supplied by the cell.

$$\eta = \frac{\text{Electrical output}}{\text{Solar energy input}}$$

The typical value is 10% to 50%

The solar cell basically consists of a single semiconductor crystal doped with both *P*-and *N*-type materials. In bright sunlight, each cell generates about 0.6 V. The amount of power which the solar cell can deliver depends upon the active surface. A cell delivers 30 mW/sq inch of the surface to a load to 5Ω.

(*c*) Selection of Materials

It consists of the selection of three suitable materials for the following:

(*i*) Semiconductor material with large absorption coefficient and high energy gap of the forbidden band for the device.

(*ii*) Impurity material for doping.

(*iii*) Material for coating the reflecting surface.

(*d*) Applications of Solar Cells

1. In calculators.
2. In solar heaters, solar cookers, solar generators etc.
3. In satellites.

17.17 BATTERY CHARGER (USING SCRs)

Figure 17.40 shows a battery cbarger. The diodes D_1 and D_2 form FW rectifier and provide input to SCR-1 and the battery to be charged. When the F.W. rectified input is sufficiently large to produce the required gate current, SCR-1 will turn ON and the battery charging starts. A Zener diode is put in the circuit for maintaining SCR-2 in OFF state, since the gate current is zero.

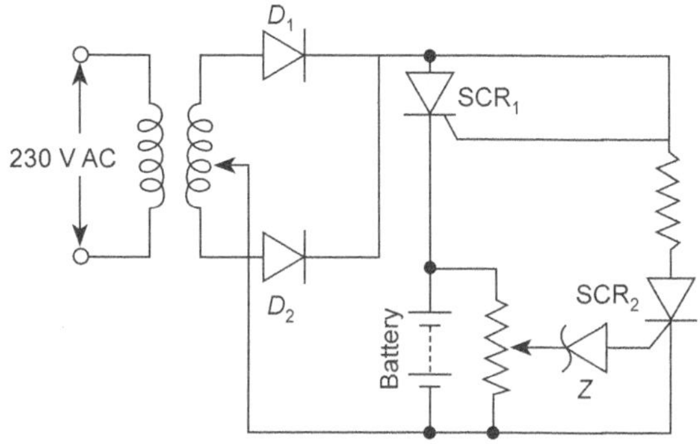

Fig. 17.40

As the charging is continued, at a point the battery voltage reaches to a value which turns ON the SCR-2 (as well as the Zener diode). Now the SCR-1 will turn OFF. When the battery is fully charged, the SCR_2 will cut-off the charging current. Thus the circuit recharges the battery whenever the voltage drops and as well as prevents overcharging of the battery.

17.18 EMERGENCY LIGHT (USING SCRs)

Figure 17.41 shows an emergency light which will maintain the charge on the battery and also provide D.C. energy to the lamp (L) in emergency. The diode D_1 and D_2 form an F.W. rectifier circuit, the output of this circuit appears across the lamp. The battery is being charged through diode D_3 and resistance R at a rate determined by R. Rest of the circuit is self explanatory.

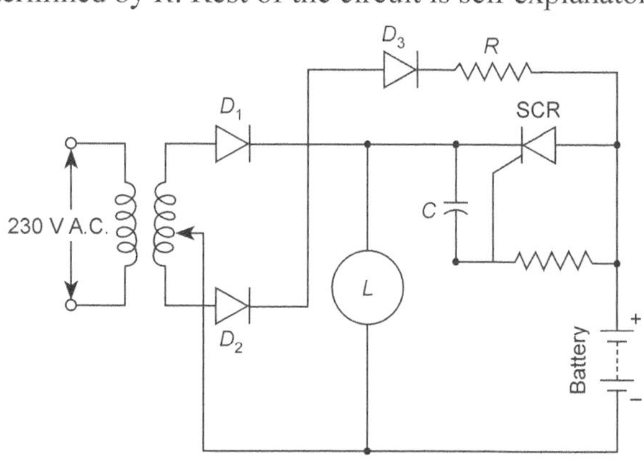

Fig. 17.41

17.19 MOBILE/CELLULAR PHONE/CELL PHONE

This is an advanced telephone system in which we can call/can be called by a person while moving. The mobile may be a handset or fitted in a vehicle. It is also called cellular phone/cellphone as the total area is divided into number of small cells. Each cell is equipped with transmitter and receiver which can operate at a given particular channel.

The working principle of this system is same as of a cordless telephone with the difference that it uses high power devices, covers a bigger range and provides service to many subscribers at the same time.

The mobile phone has replaced *pager, cordless phone* and any many other devices.

1. Terms Related to Cellular Phone Service

Below are the terms which are used regarding the cellular service.

AMPS — Advanced mobile phone service

CTS — Cellular telephone service

MTSO — Mobile telecommunication switching office

MSC — Mobile switching center

MMC —Master mobile center

NCS — Network control system

PSTN — Public switching telephone network

RSG — Remote switch group

SMC — Satellite mobile center

SIM — Subscriber identity module (mobile card)

SAT — Supervisory audio tone.

2. Distribution of Area Into Cells (See Fig. 17.42)

The total area is splitted into small parts known as "cells". Each cell has its own transmitter, receiver and other control equipment and can send/receive calls from the system, moreover each cell has a particular channel to work, however the same channel can be used for more than one conversations. Further, Each cell can handle particular frequencies for two way operation. However the adjacent cells use different frequencies to avoid interference. The cellular telephones have been assigned 666 channels (826 MHz to 845 MHz).

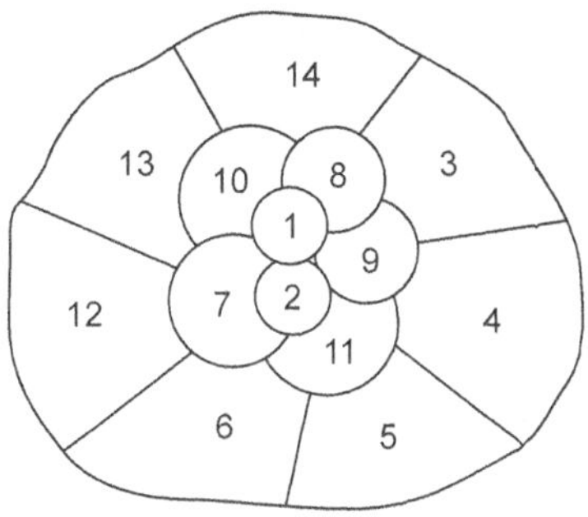

Fig. 17.42

3. Base Station

The base station has a high power transmitter, receiver, control unit and other equipment. It has also an elevated antenna. The base station may be owned by Government itself or by private companies.

The *transmitter* of the base station contains modulators, amplifiers, and the receiver contains demodulators, amplifiers etc. for processing the sound signal. The control unit has logic circuits to maintain a total control on all operations.

4. MOBILE HANDSET

Technically it is called a "cell phone" which can be used from any place within a radius of 50 km from the base station. But when a cell phone is used in a moving vehicle or it is fitted in a vehicle itself, it is called a mobile phone/unit. Note that "mobile" means "in motion". In simple words, the user always keeps the handset with him.

A mobile unit has its own transmitter, receiver and controlling equipment. It works on the same way as an ordinary/conventional telephone except that it is through the base station.

The Fig. 17.43 shows the block diagram of the hand set. Note that it is essentially a microcomputer. The blocks marked as I/O are input/ output devices.

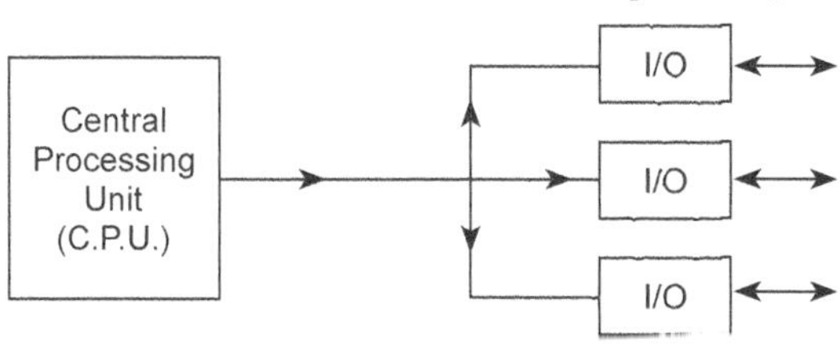

Fig. 17.43

A cell phone (handset) consists of—the mobile terminal (phone itself) and the subscriber identity module (SIM). This card is a "credit card", which fits into the handset.

Sometime if we donot have our own mobile, with this card, we can use any other person's mobile, without affecting his billing.

Note that the mobile handset should be rugged enough as it may drop on the ground. This telephone does not have a dial tone. The availability of channel is indicated by a visual indication. The mobile unit has 10 digit number like other telephones.

5. Working of a Cell Phone

For making a call, when a cell phone is switched ON, a request automatically reaches to the cellsite, which interprets the coming call and sends digital signal to MTSO (mobile telecommunication switching office) which connects the subscriber to the destination. If the user is travelling while calling, the cellular system automatically switches the cell to another cell as and when the caller moves from one area to the another.

The receipt of the call is just opposite. The central office acknowledges the caller and connects the concerned MTSO. A channel is made available and a link is established between "the caller" and "the called".

6. Block Diagram of Transmitter

The Fig. 17.44 shows block diagram of the cell site transmitter.

The frequency generator (F.G.) according to the instructions obtained from CPU of the cell phone, generates the desired channel frequency. This signal

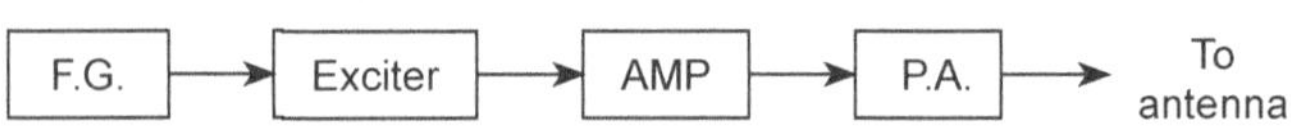

Fig. 17.44

then reaches to the exciter, where the process of modulation occurs. The output is then amplified in stages—finally it goes to the power amplifier (P.A.) and to the antenna as shown.

7. Block Diagram of Receiver

The Fig. 17.45 shows the block diagram for the receiver.

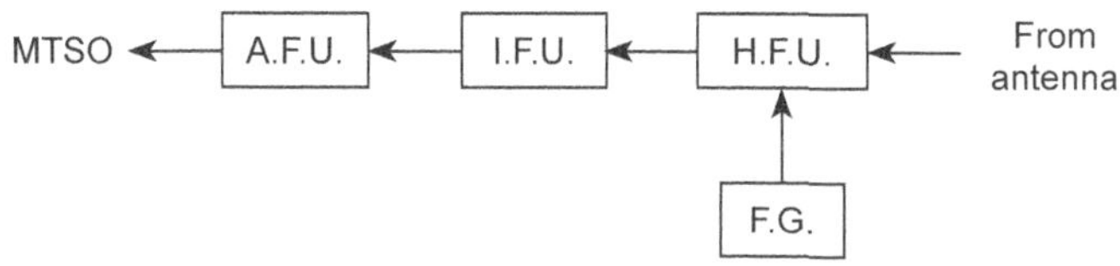

Fig. 17.45

The signal received from antenna goes to the high frequency unit (HFU) where it is preamplified and shifted to the frequenty generator (F.G.). It is then lowered by intermediate frequency unit (I.F.U.) to I.F. range and lowered by Audio Frequency Unit (A.F.U.) to the A.F. range for its further transmission to the MTSO.

8. Complete Block Diagram of a Mobile Phone

The Fig. 17.46 shows the block diagram:

1. *Duplex Filter:* Cellular phones operate on dual frequency : means they can receive and transmit signals simultaneously, Because two different frequencies are used, one for reception other for transmission. Duplex filter is used to connect transmitter, receiver circuits and antenna.
2. *RF Preamplifier:* RF preamplifier provides gain to the received signal. It amplifies the received signal to enhance its level upto a certain amplitude.
3. *Transmitter Section:* The transmitter section includes
   - Audio Per-amplifier
   - Modulator
   - Up-converter
   - RF preamplifier

   Transmitter section is basically a narrowband FM transmitter. In this section, microphone converts sound signal into electrical signal. These low noise signals are amplified by per-amplifier. Now these signal are modulated. These signals have frequency range of 300 Hz to 3 KHz with ⊥ 12KHz Frequency deviation. These signals have a bandwidth of approx. 30 KHz.

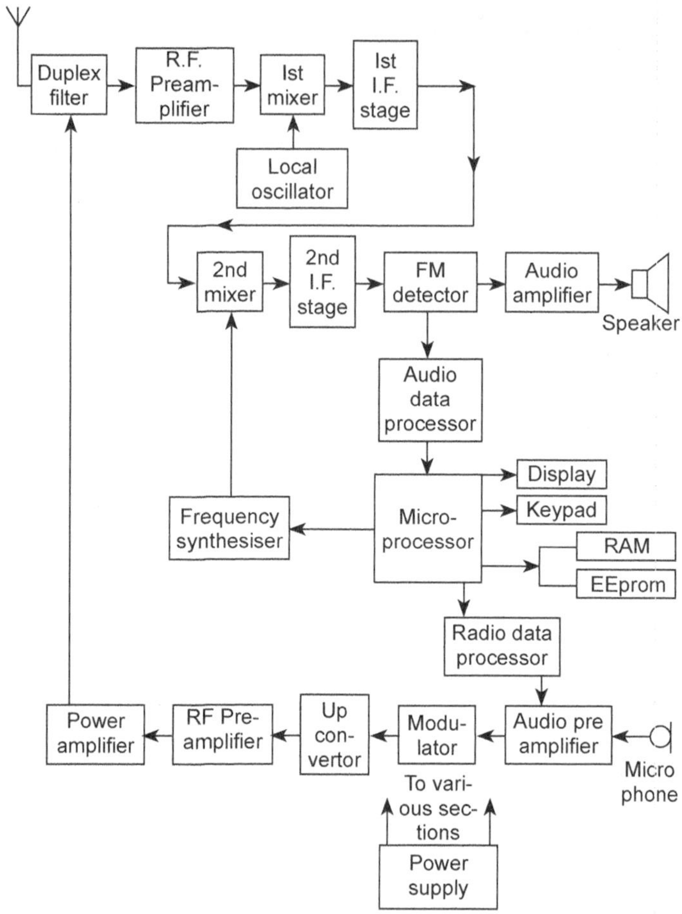

RAM = Random Access Memory
EEPROM = Electrically Erasable Programmble Read only Memory

Fig. 17.46

4. *Receiver Section:* The receiver section includes.

   - Mixers
   - Local oscillators
   - IF Amplifier
   - Amplifier
   - FM Detector

   The combination of local oscillator and mixer changes the frequency of incoming RF signal, which is called IF signal. This IF signal is amplified by IF amplifier stage and then it is again processed by a second stage of mixer and IF amplifier.

   FM detector demodulates the signal.

5. *Frequency Synthesiser.* Frequency synthesiser is an "Complementary Mosfet IC" (CMOS-IC) and used to control the frequency of the either transmitter or receiver. It synthesises the frequency in minimum time.
6. *Radio Data Processor.* Actually it is a controller used to supervise different tasks related to phone operations. It is used to make link between base station and cellular phone. There is a special frequency at which cellular phone and base station communicate with each other.
7. *Microprocessor.* System microprocessor acts as brain of cellular phone. A microprocessor:
   - Extracts the base station message from the received signals
   - Sends out status and identification signals of base station messages.
   - Directs the frequency synthesiser to set and change the transmit and receive channels.
8. *Power Amplifier.* The transmitters used in cellular phones are of low power having range of 3 watt to 6 watt. This power is adjustable and can be controlled by the microprocessor.
9. *Display.* Number of features are increasing in cell phone day by day so does the size of display. These days cellular phones have builtin camera, video GPS(Global Positioning System), phone directories etc.
10. *Power Supply.* The size and weight of power supply is very important factor. Because it is the size and weight of power supply at which cellular phone's size depends. Efforts are being made to reduce the size of battery considerably less.

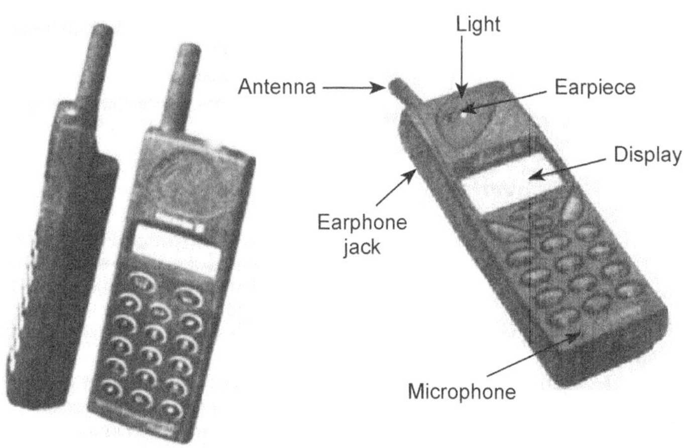

9. Modern Mobiles

The modern mobiles provide the following (and many other) facilities :

Video Camera: You can take Video clips of sports, movies, news, cartoons and a lot more.

Audio: Songs on your handset in Hindi, English and regional languages.

Surf: Access to wide range of interesting sites like Yahoo and Rediff and experience the difference.

News: The latest business, sports and weather news in 6 major Indian languages.

Messaging: Use SMS, e-mail and a range of greetings. Download from hundreds of pictures, add your text and send to loved ones.

Panchaang: Astrology, Hora, Rahukalam and Numerology.

Convenience: Access whenever you need to.

Games: Access vast library of games, download and play.

Finance: Keep a tab on you stocks through a frequently ticker.

Net: Few mobiles have internet facility also.

And video call, various channels and Apps.

etc. etc.

17.19.1 Wire Less Local Loop (WLL)

Is is a mobile with limited mobility. Generally, this does not have "roaming" and S.M.S (Short massaging service) facilities. The W.L.L. works under C.D.M.A. (Code division multiple access), while a mobile works generally under G.S.M (Global system for mobiles) technology.

17.19.2 Mobile Charger

Mobile charger is a common device being used in our home.

The Fig. 17.47 shows block diagram of a mobile charger.

Its important components are

1. *Transformer:* It is a step clown transformer which steps down the incoming a.c. supply.
2. *Full wave bridge rectifier:* It consists of four diodes D_1, D_2, D_3, D_4 in bridge form and converts input a.c. into *dc*.
3. *I.C. Regulator (5V.):* It maintain 5 V at the output precisely inspite of fluctuation at the input.

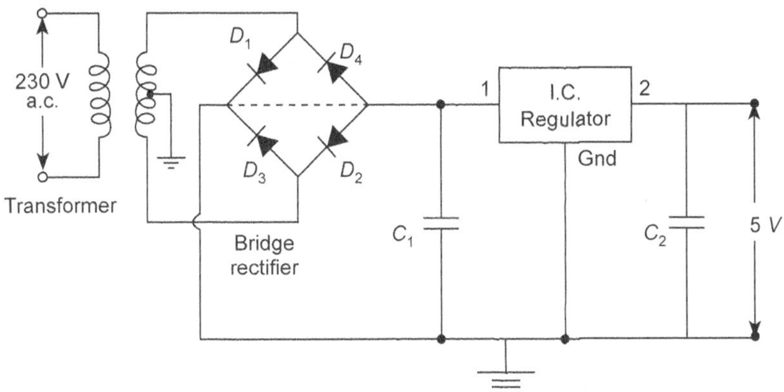

Fig. 17.47

17.20 E-MAIL

It is electronic mail often referred to as E-mail. It is the transmission of textual data from one place to another using electronic means. In simple words, an electronic mail system is a method of sending messages, mail or documents. See Fig. 17.48

Electronic Mail refers to transmission of messages at high speeds over communication channels. The simplest form of computer based mail system allows one user of the service to send a message to another. The second user at his or her convenience retrieves message on his or her display monitor.

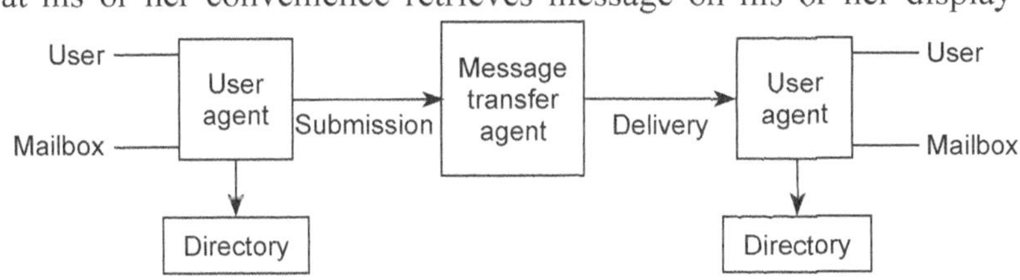

Fig. 17.48

Advantages and Services of E-mail

1. Message can be sent at whatever time or day suits to the user.
2. Messages will be in the recipient's mailbox within minutes.
3. Messages can be marked as "urgent".
4. A reply can usually by demanded.
5. Messages can be read at user's convenience.
6. Incoming messages can be saved.

Apart from the above services, E-Mail also offers Radiopaging *i.e.,* you will hear a beep (audible tone) while receiving an urgent message. Telemessages *i.e.,* replacement of old Telegram system message translation *i.e.,* message can be translated to the recipient own language and at the receiver's end.

Drawbacks of E-mail

1. Recipient must also be E-mail user.
2. Only text can be sent.
3. There is occasional use of junk-mail.
4. Until a mailbox is checked, there is no way of knowing that a message has arrived.

17.21 INTERNET

Internet is the world's Largest Computer Network (rather network of networks) in which a bunch of computers are hooked together. Somehow, all exchanging information.

The Internet is a computer network made up of thousands of networks worldwide. No one knows how many computers are connected to the internet, or how may people use it, though it is certain that both are expanding at a rapid rate. No one is incharge of the Internet. There are organizations which develop technical aspect of this network, but there is no governing body in control.

The internet is the largest computer system in the world. Any person can feed any information into thc NET.

There is no single organization which owns or controls the internet. There is no government regulation or censors governing the information. The internet is run by volunteers. It has no central headquarters, no centrally-offered services and no-link index to tell us what information is available.

(*a*) Equipment for Internet

We need the following equipment for internet. (Fig. 17.49)

1. *Computer.* A compatible PC (Personal Computer) can be used for an Internet connection.
2. *Programs.* We require special programs to use the Internet. These programs are given free of charge by most of the service providers.
3. *Modem.* A modem serves as a medium to exchange information between a computer and the Internet. Modem is device by which computers exchange information over telephone lines. The word "Modem" is short for modulator/demodulator.

4. *Telephone line.* All Internet information travels over telephone lines.
5. *ISDN line.* Integrated Services Digital Network (ISDN) is a line by which data is transmitted over digital telephone lines. ISDN is two to four times quicker than the most speedy modems.

Fig. 17.49

(*b*) Applications of Net

You think of a Job, and the NET is at your service.

Few applications are listed below:

(*i*) Sending e-mail/information/massage.

(*ii*) Receiving information on any subject such as academic, scientific, technical, sports and what not.

(*iii*) *Entertainment :* You can play games: such as chess, cricket and other electronic games

(*iv*) On line shopping.

(*v*) You can locate people, you have lost.

(*vi*) Research

(*vii*) Sales and marketing.

(*viii*) Cable TV through internet.

and so many.

(*c*) Communication Systems

The various communication systems are :

(*i*) TDMA (Time division multiple access)

(*ii*) FDMA (frequency division multiple access) and

(*iii*) CDMA (carrier division multiple access).

Table 17.4: Comparison of three communication systems

TDMA	FDMA	CDMA
1. Shares single carrier frequency with multiple users.	Multiple baseband signals are modulated on different carrier frequencies.	Uses unique spreading codes to spread the base band data before transmission.
2. Frequency/slot allocation is complex.	Number of channels required for transmission and reception are same.	Wide frequency spectrum reduces fading.
3. Pulsating power envelope helps interference with other devices	Continuous transmission implies, no framing bits needed, requires tight filtering to minimize interference.	Interference limited.

Note: GSM is also a popular communication system. It stands for "Gloal System for mobiles".

17.22 COMPUTERS

A computer is found in every home/office. Important terminology related to computers is briefly discussed below.

17.22.1. Microprocessor (μP)

A microprocessor (μP) is a LSI (Large scale Integrated) chip, that does almost all the functions of a CPU (Central Processing Unit). It is also called as a "CPU in a single chip".

The microprocessor is the heart of the microcomputer system. A variety of microprocessors are available in the market and they vary in their process¬ing capability, addressing capacity and the instruction set.

The popular microprocessors are :

8080 A (8 bit, 40 pins), 8085 (8 bit, 40 pins), 8086 (16 bit, 40 pins).

Here, we shall discuss 8085 μP:

Important points on 8085:

1. It is 8 bits general purpose microprocessor.
2. It is capable of addressing 64 K memory.
3. It has 40 pins.
4. It requires 5 V power supply.
5. It operates on 3 MHz single phase clock.
6. Its all signals are classified under six groups :
   (*i*) Address bus.

(*ii*) Data bus.

(*iii*) Control and status signals.

(*iv*) Power supply and frequency signals.

(*v*) Interrupts and peripheral initiated signals

7. It has TTL (Transistor- Transistor Logic) compatible.
8. It is developed by INTEL using NMOS technology.
9. It can interface with almost all peripherals

The Fig. 17.50 shows pin diagram of 8085 micro-processor.

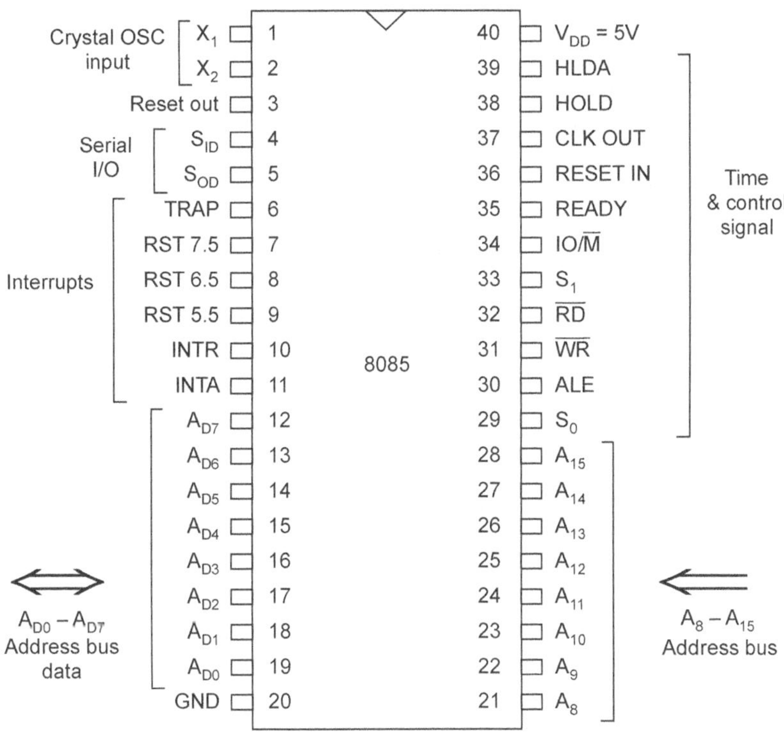

Fig. 17.50

17.22.2 Micro Controller/Mini Computer

It is a control device which incorporates a microprocessor.

A micro controller contains one or more C.P.U.s along with memory and programmable Input/output peripherals. Program memory in the form of ferroelectric RAM (Random Acesss Memory), or ROM (Read only Memory) is also included on the chip.

It is a compact I.C. designed to govern a specific operation such as displaying information in an embedded system.

Microprocessor vs micro controller

A microprocessor is an IC which has only the CPU inside it.

The micro controller possesses a CPU along with RAM, ROM and other pheripherals on the same chip. Therefore it is also termed is a "mini computer".

The 8051 is a 40 pin DIP (dual in package) micro-controller.

17.22.3 Digital Microcomputer (Fig. 17.51)

A micro-conputer is a computer which uses a microprocessor as its CPU (Central processing unit). A typical microcomputer has three types of chips: microprocessor (one chip), memory (several chips) and Input and output (I/O) (one or more chips). In some small memory microcomputers, all these three units can be fabricated on a single chip, *e.g.*, the 8048 microcomputer has an 8-bit CPU, 1088 bytes of memory and 27 I/O lines, all on a single chip.

The microcomputer contains all the blocks of a computer. The only difference being in the CPU. The CPU of the computer is a large unit, whereas in the microcomputer, the CPU is contained in the microprocessor itself.

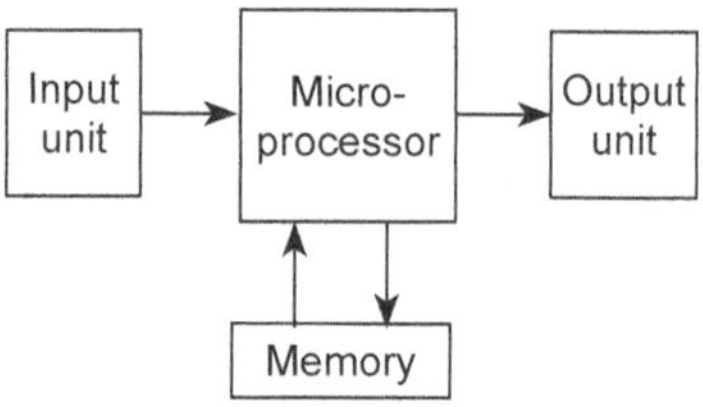

Fig. 17.51

The term micro computer does not essentially mean a small size computer though all small size computers are microcomputers. In fact all the personal computers such as PC-XT, 286, 386, 486 and Pentium fall in the category of the microcomputers. Wheras the *minicomputer*, and *super computer* fall in the category of computers.

17.22.4 Digital Computer

Figure 17.52 shows a block diagram of a digital computer. The arrows indicate the communication between the different sections.

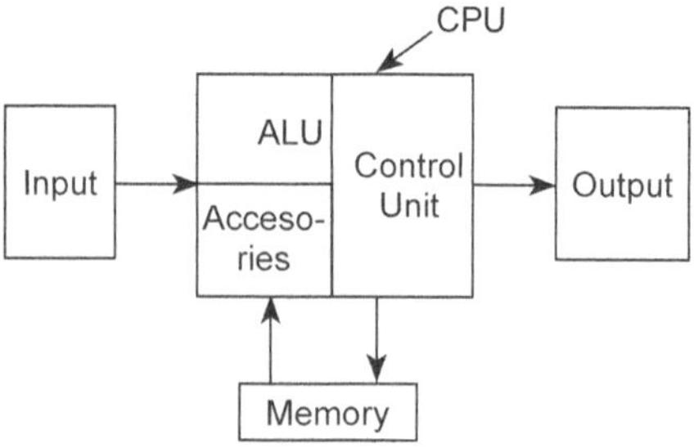

Fig. 17.52

While the central processing unit (CPU) is called heart of the computer, all devices connected to the CPU are known as peripheral devices.

Note: ALU stands for Arithmetic Logic Unit.

17.22.5 Laptop

It is a battery powered personal computer (PC), that can be easily transported and conveniently used in a small space in plane, car, etc. You can keep it in your "lap" and work comfortably. This is why it is called as "Laptop computer". Sometimes, it may have an a.c. cord also plugged into an outlet, which is also used to *"charge"* the battery.

Note that a computer designed for use by one person at a time is called personal computer (PC).

17.22.6 Desktop Computer

This is a personal computer designed for regular use at a single location on or near a desk or table due to its size and power.

Desktop publishing

The production of printed matter by means of a printer linked to a desktop computer with special software is called "desktop publishing". The quality is similar to that of typeset books.

17.22.7 Computer Software and Hardware

The computer software or simply *"software"* is a part of a computer system that consists of data or computer instructions.

The computer hardware are the physical parts or components of a computer, the examples are C.P.U., monitor, keboard, data storage, graphic card, Sound card, mother board etc.

17.22.8 Memories

A memory is a place to store an information temporarily or permanently. It also makes available the same information when needed. Each "cell" where a data is stored in the memory is assigned a "location number" called "Address." The memory in fact is equivalent to thousands of "registers" each storing a binary word. All modern computers due to more reliability and less expensive use semiconductor memories rather than magnetic memories. A typical computer may have a memory with 65, 536 locations, each with a capacity to store 1 byte (8 bits) data.

The memories can be classified as:

1. Megnetic/semiconductor memories
2. Static/dynamic memories
3. Destructive/non-destructive memories

4. Active/passive memories
5. Internal/external memories
6. Random access memory (RAM)/sequential memories
7. Volatile memories
8. Non-volatile memories such as:
   (*i*) ROM (Read only memory)
   (*ii*) PROM (Programmable ROM)
   (*iii*) EPROM (Erasable PROM)
   (*iv*) EEPROM (Electrically EPROM)

Figure 17.53 shows a block diagram of the memory unit of a microprocessor

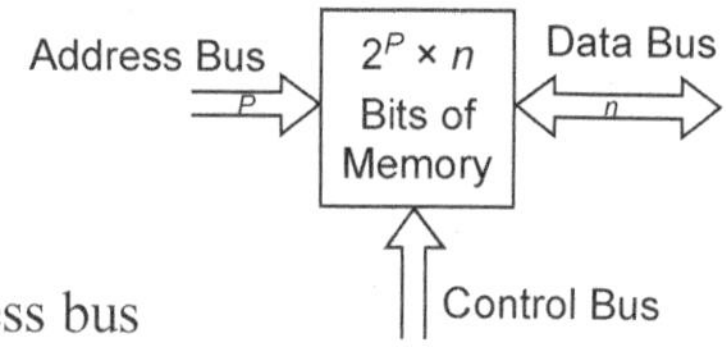

Fig. 17.53

$$\text{Capacity} = 2^P \times n \text{ bits}$$

Where p = address lines in the address bus

n = data lines in the data bus.

Terminology in Memory System

A few terms are briefly explained below:

- *Read:* To retrieve or extract information from a memory cell.
- *Write:* To enter a new data into a memory cell.
- *ROM (Read only memory):* A memory in which the data is stored permanently.
- *RAM (Random access memory):* A memory in which the data may be written or read from as and when desired.
- *Word:* A group of bits formed to give a single piece of information.
- *Capacity:* Number of bits or words a memory unit can store. A memory of capacity 1024 bits is called 1 K bit memory (1 K = 1024 bits)
- *Address:* Location of a word or bit in the memory unit.
- *Access time:* Time needed for locating a particular bit or word in the memory.
- *Random access:* It means that the user can reach any memory location, any time. The access time being equal for each location.
- *Memory cycle:* It is time required to access a particular memory location and perform an operation (Read/Write) in that location.

17.23 MEDICAL INSTRUMENTS

Below we discuss few medical instruments used is our home.

17.23.1 Blood Pressure Monitor

(*a*) Blood pressure (BP) is the force of the blood pushing against the artery walls. Each time the heart beats, it pumps blood into arteries resulting in the highest blood pressure as the heart contracts. The BP is lowest when the heart expands.

Therefore, two numbers are recorded when measuring BP. The highest number called **systolic BP** refers to the pressure inside the artery, when the heart contracts and pumps blood through the body. The lowest number called **diastolic BP**, refers to the pressure inside the artery when the heart is at rest i.e. it is expanding and taking the blood into it.

Both the "systolic" as well as "diastolic" numbers are recorded as "millimeters of mercury" (mm of Hg). This represents, how high the mercury column will be raised by the pressure of the blood.

(*b*) **Standard values of BP**

(*i*) According to the "National Heart, Lung and Blood Institute (NHBLI)". Blood pressures in adults are defined as

- 140 mm Hg : systolic
- 90 mm Hg : diastolic

It is specified as 140/90.

(*ii*) The latest guidelines issued by NHBLI are :

- 120 mm Hg : systolic
- 80 mm Hg : diastolic

It is specified as 120/80

More BP than the standard values given above come in the category of *hypertension* (high BP) which increases the risk of coronary heart diseases (heart attack) and brain stroke (brain attack). With high BP, the arteries may have an increased resistance against the flow of blood, causing the heart to pump harder to circulate the blood. So one should maintain his BP within limits (120/80).

A low BP (< 80 mm Hg) can cause fatigue, tiredness etc. Truly speaking, a low BP is more fatal than a higher

BP. We have medicines to control high BP, but unfortunately no medicine to raise a low BP. However, intake of salt (sodium), coffee (cafene) can raise the B.P.

(*c*) B.P. Measuring Techniques

The BP can be measured in two ways.

(i) Conventional Sphymo manometer: In this method, application of pressure on the upper arm causes a spurt in the blood flow in the artery (situated in the upper arm) and a typical sound is heard from the stethoscope.

In sphymo-manometer, [Fig. 17.54 (*a*) and (*b*)], there is an inflatable CUFF (hard cloth) containing bladder, which can be inflated by squeezing a rubber bulb and the pressure in the cuff is measured by the manometer. The inflated cuff can be deflated slowly by a needle valve.

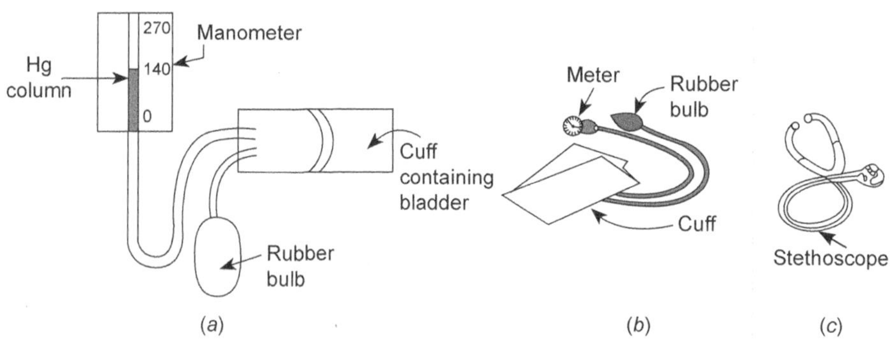

Fig. 17.54

The cuff is fastened around upper arm and with pumping of a rubber bulb, the pressure in the cuff is increased upto 220 mm Hg much above the B.P. of the "brachial" artery (situated in the upper arm). The slow release of pressure in the bulb, causes a "spurt" of blood and a typical sound is heard, now the pressure read by the manometer gives *systolic* pressure. The sound continues to be heard, till *diastolic* pressure is reached when the sound stops.

(ii) Electronic Sphymo Manometer: By the manometer described above, one cannot measure his own 'B.P'. The electronic manometer enables us to measure our own B.P. The electronic manometer instrument can also measure heart rate or pulse.

The *pulse* refers to the rhythmic expansion of an artery, that is caused by ejection of blood from the "ventricle". It can be felt, where the artery is close to the surface of the human resting on something firm.

Normal pulse rate is equal to the heart beat. It is 72 (or between 60 to 100) per minute. Our heart also beats with this rate.

The Fig. 17.55 (*a*) shows the panel of the electronic manometer and fig (*b*) shows its use

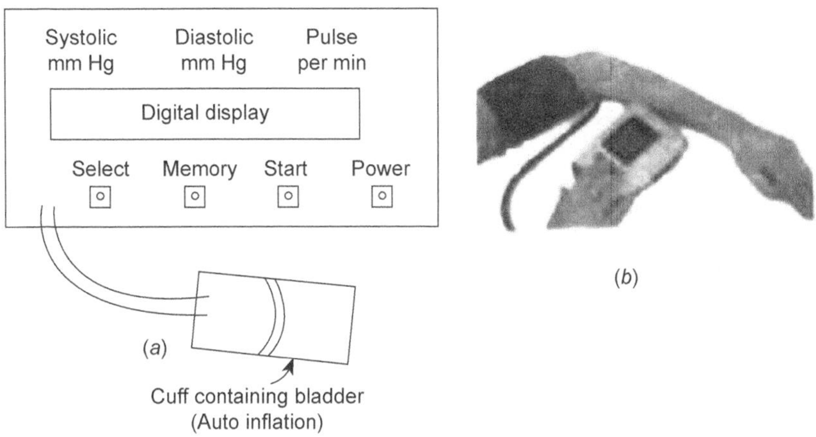

Fig. 17.55

The Electronic BP machine does not have rubber bulb. The cuff has automatic inflation. The stethescope is also not required.

Advantage: The indirect method is simple and easy to use.

Disadvantages: The indirect method has the following disadvantages.

(*i*) It does not provide a continuous recording of the blood pressure variation.

(*ii*) Its practical repetition rate is limited.

(*iii*) It can not generate waveform of the blood pressure.

(*iv*) The method is subjective and does not give proper result, if the B.P. is very low.

17.23.2 Glucometer

(*a*) **Diabetes:**

When the quantity of sugar (glucose) in the blood of a person increases beyond limit the person is known as suffering from "diabetes". The safe limit of blood sugar is given as under:

1. Fasting: When the person is empty stomach, the blood sugar should be less than 100 mg/dl.
2. Two hours after taking meals: The blood sugar may be upto 140 mg/dl.

Any value higher than this declares the person as a "Diabetic". In laymass is "language" it is called as "disease of sugar"

If the glucose level is higher than 200 mg/dl, a *urine test* becomes necessary.

High glucose level than the permissible values is very harmful. It has a very bad effect on the health of various human organs such as, kidneys, heart, lungs, eyes etc.

Diabetese is the result of the malfunctioning of an organ in human body. called *Pancreas*. It converts glucose into *insulin*, which is required by the human body.

The diabetes may be of two types:

(*i*) Type I diabetes: This occurs in children.

(*ii*) Type II diabetes: This occurs in adults.

For a diabetic person, sugar consumption is totally prohibited and he/she is to take medicine in the form of tablets or "Insulin" injection or both.

(*b*) Measurement of glucose in the blood.

The blood sugar can be easily measured by

(i) Simple laboratory test

(ii) Glucometer: This is an electronic istrument to measure glucose in the blood. The Fig. 17.56 shows the block diagram.

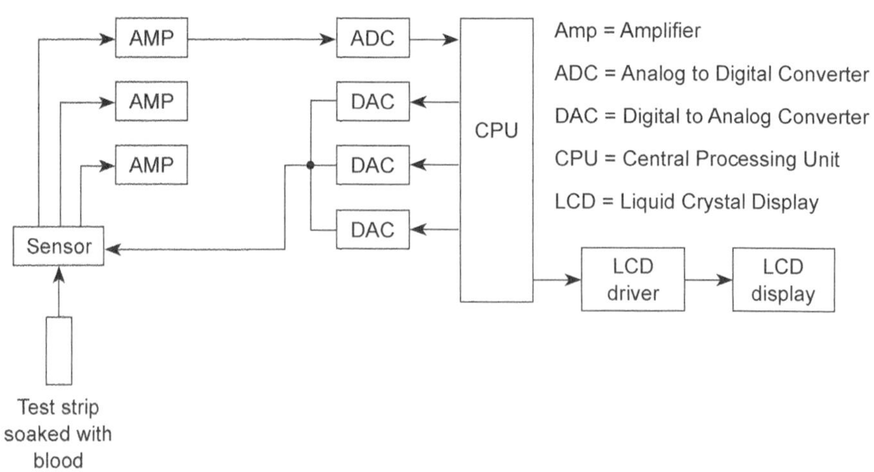

Fig. 17.56

17.23.3 Hearing Aid

(a) When our ear cannot hear properly or it has some hearing problem, we use *"hearing aid"*

The causes of hearing problems are many, few are listed below:

1. Old age
2. When the ear is subjected to high pitch sound repeatedly, the ear drum/ diaphragm gets damaged and also its internal parts lose their strength.

   The strength of sound is measured in *"decibels"* (db). Our normal conversation carries 50 db. The sound above 85 db is harmful. A repeated exposure to this sound may cause temporary or permanent loss of hearing.

(b) There are two electronic hearing aids:

(i) Digital hearing Aid: The Fig. 17.57 shows block diagram of a typical digital hearing aid.

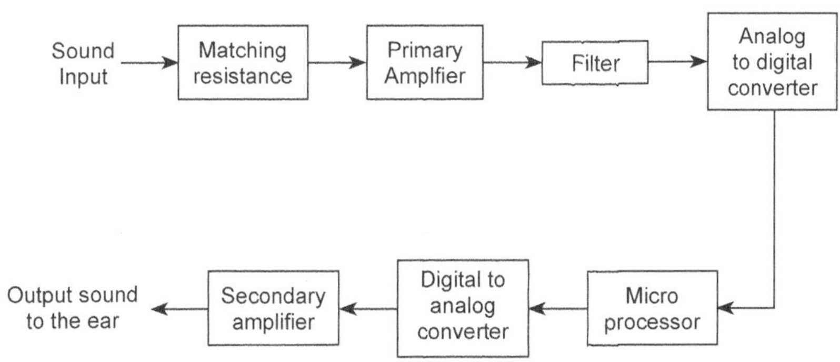

Fig. 17.57

The Weak input sound is amplified and filtered. Then it is passed through an *"Analog to digital converter"* and then through a microprocessor. The digital output is again converted into analog form, amplified finally and passed through an amplifier. The output sound reaches to the ear and can be heard.

(ii) Analog Hearing Aid: The Fig. 17.58 shows block diagram of an analog hearing aid the input sound is passes through a microphone which converts this sound into electricity. This is amplified and given to loud speaker (Receiver) which converts the electricity again into sound which become the input to the ear.

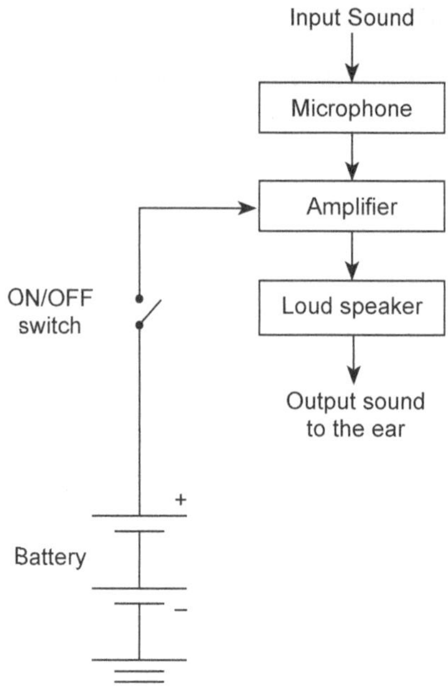

Fig. 17.58

17.23.4 Pacemaker

The electrical impulses from the heart muscles cause the human heart to beat. This electrical signal begins in the sinoatrial (SA) node located at the top of the heart's right atrium. The SA node is called the heart's *"natural pacemaker"*.

When a person's heatbeat is not normal (less than 50 per minute), he is fitted with electronic pacemaker. Electronic pacemaker is a small device, which is placed through a surgery in the chest or abdomen to control heart rhythms. This uses low energy electrical pulses to prompt the heart to beat at the normal rate.

We have two types of pacemaker.

(i) **Analog Cardiac Pacemaker:** The Fig. 17.59 shows block diagram of the analog pacemaker.

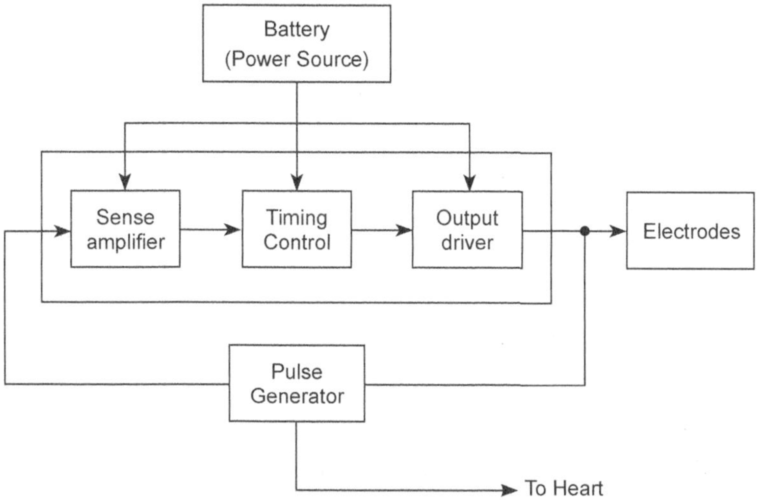

Fig. 17.59

The electrodes detect the heart's electrical activity. If the heartbeat is abnormal, a signal is sent to the pulse generator to send electrical pulses to the heart.

(ii) Digital Cardiac Pacemaker: The Fig. 17.60 shows block diagram of a digital pacemaker

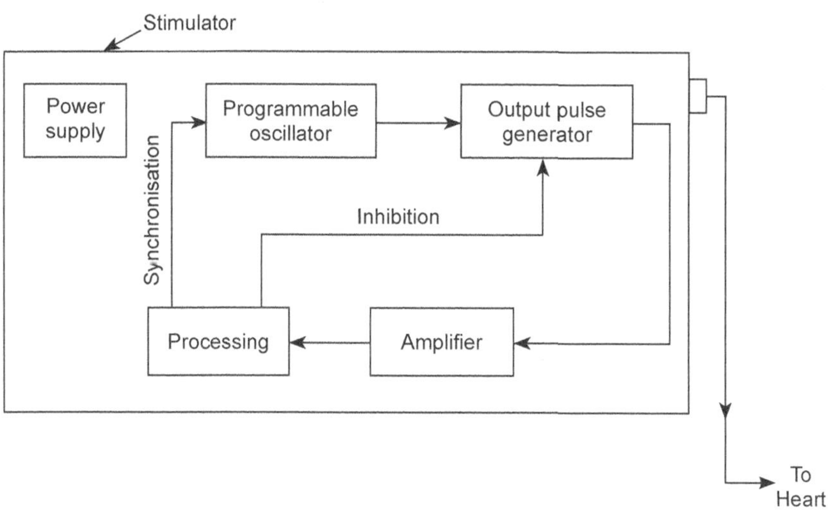

Fig. 17.60

Its functioning is same as of the analog pulsemaker.

SUMMARY

1. Digital watch uses crystal oscillator and a logic circuit.
2. The microwave oven works on the principle of high frequency heating.
3. A cordless phone uses the principle of wireless telephony.
4. By mobile, we can call a person while moving.
5. The e mail is the transmission of data or message from one place to another using electronic means.
6. Internet is a largest computer system in the world.
7. A mobile charger uses a I.C. regulator.
8. In audio video system for noise reduction, Dolby system is used.
9. Today we have μP controlled machines.
10. Binary system is used for arithmetic calculations in computers.
11. The medical instruments use electronic principles.
12. A pacemaker through surgery, is installed near the heart of a patient.

SELF TEST QUESTIONS

1. Explain construction and working of a digital watch.
2. Draw block diagram of a μ-wave oven and explain its working.
3. Draw block diagram of an automatic washing machine and explain its working.
4. Draw block diagram of a mobile phone and explain function of each block.
5. Explain important components of an Internet.
6. Explain Dolby system for noise reduction in audio video systems
7. Write note: inverter, UPS, LED, Microprocessor, Binary system, B.P. measuring instrument.

OBJECTIVE TYPE QUESTIONS

I. Fill in the Blanks

1. The button cells used in digital watch etc. chemically are __________ cells.
2. LED stands for light __________ diode.
3. LCD is the abbreviation of liquid __________ diode.
4. The microwaves are also used for treatment of various diseases such as __________.

5. Radio telephony adopts the principle of __________ phone.
6. The mobile phone is also called __________.
7. Expand CNC __________.
8. In computers __________ system is widely used.
9. Expand UPS __________
10. An example of switch mode power supply is __________
11. For noise reduction in audio video systems, __________ system is used.
12. Many medical instruments use __________ principles.

ANSWERS

1. Mercury
2. Emitting
3. Crystal
4. Cancer
5. Modulation
6. Cell
7. Computerised Numerically controlled
8. Binary
9. Uninterrupted power supply
10. inverter
11. Dolby
12. Electronic

II. State True/False

1. An LED is more superior than an LCD.
2. The microwaves have bandwidth in micrometers.
3. Todays mobiles have internet connection.
4. Inverter is more efficient than UPS
5. The important number systems used in computers are Binary, Octal and Hexa-decimal.
6. E-mail means electronic mail.
7. Many medical instruments use electronic devices
8. A pacemaker is used when a person has decreased heart beat than the normal.

ANSWERS

1. True
2. True
3. True
4. False
5. True
6. True

❑❑❑

Appendices

APPENDICES - A

Values of General Physical Constants

Constant	Symbol	Value
Charge of an electron	e	1.602×10^{-12} C
Mass of an electron	m	9.109×10^{-31} kg
Ratio of charge and mass of electron	e/m	1.759×10^{11} C/kg
Planck's constant	h	6.625×10^{-34} J.s
Boltzmann's constant	k	1.380×10^{-23} J/K
Velocity of light in vacuum	v_c	2.998×10^{3} m/s
Permeability of space	μ	1.257×10^{-4} H/m
Characteristic impedance of free space	Z_0	$120\,\pi = 377\ \Omega$
Electric permittivity of space	ε	8.854×10^{-12} F/m
Standard (noice) temperature	T_0	17°C = 290 K

APPENDICES - B

SI System of Units

	Multiplication Factor	Prefix	Symbol
Preferred Values	10^{12}	tera	T
	10^{9}	giga	G
	10^{6}	mega	M
	10^{3}	kilo	K
	10^{2}	hecta	H
	10^{1}	deca	D
	1	—	—
	10^{-1}	deci	d
	10^{-2}	centi	c
Preferred Values	10^{-3}	milli	m
	10^{-6}	micro	μ
	10^{-9}	nano	n
	10^{-12}	pico	p
	10^{-15}	femto	f
	10^{-18}	atto	a

For example 20 mega ohm = 20×10^{6} ohms

5 microfarad = 5×10^{-6} farads

APPENDICES - C

Abbreviations

AF	Audio Frequency
AM	Amplitude modulation.
Amp	Ampere or Amplifier
A3	Double-sideband, full carrier AM.
A3 A	Single-sideband, reduced-carrier AM.
A 3 B	Independent-sideband AM.
A3H	Single-sideband, full carrier AM.
A3J	Single-sideband, suppressed-carrier AM.
A5C	Vestigial-sideband AM.
AIR	All India Radio.
ADF	Automatic Direction Finder.
AFC	Automatic Frequency Correction (or conir)
AGC	Automatic Gain Control.
AOS	Atlantic Ocean Satellite
ARC	Automatic Request for Correction.
AHQ	Automatic Request for Repetition.
AVC	Automatic Volume Control.
ADM	Adaptive Delta Modulation.
ASK	Amplitude Shift Keying
ASCII	American Standard Code for Information Interchange.
BBC	British Broadcasting Corporation
BD	Breakdown
BE	Base emitter
BW	Band Width
BFO	Beat-Frequency Oscillator.
BMFW	Ballistic Missile Early Warning.
BWO	Backward-Wave Oscillator.
Baiun	Balancc-to-unbalance transformer.

Bit	Binary digit.
B & W	Black & White (monochrome) TV.
CB	Common Base or Collector Base
CE	Common Emitter or Collector Emitter
CC	Common Collector
CG	Current Gain
CT	Centre Tap or Current Transformer
Ckt.	Circuit
Ch.	Characteristic
CCRIT	Consultive Committee for Integrational Telephone & Telegraph (Code)
CFA	Crossed-field Amplifier.
CRT ,	Cathode-Ray Tube.
CRO	Cathode-Ray Oscilloscope.
CC1R	Comite Consultative International Radio.
COMSAT	Communications Satellite (Corporation).
CTE	Channel Translating Equipment.
CVV	Continuous Wave.
Codan	Carrier-operated dev ice antinoise.
Coho	Coherent Oscillator
Compander	Compressor-expander.
CW	Continuous Wave
CATV	Cable TV.
CCTV	Close Circuit TV.
DC	Direct Current or Direct Coupled
DPCM	Diff. pulse Code Modulation
DF	Direction Finding.
deci	decimal digit
Domsat	domestic satellite (system).
CM	Delta Modulation.

EB	Emitter Base
Eq	Equivalent or Equation.
ESBAR	Epitaxial Schottky Barrier (diode)
ECCM	Electronic Counter-Counter Measures.
EHT	Extra High Tension
E-mail	Electronic mail.
FB	Feedback or Forward Bias
FOM	Figure of Merit
FDM	Frequency-Division Multiplexing.
FDMA	Frequency-Division Multiple Access.
FHF	Fractional Horse Power.
FM	Frequency Modulation.
FMVFT	Frequency-Modulated Voice-Frequency Telegraph.
FSK	Frequency-Shift Keying
FCC	Federal Communications Commission.
Fax	Fascumle.
GCA	Ground Controlled Approach
Ge	Germanium
GTE	Group Translating Equipment.
HW	Half Wave
HF	High frequency
IC	Integrated Circuit
IF	Intermediate Frequency
I/P	Input
IR	Infra Red
IMP ATT	Impact Avalanche and Transit-time (diode).
INSAT	Indian National Satellite.
INTELSAT	International Telecommunication Satellite (Consortium).
ITU	International Telecommunication Union.
ISRO	Indian Space Research Organisation.
IFF	Identification Friend or Foe.
ILS	Instrument Landing System.

IOS	Indian Ocean Satellite.
ISB	Independent Sideband (modulation)
ISDN	Integrated Services Digital Network (Computer Communication).
IMU	International Measurement Unit.
LORAN	Long Range Navigation.
LE	Line Equipment.
LSA	Limited Space-charge Accumulation.
LASER	Light amplification by stimulated emission of radiation.
LAN	Local Area Network.
MEWS	Missile Early Warning System.
MFC	Multi-frequency Coding.
MTBF	Mean Time Between Failures.
MTI	Moving-Target Indicator (radar).
MUF	Maximum Usable Frequency.
MV	Multivibrator
MESFFT	Mesa Field-Effect Transistor.
MARISAT	Maritime Satellite (system)
Maser	Microwave Amplification by Stimulated Emission of Radiation.
MODEM	Modulator-demodulator.
NASA	National Aeronautics and Space Administration.
NTSC	National Television Standards Committee.
OFC	Optical Fibre Cable.
OOK	On-Off Keying.
OC	Open Circuit
OSC	Oscillator
OP	Operation or Operating
O/P	Output
OCS	Overseas Communication Services (India).

OTC (A)	Overseas Telecommunications Commission (Australia).
PLCC	Power Line Carrier Communication.
PAM	Pulse-Amplitude Modulation.
PCM	Pulse-Code Modulation.
PG	Power Gain
PA	Power Amplifier or Public Address System
PT	Potential Transformer / Picture Tube
PM	Phase Modulation.
PPM	Pulse-Position Modulation.
PSK	Phase-Shift Keying.
PTM	Pulse-Time Modulation.
PWM	Pulse-width Modulation.
PFN	Pulse-Forming Network.
PLL	Phase-Locked Loop.
POS	Pacific Ocean Satellite.
PPI	Plan Position Indicator.
PPM	Periodic-Permanent Magnet/Parts Per million.
PRF	Pulse-Repetition Frequency.
PIN	P-Intrinsic-N (diode).
PRR	Pulse Repetition Rate.
PSTN	Public Switching Telephone Network.
RB	Reverse Bias
NRF	Radio Frequency or Rupple Factor
RDF	Radio Direction Finding.
RF	Radio Frequency.
Radar	Radio Detection and Ranging.
SC	Shori Circuit
Si	Silicon
SLV	Satellite Launch Vehicle.
SAGE	Semiautomatic Ground Environment.
Stalo	Stable Oscillator (in radar).

SMG	Stabilized Master Oscillator.
SWR	Standing Wave Ratio.
SSB	Single Sideband
SIM	Subscriber Identity Module (mobile card).
TE	Transverse Electric.
TEM	Transverse Electromagnetic.
TM	Transverse Magnetic.
Tr	Transistor
T/F	Transformer
TR	Transmit-Receiver.
TRF	Tuned Radio Frequency.
TWS	Track While Scan.
TDM	Time-Division Multiplexing.
TDMA	Time-Division Multiple Access.
TED	Transferred Electron Device.
TEO	Transferred Electron Oscillator.
TRAPATT	Trapped Plasma Avalanche Triggered Transit (diode).
TWT	Travelling Wave Tube.
USB	Upper Side Band.
UHF	Ultra High Frequency.
UV	Ultra Violet
VG	Voltage Gain
VAR	Visual Aural Range,
VOR	Visual Omni Range.
VTM	Voltage-Tunable Magnetron.
VCO	Voltage-ControUed Oscillator.
VFO	Variable-Frequency Oscillator.
VHF	Very High Frequency.
VSB	Vestigial Side Band (transmission).
WB	Wein Bridge
WLL	Wireless Local Loop (Mobile phone)

APPENDICES - D: Glossary

ACTIVE HOMING

In active homing, the missile has its own radar. The missile homes on to the reflected energy from the target.

ACCURACY

The accuracy of a D/A converter is the difference between the actual analog output and idea' expected output when a given digital input is applied.

ADAPTIVE RADAR

In an adaptive radar, the operational parameters of the radar such as PRR, Pulse width etc. varied within limits as a function of radar's performance on variety of targets.

AGC

AGC stands for Automatic Gain Control. In this technique, a control input to an amplifier is used to control its gain to keep the amplifier output constant irrespective of the changes in the input signal amplitude.

AMPLITUDE MODULATION (AM)

In amplitude modulation, amplitude of carrier is varied in accordance with the amplitude variation of the intelligence to be transmitted. The carrier frequency remains constant.

AMCW RADAR

The transmitted signal is amplitude modulated CW signal. It is a CW (Continous Wave) radar that makes use of two slightly differing CW transmissions. The range is determined from the phase difference between the two received doppler signals. The phase difference is linearly proportional to the distance and is unambiguous upto a phase difference of 2π radians.

AM DEMODULATOR

It is an electronic circuit that demodulates an amplitude modulated signal fed to it. It recovers the modulating signal from the modulated signal.

AND-GATE

The output of an AND-GATE is logic 'I' only when all of its input are to logic 'I' state. For all other possible input combinations, the output is a logic 'O'.

ANTENNA GAIN

The antenna gain is defined as the ratio of voltages produced at a given point by the actual and the hypothetical antennas. The hypothetical antenna is nothing but an omni-directional antenna.

ANTENNA RECIPROCITY

The phenomenon of using the same antenna for transmission as well as for reception is known as antenna reciprocity.

ASCII

Stands for American Code for Information Interchange. This is a 7-bit code and is the most commonly used code all over the world. It can be used to represent up to 128 characters including all alphabet !'both lower case and upper case), numeric digits and a number of special characters.

ASPECT RATIO

The ratio of width to heigh of the picture frame is called aspect ratio.

ASIC

ASIC (an abbreviation for Application Specific Integrated Circuit) is an IC belonging to LSI or VLSI category that has been designed for specific applications.

ASYNCHRONOUS TRANSMISSION

It is a transmission mode in which each information character is individually synchronised by the use of start and stop elements.

ATM

ATM stands for Automatic Teller Machine. An ATM is a computer terminal that allows individual customers to carry out transactions without any human intervention.

AUDIO TRANSFORMER

It is a transformer used to transform electrical signals spread out in. the audio frequency range from one circuit to another.

AVC

AVC stands for Automatic Volume Control. It is used to keep the strength of the received signal at the output of the detector as constant irrespective of the strength of signal received.

BAUD

It is a unit of measurement of signalling speed and is equal to the number of signal events per second, or bits per second.

BCD

Stands or Binary Codcd Decimal representation. This is a system of representing decimal numbers in the form of binary numbers.

BISTATIC RADAR

In the bistatic radar system, separate antennas are used for transmit and receive functions. In a Typical bistatic radar set up, these antennas may even be miles apart.

BIT

BIT stands for Binary Digit and is the smallest unit of information. It is either 0 or 1.

BLIND RANGES (Radar)

When a target lies at such a distance from radar that the radar receives the echo while it is still transmitting, it fails to detect the target. In such a case, the target is said to be at a blind range.

BODY STABILIZED (Satellite)

A satellite whose solar panels are fixed always to face the sun while antenna and body stabilization is provided through a spinning wheel.

BOOLEAN ALGEBRA

It is an algebra which has been named after its inventor, George Boole. This algebra is quite similar to ordinary algebra but handles only two valued variables, known as Boolean variables, which can take on one of the two values TRUE or FALSE.

BPI

Stands for Bits Per Inch. On a disc, data are recorded serially on tracks. On a tape, data are recorded in parallel on several tracks. BPI tells us about the storage density of any mass storage device.

BYTE

A string of 8 bits is called a byte. Byte is the basic unit of data operated on as a single unit in computers.

CAD/CAM SYSTEMS

CAD/CAM stands for Computer-Aided-design and Computer-Aided-Manufacturing systems. These systems usually have a CRT display, a keyboard, plotter and some graphics I/O devices.

CAPTURE RANGE

Capture range is frequency range over which a VCO output can acquire lock with the input frequency.

CARD READER

Card reader is an input device which is used to read the punched cards and the data read is then transferred to the computer for processing.

CAT (or CT)

CAT stands for Computerised Axial Tomography. It is a type of non-invasive testing that combines X-Ray techniques and computerised processing to carry out medical diagnosis.

CCD

The CCD (Charge Coupled Device.) is a storage element structured to offer a serial access. It can be regarded as a stretched enhancement MOSFET with a string of gales between source and dram.

CHIRP RADAR

See Pulse Compression Radar.

CHROMATICITY

The chromaticity diagram is a graphical representation in space co-ordinates X, Y, Z, of all spectral colours. The colours which are produced as a result of additive mixing of these colours based on values i>f the component colours are also represented in this diagram.

CHROMINANCE SIGNAL

Chrominance signal is the electrical signal that represents colour information of the scene to be televised.

CITIZEN BAND

The range of frequencies from 26.96 MHz to 27.41 MHz allocated to private citizens for short-range radio communication.

CO-AXIAL CABLE

A co-axial cable is a two conductor transmission line with a center conductor surrounded by a braided shield. The inner conductor is supported by some form of dielectric insulation.

CO-AXIAL RELAY

Co-axial relay is terminated into some kind of .... operations of equipment interconnected by cables.

CLOCK

It is an electronic device that generates periodic signals and is used to control the timing of ail the CPU operations.

COMMAND GUIDANCE

It is a technique for missile guidance. In command guidance of missiles, there are two separate tracking radars, one tracking the missile and the other tracking the target. The computer sends guidance commands to the missile continuously till it intercepts the target.

COMPANDING D/A CONVERTER

Companding type D/A converters are so constructed that the more significant bits of the digital input have a larger binary relationship than to the less significant bits

COMPANDING (PCM)

Companding mean compressing the signal at the transmitter and expanding it at the receiver. The signal is modified at the transmitter by artificially making the smaller signals immune to quantizing noise by increasing the amplitudes of smaller signals. The original amplitudes are restored at the receiver.

CONVERSION TIME (A/D Converters)

It is the time that elapses from the time instant of start conversion signal until the conversion of complete signal occurs. It ranges from a few nanoseconds to microseconds.

COPROCESSOR

It is a microprocessor that can be plugged into a microcomputer to replace or work with microcomputer original (or main) microprocessor

CROWBARING

Crowbaring is a type of over voltage protection for the power supplies. Whenever the output voltage reaches a certain preset limit, the power supply output is short circuited by an SCR which takes the short circuit current through it.

CRT

Stands for Cathode Ray Tube. It is an electronic device, upon which the information can be displayed.

dBM

(IBM is nothing but decibels referred to 1 milliwatt. It is used as a measure of signal power in communication circuits. Zero dBM equal 1 milliwatt into a specified impedance, often 600 ohms.

DECIBEL (DB)

A logarithmic measure of ratio between two powers.

DECODER

A decoder is a circuit that converts information from a-inputs to a minimum of 2 output lines.

DIFFERENTIAL PCM

In differential PCM, as compared to the conventional PCM. only the relative amplitudes of various sample., and not the absolute amplitudes are indicated.

DELTA PCM

It is similar to differential PCM. In this technique, only one bit is transmitted per sample just to indicate whether the sample in question is larger or smaller in amplitude than the immediately preceding sample.

DEMULTIPLEXER

It is exactly the reverse of a multiplexer. It receives data from one high speed line and distributes it to one of the low speed output line

DIFFERENTIAL NON-LINEARITY(DNL)

It indicates the worst case difference between the actual analog voltage change and the ideal LSB (Lower Side Band) voltage change. DNL is also expressed in percentage of full scale.

DIGITIZER

A digitizer (oc scanner) is a direct entry input device that can be moved over text, graphics, maps, drawings, pictures etc. to convert them into digital data. This is used to find coded price of books etc.

DOPPLER EFFECT

Doppler effect implies that the frequency of a wave when transmitted by a source is not the same as the frequency of the same wave when picked up by a receiver if there is a relative motion between the transmitter and the receiver.

ECHO SUPPRESSOR

Echo suppressor is a device or a circuit that allows transmission only in one direction at a time. Echo suppressors are used in telephone circuits to attenuate echo on long distance circuits.

EBCDIC

Stands for Extended Binary Coded Decimal Interchange Code. It is an eight-bit code.

EDTV

EDTV stands for Enhanced Definition Television. It employs newer encoding techniques and a compatible transmission to achieve a resolution better than the conventional system.

ENCODER

Encoder is any device which modifies information into the desired pattern or form for a specific method of transmission.

EQUALIZATION

Equalization is the process of reducing the effects of amplitude, frequency and/or phase distortion of a circuit.

FACSIMILE (Or Fax)

Facsimile is a method of transmitting pictures, printed pages or film to a remote location, when: mt transmitted information may be reproduced in the hard copy form.

FADING

Fading phenomenon is the result of electromagnetic waves reaching the receiving antenna via two different paths.

FIDELITY

The fidelity of a receiver represents the variation of output with modulation frequency when the output load impedance is a pure resistance.

FLAT PANEL DISPLAY

It is an alphanumeric or graphic terminal that uses an LCD type or an electroluminescent type display.

FLOPPY DISKETE

It is a thin, flexible circular plastic plate, coated with magnetic material and enclosed in a card board jacket, it is used as a secondary storage device.

FM ALTIMETER

In the modulated signal, the instantaneous frequency depends upon the time that elapses after the start of (he pulse. The frequency of the received echo thus gives an indication of the height of the transmitter above ground.

FMCW RADAR

In an FMCW radar, the transmitted signal is a frequency modulated CW (Continous Wave) signal. It can be used for measurement of both the target velocity and range.

FREQUENCY MODULATION

It is the modulation technique in which frequency of the carrier varies in accordance with the amplitude of the modulating signal. The rate of change of frequency is proportional to the frequency of the modulating signal.

FREQUENCY SHIFT KEYING

It is a form of frequency modulation commonly used in low speed MODEMS in which the two states of the signal are transmitted as two separate frequencies.

FULL-DUPLEX

Full duplex refers to a communication system or equipment that is capable of simultaneous two-way communications.

FM DISCRIMINATOR

It is a circuit that converts signal frequency variation to corresponding amplitude variation fed at its input. It demodulates a frequency modulated signal.

FREQUENCY DIVISION MULTIPLEXING

Frequency Division Multiplexing is the process of utilising the frequency scale for simultaneous transmission of more than one signals on the same carrier.

GEOSTATIONARY

It is a satellite orbit where the speed of rotation of satellite is equal to the speed of earth's rotation with the result that any spot on earth within the coverage of the satellite remains in the same relative position with respect to the satellite.

GPS

The GPS (Global Positioning System) is a navigational aid implemented with a series of satellites. System can he used to fix user's position on earth with an accuracy of few meters.

HALF DUPLEX

Half duplex refers to a communication system or equipment that is capable of providing communication in both directions but in only one direction at a time.

HDTV

HDTV stands for High Definition Television. It is the new generation TV technology, displaying images with much higher resolution on screens having an aspect ratio of 16 : 9 instead of 4 : 3 at present.

HOLLERITH CODE

A type of code used to represent alphanumeric data on punched cards.

DF ORMATION

DF Ormation is data that is organised and is meaningful to the user. Data processor transforms data into deformation.

INTERTIAL NAVIGATION

Intertial navigation system is a self contained navigation system and is associated with the use of an Intertial Measurement Unit (IMU) to help determine slate propagation.

INSTRUMENT LANDING SYSTEM

U is the system that is used by aeroplanes to assist the pilots to make proper landing. The system comprises of two transmitters, the localizer and the Glide path.

INTERACTIVE VIDEO

Interactive video refers to multimedia concept that merges computer text, audio and graphics by using a video disk, video disk player, micro computer and computer software.

INTERLACED SCANNING

In the interlaced scanning, all the odd lines are scanned first followed by scanning of all the even lines dividing one complete frame into two parts, y

IRRIDIUM PROJECT

This is a satellite based global communication system.

ISDN

ISDN stands for "Integrated Services Digital Network." it integrates different data and voice services so as to transmit them over a single communication channel.

ISC-9000

The term ISO-9000 broadly refers to a series of international standards laid down by the International Organisation for Standardisation.

LORAN

LORAN stands for Long Range Navigation. It is used to provide navigational aid to ships and planes.

MAXIMUM FREQUENCY DEVIATION

It is the range of frequencies between the lowest carrier frequency and the unmodulated carrier frequency.

MODULATION INDEX (AM)

Modulation index in amplitude modulation is defined as the ratio of peak amplitude of the modulating signal to the peak amplitude of unmodulated carrier signal.

MODULATION INDEX (FM)

Modulation index in FM is defined as the ratio of maximum frequency deviation and the highest modulating signal frequency.

MODEM

MODEM stands for Modulator-Demodulator. It is a type of data communication equipment that converts digital data to an analog signal (and vice-versa) for transmission on telephone circuits.

MONOTONOCITY

The D/A converter is said to exhibit monotonocity if its analog output either increases or remains same hut does not decrease.

MTI RADAR

MTI radar is used to determine both range as well as velocity of a moving target. Velocity determination is based on the doppler principle while range is determined as it is done incase of pulse radar

MULTIPLEXER

It is an electronic device that receives input from many low speed lines and then multiplexes or concentrates and transmits a compressed data stream on a high speed and a much more efficient data transmission channel.

MULTIPLYING D/A CONVERTER

In a multiplying D/A converter, the converter multiplies an analog reference by a digital word.

NAVIGATION

It is a proccs.. or a technique used for directing the movement of a vehicle (aeroplane, ship, missile etc.) from one position to another.

NEGATIVE LOGIC

In this, the more positive of the two voltage or current levels represents a logic 0 and the less positive of the two voltage or current levels represents a logic 1.

NOISE FIGURE

Noise Figure (NF) of a receiver is a measure of the extent to which the noise appearing in the receiver output in the absence of signal is greater than the noise that could be present if the receiver were a perfect one.

NON-LINEARITY

Non-linearity specification of an A/D converter describes its departure from a linear transfer curve. Non-linearity error does not include Gain, Offset and Quantization errors.

OMNIDIRECTIONAL ANTENNA

An antenna that radiates in or receives equally from all directions is called on omnidirectional antenna.

OVER MODULATION

Modulation percentage greater than 100 per cent is called over modulation.

PASSIVE HOMING

Passive homing depends for its functioning on the natural energy emitted by or reflected from the target. The sensor in the missile makes use of this energy to track the target. Anti-radar missiles for instance may home onto the transmitted power produced by the radar.

PERCENTAGE OF MODULATION (FM)

The percentage of modulation in an FM signal is defined as the ratio of maximum frequency deviation to the standard deviation (75 kHz) fixed for the service.

PHASED ARRAY RADAR

A phased array radar produces a beam that can be electronically scanned over a wide angle without any physical movement of the antenna.

POINT CONTACT DIODE

Point contact diode is primarily intended for RF applications due to its extremely small internal capacitance.

POLARISATION

Polarisation is that characteristic of the electromagnetic wave which gives direction of electrical component of the wave with respect to ground.

PRIMARY COLOURS

Red, Green and Blue have been identified as the primary colours.

PULSE AMPLITUDE MODULATION (PAM)

In a pulse amplitude modulated signal, the amplitude of the unmodulated pulse train varies in accordance with the instantaneous amplitude of the modulating signal.

PULSE CODE MODULATION

In Pulse Code Modulation (PCM), the signal be ....samples are then transmitted.

PULSE COMPRESSION RADAR

A pulse compression radar (also known as chirp radar) has the increased range detection capability of long-pulse radar system while retaining better range resolution

PULSE POSITION MODULATION (PPM)

In a pulse position modulated signal, the time of occurrence of pulses in the unmodulated pulse train changes in accordance with the instantaneous amplitude of the modulation signal.

PULSE WIDTH MODULATION (PWM)

In a pulse width modulated signal, the width of the pulses is changed in accordance with the instantaneous amplitude of the modulating signal.

PULSE TRANSFORMER

A pulse transformer is used to transform pulses from one circuit to another.

PUNCH CARDS

These are cards made up of thick paper (or card-board) and have 80 columns and 12 rows on which holes can be punched. Each column refers to a character and characters are stored by punching holes into the card.

QUANTIZATION ERROR-(A/D Converter)

The error resulting from the resolution limit is known as the Quantization Error.

RADAR ALTIMETER

Radar altimeter is similar to a pulse radar and is used to determine height of the aircraft above ground

RADAR BEACON

Radar beacons are stations to provide signals for use as navigational aids.

RADAR CLUTTER

All those objects which reflect the radar signal and which are other than the target are termed as clutter.

RADIX

Radix of a number system is defined as the number of different symbols used in the number system. The radix of binary number system, for instance is 2.

RASTAR

Rastar is illumination on screen in the absence of any video signal.

RDF

RDF stands for "Radio Direction Finding." It is the process of using radio signals to find out the relative position of the aeroplane

RECIPROCITY THEOREM

In a linear, active and bilateral network with a single source, the ratio of excitation to response remains unchanged when the excitation and response interchange their positions.

RESOLUTION

The resolution of a A/D converter is the number of states that the full scale range is divided or resolved into.

RESOLUTION

It is the ability of the radar to distinguish between two closely spaced targets.

RF AND IF TRANSFORMERS

These are narrow band transformers that are used to couple signals in a narrow band around a center frequency from one stage to another.

SCANNING (HORIZONTAL/VERTICAL)

Picture elements in that line (called horizontal scanning) and then various lines may be scanned sequentially in order to cover the complete picture frame (called vertical scanning).

SECONDARY COLOURS

Yellow (= Red + Green), Cyan (= Blue + Green) and Magneta (= Red + Blue) are known as secondary colours.

SENSITIVITY

Sensitivity of a communication receiver is the ability of the receiver to gel a standard output

SMART TERMINAL

A smart terminal has a microprocessor and primary storage to enable the user to do some processing before data is actually sent to the computer.

SVGA-(Super Video Graphics Array)

Is a type of video system offering upto 1024 × 768 resolutions producing a total of 7,86432 pixels.

SYNTHETIC APERTURE RADAR (SAR)

SAR coherently combines signals obtained sequentially from different locations of a moving antenna for mapping or target identification purposes.

TELECONFERENCE

Teleconference refers to a meeting that occurs via telephone/electronic facilities thereby eliminating need tor travel.

TELECOMMUTING

Telecummuting refers to working at home and communicate with the office or send data to the office via electronic machines.

TELEX

Telex is a communication service used for transmission of plain messages, it makes use of a teleprinter as terminal and telephone line as communication medium.

TELETEXT

Teletext service is the use of television sets for display of data, usually alphanumeric, in addition to receiving television broadcast.

TRUTH TABLE

A truth table lists all possible combinations of input and the corresponding outputs of a logic system.

TURN-AROUND TIME

It is the actual time required to reverse the direction of transmission from sender to receiver or vice versa when using a half duplex circuit.

TRACKING

Tracking in a communication receiver is the ability of different tuned circuits to change their resonant frequencies in the same proportion when a common shaft is rotated.

TRACKING RADAR

A tracking radar is used to continuously track a moving target.

UART

UART stands for Universal Asynchronous Receiver/Transmitter

UPS-(Uninterrupted Power Supply)

UPS is a power supply that is not interrupted by mains failure. When the mains power failure occurs the UPS acts as an auxiliary power source for the equipment. The UPS system draws its power source for storage a source such as batteries.

VAR

VAR stands for Visual Aural Range. VAR system can be used by aeroplanes and ships to determine their location.

VIDEO TRANSFORMER

It is a wideband transformer with a frequency response spread over frequency range from a few Hertz to a few Megahertz.

VIDEO DISPLAY ADAPTER

It is located it) one of the expansion slots in the computer and acts as an interface between computer and the Video Display Unit.

VIDEOTEXT

It is a general name given to a two way interactive information system which allows collection and retrieval of information through terminals which are specially adapted TV sets.

VOR

VOR stands for Visual Omni Range. VOR can be used to give exact bearing of the plane.

WAVEGUIDE

It is a specially designed and constructed metallic pipe for transmission of microwave electromagnetic radiations.

❑❑❑